The Joy of Finite Mathematics

The Language and Art of Math

The Joy of Finite Mathematics
The Language and Art of Math

Chris Tsokos

University of South Florida,
Department of Mathematics and Statistics,
Tampa, FL 33620

Rebecca Wooten

The Pedagogue, LLC,
Developing Educational Materials,
Tampa, FL

AMSTERDAM • BOSTON • HEIDELBERG • LONDON
NEW YORK • OXFORD • PARIS • SAN DIEGO
SAN FRANCISCO • SINGAPORE • SYDNEY • TOKYO
Academic Press is an imprint of Elsevier

Academic Press is an imprint of Elsevier
125 London Wall, London, EC2Y 5AS, UK
525 B Street, Suite 1800, San Diego, CA 92101–4495, USA
225 Wyman Street, Waltham, MA 02451, USA
The Boulevard, Langford Lane, Kidlington, Oxford OX5 1GB, UK

British Library Cataloguing in Publication Data
A catalogue record for this book is available from the British Library

Library of Congress Cataloging-in-Publication Data
A catalog record for this book is available from the Library of Congress

ISBN: 978-0-12-802967-1

For information on all Academic Press publications
visit our website at http://store.elsevier.com/

Working together
to grow libraries in
developing countries

www.elsevier.com • www.bookaid.org

Dedications

"We think faster than we speak and speak faster than we write; therefore, when creating great things we abbreviate everything. This abbreviated language is called Math."

RDW

This text is dedicated to my wife Debbie and my children Matthew, Jonathan and Maria.
Chris P. Tsokos

This text is dedicated to my husband Dana Miller, and all my family.
Rebecca D. Wooten

Contents

About the Authors

Chris P. Tsokos is Distinguished University Professor of Mathematics and Statistics at the University of South Florida. Dr. Tsokos received his B.S. in Engineering Sciences/Mathematics, his M.A. in Mathematics from the University of Rhode Island, and his Ph.D. in Statistics and Probability from the University of Connecticut. Professor Tsokos has also served on the faculties at Virginia Polytechnic Institute and State University and the University of Rhode Island.

Dr. Tsokos' research has extended into a variety of areas, including stochastic systems, statistical models, reliability analysis, ecological systems, operations research, time series, Bayesian analysis, and mathematical and statistical modeling of global warming, both parametric and nonparametric survival analysis, among others. He is the author of more than 300 research publications in these areas.

For the past four years Professor Tsokos' research efforts have been focused on developing probabilistic models, parametric and nonparametric statistical models for cancer and GLOBAL WARMING data. Specifically, his research aims are data driven and are oriented toward understanding the behavior of breast, lung, brain, and colon cancers. Information on the subject matter can be found on his website.

Professor Tsokos has more than 300 publications in his research areas of interest. He is the author of several research monographs and books, including *Random Integral Equations with Applications to Life Sciences and Engineering*, *Probability Distribution: An Introduction to Probability Theory with Applications*, *Mainstreams of Finite Mathematics with Applications*, *Probability with the Essential Analysis*, *Applied Probability Bayesian Statistical Methods with Applications to Reliability*, and *Mathematical Statistics with Applications*, among others.

Dr. Tsokos is the recipient of many distinguished awards and honors, including Fellow of the American Statistical Association, USF Distinguished Scholar Award, Sigma Xi Outstanding Research Award, USF Outstanding Undergraduate Teaching Award, USF Professional Excellence Award, URI Alumni Excellence Award in Science and Technology, Pi Mu Epsilon, election to the International Statistical Institute, Sigma Pi Sigma, USF Teaching Incentive Program, and several humanitarian and philanthropic recognitions and awards.

Professor Tsokos is a member of several academic and professional societies. He is serving as Honorary Editor, Chief-Editor, Editor or Associate Editor of more than twelve academic research journals.

Rebecca D. Wooten is Assistant Professor of Mathematics and Statistics at the University of South Florida. She received her M.A./B.A. in Mathematics and her Ph.D. in Statistics from the University of South Florida. She has worked for 15 years in teaching and has been recognized for her excellence in teaching; teaching courses such as Liberal Arts Math, Finite Mathematics, Basic Statistics, Introduction to Statistics, and Applied Statistics Methods.

Her research interests are concentrated in Applied Statistics with emphasis on Environmental Studies. Her research publications span a variety of areas such as Global Warming (carbon dioxide and temperature), Atmospheric Sciences and Geography (hurricanes), Geology (volcanic ash fall), Marine Biology (red tide), among others.

Professor Wooten is extensively involved in activities to improve education not only in Mathematics and Statistics, but Education in general. She is the Academic Coordinator for two free-educational assistance program which offer opportunities for students to volunteer and the local community to get the assistance in their studies that they would otherwise be unable to afford.

Preface

This book has been written to present certain aspects of modern finite mathematics from an elementary point of view, with emphasis on relevance to real-world problems. The objective is to create a positive attitude toward mathematics for the non-science-orientated college student and to demonstrate its usefulness in solving problems that we frequently encounter in our complex society.

Throughout the text, the aim has been to de-emphasize difficult theoretical concepts; thus, an intuitive treatment leads to practical applications of the various subject area topics. We believe that with such an approach, the modern college student will complete this course with the good feeling that mathematics is not only useful but enjoyable to work with.

The *Joy of Finite Mathematics* has several distinguishing features:

▣ The text has been written for students with only high school mathematics.

▣ Diagrams and graphs are used to illustrate mathematical concepts or thoughts.

▣ Step-by-step directions are given for the implementation of mathematical methods to problem solving.

▣ Emphasis has been placed on usefulness of mathematics to real-world problems.

▣ To provide motivation to the reader, most chapters are preceded by a short biography of a scientist who made important contributions to the subject area under consideration.

▣ Mathematical concepts are introduced as clearly and as simply as possible, and they are followed by one or more examples as an aid to thorough understanding.

▣ Each chapter ends with a complete summary that includes the definitions, properties, and rules of the chapter, followed by a Review Test.

▣ Each chapter contains numerous critical thinking and basis exercises with problems that reflect on the mainstream of the chapter.

The book has been designed to give the instructor wide flexibility in structuring a one or two-semester course, or a full-year course. Although some chapters are dependent on other others, many options are allowed (see accompanying diagraph).

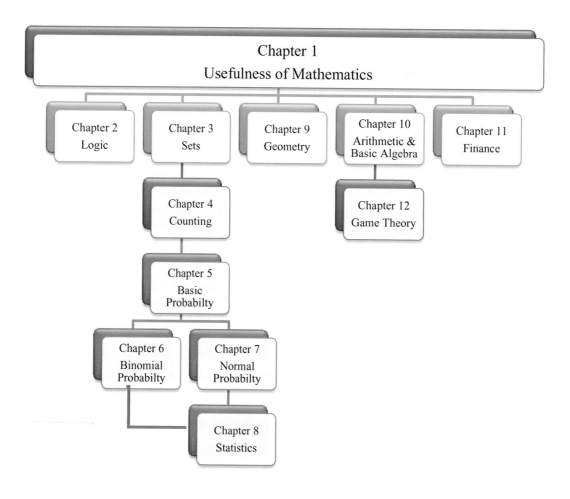

Very Special Acknowledgment

We wish to express our appreciation to the following academic educators for having reviewed our book and expressed their opinions.

An outstanding book for students to obtain basic knowledge of the usefulness of mathematics. Excellent motivation strategies throughout the book. It will inspire the student to learn the importance of mathematics.

Dr. Ram Kafle, Department of Mathematics and Statistics, Sam Houston University.

A very constructive and motivating book of finite mathematics. Special emphasis on the applications of math to real-world problems. The interactive approach of presenting their material is excellent. The student will acquire a very good understanding of what mathematics is all about.

Dr. Bong-jin Choi, Lineberger Comprehensive Cancer Center, The University of North Carolina at Chapel Hill.

This book is a masterful treatment of finite mathematics for undergraduate students who are afraid of mathematics. It will enlighten the student of the interdisciplinary use of mathematics at the very basic level. The book provides excellent illustrations of the use of mathematics/statistics to solve important problems.

Dr. Yong Xu, Department of Mathematics and Statistics, Radford University.

The Joy of Finite Mathematics *provides an excellent treatment of the subject. Unique emphasis on the importance of mathematical sciences to our society. The non-mathematics-oriented undergraduate student will find the contents of the book easy to read and very inspiring to learn more of the subject matter.*

Dr. K. Pokhrel, Department of Mathematics & Computer Systems, Mercyhurst University.

This is an excellent book of finite mathematics. It offers a justifiable, useful, and motivating approach to what mathematics is all about to the undergraduate student with minimum prior knowledge of the subject. The selection of the contents of the book, examples, and exercises is outstanding.

Dr. N. Khanal, Department of Mathematics, University of Tampa.

Several Options for a Semester Course in Finite Mathematics

Five possible options in designing a basic course in **finite mathematics** are given below, along with some remarks for each selection.

Options 1 and 2 offer a detailed coverage of specific topics in **math**, each spanning seven chapters:

Option 1:

Chapters Covered	Title
Chapter 1	The Usefulness of Mathematics
Chapter 2	Logic
Chapter 3	Sets
Chapter 4	Counting Techniques
Chapter 5	Probability
Chapter 8	Statistics
Chapter 9	Geometry

Covering materials necessary for the **CLAST (College Level Academic Skill Test)** exam, excluding algebra, these six topics are often taught collectively. In addition to the necessary high school algebra, these topics prepare a student well for the **CLAST** exam.

Option 2:

Chapters Covered	Title
Chapter 1	The Usefulness of Mathematics
Chapter 2	Logic
Chapter 3	Sets
Chapter 5	Probability
Chapter 6	Bernoulli Trials
Chapter 7	The Bell-shaped Curve
Chapter 8	Statistics

Option 2 provides the materials necessary for a comprehensive understanding of basic **probability** and **statistics**. This option is a broad introduction, including the underlying probabilities necessary to compute basic descriptive statistics, as well as inferential statistics in terms of interval estimates and tests of hypothesis.

Options 3-5 offer a more detailed coverage of specific topics in **math, each spanning** six chapters:

Option 3:

Chapters Covered	Title
Chapter 1	The Usefulness of Mathematics
Chapter 3	Sets
Chapter 4	Counting Techniques
Chapter 5	Probability
Chapter 6	Bernoulli Trials
Chapter 7	The Bell-shaped Curve

These topics enhance the study of **probability**. Option 3 begins with the basic concepts of categorization into **sets, counting** sets, and measuring **basic probabilities** empirically. It then continues with measuring basic probabilities hypothetically using either the discrete **binomial probability distribution, or** the **continuous normal probability** distribution.

Option 4:

Chapters Covered	Title
Chapter 1	The Usefulness of Mathematics
Chapter 4	Counting Techniques
Chapter 5	Probability
Chapter 6	Bernoulli Trials
Chapter 7	The Bell-shaped Curve
Chapter 8	Statistics

Option 4 covers materials necessary for the study of the basic aspects of **statistics**. This option includes **counting basic** empirical and hypothetical **probabilities** empirically. It also includes the basic necessities of **statistics**, descriptively and inferentially, for means and proportions.

Option 5:

Chapters Covered	Title
Chapter 1	The Usefulness of Mathematics
Chapter 2	Logic
Chapter 3	Sets
Chapter 4	Counting Techniques
Chapter 9	Geometry
Chapter 11	Arithmetic and Algebra

Option 5 covers materials necessary to gain a basic understanding of the language of deterministic **math**. This option provides a basic understanding of **logic, sets, counting, geometry,** and **algebra**.

Note: Game theory can be included in any scheme that includes the algebra and arithmetic.

A SUMMARY OF THE PROPOSED OPTIONS

Depending on which option you choose (1, 2, 3, 4, or 5), the purple indicates which chapters should be included; the green indicates optional chapters in each scheme.

Options ➲ Chapter ⟳	1	2	3	4	5
1	purple	purple	purple	purple	purple
2	purple	purple			purple
3	purple	purple	purple		purple
4	purple		purple	purple	purple
5	purple	purple	purple	purple	
6		purple	purple	purple	
7		purple	purple	purple	
8	purple	purple		purple	
9	purple				purple
10	green	green	green	green	green
11					purple
12	green	green	green	green	green

Joy of Finite Mathematics

Special Features

Motivation

◙ The usefulness of mathematics, especially those branches that constitute **finite math**, is illustrated both from a historical perspective, and by the role it plays in our daily lives.

◙ We emphasize an interactive approach to teaching **finite mathematics**.

The Language

◙ Teaching any student basic finite mathematics requires a basic understanding of the underlying symbolic language. Mathematics has many dialects: **logic**, **set theory**, **combinatorics (counting)**, **probability**, **statistics**, **geometry**, **algebra**, and **finance**, for example. Learning through relevance and interpretation of symbolism is vital.

The Relevant Questions

◙ A complete introduction of mathematics in a finite world, the notation used, and the underlying interpretation is presented. Relevant and useful questions associated with each dialect are posed, which will be answered through the process of learning finite mathematics.

The Review

◙ Reviews of each basic concept are given at the end of each chapter. The reviews enhance the learning of the basic aspects of each topic and their usefulness.

Step-by-Step

◙ Clear and concise **step-by-step** procedures are used in the development of various methodologies. Procedures are easy to follow, comprehend, and use to solve problems.

Highlights

◙ **Definitions**, **rules**, **methods** and **procedures** are highlighted with boldface and their meanings and usefulness follow with an abundance of relevant examples and applications.

Graphs and Tables

◙ Throughout the book, emphasis is placed on the extensive use of **tables**, **diagrams**, and **graphs** to clearly illustrate definitions, outlined methods, comparisons, etc. These visual aids invite clear interpretation of what they represent and their relevance to the text.

Applications and Interpretation

◙ We utilize a **step-by-step** approach in the illustrated examples (applications) that relate to the various dialects and their interpretations that have been introduced. Emphasis is placed on properly denoting the problems symbolically, interpreting the argument, outlining the defined set, measuring the probability, or, in general, finding the solution. Then, we encourage the student to clearly state any conclusions that can be drawn from the application.

Critical Reviews

◨ Each chapter ends with a review of: the new mathematical vocabulary, the most important concepts and methods, an abundance of review exercises, and a practice test that is based on the material from the preceding chapter.

Inspiration

◨ Throughout the book, we utilize important **historical facts** and **pose interesting** and **relevant questions**. We also include **humorous events**, **pictures**, **graphs**, **tables**, **biographical sketches of famous scientists**, **popular** and **classical quotes**, and more. These are all tools to **challenge**, **inspire** and **motivate** students to learn the mathematical thinking and to illustrate the absolute relevance of math to our society.

Challenging Problems

◨ Throughout the book, there are sections and challenging problems that are somewhat more advanced for a basic course in **finite mathematics** and are left to the discretion of the instructor.

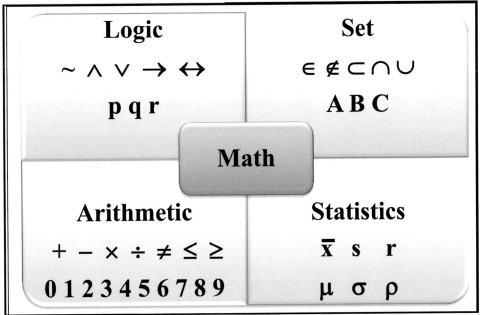

Dialects of Mathematics

Number is the within of all things.

Chapter 1   The Usefulness of Mathematics

If people do not believe that mathematics is simple, it is only because they do not realize how complicated life is.

John Louis von Neumann

The essence of mathematics is not to make simple things complicated, but to make complicated things simple.

S. Gudder

Go down deep enough into anything and you will find mathematics.

Dean Schlicter

The man ignorant of mathematics will be increasingly limited in his grasp of the main forces of civilization.

John Kemeny

Pure mathematics is, in its way, the poetry of logical ideas.

Albert Einstein

Mathematics is a more powerful instrument of knowledge than any other that has been bequeathed to us by human agency.

Descartes

Mathematics is the Queen of the Sciences.

Carl Friedrich Gauss

Mathematics is the science of definiteness, the necessary vocabulary of those who know.

W.J. White

Mathematics is the science which uses easy words for hard ideas.

Edward Kasner and James R. Newman

***Philosophy** is a game with objectives and no rules.*

***Mathematics** is a game with rules and no objectives.*

Goals and Objectives

The main objective of this chapter is to give an overview and motivate the non-mathematically oriented student about the usefulness of mathematics in several important fields. We begin with a brief historical perspective of the subject and proceed to discuss the importance and usefulness of all the areas that we believe constitute a course in Finite Mathematics. The diagram below illustrates the areas covered. Although not all the chapters of the textbook need to be covered in a one semester or two quarter course, we believe that the student can gain some basic knowledge by studying this chapter.

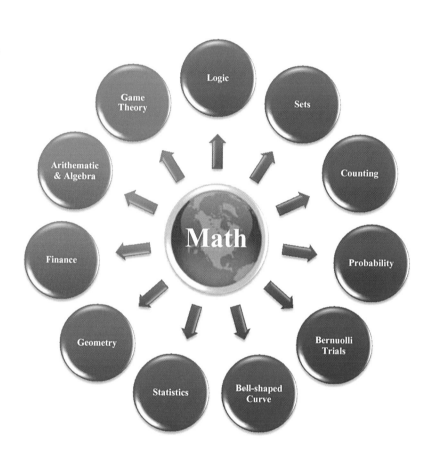

Thus, our goal here is to familiarize you with different areas and "dialects" of Math and:

➢ Learn about the history of Math
➢ Learn about the Math that is the foundation of logic
➢ Learn about the interplay of Math and sets
➢ Learn about counting techniques
➢ Learn about Math used to obtain probabilities of events
➢ Learn about binomial trials that leads to the Bernoulli probabilities
➢ Learn about the paramount importance of the Bell-Shaped Curve
➢ Learn about using Math to develop useful statistical methods

> Learn how Math is used to obtain measures of the earth or geometry
> Learn the Math that is arithmetic and algebra
> Learn how we use Math to answer basic financial questions
> Learn how we use Math in Game Theory, solving systems of equations to optimize strategies.

Pythagoras of Samos was the first to call himself a philosopher, Greek for "lover of wisdom." Pythagorean ideas greatly influenced western philosophy. Best known for the theorem which carries his name, Pythagoras was also a mathematician, scientist, musician, and mystic.

He founded the religious movement called Pythagoreanism. The Pythagoreans first applied themselves to mathematics, a science which they improved, and penetrated within; they fancied that the principles of mathematics were the principles of all things. A younger contemporary, Eudemus, shrewdly remarked that "they changed geometry into a literal science; they diverted arithmetics from the service of commerce" … Aristotle.

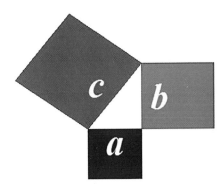

1.1   INTRODUCTION TO MATH

Mathematics played a very significant role in all our technological, scientific, medical, educational and economic accomplishments in our global society. However, just as important is the fact that mathematics indirectly interweaves every aspect of our daily lives; **mathematics** is the most powerful interdisciplinary language in almost all fields of **engineering**, every aspect of **health sciences, education, social** and **physical sciences, economics, finance, environmental sciences, Global Warming**, and of course **music** and **art**, among many other disciplines.

The word mathematics comes from the Greek word ***matheno*** which means *I learn*. Historically mathematics has its origin in the Orient when the Babylonians, in about 2000 BC, collected a lot of materials on the subject that we identify as **elementary algebra**. However, the modern concept of mathematics started in Greece around the fifth and fourth centuries BC. At this time, mathematics was subjected to philosophical discussion that was a unique priority in the Greek city states. The Greek philosophers were quite aware of the mathematical difficulties involved in understanding continuity, infinity, motion and the problems of making measurements of arbitrary quantities. Eudoxus' theory was very significant in geometrically understanding these concepts that were later significantly improved by Euclid's elements. Thus, the Greeks have an enormous influence on the tremendous development of today's mathematics.

Math is the language of thought. We think faster than we speak and we speak faster than we write… therefore, to convey our thoughts quickly, Mathematicians abbreviate everything.

Rebecca D. Wooten

Mathematics is the "brain" for
- Engineering
- Health Sciences
- Education
- Social Science
- Physical Sciences
- Economics
- Finance
- Environmental Sciences
- Global Warming
- Music
- Art
- Among others…

Discrete

 Apart or detached from others; separate; distinct

Absolute

 Not mixed or adulterated; pure

Relative

 Something having, or standing in, some relation to something else

Continued

 To go on with or persist in; to continue an action

Stable

 Not likely to fall or give way; firm, steady

Moving

 To pass from one place or position to another

H. Weyl, one of the truly great mathematicians of the twentieth century, stated *"without the concepts, methods, and results found and developed by previous generations right down to the Greek antiquity, one cannot understand either the aim or the achievements of mathematics in the last 50 years"* (American Math Monthly, Vol. 102, 1995).

Historically mathematics was defined as the logical study of shape, arrangement, and quantity. Furthermore, attempts have been made to think of mathematics as two branches: **Applied Mathematics** and **Pure** or **Abstract Mathematics**. The branches of applied mathematics are concerned with the study of physical, biological, medical or sociological worlds. Pure or Abstract Mathematics is concerned with the study and development of the principles of mathematics as such and is not concerned with their immediate usefulness.

In addition, we also had a divide of mathematics in the **discrete** and the **continuous**. Herbert W. Turnbull, in his essay on the "**World of Mathematics**" states: *To Pythagoras we owe the very word mathematics and its double fold branches*; that is,

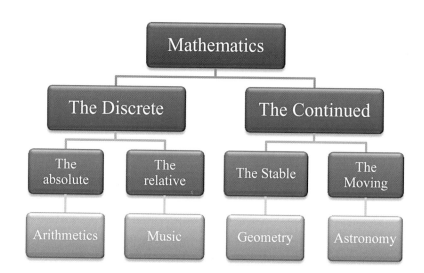

This double fold of mathematics played a major role in the development and usefulness of mathematics. In fact, **Aristotle** summarizes this historical divide as follows:

"The Pythagoreans first applied themselves to mathematics, a science which they improved; and, penetrated with it, they fancied that the principles of mathematics were the principles of all things." And a younger contemporary, Eudemus, shrewdly remarked that *"they changed geometry into a literal science; they diverted arithmetic from the service of commerce"*.

The *Joy of Finite Mathematics* is written to show at a very basic level, that mathematics is useful to virtually everyone, especially those students who do not like mathematics as we approach mathematics as a language used to describe simple and complex problems that we encounter in our daily lives.

In the essay on *"The Nature of Mathematics,"* by Philip E B Jourdain, he begins with *"An eminent mathematician once remarked that he was never satisfied with his knowledge of a mathematical theory until he could explain it to the next man that he met in the street."* This is so very true and we believe it is our responsibility in writing this text to explain to our students the usefulness of mathematical methods and theories using real world problems. Thus, the student has the right to ask *"what is the usefulness of mathematics?"*

We have taken that aspect of the student asking such questions as our responsibility in positively responding. We proceed to address this important issue by raising several relative questions in the interdisciplinary structure of mathematics that constitute the areas of the subject that we have identified as "Finite Mathematics." Thus, in what follows is the main thrust of the basic dialects of mathematics for students whose primary interest is not the subject matter, but how to enhance their understanding of the usefulness of mathematics. For motivating the students we begin each branch of mathematics by stating several real world questions, the answers to which will lead to the importance and usefulness of mathematics. We believe that this interactive approach will motivate the learning process and take our students on a very "joyful ride" to learn finite mathematics.

Intelligence is the ability to adapt to change.
Stephen Hawking

Number is the within of all things
Pythagoras

My goal is simple. It is a complete understanding of the universe, why it is as it is and why it exists at all.
Stephen Hawking

1.2 WHAT IS LOGIC?

Logic is derived from the Greek *λογική* meaning *conforming to laws of reasoning*. The branch of philosophy that treats forms of **thinking**, **reasoning** or **arguing** is also referred to as **Logic**. Averroes defines logic as *"the tool for distinguishing between the true and the false."* **Logic** is divided into two parts: inductive and deductive reasoning. **Inductive reasoning** draws conclusions based on specific examples whereas **deductive reasoning** draws conclusions from definitions and axioms.

Thus, our goal in learning **Logic** is to be in a position to make logical decisions regarding such questions as:

□ **Politics**: A politician claims *"if you don't vote for me, then you will not get the tax cuts"*—does this imply that if you do vote for him, that you will get the tax cuts?

□ **Health**: *If you work out more, then you will lose weight and tone your muscles, and if you watch your calorie intake, you will lose weight.* Does this mean that *if you lost weight that you must have both worked out and watched your calorie intake*?

□ **Travel**: *If Athens is in Greece and Berlin is in Germany*, then *when I visit Germany and not Athens*, then does it follow that *I went to Berlin*?

□ **Lottery**: *If Frank wins the lottery, then Frank will take you to dinner. Frank did not win the lottery and did not take you to dinner*. Did Frank lie?

□ **Law**: *If you are 17, then you are a minor. Jordan is not 17*; therefore, can we conclude that *Jordan is not a minor*?

Archival Note
Averroes as he is known in Greek is **Abū 'l-Walīd Muḥammad bin Aḥmad bin Rushd**, and he defined **logic** and is the founder of **Algebra**.

Debate between Averroes and Porphyry
Monfredo de Monte Imperiali Liber de herbis, *14th century*

It is the aim to teach logical reasoning to enable students to reason using the art of deduction and to draw correct conclusions when confronted with facts.

1.3 USEFULNESS OF SETS

The word "**set**" has more definitions than any other word in the English language due to its many origins. One origin of the word set is from the Old English *settan* meaning *cause to sit, put in some place, fix firmly*, and another is from the Old French *setta*, meaning *collection of things*. The branch of mathematics which deals with the study of sets is called **Set Theory**. The modern study of Set Theory was begun by **Georg Cantor** and **Richard Dedekind** in the 1870s. The language of sets can be used to define nearly all mathematical "**objects**" such as functions.

Set Theory begins with a fundamental binary relationship (similar to logic) between objects $\mathcal{O}$ and a set S, namely that of membership. Either an object (element) belongs to a set, or it does not.

Usefulness of Sets: To present data, relevant information in a systematic manner so that it will be visually attractive and easily understood and so that it can be used effectively to address various questions of interest. Thus, our goal in learning about sets is to be in a position to make categorical decisions regarding such questions as:

☐ Business: A store owner notes that more people like chocolate muffins than blueberry muffins. With this information, how many of each should be made? How many customers are expected to purchase both?

☐ Cancer: If survival is a function on the type of treatment(s) received, then which treatment is better or is a combination of treatments better?

☐ Meteorology: Given 20 readings of temperatures over a period of 14 days taken at two relatively close stations, when comparing these temperatures; do they appear to fall in the same temperature range?

It is our aim in learning **Set Theory** that students will be able to describe information **categorically** as well as to be able to display information graphically in **Venn diagrams** and use these graphics to support any inferences made regarding relationships among the various sets.

1.4 COUNTING TECHNIQUES

Count is from the Old French counter meaning *add up*, but also *tell a story*. Some of the first known use of counting was with shepherds who, when tending their sheep, would tie **knots in a rope** as they sent their sheep out to graze. In the evening, when the sheep returned to the fold for safety during the night, the shepherd would untie a knot and if there were any knots in the rope, they knew there were sheep that needed to be found. The branch of **mathematics** dealing with counting can extend from **tally marks**—making a mark for each number and then tallying these marks, **enumeration**—counting aloud

Archival Note

Georg Ferdinand Ludwig Philipp Cantor was a German mathematician, the inventor of **Set Theory**, and the first to establish the importance of one-to-one correspondence between sets.

Archival Note

Julius Wilhelm Richard Dedekind was another German mathematician who worked with Cantor and is also well known for his work in **abstract algebra**.

Categorical

 Unambiguously explicit and direct

 Data consisting of nominal information

 Qualitative data organized in a contingency table

Venn Diagram

 A set diagram that shows all possible relations between finite collections

Archival Note

Counting only involves the whole (counting) numbers, 0, 1, 2, 3... It first started with the natural counting numbers, and then we introduced the number zero to represent "nothing" or the number of elements in the empty set. It was **Jiu zhange suan-shu** who first used red rods to denote "**positive**" values and black rods to denote "**negative**" values in his writing *Nine Chapters on the Mathematical Art*. For a long period of time, negative solutions to problems where considered "false."

Diophantus, in the third century AD, referred to the idea of "$2x + 10 = 0$" as absurd.

or on your fingers to more complex counting techniques such as **combinations** and **permutations**.

Thus, our goal in **Counting Techniques** is to be in a position to count and discern the implication of such questions as:

☐ **Social**: At a social get-together of ten individuals, how many introductions will be needed to ensure everyone has met face-to-face?

☐ **Coding**: When coding a confidential letter using only the letters in the English alphabet, how many distinct codings are there? How many are needed such that no letter is mapped to itself in the coding?

☐ **Civics**: A board of directors consisting of ten women and 15 men need to form a five member committee to oversee next year's fund raiser. How many possible committees consist of exactly three men and two women?

☐ **Job Assignment**: A real estate agency has ten realtors and only nine new property listings. How many possible assignments of realtors to a house?

☐ **Diet**: There is a list of ten fruits you are willing to eat and your goal is to eat four fruits a day. To mix it up each day you create a meal plan that covers all possible combinations of four out of ten fruits. How many options are there for fruits in this meal plan?

The purpose in learning various **counting techniques** is to enable the student to determine the logistics necessary in such detailed coordination of a complex operation. This ranges from counting people or supplies to organizing committees and daily life.

Combination

 When r out of n objects are taken without replacement and without distinction in ordering

Permutation

 When r out of n objects are taken without replacement and with distinction in ordering

Counting is the religion of this generation. It is its hope and its salvation.

        Gertrude Stein

It's not the voting that's democracy, it's the counting.

        Alfred Emanuel Smith

Innumerable actions are going on through us all the time. If we started counting them, we should never come to an end.

        Vinoba Bhave

Music is the pleasure the human mind experiences from counting without being aware that it is counting.

        Gottfried Leibniz

Probability is the very guide to life.

        Cicero

1.5 PROBABILITY

Probability is from the French *probabilite'* meaning *quality of being probable* or *something likely to be true*. **Probability** is the branch of mathematics which is a way of expressing knowledge (or belief) that an event will or will not occur numerically, a form of empirical inductive reasoning leading to statistical inferences.

The idea of **probability** originated with games of chance in the seventeenth century. The earliest writings in the area were the result of the collaboration of the eminent mathematicians Blaise Pascal and Pierre Fermat, and a gambler, Chevalier de Mere. To them, there seem to be contradictions between mathematical calculations and the events of actual games of chance involving throwing dice, tossing a coin, spinning a roulette wheel, or playing cards.

Thus, our goal in learning about **Probability** is to be in a position to compute and interpret the relevance of probabilities and address such questions as:

☐ **Breast Cancer**: A patient goes to the doctor with a lump in her breast. What is the probability that it is a tumor? What is the probability that it is cancerous?

☐ **Finance**: What is the probability that the value of the Dollar will be higher than the Euro in 2015?

☐ **Sports**: What is the probability that the USF quarterback will complete half of his passes in a given game?

☐ **Engineering**: What is the probability that a computer software package will fail?

Probability is expectation founded upon partial knowledge.

        George Boole

☐ **Sociology**: What is the probability that a disadvantaged child in an urban area will pass the Florida Comprehensive Assessment Test (FCAT)?

☐ **Meteorology**: What is the probability that a hurricane will obtain hurricane status category 3 or more?

☐ **Statistics**: What is the probability that the mean number of accidents during New Year's Eve will exceed that of the previous year's number of accidents?

☐ **Physiology**: What is the probability that an experimental animal will convulse upon administration of a certain pharmacological agent?

☐ **Education**: What is the probability that an individual's score on an intelligence test will show significant improvement following a refresher course in verbal skills?

It is our aim to learn some of the very basic aspects of probability so that we not only answer questions such as those given above, but also to understand the role the subject plays in our daily lives. Learning **probability** is intended to put the student in a position to apply probability to any area of study that they are interested in: **statistics, engineering, operations research, physics, medicine, business, economics, accounting, education, sociology, physiology, agriculture, meteorology, linguistics** and **political science**, among others, and to use this information to make knowledgeable decisions.

Discrete

 A type of measure such that the outcomes are separate and distinct

Random

 Taken such that each individual is equally likely to be selected

Variable

 A distinct characteristic of an individual to be observed or measured

How dare we speak of the laws of chance? Is not chance the antithesis of all law?

Joseph Bertrand

1.6   BERNOULLI TRIALS

Trial is Anglo-French meaning *act or process of testing*. A **Bernoulli trial** is an experiment whose outcome is random, but has one of only two possible outcomes: *success* or *failure*. The discrete probability distribution that we use to answer such questions, among others, is the **binomial** or **Bernoulli probability distribution;** a mathematical expression that generates the actual probability for specific inputs that relate to a given question. We encounter many important situations that can be characterized by a **discrete random variable** with this developed distribution.

It is our goal in studying **Bernoulli trials** to put ourselves in a position to compute binomial probabilities and address such questions as:

☐ **Births**: A baby born less than 36 weeks is consider premature. What is the probability that a baby will be born premature?

☐ **Medicine**: What is the probability that a given drug will be effective to cure a specific disease?

☐ **Politics**: What is the probability that Candidate A will be elected president of the US?

☐ **Gambling**: What is the probability that I will obtain an odd number in a single roll of a fair die?

☐ **Computers**: What is the probability that the computer you purchased online will be operable (non-defective)?

We will learn how to use this very important probability distribution to answer the above questions, among others.

1.7   THE BELL-SHAPED CURVE

The **Bell-shaped Curve** is commonly called the **normal curve** and is mathematically referred to as the **Gaussian probability distribution**.

Unlike **Bernoulli trials** which are based on discrete counts, the **normal distribution** is used to determine the probability of a continuous random variable.

The **normal** or **Gaussian probability distribution** is the most popular and important distribution because of its unique mathematical properties, which facilitate its application to practically any physical problem in the real world; if not for the data's distribution directly, then in terms of the distribution associated with sampling. It constitutes the basis for the development of many of the statistical methods that we will learn in the following chapters.

Thus, our goal in studying the **Bell-Shaped Curve** is to put ourselves in a position to compute and interpret probabilities associated with continuous random variables and address such questions as:

- ☐ **Cancer**: What is the probability that in a given group of lung cancer patients, an individual selected at random is Asian?
- ☐ **Education**: What is the probability that a student will have a final grade in finite mathematics between 85 and 95?
- ☐ **Sports**: What is the probability that a given lineman's weight on the USF football team will be between 275 and 325 pounds?
- ☐ **Rainfall**: What is the probability that the average rainfall in the State of Rhode Island in the year 2012 will be between 16 and 24 inches?
- ☐ **Chemistry**: What is the probability that an acid solution made by a specific method will satisfactorily etch a tray?

The objective in learning the mathematical properties of the **normal probability distribution** is to realize its usefulness in characterizing the behavior of continuous random variables that frequently occur in daily experience.

1.8 STATISTICS

The branch of *Statistics*, meaning *quantitative fact or statement*, is becoming more widely accepted as a necessity for understanding all aspects that influence our daily lives. In almost every field of study, **statistics** is used to estimate the unknown, a characteristic of the individual we would like to know about in a given population. It is similar to the **Scientific Method**, in that we must first understand and clearly state the problem, gather the relevant information, formulate a hypothesis and test this hypotheses by recording and analyzing the data, before we can interpret the data and state our conclusion.

The basic idea behind **descriptive statistics** is to reduce a set of data down to one piece of information that describes some aspect of the data—an estimate of the population mean, or its central tendency, deviation, range, extremes, etc. Thus, our goal in studying **Statistics** is to be able to analyze and interpret real world data so that we will better understand the phenomenon that we are studying and address such questions as:

- ☐ **Business**: What is the mean profit made per hours of production time as a function of employees on the floor?
- ☐ **Politics**: What percentage of the people truly desire a tax increase given only 40% of individuals vote?
- ☐ **Chemistry**: What is the point of saturation for carbon dioxide in the atmosphere?
- ☐ **Medicine**: What is the mean tumor size in a patient with brain cancer?

Continuous

 A type of measure such that the outcomes are dense, that is, between any two outcomes, other possible outcomes exist.

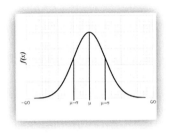

The graph of the normal probability distribution is a "**bell-shaped**" curve, as shown in the figure above. The constants μ and σ are the parameters.

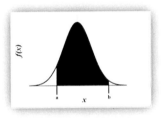

The area under the curve represents the underlying probability of the situation.

Statistics

 The art of decision making in the presence of uncertainty

As opposed to **statistic**—
a numerical datum

Hypothesis

 Greek meaning "to suppose"

A statistical analysis, properly conducted, is a delicate dissection of uncertainties, a surgery of suppositions.

M.J. Moroney

Statistics may be defined as "a body of methods for making wise decisions in the face of uncertainty."

W.A. Wallis

□ **Engineering**: What is the mean maximum load (kN) for a fishing line? What is the mean elongation?

□ **Astronomy**: What is the mean temperature fluctuation in the Sea of Tranquility on the moon?

□ **Agriculture**: What is the mean yield of corn per acre given the number of acres planted?

□ **Sociology**: What is the mean number of texts sent by a cellular phone user in a given month? Is there a difference in usage between teens and adults?

The point in learning basic statistics is to be able to efficiently gather, organize, analyze and interpret data in order to address questions that arise from every field of study and that apply to everyday living in a growing global society.

1.9   GEOMETRY

Geometry is from the Ancient Greek word γεωμετρια meaning *measurement of earth or land*. This branch of mathematics is concerned with questions regarding the **shape**, **size**, **relative positions** and properties of **space**. **Euclidean geometry** is a mathematical system that assumes a small set of axioms and deductive propositions and theorems that can be used to make accurate measurement of unknown values based on their geometric relation to known measures.

Thus, our goal in studying **Geometry** is to be able to accurately measure the world around us, perform basic calculations that address such questions as:

□ **Agriculture**: Using similar triangles, given the height of a stick and the length of its shadow at 2:00 PM, measuring the shadow of the tree at the same time, determine the height of the tree.

□ **Carpentry**: If two boards, mitered at a 60° angle, are reversed and attached to create a frame, what is the angle formed by the joint?

□ **Playground**: How large should a sandbox be if only 5 ft of wood is available and how much sand is needed if the wood is 6 in. tall?

□ **Rubik Cube**: How many squares are there on the surface of a Rubik Cube?

□ **Business**: If a showroom has 10,000 square feet of space to be converted into offices, but must leave 5000 square feet for the showroom floor and each office must be 200 square feet of space, how many offices can be created at most?

□ **Chemistry**: What is the shape of a sugar molecule? How does this differ between mono-dextrose and poly-dextrose sugars? What is the difference in volume?

The intention behind learning **Geometry** is to enable the student to be proficient in both the art and science of **geometry**. **Geometry** is used in areas ranging from **graphic design** to **Einstein's theory of general relativity**; when a surveyor plots land, a manufacturer determines the best packaging for a stack of spherical oranges to be shipped or a car manufacturer redesigns a parabolic headlight, for example.

1.10   ARITHMETIC AND ALGEBRA

Arithmetic means *the art of counting*, and *Algebra* means *reunion of broken pieces*. **Arithmetic** is the oldest and most elementary branch of mathematics and deals with the study of quantity such as those that result from combining other quantities, which leads directly to **Algebra**. **Algebra** is a branch of

Euclid is the **Father of Geometry** best known for his book *Elements which consist of 13 books covering Euclidian Geometry.*

Title page of Sir Henry Billingsley's first English version of **Euclid's Elements**, 1570

I've always been passionate about geometry and the study of three-dimensional forms.

Erno Rubik

There is geometry in the humming of the strings; there is music in the spacing of the spheres.

Pythagoras

Give me a lever long enough and a fulcrum on which to place it and I shall move the world.

Archimedes

mathematics outlining arithmetics, the rules of **operations** such as **addition**, **subtraction**, **multiplication** and **division**, but also **relations** such as **equalities**, **inequalities** and **functions**.

Arithmetic and **Algebra** are the building blocks of most areas in **mathematics**, usually taught as part of the curriculum in primary and secondary education. However, even at university level, these topics are extremely useful allowing general formulations to be the first step in the systematic exploration of more complex problems that can be solved using **Math**. Thus, our goal in studying **Arithmetic & Algebra** is to be able answer such questions as:

- ☐ **Health**: Based on the nutritional information for three dog foods based on three required nutrients, how much of each type of dog food should be included in a single serving to optimize the nutritional intake?
- ☐ **Farming**: Given 100 feet of fencing, how should the length of a pen be related to the width, if the fence is to create two adjacent pens sharing a common size with maximum area?
- ☐ **Business**: If you sell tickets for $20 each and you sell as many as you can, which beforehand is an unknown quantity, x, how does your profit relate to this unknown value x?
- ☐ **Social**: If you know that you have x adults coming for dinner and one child, and each adult eats three manicotti shells and the child eats one, how many shells must be made, y, as a function of the number of adults invited, x.

The aim of learning **Arithmetic** and **Algebra** is to refresh the student's understanding of the subject matter and to introduce more relevant uses of this dialect of **Math**. Remember: *we think faster than we speak and we speak faster than we write*. Therefore, to address large complex problems such as building a bridge, we need a very short handed language. This universal language is **Math**, and **Arithmetic** and **Algebra** are a large part of this language.

1.11 FINANCE

Finance means *to ransom*, or **to manage money**. This science of funds management includes **business finance**, **personal finance** and **public finance**. Our goal is to use mathematics to teach the student to have a better understanding of **basic personal finance**; such finances will include savings and loans in terms of time, money, risk and how they are interrelated in addition to spending and budget.

Thus, our goal in studying the **Basics of Finance** is to be able to understand and manage personal finances and address such questions as:

- ☐ **Personal Budget**: How much do you spend each month on Rent, Electricity, Phone, Internet, Food, Gas, Insurance, etc.
- ☐ **Wedding**: How much can I afford to spend? If I finance a wedding on credit, how long will it take to pay off this debt and how much will it eventually cost?
- ☐ **Transportation**: What should be the maximum payment I should agree to in order to ensure my vehicle is not repossessed.
- ☐ **Housing**: Can you afford to move out of your apartment and into a house? What is the expected down payment? Inspection fees? The expected property taxes?

Diophantus is traditionally known as the **Father of Algebra**, but this has recently put up to debate in that **Al-Jabr**, the author of **Arithmetic** gives the elementary algebra before **Diophantus** in 200-214 CE.

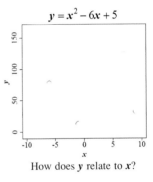

$y = x^2 - 6x + 5$

How does y relate to x?

In the business world, the rearview mirror is always clearer than the windshield.

Warren Buffett

Budget

 An estimate, often itemized, of expected income and expense for a given period of time in the future

 A sum of money set aside for a specific purpose

In the absence of the gold standard, there is no way to protect savings from confiscation through inflation. There is no safe store of value.

Alan Greenspan

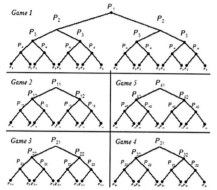

If thou dost play with him at any game,
Thou art sure to lose, and, of that natural luck,
He beats thee 'gainst the odds.

Shakespeare

☐ **Home Repairs**: How long would it take to save enough to replace an air conditioning unit? How much can be saved be investing into a sinking fund versus using credit?

☐ **Credit**: If you make the minimum required payment and have a minimum purchase each month, how long will you be indebted to the creditor?

☐ **Christmas Funds**: How much needs to be put into a sinking fund in order for you to have saved up $1000 in a Christmas fund over a period of 11 months starting in January?

In is imperative in today's economy that everyone has a basic understanding of finance. Many individuals are overwhelmed when confronted with mounting bills or credit; however, it is important that they budget, even if the final amount is negative. Once we are aware of the problem, we can begin to work out the solution. Understanding the basic **mathematics** behind **Basic Finance**, we will be in a position to make positive changes in our present financial state and better plan for the future.

1.12   GAME THEORY

Game Theory is a study of strategic decision making between two rational decision-makers. Here, we address **two-person zero-sum games**; games designed such that one players gain equals the second player's loss.

☐ Strictly Determined Games: The Saddle Point
☐ Games with Mixed Strategies
☐ Reducing Matrix Games to a System of Linear Equations

CRITICAL THINKING AND BASIC EXERCISE

1.1. Who was the first to call himself a philosopher?
1.2. What is Pythagoreanism?
1.3. What does the word "matheno" mean?
1.4. Mathematics can be divided into what two branches?
1.5. Who wrote "The Nature of Mathematics"?
1.6. What word is derived from the Greek meaning conforming to laws of reason?
1.7. Distinguish between the two types of reasoning.
1.8. In Logic, which type of reasoning is used to draw correct conclusions when confronted with facts?
1.9. What does the word "setta" mean?
1.10. The modern study of sets began with two mathematicians; name them and state where they are from.
1.11. Sets are used for what type of measure: numerical or categorical?
1.12. What diagram is used to graph categorical information and to support any inferences made regarding relationships among the various sets?
1.13. What are some of the first known uses of counting?
1.14. Name three counting techniques.
1.15. Name the area of study that is a form of empirical inductive reasoning leading to statistical inferences.
1.16. In what area of study are Blaise Pascal, Pierre Fermat and Chevalier de Mere known to have collaborated?
1.17. Who said "Probability is expectation founded upon partial knowledge"?
1.18. A binomial experiment is also known by what other name?
1.19. In a Bernoulli trial, there are exactly how many possible outcomes?
1.20. The binomial probability distribution is characterized by what type of random variable? Continuous or Discrete.
1.21. The normal probability distribution is also known by what other name?
1.22. Outline the steps associated with the Scientific Method.
1.23. Distinguish between Descriptive and Inferential Statistics.

1.24. List the points you need to learn in basic statistics to be an efficient researcher.

1.25. What branch of mathematics is concerned with questions regarding shape, size, relative positions and properties of space?

1.26. Name the mathematical system that assumes a small set of axioms and deductive propositions and theorems that can be used to make accurate measurement,

1.27. The art of counting is better known by what name?

1.28. Name the rules of operations in Algebra.

1.29. Why are Arithmetic and Algebra important?

1.30. Name the science of funds management.

1.31. Name the study of strategic decision making between two rational decision-makers.

SUMMARY OF IMPORTANT CONCEPTS

The first chapter, *The Usefulness of Mathematics*, introduces Mathematics and its history. This motivational chapter answers the question "what is logic"; outlining the usefulness of Sets; the start of Counting Techniques and how counts are the foundation of empirical probabilities. The first chapter also includes the usefulness of Mathematics in basic Probability and Statistics, Geometry, and Finance; an overview of basic Arithmetic and Algebra along with Game Theory.

The second chapter on *Logic* covers statements and their truth values; outlines the symbolisms used to express statements in the short hand language of Math including logical operators: conjunction, disjunction, negation and implication; how to construct truth tables and determine equivalent statements. This chapter helps the student understand logical reasoning by interpreting logical symbolism by giving their English translation. Properties of Logic covered include Tautologies, Self-Contradictions, Paradox, Equivalence, and Algebra of Statements; Variations on the Conditional Statement; Quantified Statements; Testing the Validity of an Argument and Applications of Logic.

The third chapter on *Sets* gives an introduction to Set Theory, covers collections of objects, the symbolisms used to express these collections (sets) in the short hand language of Math including set operators: intersection, union, complement and subset; and how they relate to logical operators. This chapter covers the Algebra of Sets, some basic counting principles applied to sets.

The fourth chapter on *Counting Techniques* introduces counting principles beyond that of simple sets to that of the Multiplication Principle, Permutations and Combinations, Distinct Orderings, and other counting techniques such as the Binomial Theorem, and Pascal's Triangle.

The fifth chapter on *Basic Probabilities* gives an introduction to probability and various definitions: personal probability, empirical and theoretical. This chapter covers the experimental probabilities using sample spaces, the basic laws of probability, conditional probability and Bayes rule.

The sixth chapter on *Binomial Probability* introduces discrete random variables, discrete probability distribution in general including expected value and variance followed by the Binomial Probability Distribution and the expected value and variance for the Binomial random variable.

The seventh chapter on *Normal Probability* introduces continuous random variables and the Normal probability distribution. This chapter also ties back in with discrete random variables covering Normal Approximation to the Binomial.

The eighth chapter on *Descriptive Statistics* covers gathering and organizing data; graphical representations of qualitative information and quantitative information; and measuring central tendencies and deviations from the center.

The ninth chapter on *Geometry* covers rounding and types of measurement; properties of lines: linear, linear pairs, two lines and three lines; properties of angles: categorization and additive principles; properties of triangles: categorization and similar/equivalent triangles; and properties of quadrilaterals and polygons. This chapter also covers area, surface area and volume.

The tenth chapter on *Arithmetic and Basic Algebra* covers the real number system, basic arithmetic: addition, subtraction, multiplication and division; pattern recognition: sequences and series; algebraic expressions and relationships; equations: equalities and systems of equations; and functions: linear and quadratic equation.

The eleventh chapter on *Finance* covers basic financing including sinking funds and amortization: various savings situations and comparison shopping: credit versus cash, leasing versus purchasing, and renting versus owning. This chapter also covers effective rates and uses them to compare CD versus credit and comparisons of credit cards. There is also a section on personal finance: how to create a monthly budget; insurance: what every homeowner should know and your credit report.

The twelfth chapter on *Game Theory* covers two-person zero-sum games. This chapter covers the Matrix Game; strictly determined games, games with mixed strategies and instructions on how to reduce Matrix Games to Systems of Equations.

Propositions

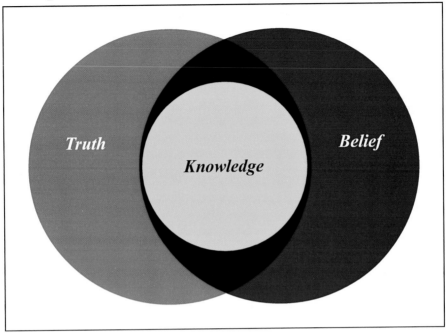

Number is the within of all things.

PYTHAGORAS

Chapter 2	Logic

In the field of mathematics, Aristotle is probably best noted for his contributions to the methods of proofs. He was the first to provide clear distinctions between axioms, postulates, and definitions. In addition, he contributed theorems of geometry, infinity, and continuity. He was considered to be a philosopher, but the philosophy of his time included what is now classified as natural sciences. Among Aristotle's writings on logic (called later the Organon) are Prior Analytics, Posterior Analytics, and Sophisms. In these and in his other works, he systematized the formal rules of logic and introduced syllogism, a form of deductive reasoning.

Aristotle was a student of Plato's Academy and later became a teacher. When he was forty-one, he began to supervise the education of Alexander the Great, for which he received the beginnings of his fortune. He later taught in the Lyceum in Athens and began amassing a book and map collection for a museum of natural history. This arrangement eventually led to a "school" after his death at the age of sixty-two.

Aristotle's influence was so encompassing and pervasive that many of his contributions were not even questioned until the middle of the nineteenth century even though many of his theories were incorrect. His writings, including accurate as well as misdirected ideas, were accepted as the ultimate authority during the medieval period and were upheld by the Roman Catholic Church beyond the time of Galileo (Aristotle had believed in the geocentric; i.e., earth-centered solar system, which Galileo unsuccessfully argued against in the Inquisition). Although his dogmatic followers deterred further advances for many centuries, Aristotle did much to advance science in his time. His many fields of study included biology (he devised classifications for all kinds of plants and animals), metaphysics and logic, ethics and politics, rhetoric and poetics, weather, and the other physical sciences.

Logic

 Derived from the Greek word logos which means reason or discourse

Syllogism

 An argument supported by two premises; deductive reasoning

 An extremely subtle, but sophisticated argument

Aristotelian

 Of, pertaining to, based on, or derived from Aristotle or his theories.

1. *To be able to apply logic to analyze problems, and*
2. *To obtain proficiency in the correct methods of logical reasoning.*

2.1   LOGIC

Why should we have chosen to begin the study of **finite mathematics** with a chapter on logic? The following argument is offered by way of illustration: Mathematics must be based on logic. This is a basic course in mathematics with emphasis on its usefulness. Therefore this course must be based on logic. An argument of this form is known as a **syllogism**. A **syllogism** is a typical **Aristotelian** argument. **Aristotle** gave the first systematic treatment of the principles of logical reasoning which earlier **Greeks** had begun to formulate. **Aristotelian logic** is the fundamental form of logical reasoning which is still utilized today.

The assertion that mathematics must be based on logic is justifiable because virtually all mathematical results are obtained by logical deductions from other previously obtained results now generally accepted as true, or from assertions which have been assumed without proof. Sometimes, making logical deductions is not as straightforward and simple as we might like; thus, it is true here as in many other situations that possession of a set of rather specific rules makes the task much easier.

George Boole (1815-1864), an English mathematician, integrated logic into algebra and essentially founded the field of mathematical logic. He introduced the use of symbols to represent statements or assertions, which greatly increases the ease and speed of manipulation of concepts in deductive logic. **Mathematical logic** is also known as **symbolic logic** for this reason. In this chapter some of the fundamental concepts of mathematical logic will be discussed with a view toward enabling the reader:

… no general method for the solution of questions in the theory of probabilities can be established which does not explicitly recognize … those universal laws of thought which are the basis of all reasoning …

George Boole

To this end, we must consider that we think faster than we speak, we speak faster than we write, therefore to think quickly and communicate these ideas, we must learn to abbreviate almost everything. A summary of modern symbolic logic can be found in the summary, at the end of the chapter.

2.2 STATEMENTS AND THEIR TRUTH VALUES

In this section we shall discuss one of the basic concepts of logic; namely, that of a statement. We shall also introduce some other important terms and symbols. We begin with a definition.

Definition 2.2.1 Statement

A **statement** is a declarative sentence which is either true or false, but not both. We shall denote statements symbolically by lower case letters p, q, r…

We judge statements with respect to their truth value. That is,

---

Definition 2.2.2 Truth Value

The **truth value** of a statement is the truth or falsity of the statement. We shall denote *true* by **T** and *false* by **F**.

Example 2.2.1 Classify Sentences

Consider the following sentences; classify each as statement, question or command:
(a) London is in France
(b) $3 + 5 = 8$
(c) Who is here?
(d) Put the book on the shelf.
(e) Sometimes it rains.

Solution

Sentence (a) is a false statement, and sentences (b) and (e) are true statements. However, sentences (c) and (d) are not statements because neither can be assigned a truth value of true or false. Sentence (c) is a question and sentence (d) is a command.

True Statements:
- *Monday is a day of the week*
- **5** *is a natural number*

False Statements:
- *January is a day of the week*
- **5** *is a negative integer*

Facts:
(a) London is, in fact, not in France
(b) **3 + 5** is equal to **8**
(c) Not a statement
(d) Not a statement
(e) Sometimes it does rain.

Simple Statements:
- *Today is Monday*
- *Tomorrow is Christmas*
- *The sun is shining*
- *There are rain clouds in the sky*
- *I will study English*
- *I will study Math*

The statements given in the preceding example are composed of terms in a certain relation to each other.

Compound Statements:

- *Today is Monday and it is the day before Christmas*
- *The sun is not shining and there are rain clouds in the sky.*
- *I will study English or Math*

Definition 2.2.3 Compound Statement

A statement consisting of a single such relationship is called a **simple statement**.

However, consider the statement "The square root of thirty-six is six and six is an even number." This statement is a combination of two components; namely, "The square root of thirty-six is six" and "six is an even number." Thus, we have a compound statement.

Note: iff reads "if and only if"

Definition 2.2.4 Compound Statement

A **compound statement** is a statement composed of two or more statements connected by the logical connectives, "and," "or," "if then," "not," and "if and only if." A statement which is not compound is said to be a **simple statement**.

Let b represent **Deb is beautiful** *and s represent* **Deb is smart**.

(a) $b \wedge s$

   $\wedge$ *reads "and"*

Let h represent **Matthew is here**, w represent **Washington is in the United States** and s represent **Sugar is sweet**.

(b) $(h \rightarrow w) \vee s$

   $\rightarrow$ *reads "implies"*

   $\vee$ *reads "or"*

Example 2.2.2 Simple/Compound Statement

Consider the following statements:

(a) "**Deb is beautiful and Deb is smart**." This statement is a compound statement composed of the simple statements "**Deb is beautiful**" and "**Deb is smart**" linked by means of the logical connective "**and**."

(b) "**If Matthew is here, then Washington is in the United States, or sugar is sweet**." This is a compound statement composed of the compound statement "**If Matthew is here, then Washington is in the United States**" and the simple statement "**Sugar is sweet**" by means of the logical connective "or."

Good, too, Logic, of course; in itself, but not in fine weather

__A. H. Clough__
English Poet (1848)

The truth value of a compound statement is completely determined by the truth values of the simple statements that form the compound statement.

We shall now study some of the most important **connectives** and illustrate their meanings by various examples. In logic, connectives are referred to as **operators**.

$\wedge$
reads
"and"

"and"
means
"both"

Definition 2.2.5 Conjunction: $p \wedge q$

The **conjunction** of two statements p and q is the compound statement "p and q"; written symbolically, $p \wedge q$, where the symbol $\wedge$ is read "and" or "but."

The truth value of $p \wedge q$ is determined using property 2.2.1.

Property 2.2.1 Conjunction: $p \wedge q$

If p is true and q is true, then $p \wedge q$ is true; otherwise $p \wedge q$ is false.

Row 1: When both p and q are true, the statement "p and q" is true.

Row 2: When p is true, but q is not true, the statement "p and q" is false.

Row 3: When p is false, and only q is true, the statement "p and q" is false.

Row 4: When both p and q are false, the statement "p and q" is false.

Columns 1 and 2 give all possible truth value combinations of the statements p and q. Column 3 gives the truth value of the statement $p \wedge q$ for each of the four combinations of the individual truth values of p and q. We observe that p and q is true only when p and q are both true; otherwise $p \wedge q$ is false. This table defines the truth value of the compound statement $p \wedge q$ as determined by the truth values of p and q separately. The representation shown by Table 2.1 is called the truth table for the conjunction $p \wedge q$.

Facts:

(a) *Athens is in Greece   Rhodes is an island*

(b) *Berlin is in Germany   Casablanca is in Morocco*

(c) *Roses are red   Violets are blue*

Example 2.2.3 Truth Values

Determine the truth value of each of the following conjunctions:

(a) Athens is in Greece and Rhodes is an island.

(b) Berlin is in Germany and Casablanca is in Williamsburg.

(c) Roses are red and violets are blue.

Solution

By Property 1, "$p \wedge q$" is true only when p and q are both true. Thus, we have (a) true, (b) false, and (c) true.

Definition 2.2.6 Disjunction: $p \vee q$

The **disjunction** of two statements p and q is the compound statement "p or q"; written symbolically is, $p \vee q$, where the symbol $\vee$ is read "or."

$\vee$
reads
"or"

Example 2.2.4 Symbolism in Logic

Let p and q represent the statements "George teaches mathematics" and "George lives in Greece"; then $p \vee q$ denotes the disjunction "George teaches mathematics or George lives in Greece."

TABLE 2.1 Truth Table for the conjunction $p \wedge q$

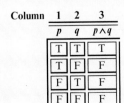

Column	1	2	3
	p	q	$p \wedge q$
	T	T	T
	T	F	F
	F	T	F
	F	F	F

The truth value of $p \vee q$ is determined by:

Property 2.2.2 Disjunction: $p \vee q$ (Inclusive)

When p is true or q is true, or if both p and q are true, then $p \vee q$ is true; otherwise $p \vee q$ is false. Thus, the disjunction $p \vee q$ of the two statements p and q is false only when both p and q are false.

That is, the standard disjunction $\vee$ is the **inclusive disjunction** which is true if at least one of the statements is true. The **exclusive disjunctive**, symbolized $\underline{\vee}$ is only true if one or the other statements are true, but not both. Thus the truth value of $p \underline{\vee} q$ is determined by: p is true and q is false, or p is false and q is true. In this text "or" will be inclusive. The truth value of $p \vee q$ is determined from the truth values of p and q by the following truth table (Table 2.2):

Column 3 tells us that the disjunction, $p \vee q$, is false only when both p is false and q is false; in other words, when p and q are *both false*.

Example 2.2.5 Disjunctions

Obtain the truth value of each of the following disjunctions:
(a) $3 + 5 = 9$ or $4 + 10 = 14$.
(b) Atlanta is in Florida or Armstrong landed on the moon.
(c) The earth is square or football is a gentle game.

Solution

By Property 2, "$p \vee q$" is false only when p and q are both false. Thus, we have (a) true, (b) true, and (c) false.

A good decision is based on knowledge and not on numbers.

Plato

Inclusive
"or"
means
"at least one"

Row 1: *When both p and q are true, the statement "p or q" is true.*
Row 2: *When p is true, but q is not true, the statement "p or q" is true.*
Row 3: *When p is false, and only q is true, the statement "p or q" is true.*
Row 4: *However, when both p and q are false, the statement "p and q" is false.*

*A mind all **logic** is like a knife all blade, it makes the hand bleed that uses it's*

Tagore

TABLE 2.2 Truth Table for the inclusive disjunction $p \vee q$

Column	1	2	3
	p	q	$p \vee q$
	T	T	T
	T	F	T
	F	T	T
	F	F	F

TABLE 2.3 Truth Table for the exclusive disjunction $p \veebar q$

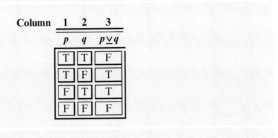

Row 1: When both **p** and **q** are true, the statement "**p** or **q**" is false.

Row 2: When **p** is true, but **q** is not true, the statement "**p** or **q**" is true.

Row 3: When **p** is false, and only **q** is true, the statement "**p** or **q**" is true.

Row 4: However, when both **p** and **q** are false, the statement "**p** and **q**" is false.

Exclusive
"or"
means
"exactly one"

Not true is false
Not false is true

For the exclusive "or," that is "**p** or **q**, but *not* both," the truth value is determined by:

Property 2.2.3 Disjunction: $p \veebar q$ (Exclusive)

When **p** is true and **q** is false, or when **p** is false and **q** is true, then $p \veebar q$ is true; otherwise $p \veebar q$ is false. Thus, this alternative disjunction $p \veebar q$ of the two statements; **p** and **q** are false only when both **p** and **q** are false and when both **p** and **q** are true.

In Table 2.3, Column 3 tells us that the exclusive disjunction $p \veebar q$ is true only when exactly one of the statements **p** and **q** are true. If **p** and **q** have the same truth value then $p \veebar q$ is false.

Many times it is necessary to negate a given statement, **p**, forming the "negation of **p**." This is accomplished by writing "It is false that" before **p**, or, if possible, using the word "**not**" in the statement **p**.

Example 2.2.6 Negations

Following is the statement **p**, give its corresponding negation $\sim p$:

(a) $4 + 6 = 10$.
(b) London is in England.
(c) Maria is pretty.
(d) Jonathan is not here.

Solution

(a) $\sim (4 + 6 = 10)$ is $4 + 6 \neq 10$, (b) It is not the case that London is in England *or* London is not in England, (c) Maria is not pretty, (d) Jonathan is here.

Definition 2.2.7 Negation: $\sim p$

The **negation** of **p** is denoted by $\sim p$. The truth value is obtained from the property: If **p** is true, then $\sim p$ is false and if **p** is false, then $\sim p$ is true. The symbol $\sim$ is read as not; that is, $\sim p$ is the statement "not **p**." Other common denotations of the negation of **p** are $\bar{p}$, $\neg p$ or even $-p$; however, we shall use the notation $\sim p$.

TABLE 2.4 Truth Table for the negation $\sim p$

Column	1	2
	p	$\sim p$
	T	F
	F	T

The truth value of the negation $\sim p$ is determined by the following property:

Property 2.2.4 Negation: $\sim p$

When p is true, then $\sim p$ is false; if p is false, then $\sim p$ is true.

We observe that the truth value of the negation of any statement is always the opposite of the truth value of the original statement.

The following truth table defines the truth value of $\sim p$ as it depends on the truth values of p, Table 2.4.

It is clear that the first column in Table 2.4 gives the original statement while Column 2 gives the corresponding negation.

Just as in English, a double negative statement is the positive statement; that is, not "not p" is p; symbolically $\sim(\sim p)$ is p.

Definition 2.2.8 Conditional Statement: $p \rightarrow q$

Statements of the form "if p then q" are called **conditional statements** and are denoted by $p \rightarrow q$. The statement p is called the **hypothesis** (**condition/antecedent**) while statement q is called the **conclusion** (**consequence**). The truth value of $p \rightarrow q$ is given by the property: $p \rightarrow q$ is false only when p is true and q is false; otherwise $p \rightarrow q$ is true.

It is not the statements' truth value we are concerned with, but whether the implication "$\rightarrow$" fails to be true. The key word here is "if," this condition is often misinterpreted. A mother say to the child, "if you misbehave, you will not get to go outside and play" and the child is horrible but the mother says, "Just go and play." This teaches the child that either mom is lying or "if" has no meaning. Unfortunately, children often assume the latter. However, this is not true, "if" is a very powerful word; only "if" the premise is true does the implication have to lead to the conclusion. Without the condition, the implication is vacuously true. Given the statement "if I win the lottery, then I will buy you dinner" and then this person does not win the lottery, but does buy me dinner, this does not make the original statement a lie (false), it simply has not come to pass.

Example 2.2.7 Conditional Statements

Consider again the following statements:

   p: I win the lottery.
   q: I will buy you dinner.

Statements:

- *I like vanilla ice cream*
- *I am not a Leo*
- *My favorite color is green*

Their Negations:

- *I do not like vanilla ice cream*
- *I am a Leo*
- *My favorite color is not green*

$\rightarrow$
reads
"implies"

Hypothesis

 From the Greek—basis or supposition

 An assumption or concession made for the sake of argument

 A preceding event, condition or cause

Conclusion

 The necessary consequence of a preceding event, condition or cause

 Something necessarily following from a set of conditions

Example 2.2.7 Conditional Statements—cont'd

Hence, $p \rightarrow q$ represents "If I win the lottery, then I will buy you dinner." If I win the lottery and I buy you dinner, then I have kept my word. Thus, $p \rightarrow q$ is true when p is true and q is true. However, if I did not win the lottery and do not buy you dinner, or I did not win the lottery but do take you to dinner, I did *not* break my word. Hence, $p \rightarrow q$ is true when p is false whether q is true or false. Only when I win the lottery and do not take you to dinner has my word been broken. Hence, the only time that $p \rightarrow q$ is false is when p is true, but (and) q is false.

Under what conditions is this conditional statement $p \rightarrow q$ is true.

This example leads us to the following property for the conditional statement $p \rightarrow q$:

Property 2.2.5 Conditional Statement: $p \rightarrow q$

The **conditional statement** $p \rightarrow q$ is false only when p is true and q is false; otherwise $p \rightarrow q$ is true.

Table 2.5 defines the truth value of $p \rightarrow q$, dependent on the truth values of p and q.

Column 3 tells us that the conditional statement $p \rightarrow q$ is false only when p is true and q is false; in other words, when the hypothesis is true and the conclusion is false.

Example 2.2.8 Conditional Truth Values

Determine the truth value of each of the following conditional statements:
(a) If $2 + 3 = 5$, then $6 + 9 = 15$.
(b) If $6 + 8 = 10$, then $1 + 1 = 2$.
(c) If $2 + 1 = 4$, then $2 + 3 = 6$.
(d) If $3 + 7 = 10$, then $4 + 7 = 14$.

Solution

We observe that (d) is the only conditional statement having the hypothesis "$3 + 7 = 10$" true and the conclusion "$4 + 7 = 14$" false. Thus, (d) is a false statement. Statements (a), (b), and (c) are all true. Note that even though the hypothesis in both (b) and (c) is false, the resulting conditional statement is true because of Property 5.

TABLE 2.5 **Truth Table for the conditional statement** $p \rightarrow q$

Column	1	2	3
	p	q	$p \rightarrow q$
	T	T	T
	T	F	F
	F	T	T
	F	F	T

Row 1: When both p and q are true, the statement "if p, then q" is true.

Row 2: When p is true, but q is not true, the statement "if p, then q" is false.

Row 3: When p is false, and only q is true, the statement "if p, then q" is not false; that is, true.

Row 4: However, when both p and q are false, the statement "if p, then q" is not false; that is, true.

(a) $T \rightarrow T$ is T
(b) $F \rightarrow T$ is T
(c) $F \rightarrow F$ is T
(d) $T \rightarrow F$ is F

TRUE WHEN

$$T \rightarrow T$$
$$F \rightarrow T$$
$$F \rightarrow F$$

If p true, it is suffice to say, then q is true.
However, q only needs to be true when p is true.

(a) $2 + 3 = 5$ T
    $6 + 9 = 15$ T
(b) $6 + 8 = 10$ F
    $1 + 1 = 2$ T
(c) $2 + 1 = 4$ F
    $2 + 3 = 6$ F
(d) $3 + 7 = 10$ T
    $4 + 7 = 14$ F

(a) $F \rightarrow T$ is T
(b) $T \rightarrow T$ is T
(c) $F \rightarrow F$ is T
(d) $T \rightarrow F$ is F

Common: Conditional Statement

There are four common ways in which the **conditional statement** $p \rightarrow q$ can be expressed.
1. *p implies q* (Direct statement)
2. *p only if q*
3. *p* is **sufficient** for *q*
4. *q* is **necessary** for *p*

*If science and **logic** chatters as fine and as fast as he can; though I am no judge of such matters, I'm sure he's a talented man*

W. M. Praed
Write of "The Talented Man" (1830)

The following example depicts the usage of the preceding expressions. However, it should be noted that a frequent usage of "*p* implies *q*" is the meaning that *q* is true whenever *p* is true. The conditional "$p \to q$" is a new statement compounded from two given statements, while the implication "*p* implies *q*" is a relation between two statements. The connection is the following: "*p* implies *q*" means the conditional statement $p \to q$ is always true.

Necessary

 Being essential, indispensable, or requisite

 Existing by necessity

Sufficient

 Adequate for the purpose; enough

 Of a condition

Example 2.2.9 Necessary and Sufficient

Obtain the truth value of each of the following conditional statements:
(a) **3 + 7 = 11** is a sufficient condition for **4 + 8 = 12**.
(b) **4 + 11 = 15** is a necessary condition that **3 × 3 = 9**.
(c) Sugar is sour implies vinegar is sweet.
(d) **4 + 10 = 14** only if **5 × 4 = 23**.

Solution

Statement (a) can be expressed as "If **3 + 7 = 11**, then **4 + 8 = 12**," which is true because the hypothesis "**3 + 7 = 11**" is false and the conclusion "**4 + 8 = 12**" is true. Statement (b) can be expressed as "If **3×3=9**, then **4 + 11 = 15**," which is true. Statement (c) can be restated in the form "If sugar is sour, then vinegar is sweet." We observe that since both hypotheses and the conclusion are false, the given conditional statement is true. Statement (d) can be restated as "If **4 + 10 = 14**, then **5×4= 23**," which is false because the hypothesis "**4 + 10 = 14**" is true but the conclusion "**5×4= 23**" is false.

$$p \leftrightarrow q$$
A bi-conditional statement

Definition 2.2.9 Bi-conditional Statement: $p \leftrightarrow q$

A statement of the form "*p* if and only if *q*" is called a **bi-conditional statement**, written in shorthand, you will often write "*p* iff *q*," symbolically, $p \leftrightarrow q$.

Bi

 *From the Latin "**bis**" meaning twice*
$$p \leftrightarrow q$$

Its truth value is obtained from the property 2.2.6. If *p* and *q* are both true or if *p* and *q* are both false, then $p \leftrightarrow q$ is true; if *p* and *q* have opposite truth values, then $p \leftrightarrow q$ is false. This is due to the fact that the statement $p \leftrightarrow q$ is equivalent to the conjunction of conditional statements, $(p \to q) \wedge (q \to p)$, hence if both *p* and *q* are both true then both $p \to q$ and $q \to p$ are true and if both *p* and *q* are both false then both $p \to q$ and $q \to p$ are vacuously true; thus $(p \to q) \wedge (q \to p)$ is true.

A convenient way to analyze logically compound statements formed by the connectives $\wedge$, $\vee$, $\sim$, $\to$ and $\leftrightarrow$ is by structuring their truth tables. Truth tables will be introduced in Section 2.4.

Example 2.2.10 Bi-Conditional Statement

Let *p* represent the statement "Maria is happy" and *q* "Maria looks beautiful"; then $p \leftrightarrow q$ denotes the bi-conditional statement "Maria is happy if and only if she looks beautiful."

The following property is used to determine the truth value of the bi-conditional statement:

TABLE 2.6 Truth Table for the bi-conditional statement $p \leftrightarrow q$

Column 1	2	3
p	q	$p \leftrightarrow q$
T	T	T
T	F	F
F	T	F
F	F	T

Property 2.2.6 Bi-Conditional Statement: $p \leftrightarrow q$

If p and q are either both statements are true or both statements are false, then $p \leftrightarrow q$ is true; if p and q have opposite truth values, then $p \leftrightarrow q$ is false.

Row 1: When both **p** and **q** are true, the statement "**p** iff **q**" is true.

Row 2: When **p** is true, but **q** is not true, the statement "**p** iff **q**" is false.

Row 3: When **p** is false, and only **q** is true, the statement "**p** iff **q**" is false.

Row 4: However, when both **p** and **q** are false, the statement "**p** iff **q**" is true.

The truth value of the biconditional statement $p \leftrightarrow q$ is given in Table 2.6.

In Table 2.6, Column 3 tells us that the bi-conditional statement $p \leftrightarrow q$ is true only if p and q have the same truth value, as in the first and fourth lines of the table.

Example 2.2.11 Bi-Conditional Statement Truth Value

Determine the truth value of each of the following bi-conditional statements:

(a) $4+8=12$ if and only if $3+7=10$.
(b) London is in France if and only if "Sharks" live on the moon.
(c) $14+5=20$ if and only if $3+8=15$.
(d) $5+3=8$ if and only if **6** is greater than **8**.

(a) $\text{T} \leftrightarrow \text{T}$ is T
(b) $\text{F} \leftrightarrow \text{F}$ is T
(c) $\text{F} \leftrightarrow \text{F}$ is T
(d) $\text{T} \leftrightarrow \text{F}$ is F

Solution

We observe that only in Statements (a), (b), and (c) have p and q have the same truth value, both being true in (a) and both false in (b) and (c). Thus, we can conclude that Statements (a), (b), and (c) are true, whereas Statement (d) is false.

We should also mention here that a bi-conditional statement, $p \leftrightarrow q$, is often stated in the form "p is necessary and sufficient for q."

If it was so, it might be; and if it were so, it would be: but as it isn't, it ain't. That's logic

Lewis Carroll
Logician in **Through the Looking Glass** (1872)

Example 2.2.12 Bi-Conditional Statement Truth Value

Obtain the truth value of each of the following bi-conditional statements:

(a) $4+5=9$ is a necessary and sufficient condition for $8+7=15$.
(b) Honey is sweet is necessary and sufficient for $8+10=25$.

Solution

We can rewrite Statement (a) as "$4 + 5 = 9$ if and only if $8 + 7 = 15$," which is true. Statement (b) can be restated in the form "Honey is sweet if and only if $8 + 10 = 25$," which is a false bi-conditional statement since p and q have opposite truth values.

EXERCISES

Critical Thinking

Indicate which phrases are acceptable statements; and for those which are, indicate whether they are true or false.

2.2.1. Sparta is in Greece.
2.2.2. $11 + 1 = 10$.
2.2.3. Who is coming?
2.2.4. Put the fish in the water.
2.2.5. Florida has a cold climate.
2.2.6. Tennis is fun.
2.2.7. $-3 + 5 = 2$.
2.2.8. Goodbye Columbus.
2.2.9. Put your shoes on.
2.2.10. Casablanca is in North Africa.
2.2.11. Tampa is the capital of Florida.
2.2.12. Answer the phone.
2.2.13. This statement is false.

Analyze the given statements and indicate the appropriate connectives and the simple statements.

2.2.14. $4 + 1 = 5$ and Maria is pretty.
2.2.15. If you are swimming in the Gulf, then Washington is in England or $4 + 7 = 10$.
2.2.16. If Columbus was from Italy or Columbus was from Portugal, then 1492 was an Italian year.
2.2.17. If Jonathan is thirsty, then Maria needs water, but 1977 is a dry year.
2.2.18. If $4 - 11 = 7$, then $-4 + 11 = 10$ and $6 - 7 = -1$.
2.2.19. Deb is athletic if Maria and Jonathan are, but Mathew is a basketball player.
2.2.20. It is not the case that it is raining and the sun is shining.

Basic Problems

Write each statement in symbolic form using the indicated letters to represent the corresponding simple statement.

2.2.21. Roses are red (*r*), Violets are blue (*b*)
2.2.22. I will travel by train (*t*), plane (*p*) or automobile (*a*).
2.2.23. I will study art (*a*) and music (*m*) as a minor.
2.2.24. You may go to the movies (*m*) if and only if you clean your room (*r*).
2.2.25. If I win the lottery (*l*), then I will buy you dinner (*d*).

2.2.26. Give a verbal translation of each compound statement given *p* represents "I love Lucy" and *q* represents "Lucy is a research scientist."

a. $p \wedge q$ e. $\sim p \vee \sim q$
b. $p \vee q$ f. $p \wedge \sim q$
c. $\sim p \wedge q$ g. $\sim p \vee q$
d. $\sim (p \wedge q)$

Write the negative of each statement.

2.2.27. Our coffee shop showed profit this year.
2.2.28. My cats name is Snowflake.
2.2.29. My cat is not a Siamese.
2.2.30. Peggy loves chocolate cake.
2.2.31. All positive integers are even
2.2.32. All rhombi are rectangles.
2.2.33. Some democrats are politicians.
2.2.34. The number 5 is a whole number.

2.2.35. Everyone likes ice cream.

2.2.36. There are no absolutes; that is nothing is certain.

2.2.37. All circles are round.

Determine the truth value of the given statements or give the conditions that will make the statement true.

2.2.38. $4 + 16 = 19$ and Florida is in Canada.

2.2.39. Peter is a swimmer and Alex a boxer.

2.2.40. $4 - 6 = 0$ and $5 + 5 = 10$.

2.2.41. Sopoto is in Greece and Chris is from Sopoto.

2.2.42. $3 - 3 = 0$ or $9 + 1 = 8$.

2.2.43. Miami is in Mississippi or Newport is in Rhode Island.

2.2.44. The earth is round or the moon is square.

2.2.45. If Houston is in Texas, then Boston is in New York.

2.2.46. If $6 + 3 = 8$ then $6 - 3 = 8$.

2.2.47. If $5 + 3 = 8$, then $6 + 2 = 8$.

2.2.48. If the sun is shining, then it is cold.

2.2.49. If $6 - 11 = -5$, then Chris is Greek.

2.2.50. $6 + 6 = 12$ if and only if $3 + 6 = 9$.

2.2.51. Chicago is in Michigan if Tampa is in Florida.

2.2.52. Patras is in the Corinthian Gulf if and only if Peter is from Sopoto.

2.2.53. $11 + 4 = 15$ if and only if $6 - 11 = -5$.

2.2.54. $6 + 11 = 17$ is a sufficient condition for $6 + 8 = 12$.

2.2.55. Honey is sweet implies Jan is sour.

2.2.56. $6 + 14 = 20$ is a necessary condition for $3 + 8 = 11$.

2.2.57. $9 + 22 = 31$ only if $17 + 6 = 21$.

2.2.58. If $16 + 22 = 38$, then Alex is pretty.

2.2.59. San Francisco is in California.

2.3 TRUTH TABLES

Often, in logical reasoning, complex compound statements are formed by the logical connectives such as conjunctive, disjunctive, negation, implication and bi-implications written symbolically:

$$\wedge, \vee, \sim, \rightarrow \text{ and} \leftrightarrow$$

along with simple statements $p, q, r\ldots$ Our aim in this section is to learn how to determine the truth value of such a compound statement when the truth values of its components $p, q, r\ldots$ are known. An effective way of doing this is by using truth tables.

Given n simple statements, then there are 2^n different possible combinations of true values. This can be seen in previous examples; when there is a single statement as in Table 2.4 when the negation is considered, there are only $2^1 = 2$ possibilities, either p is true or p is false. When there are two simple statements as in Table 2.1 when the conjunction of two simple statements is considered, there are $2^2 = 4$ possibilities: **TT, TF, FT,** and **FF**. Notice, half the first are **T** and half are **F**; this is also true for the second, however they are "half"; that is, half the **T**'s for the first statement are **T** for the second statement and half for the second statement are **F** as well as half the **F**'s for the first statement are **T** for the second statement and half for the second statement are **F**.

Extending this "halving" technique, for three simple statements, there are $2^3 = 8$ different possible combinations; for the first statement, four are **T** and four are **F**; for the four **T**'s statements, two are **T** and two are **F** which is

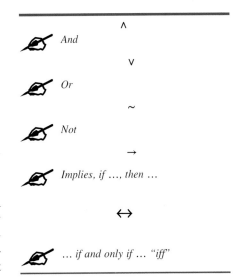

		$\wedge$
☞	*And*	
		$\vee$
☞	*Or*	
		$\sim$
☞	*Not*	
		$\rightarrow$
☞	*Implies, if ..., then ...*	
		$\leftrightarrow$
☞	*... if and only if ... "iff"*	

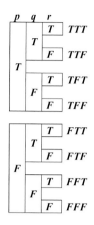

If I do not have to work today, then I will go to the store.

Not true is false and not false is true:
(a)   ~T≡F
(b)   ~T≡F
(c)   ~F≡T
(d)   ~F≡T

Fails only when premise is true and the conclusion does not follow:
(a)   ~T→T is **T**
(b)   ~T→F is **T**
(c)   ~F→T is **T**
(d)   ~F→F is **F**

TABLE 2.7 Truth values for three simple statements

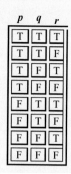

the same for the **F**'s. Then, for each of the given possibilities, the last statement is either true or false (Table 2.7).

Example 2.3.1

Determine the truth table for $\sim p \to q$.

Solution

First we observe that this statement is composed of the simple statements $\sim p$ and q along with the connectives $\sim$ and $\to$. Since there are just two simple statements, p and q, involved, each one being either true or false, the truth table will have 4 rows, giving all possible truth-value combinations of p and q. We present this procedure in five steps, as illustrated below:

  STEP 1: Construct the truth table for the given number of simple statements

  STEP 2: We determine the truth values for $\sim p$ by using Table 2.4

p	q	~p
T	T	F
T	F	F
F	T	T
F	F	T

  STEP 3: Lastly, we determine the truth values for $\sim p \to q$ using statement builder tables.

p	q	~p	~p → q
T	T	F	T
T	F	F	T
F	T	T	T
F	F	T	F

Solution—cont'd

The last column was obtained by using the truth values of $\sim p$ and q along with the truth table in Property 2.2.5, defining the truth value of a conditional statement.

An alternative way of constructing the truth table is to reproduce this information as in the chart below allowing a column for each connective.

STEP 4: Reproduce information need in final statement.

$\sim p$	$\to$	q
F		T
F		F
T		T
T		F

STEP 5: Then use this information to construct the conditional statement $\sim p \to q$

$\sim p$	$\to$	q
F	T	T
F	T	F
T	T	T
T	F	F

The entire problem presented in a single table as follows:

1	2	3	4	6	5
p	q	$\sim p$	$\sim p$	$\to$	q
T	T	F	F	T	T
T	F	F	F	T	F
F	T	T	T	T	T
F	F	T	T	F	F

Example 2.3.2

Construct a truth table for the $(p \lor q) \to p$

Solution

p	q	$p \lor q$	$(p \lor q) \to p$
T	T	T	T
T	F	T	T
F	T	T	F
F	F	F	T

Or, alternatively

1	2	3	5	4
p	q	$p \lor q$	$\to$	p
T	T	T	T	T
T	F	T	T	T
F	T	T	F	F
F	F	F	T	F

We note from the answer column indicates that $(p \lor q) \to p$ is false only when p is false and q is true; otherwise $(p \lor q) \to p$ is true.

If you have pride or money, then you have pride.

*Hence, when you **have pride** and **have money**, then you do have **pride**, this is true.*

*Then **when you don't have pride** but **you do have money**, then stating this implies **pride** is fallacious.*

*When you **have pride**, but **do not have money**, then you can conclude you have **pride**.*

*However, when you **don't have pride** and you **don't have money**, then stating "**If you have pride or money, then you have pride**" is not false, and is therefore true.*

*If we may believe our **logicians**, man is distinguished from all others creatures by the faculty of laughter*

The Spectator *(1712)*

If you know Math, English and Science, then you know Math or English.

Table 2.1

1	2	3
p	q	$p \wedge q$
T	T	T
T	F	F
F	T	F
F	F	F

Table 2.2

1	2	3
p	q	$p \vee q$
T	T	T
T	F	T
F	T	T
F	F	F

Table 2.5

1	2	3
p	q	$p \rightarrow q$
T	T	T
T	F	F
F	T	T
F	F	T

Regardless of the validity of each simple statement, the statement is always true; hence, a tautology.

Example 2.3.3

Construct a truth table for the statement $[p \rightarrow (p \wedge q)] \vee \sim p$

Solution

p	q	$p \wedge q$	$p \rightarrow (p \wedge q)$	$\sim p$	$[p \rightarrow (p \wedge q)] \vee \sim p$
T	T	T	T	F	T
T	F	F	F	F	F
F	T	F	T	T	T
F	F	F	T	T	T

Or, alternatively

1	2	3	5	4	7	6
p	q	$[p$	$\rightarrow$	$p \wedge q]$	$\vee$	$\sim p$
T	T	T	T	T	T	T
T	F	T	F	F	F	F
F	T	F	T	F	T	T
F	F	F	T	F	T	T

➤ *Step 5 is obtained from Steps 3 and 4, giving the truth value of $p \rightarrow (p \wedge q)$. We see that $p \rightarrow (p \wedge q)$ is false only when p is true and $(p \wedge q)$ is false. Table 2.4 is used to complete the column.*

➤ *Step 7 is obtained by using the truth values in Steps 5 and 6 along with Table 2.2 for the disjunction. Note that the disjunction $[p \rightarrow (p \wedge q)] \vee \sim p$ is false only in the second line, where both $p \rightarrow (p \wedge q)$ and $\sim p$ are false.*

Example 2.3.4

Determine the truth table for $[(p \wedge q) \wedge r] \rightarrow (p \vee q)$

Solution

Since there are three simple statements forming this compound statement; namely, p, q, and r, the truth table will have $2^3 = 8$ rows.

Following the indicated steps carefully, using Table 2.1 for the conjunction $\wedge$, Table 2.2 for the disjunction $\vee$ and Table 2.5 for the conditional $\rightarrow$, we have

p	q	r	$p \wedge q$	$(p \wedge q) \wedge r$	$p \vee q$	$[(p \wedge q) \wedge r] \rightarrow (p \vee q)$
T	T	T	T	T	T	T
T	T	F	T	F	T	T
T	F	T	F	F	T	T
T	F	F	F	F	T	T
F	T	T	F	F	T	T
F	T	F	F	F	T	T
F	F	T	F	F	F	T
F	F	F	F	F	F	T

Alternatively,

			1	2	3	4	6	5	8	7
			p	q	r	$[(p \wedge q)$	$\wedge$	$r]$	$\rightarrow$	$(p \vee q)$
			T	T	T	T	T	T	T	T
			T	T	F	T	F	F	T	T
			T	F	T	F	F	T	T	T
			T	F	F	F	F	F	T	T
			F	T	T	F	F	T	T	T
			F	T	F	F	F	F	T	T
			F	F	T	F	F	T	T	F
			F	F	F	F	F	F	T	F

An analysis of Step 8 reveals that the compound statement $[(p \wedge q) \wedge r] \rightarrow (p \vee q)$ is true, regardless of the truth values of p, q, and r. Step 8 was computed from Steps 6 and 7 by use of Table 2.5, which defines the conditional.

Example 2.3.5 _____

Construct a truth table for $(p \wedge \sim q) \leftrightarrow r$.

Solution _____

Following the indicated steps carefully, we can write

p	q	r	$\sim q$	$p \wedge \sim q$	$(p \wedge \sim q) \leftrightarrow r$ OR $r \leftrightarrow (p \wedge \sim q)$
T	T	T	F	F	F
T	T	F	F	F	T
T	F	T	T	T	T
T	F	F	T	T	F
F	T	T	F	F	F
F	T	F	F	F	T
F	F	T	T	F	F
F	F	F	T	F	T

Alternatively,

1	2	3	4	6	8	7
p	q	r	$\sim q$	$p \wedge \sim q$	$\leftrightarrow$	r
T	T	T	F	F	F	T
T	T	F	F	F	T	F
T	F	T	T	T	T	T
T	F	F	T	T	F	F
F	T	T	F	F	F	T
F	T	F	F	F	T	F
F	F	T	T	F	F	T
F	F	F	T	F	T	F

Note that **Step 7** was obtained from **Steps 5** and **6**, making use of **Table 2.6** for the bi-conditional statement.

Example 2.3.6 _____

Construct a truth table for the statement $(p \underline{\vee} q) \rightarrow (p \wedge \sim q)$

There is rain and not sunshine if and only if there are clouds in the sky.

*Let **p** represent "**there is rain**," **q** represent "**the sun is shining**" and **r** represent "**there are clouds**."*

This statement is not true when:

*There is **sunshine** with the **rain**, regardless of the **clouds**; or, there is **no rain** when there are **clouds**.*

Table 2.6

1	2	3
p	q	$p \leftrightarrow q$
T	T	T
T	F	F
F	T	F
F	F	T

*"**You can have ice cream and not cake if and only if you finish your dinner**"*

*This statement is false when: I have **cake** or ice cream **without finishing my dinner**; this includes **cake and ice cream**; or I **don't have cake** and I **don't have ice cream** when I do **finish my dinner**.*

*"**If you want to drink either chocolate milk or soda, then you may have chocolate milk, but not a soda.**" This statement is false when: I have the **soda** and **not chocolate milk**.*

*If there is either a **cheaper guitar** or a **payment plan** r, then I will **buy a cheaper guitar and not take the payment plan.** This statement is false when: There is **not a cheaper guitar** and I **do take the payment plan.***

Solution

Following the indicated steps we have

p	q	$p\underline{\vee}q$	$\sim q$	$p\wedge\sim q$	$(p\underline{\vee}q)\to(p\wedge\sim q)$
T	T	F	F	F	T
T	F	T	T	T	T
F	T	T	F	F	F
F	F	F	T	F	T

Or, alternatively

1	2	3	7	4	6	5
p	q	$(p\underline{\vee}q)$	$\to$	$(p$	$\wedge$	$\sim q)$
T	T	F	T	T	F	F
T	F	T	T	T	T	T
F	T	T	F	F	F	F
F	F	F	T	F	F	T

$p\Delta q\equiv\ \sim(p\to q)$

*Let **p** represent "**John is smart**," **q** represent "**Leroy is tall**" and **r** represent "**Sherrie is bashful**."*

It is the logic of our times, no subject for immortal verse – that we who lived by honest dreams. Defend the bad against the worse

C. Day-Lewis
Anglo-Irish Poet (1943)

Example 2.3.7

Let the connective Δ be defined by the truth table,

p	q	$p\Delta q$
T	T	F
T	F	F
F	T	T
F	F	F

Find the truth value for $p\to[\sim p\Delta(\sim q\wedge p)]$

p	q	$\sim p$	$\sim q$	$\sim q\wedge p$	$\sim p\Delta(\sim q\wedge p)$	$p\to[\sim p\Delta(\sim q\wedge p)]$
T	T	F	F	F	F	F
T	F	F	T	T	T	T
F	T	T	F	F	F	T
F	F	T	T	F	F	T

Alternatively,

1	2	3	9	4	8	5	7	6
p	q	p	$\to$	$[\sim p$	Δ	$(\sim q$	$\wedge$	$p)]$
T	T	T	T	F	F	F	F	T
T	F	T	T	F	T	T	T	T
F	T	F	F	T	F	F	F	F
F	F	F	T	F	T	T	F	F

Step 7 giving the truth values of $\sim q\wedge p$ was obtained from **Steps 5 and 6**, making use of Table 2.1 for $\wedge$. The truth values for $\sim p\Delta(\sim q\wedge p)$ were obtained from **Steps 4 and 7** and the truth table for the connective $\wedge$ given at the beginning of this problem. Finally, the answer in **Step 9** was derived from **Steps 3 and 8**, making use of Table 2.5 for the conditional statement.

Example 2.3.8

Determine the truth value of the following statement: "**John** is smart or **Leroy** is tall if and only if **Sherrie** is bashful implies **John** is not smart." Given that **John** is smart, **Leroy** is not tall and Sherrie is bashful.

Solution

First, we note that "**John** is smart or **Leroy** is tall" can be written symbolically as

$$p \lor q$$

Secondly, we can write the statement "**Sherrie** is bashful implies **John** is not smart" as

$$r \to \sim p.$$

Thirdly, joining the two statements with the bi-conditional "if and only if," we can represent the entire statement as

$$(p \lor q) \leftrightarrow (r \to \sim p)$$

Now, since $(p \lor q)$ is true when p is true and q is false and $(r \to \sim p)$ is false when r is true and p is true, we can write

$$(T \lor F) \leftrightarrow (T \to \sim T)$$
$$(T) \leftrightarrow (T \to F)$$

This leads us to $T \leftrightarrow F$, which means that the given statement is *false*.

EXERCISES

Critical Thinking

2.3.1. How many rows are required in a truth table for a compound statement that contains four simple statements, *p, q, r,* and *s*?

2.3.2. How many rows are required in a truth table for a compound statement that contains five simple statements, *p, q, r, s* and *t*?

Basic Problems

Construct the truth table for the following statements:

2.3.3. $\sim q \land p$

2.3.4. $\sim (p \land q)$

2.3.5. $\sim (\sim p \lor \sim q)$

2.3.6. $p \land (q \lor p)$

2.3.7. $(p \lor q) \to p$

2.3.8. $(p \land q) \to p$

2.3.9. $(p \lor q) \to (p \land q)$

2.3.10. $[(p \to q) \lor p]$

2.3.11. $[(p \land q) \lor q] \to p$

2.3.12. $[(p \to q) \lor \sim p] \to q$

2.3.13. $p \lor (q \land r)$

2.3.14. $(p \lor q) \land (p \lor r)$

2.3.15. $p \Delta \sim q$

2.3.16. $\sim p \Delta q$

Determine the truth value of the following statements given *p* is true, *q* is false and *r* is true.

2.3.17. $p \to (q \lor r)$

2.3.18. $p \land (q \lor r)$

2.3.19. $p \lor \sim p$

2.3.20. $p \land \sim (q \to r)$

2.4 PROPERTIES OF LOGIC

In certain cases of logical reasoning we are concerned with statements that are always true; for example, $p \vee \sim p$ is always *true*. If p is true then we have $T \vee F$, which is true; however, if p is false then we have $F \vee T$, which is true. More specifically, consider the following definition:

Definition 2.4.1 Tautology

A compound statement, τ, is said to be a **tautology** or **logically true** if it is true for all possible truth values of its components.

In other words a **tautology, τ,** is a compound statement which has only true, **T,** in the last column of its truth table. Note that in **Example 2.16** the $[(p \wedge q) \wedge r] \rightarrow (p \vee q)$ is a **tautology**.

Example 2.4.1 Tautology

Show that the statement $\sim (p \wedge q) \vee p$ is a tautology.

Tautology

From the Greek word ταυτολογία *meaning a formula which is* **unconditional true**, *tautology was first applied by* **Ludwig Wittgenstein** *to redundancies of propositional logic in 1921.*

A tune is a kind of tautology ... complete in itself
Wittgenstein

*If you do **not** have both **p** and **q**, then you are missing at least one; hence, you either **don't have** **p** (or **not q**), or you do have **p**—this is always true. With anything, you either you have it, or your do not, this is an unconditional truth.*

If you don't know, then you know.
You must clean both your room and the kitchen, but you don't have to do at least one of these.

Solution

Constructing the truth table for the statement $\sim (p \wedge q) \vee p$, we have

p	q	$p \wedge q$	$\sim (p \wedge q)$	$\sim (p \wedge q) \vee p$
T	T	T	F	T
T	F	F	T	T
F	T	F	T	T
F	F	F	T	T

Since the truth table for the statement $\sim (p \wedge q) \vee p$ is true, **T,** for all possible truth value combinations of p and q, the statement is a **tautology**.

Analogous to **tautologies**, there are statements that are logically false; for example, $p \wedge \sim p$ is always *false*. If p is true then we have $T \wedge F$, which is false; however, if p is false then we have $F \wedge T$, which is false.

That is,

Definition 2.4.2 Contradiction

A compound statement, ϕ, is said to be a **self-contradiction** or **logically false** if it is false for all possible truth values of its components.

This means that a self-contradiction, φ, is a compound statement that has only false, **F,** in the last column of its truth table. Consider the following example:

Example 2.4.2 Self-Contradiction

Show that the statement $(p \wedge q) \wedge \sim (p \vee q)$ is a self-contradiction.

1	2	3	4	5	6
p	q	$p \wedge q$	$p \vee q$	$\sim(p \vee q)$	$(p \wedge q) \wedge \sim(p \vee q)$
T	T	T	T	F	F
T	F	F	T	F	F
F	T	F	T	F	F
F	F	F	F	T	F

Observe that Step 6 was obtained from Steps 3 and 5, making use of Table 2.1 for the conjunction. Since only **F** appears in Step 6, we conclude that the statement $(p \wedge q) \wedge \sim (p \vee q)$ is a self-contradiction.

Definition 2.4.3 Paradox

A **paradox** is an apparently true statement or group of statements that leads to a contradiction.

For example, the liars' **paradox**: "this statement is false"; when this statement is true, this implies it is false or vice versa.

In logical reasoning we often encounter statements that are the same or equivalent.

That is,

Definition 2.4.4 Equivalent

Two statements *r* and *s* are said to be **logically equivalent** or simply **equivalent** if they have identical truth tables; that is, if $r \leftrightarrow s$ is a tautology. To symbolize two equivalent statements *r* and *s*, we write $r \equiv s$ or $r \Leftrightarrow s$.

Similarly, we write $p \Rightarrow q$ when the statement $p \to q$ is a **tautology**.

Example 2.4.3 Equivalent Statements

Show that the statement $\sim p \vee q$ is equivalent to the statement $p \to q$.

Solution

We begin by constructing the truth table for $(\sim p \vee q) \leftrightarrow (p \to q)$.

p	q	$\sim p$	$\sim p \vee q$	$p \to q$	$(\sim p \vee q) \leftrightarrow (p \to q)$
T	T	F	T	T	T
T	F	F	F	F	T
F	T	T	T	T	T
F	F	T	T	T	T

Since $(\sim p \vee q) \leftrightarrow (p \to q)$ is a tautology, we see that $(\sim p \vee q) \equiv (p \to q)$.

*When you do both, then you did not listen when I stated you **don't have to do at least one**, and when you don't do both, you did not listen to the first statement which stated do **both your room and the kitchen**.*

Archival Note
*The **Epimenides paradox** (circa 600 BCE), is a liar's paradox; what do you think form a man from **Crete** when he states "Cretans are always liars."*

Syntactically

$$r \equiv s$$

Semantically

$$r \Leftrightarrow s$$

*I will not win the lottery ticket or I will take you to dinner is equivalent to stating **if I win the lottery, then I will take you to dinner**.*

Lottery

 *From the Italian **lotteria** meaning arrangement for a distribution of prizes by chance*
If the sun is shining, then I will go swimming is equivalent to stating if I don't go swimming, then the sun was not shining.
If you are seventeen, then you are a minor is equivalent to stating if you are not a minor, then you are not seventeen.
If you have an A in this course, then you understand logic is equivalent to stating if you do not know logic, then you will not have an A in this course.
***If you are seventeen, then you are a minor** is NOT equivalent to stating **if you are a minor, then you are seventeen**. There are many ways to be a minor and not be seventeen.*

Minor

 A person under the age of full legal responsibility

Example 2.4.4 Contrapositive: $\sim q \to \sim p$

Show that $(p \to q) \equiv (\sim q \to \sim p)$.

Solution

It suffices to show that the statement $(p \to q) \leftrightarrow (\sim q \to \sim p)$ is a tautology. We can show this by structuring the appropriate truth table.

That is,

p	q	$\sim p$	$\sim q$	$p \to q$	$\sim q \to \sim p$	$(p \to q) \leftrightarrow (\sim q \to \sim p)$
T	T	F	F	T	T	T
T	F	F	T	F	F	T
F	T	T	F	T	T	T
F	F	T	T	T	T	T

Example 2.4.5 Converse: $q \to p$

Show that the statement $p \to q$ is not logically equivalent to the statement $q \to p$.

If the figure is a square, then the rectangle is NOT equivalent to stating if the figure is a rectangle, then the figure is a square. A rectangle has equal angles whereas a square has both equal angles and equal sides.

Given it is not the case that if you are an adult, then you do drive, this is equivalent to stating there are adults out there that do not drive.

Solution

We begin by constructing the truth table for the statement $(p \to q) \leftrightarrow (q \to p)$.

p	q	$p \to q$	$q \to p$	$(p \to q) \leftrightarrow (\sim q \to \sim p)$
T	T	T	T	T
T	F	F	T	F
F	T	T	F	F
F	F	T	T	T

It is clear from the results of Step 5 that the statement $(p \to q) \leftrightarrow (\sim q \to \sim p)$ is not a tautology; thus

$$(p \to q) \not\equiv (\sim q \to \sim p),$$

where $\not\equiv$ is read "not equivalent."

Example 2.4.6 Not Implied: $\sim (p \to q)$

Is the statement $\sim (p \to q)$ is equivalent to the statement $p \wedge \sim q$?

Solution

We begin by constructing the truth table for the statement $\sim (p \to q) \leftrightarrow (p \wedge \sim q)$.

1	2	3	4	5	6	7
p	q	$\sim q$	$p \to q$	$\sim (p \to q)$	$p \wedge \sim q$	$\sim (p \to q) \leftrightarrow (p \wedge \sim q)$
T	T	F	T	F	F	T
T	F	T	F	T	T	T
F	T	F	T	F	F	T
F	F	T	T	F	F	T

We observe that the truth table for the statement $\sim(p \to q)$ in **Step 5** is the same as the truth table for $p \land \sim q$ in Step 6. Thus, we conclude that $\sim(p \to q) \equiv (p \land \sim q)$. This means that the negation of the conditional statement, $p \to q$, is equivalent to $p \land \sim q$. Moreover, using this equivalence in conjunction with the equivalence in Example 2.4.3, $\sim(\sim p \lor q) \equiv (p \land \sim q)$; this distributive property called De Morgan's Laws.

Example 2.4.7 Equivalent Statements

Show that $(p \land q) \land r \equiv p \land (q \land r)$.

Solution

The equivalence relation of these two statements can be seen by constructing the following truth table:

1	2	3	4	5	6	7	8
p	q	r	$p \land q$	$(p \land q) \land r$	$q \land r$	$p \land (q \land r)$	$(p \land q) \land r$ $\leftrightarrow$ $p \land (q \land r)$
T	T	T	T	F	T	F	T
T	T	F	T	F	F	F	T
T	F	T	F	T	F	T	T
T	F	F	F	T	F	T	T
F	T	T	F	F	T	F	T
F	T	F	F	F	F	F	T
F	F	T	F	F	F	F	T
F	F	F	F	F	F	F	T

Since the truth table for $(p \land q) \land r$ in Step 5 is identical to that $p \land (q \land r)$ in **Step 7**, we conclude that, indeed, the two statements are equivalent. This is further seen in **Step 8** which shows that $(p \land q) \land r \leftrightarrow p \land (q \land r)$ is a **tautology**.

The statements that we have studied under the equivalence relation, $\equiv$, satisfy some very basic laws of algebra listed below. We shall state these laws and illustrate their usefulness with some examples. In order to prove any of these laws of algebra it is sufficient to construct a truth table to show that the given equivalence statement is indeed true.

Let p, q, and r represent given statements.

Rule 2.4.1 Idempotent

$$p \lor p \equiv p \text{ and } p \land p \equiv p$$

Idempotent describes the property of operations, as in mathematics and computer science, which yield the same result after the operation is applied multiple times.

*Given it is not the case that if you try to get pregnant that you will get pregnant is equivalent to stating **there are those out there who try to get pregnant and do not get pregnant**.*

*Given it is not the case that having money implies happiness is equivalent to stating **there are those who have money, but are not happy**.*

Commutative

 *From Latin **commutate** to change*

 Involving the quality that quantities connected by operators give the same result when commuted

$$a \times b = b \times a$$

Tautology

 *From the Greek **tautologia**—the same saying*

A statement that is true by necessity or by virtue of its logical form

Idempotence

 Applied a multiple number of times without change

Example:

$$f(x) = x$$
$$f(f(x)) = f(x) = x$$
$$\vdots$$

Comparable to Association in Addition and Multiplication

$$(a+b)+c=a+(b+c)$$
$$(a\times b)\times c=a\times(b\times c)$$

Comparable to Commutative in Addition and Multiplication

$$a+b=b+a$$

$$a\times b=b\times a$$

Not directly comparable to Distributive in Addition and Multiplications *multiplication is distributive over addition, but not the reverse.*

Whereas in Logic, conjunction distributions over disjunction and disjunction distributions over conjunction.

Not exactly comparable to the Identity in Addition and Multiplication *as the last statement would not follow in addition*

$$a + 0 = a \checkmark$$
$$a \times 1 = a \checkmark$$
$$a \times 0 = 0 \checkmark$$
$$a + 1 = a \; \text{✗}$$

Directly comparable to maximum of 0/1 versus minimum 0/1; that is, $a=0$ (false) and $a=1$ (true):

$$max\{a, 0\} = a \checkmark$$
$$min\{a, 1\} = a \checkmark$$
$$min\{a, 0\} = 0 \checkmark$$
$$max\{a, 1\} = 1 \checkmark$$

or $\equiv$ *addition*
and $\equiv$ *multiply*

> **Rule 2.4.2 Associative**
> $$(p \vee q) \vee r \equiv p \vee (q \vee r) \text{ and } (p \wedge q) \wedge r \equiv p \wedge (q \wedge r)$$

Associative describes the property of operations that enables statements to re-associate while yielding the same result. The associative law in logic is comparable to the associative law of addition and multiplication in algebra; that is, $(a+b)+c=a+(b+c)$ and $(a\times b)\times c=a\times(b\times c)$.

> **Rule 2.4.3 Commutative**
> $$p \vee q \equiv q \vee p \text{ and } p \wedge q \equiv q \wedge p$$

Commutative describes the property of operations that enables statements to move or commute while yielding the same result. The commutative law in logic is comparable to the commutative law of addition and multiplication in algebra; that is, $a+b=b+a$ and $a\times b=b\times a$.

> **Rule 2.4.4 Distributive**
> $$p \vee (q \wedge r) \equiv (p \vee q) \wedge (p \vee r) \text{ and } p \wedge (q \vee r) \equiv (p \wedge q) \vee (p \wedge r)$$

Distributive describes the property of one operator to be expanded in a particular way which yield the same result; that is, an equivalent expression. The distributive law in logic is more extensive than the distributive property in algebra. In logic, the operator for "and" is distributive over the operator for "or" and vise verse, whereas in algebra, this relation only works for "multiplication" over "addition" but not "addition over multiplication." However, the idea of distribution of one operator over a second operator is comparable; for multiplication over addition we have $a(b+c)=ab+ac$ and for power over multiplication we have $(a\times b)^n=a^n\times b^n$.

> **Rule 2.4.5 Identity**
> If τ is a **tautology** and φ is any **contradiction**, then
> $$p \vee \phi \equiv p, \; p \wedge \tau \equiv p, \; p \wedge \phi \equiv \phi \text{ and } p \vee \tau \equiv \tau$$

Identity is the state or fact of remaining the same one or ones, as under varying aspects or conditions, to identify. Hence, a contradiction in a disjunction identifies the truth value of the statement p, but in conjunction identifies the contradiction. A tautology in disjunction identifies the tautology, whereas in conjunction identifies the true value of the statement p. The first two are comparable to the identity property of addition and multiplication: $a+0=a$ and $a\times 1=a$, where 0 is compared to the contradiction and 1 is compared to the tautology; therefore, "or" is comparable to "addition" and "and" is comparable to "multiplication." The third is comparable to zero property, $a\times 0=0$; however, the four as will the law of distribution is not exactly comparable to any given algebraic property or law. These comparisons will come back into play in Chapter 3, which introduces the fundamental principle of counting; basically "and" means "multiply" (when there are more than one independent events) and "or" means "add" (minus overlap).

Rule 2.4.6 Complementary

If τ is a **tautology** and ϕ is any **contradiction**, then

$$p \vee \sim p \equiv \tau, \; p \wedge \sim p \equiv \phi, \; \sim(\sim p) \equiv p, \; \sim\tau \equiv \phi \text{ and } \sim\phi \equiv \tau$$

Directly comparable to maximum of 0/1 versus minimum 0/1; that is, a=0 (false) and a=1 (true); if p is a, then ~p is 1−a:

$$max\{a, 1-a\} = 1$$
$$min\{a, 1-a\} = 0$$
$$1 - 1 = 0$$
$$1 - 0 = 1$$

A **complement** is the part needed to make complete or perfect; in **logic**, this is the relationship between *true* and *false*, an event and not the event, etc. In **logic**, the statement is true or the statement is false and thus $p \vee \sim p$ is a **tautology**, τ; however, a statement cannot be both true and false; hence, $p \wedge \sim p$ is a contradiction, ϕ. Furthermore, as in English, a double negative is the positive statement; not "not p" then p. Finally, logically speaking, the negation of a **tautology** is a **contradiction** and the negation of a **contradiction** is a **tautology**. They complement each other, without tautologies, contradictions would not exist.

Rule 2.4.7 De Morgan's Rule

$$\sim(p \vee q) \equiv \sim p \wedge \sim q \text{ and } \sim(p \wedge q) \equiv \sim p \vee \sim q$$

De Morgan's Rule illustrates the fact that "or" is the complement of "and" and vice versa. In logic, "or" means at least one and "and" means both; hence, when *you do not have at least one*, $\sim(p \vee q)$, this is equivalent to you do not have either, or you are missing both, $\sim p \wedge \sim q$; *you do not have the first and you do not have the second*. Similarly, when *you do not have both*, $\sim(p \wedge q)$, this is equivalent to you are missing at least one, $\sim p \vee \sim q$; *you do not have the first or you do not have the second*. The idea of "negation" will be discuss further in Section 2.6.

It is left to the student as an exercise to construct the truth tables to show the stated equivalence of the preceding laws of algebra. However, we use these laws to simplify compound statements.

De Morgan was a British mathematician and logician who formulated De Morgan's laws and introduced rigor to mathematical induction.

This will be similar in sets or events:
When you do not have at least one, then you do not have the first and you do not have the second.
   When you do have both, you are missing at least one; either the first one or the second one (or you are missing both), you just do not have both.

Let's use the following laws:
➲ *Commutative*
➲ *Distribution*
➲ *Complement*
➲ *Identity*
➲ *De Morgan's Law*

Example 2.4.8 Properties of Logic

Simplify each of the following statements by using the laws of statements and using t for **tautologies** and ϕ for **contradictions**:

(a) $(p \vee q) \wedge \sim p$ (b) $p \vee (p \wedge q)$

(c) $[\sim(p \vee q)] \vee (\sim p \wedge q)$ (d) $[\sim(p \wedge q)] \vee (p \wedge \sim q)$

I will do the dishes or wash the car, but I will not do the dishes is equivalent to saying I will not do the dishes and I will wash the car.

Solution

	Statement (a)	Reason
1.	$(p \vee q) \wedge \sim p \equiv \sim p \wedge (p \vee q)$	**Commutative Law**
2.	$\equiv (\sim p \wedge p) \vee (\sim p \wedge q)$	**Distributive Law**
3.	$\equiv \phi \vee (\sim p \wedge q)$	**Complement Law**
4.	$\equiv (\sim p \wedge q)$	**Identity Law**

*I have ice cream, or I have ice cream and cake is equivalent to **I have ice cream, and I have ice cream or cake**.*

*Either I don't have at least one (p or q) or I am missing the first (p) and have the second (q) is equivalent to **I don't have the first (p)**. As for the second, I either have it or I do not—it is irrelevant.*

*Either I do not have both (p and q) or I have the first (p) and not the second (q) is equivalent to stating **I do not have both the first and second**.*
*   **I lost my pen and paper, or I have the pen and not the paper** is equivalent to stating **I lost either the pen or I lost the paper**; that is, **I do not have both the pen and paper**.*

	Statement (b)	Reason
1.	$p \vee (p \wedge q) \equiv (p \vee p) \wedge (p \vee q)$	**Distributive Law**
2.	$\equiv p \wedge (p \vee q)$	**Identity Law**

	Statement (c)	Reason
1.	$[\sim(p \vee q)] \vee (\sim p \wedge q)$	**De Morgan's Law**
	$\equiv (\sim p \wedge \sim q) \vee (\sim p \wedge q)$	
2.	$\equiv \sim p \wedge (\sim q \vee q)$	**Distributive Law**
3.	$\equiv \sim p \wedge \tau$	**Complement Law**
4.	$\equiv \sim p$	**Identity Law**

	Statement (d)	Reason
1.	$[\sim(p \wedge q)] \vee (p \wedge \sim q)$	De Morgan's Law
	$\equiv (\sim p \vee \sim q) \vee (p \wedge \sim q)$	
2.	$\equiv [(\sim p \vee \sim q) \vee p] \wedge [(\sim p \vee \sim q) \vee \sim q]$	Distributive Law
3.	$\equiv [p \vee (\sim p \vee \sim q)] \wedge [(\sim p \vee \sim q) \vee \sim q]$	Commutative Law
4.	$\equiv [(p \vee \sim p) \vee q] \wedge [\sim p \vee (\sim q \vee \sim q)]$	Associative Law
5.	$\equiv [\tau \vee \sim q] \wedge [\sim p \vee (\sim q \vee \sim q)]$	Complement Law
6.	$\equiv [\tau \vee \sim q] \wedge [\sim p \vee \sim q]$	Idempotent Law
7.	$\equiv \tau \wedge [\sim p \vee \sim q]$	Identity Law
8.	$\equiv \sim p \vee \sim q$	Identity Law
9.	$\equiv \sim(p \wedge q)$	De Morgan's Law

EXERCISES

Summary

⮑ **Idempotent Law**

$$p \wedge p = p \qquad p \vee p = p$$

⮑ **Associative Law**

$$p \wedge (q \wedge r) = (p \wedge q) \wedge r \qquad p \vee (q \vee r) = (p \vee q) \vee r$$

⮑ **Commutative Law**

$$p \wedge q = q \wedge p \qquad p \vee q = q \vee p$$

⮑ **Distributive Laws**

$$p \vee (q \wedge r) = (p \vee q) \wedge (p \vee r) \qquad p \wedge (q \vee r) = (p \wedge q) \vee (p \wedge r)$$

⮑ **Identity Law**

$$p \vee \phi = p, \quad p \wedge \tau = p, \quad p \wedge \phi = \phi, \quad p \vee \tau = \tau$$

⮑ **Complement Law**

$$p \vee \sim p = \tau, \quad p \wedge \sim p = \phi, \quad \sim(\sim p) = p, \quad \sim \tau = \phi, \quad \sim \phi = \tau$$

⮑ **De Morgan's Laws**

$$\sim(p \vee q) = \sim p \wedge \sim q, \quad \sim(p \wedge q) = \sim p \vee \sim q$$

Critical Thinking

2.4.1. Is the statement $\sim(p \vee q) \wedge p$ a tautology?

2.4.2. Show that $p \vee \sim(p \wedge q)$ is a tautology.

2.4.3. Determine whether or not the statement $(\sim p \wedge \sim q) \vee (p \vee q)$ is a tautology.

2.4.4. Is the statement $(p \vee q) \vee \sim(p \wedge q)$ a tautology?

2.4.5. Determine whether or not the statement $(p \vee q) \rightarrow \sim(p \wedge q)$ is a self-contradiction.

2.4.6. What can you say about the two statements $q \rightarrow p$ and $\sim p \rightarrow \sim q$

2.4.7. Show that the statement $(p \wedge q) \vee \sim (p \vee q)$ is a self-contradiction.

2.4.8. Is the statement $p \rightarrow q$ logically equivalent to the statement $q \rightarrow \sim p$?

2.4.9. Prove that $(p \vee q) \vee r \equiv p \vee (q \vee r)$.

2.4.10. Show that $(p \vee q) \wedge \sim p \equiv \sim p \wedge q$.

2.4.11. Is the statement $p \vee q$ equivalent to $\sim (\sim p \wedge \sim q)$?

2.4.12. Verify the associative law $(p \wedge q) \wedge r \equiv p \wedge (q \wedge r)$ by constructing the appropriate truth table.

2.4.13. Prove the distributive law $p \vee (q \wedge r) \equiv (p \vee q) \wedge (p \vee r)$.

Basic Problems

2.4.14. Show that $\sim (p \wedge q) \equiv \sim p \vee \sim q$.

2.4.15. What can you say about the statements $\sim (\sim p)$ and p?

2.4.16. Simplify the statement $(p \wedge q) \wedge \sim p$.

2.4.17. Using the laws of algebra simplify the statement $p \wedge (p \wedge q)$.

2.4.18. Simplify $\sim (p \wedge q) \wedge (\sim p \wedge q)$.

2.4.19. Simplify $\sim (\sim p \wedge q)$.

Determine if the following statement is a **tautology**, **self-contradiction** and **paradox**.

2.4.20. There is an exception to every rule; except this rule.

2.4.21. To be or not to be.

2.4.22. You travel back in time and kill your grandfather before he meets your grandmother. Hence, you are never born and, therefore, you couldn't go back in time and kill your grandfather.

2.4.23. Yes and no, and I don't mean maybe.

2.5 VARIATIONS OF THE CONDITIONAL STATEMENT

We have seen that equivalent statements have identical truth tables and may be thought of as different forms of the same statement. In this section we shall be concerned with some of the different forms by which the conditional statement $p \rightarrow q$ can be expressed. That is, given a conditional statement $p \rightarrow q$, we shall study three variations formed from statements p, q, and the logical connectives, $\rightarrow$ and $\sim$.

Statement	Name
$p \rightarrow q$	Conditional Statement
$q \rightarrow p$	**Converse** of $p \rightarrow q$
$\sim p \rightarrow \sim q$	**Inverse** of $p \rightarrow q$
$\sim q \rightarrow \sim p$	**Contra-positive** of $p \rightarrow q$

A comparison of the truth tables for these statements is given by Table 2.8.

TABLE 2.8 Conditional variations

	1	2	3	4	5	6	7	8
	p	q	$\sim p$	$\sim q$	$p \rightarrow q$	$q \rightarrow p$	$\sim p \rightarrow \sim q$	$\sim q \rightarrow \sim p$
	T	T	F	F	T	T	T	T
	T	F	F	T	F	T	T	F
	F	T	T	F	T	F	F	T
	F	F	T	T	T	T	T	T

Conditional

 Subject to one or more requirements

Sentences that discuss factual implication

Example:

If you are seventeen then you are a minor.

If you are a minor, then you are seventeen.

If you are not seventeen, then you are not a minor.

If you are not a minor, then you are not seventeen.

Given the conditional statement is true, the inverse and the converse need not follow.

If the converse and the inverse do hold, this would be a bi-conditional.

Common: Equivalent Statements

Analyzing Table 2.8 on the previous page we observe that

1. $p \rightarrow q$ is not equivalent to $q \rightarrow p$ because Columns 5 and 6 are **not identical**.

2. $p \rightarrow q$ is not equivalent to $\sim p \rightarrow \sim q$ because Columns 5 and 7 are **not identical**.

3. $p \rightarrow q$ is equivalent to $\sim q \rightarrow \sim p$ because Columns 5 and 8 are **identical**.

4. $q \rightarrow p$ is equivalent to $\sim p \rightarrow \sim q$ because Columns 6 and 7 are **identical**.

If I win the lottery, then I will buy you dinner.

If I buy you dinner, then I will win the lottery? This is not equivalent.

If I do not win the lottery, then I cannot buy you dinner? This does not follow.

If I do not take you to dinner, then I have not won the lottery. This is true.

Among the three different forms statements, the first two are equivalent to each other.

Conditional Statement
$$p \to q$$
Converse
$$q \to p$$
Inverse
$$\sim p \to \sim q$$
Contrapositive
$$\sim q \to \sim p$$

Original Statement:
A well supported figure.

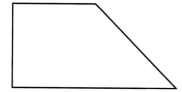

Inverse Statement:

Not the same support

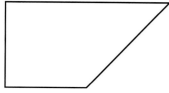

Converse Statement:

Not the same support as the original, but the same as the inverse.

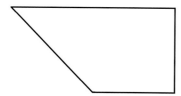

Contrapositive Statement:

Same support as the original figure.

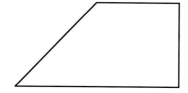

Therefore, there are two equivalences: $p \to q \equiv \sim q \to \sim p$ and similarly $q \to p \equiv \sim p \to \sim q$. Hence, a conditional statement is equivalent to the counter of the positive statement, that is, the contra-positive.

Example 2.5.1 Contra-positive

Given the conditional statement "If you are seventeen, then you are a minor," find the **converse**, **inverse**, and the **contra-positive** of the statement.

Solution

Converse: If you are a minor, then you are seventeen.
Inverse: If you are not seventeen, then you are not minor.
Contra-positive: If you not a minor, then you are not seventeen.

By definition of a minor, both the original conditional statement and the contra-positive statement both true (that is, equivalent statements) whereas, the converse and inverse are not true. If you are a minor, you might be twelve and not seventeen; and if you are not seventeen, then you might be sixteen which is still a minor. Hence, in this case the conditional statement and the contra-positive statement are true and the converse and the inverse are false; however, this need not be the case. If all four statements are true, then the conditional statement is actually a bi-conditional statement. However, in general, $\to$ is not commutative; that is, $p \to q \not\equiv q \to p$.

Example 2.5.2 Conditional Statements

Suppose that p is true and q is false. What is the truth value of the following statements?
(a) $p \to q$
(b) The inverse of $\sim q \to p$
(c) The contra-positive of $\sim q \to \sim p$
(d) q is sufficient for $\sim p$
(e) $\sim p$ is necessary for $\sim q$
(f) The converse of $\sim q$ only if p

Solution

(a) For $p \to q$ we have $T \to F$, which is false by **Table 2.8**.
(b) The **inverse** of $\sim q \to p$ is $\sim(\sim q) \to \sim p$ or **equivalently** $q \to \sim p$. Thus, we have $F \to \sim T$ or $F \to F$, which is vacuously true.
(c) The **contra-positive** of $\sim q \to \sim p$ is $\sim(\sim q) \to \sim(\sim p)$ or equivalently $q \to p$. Hence, $F \to T$, which is vacuously true.
(d) The statement "q is sufficient for $\sim p$" is another way of saying "if q then $\sim p$" or, symbolically $q \to \sim p$. Thus, we have $F \to \sim T$, which is equivalent to $F \to F$, which is vacuously true.
(e) The statement "$\sim p$ is necessary for $\sim q$" is another way of saying "if you have $\sim q$, then it was necessary that you had $\sim p$"; in other words, "if $\sim q$ then $\sim p$" or written symbolically $\sim q \to \sim p$. Hence, $\sim F \to \sim T$, which gives us $T \to F$, which is false. Alternatively, $\sim q \to \sim p$ is the contra-positive of $p \to q$ which, by (a) is false.
(f) "$\sim q$ only if p" can be written as $\sim q \to p$ and therefore the converse is $p \to \sim q$. Since p is true and q is false, we have $T \to \sim F$, that is, $T \to T$, which is true.

Example 2.5.3 Associative

Show that the associative rule does not hold for $\rightarrow$; that is, $(p \rightarrow q) \rightarrow r \not\equiv p \rightarrow (q \rightarrow r)$.

Solution

Here, to prove that $(p \rightarrow q) \rightarrow r$ is not equivalent to $p \rightarrow (q \rightarrow r)$.

This is shown by structuring the following truth table:

1	2	3	4	5	6	7	8
p	q	r	$p \rightarrow q$	$q \rightarrow r$	$(p \rightarrow q) \rightarrow r$	$p \rightarrow (q \rightarrow r)$	$(p \rightarrow q) \rightarrow r \leftrightarrow p \rightarrow (q \rightarrow r)$
T	T	T	T	T	T	T	T
T	T	F	T	F	F	F	T
T	F	T	F	T	T	T	T
T	F	F	F	T	T	T	T
F	T	T	T	T	T	T	T
F	T	F	T	F	F	T	F
F	F	T	T	T	T	T	T
F	F	F	T	T	F	T	F

Since **Steps 6** and **7** do not yield identical truth tables for $(p \rightarrow q) \rightarrow r$ and $p \rightarrow (q \rightarrow r)$, we observe that these two statements are not equivalent; that is, $(p \rightarrow q) \rightarrow r \leftrightarrow p \rightarrow (q \rightarrow r)$ is not a tautology. Thus $\rightarrow$ is not associative.

Example 2.5.4 Equivalency

Show that the following statements are equivalent without the use of truth tables:
(a) $p \rightarrow \sim q$ and $\sim q \rightarrow p$
(b) $p \rightarrow (q \vee p)$ and $(\sim q \wedge \sim r) \rightarrow \sim p$

Solution

(a) The contra-positive of $p \rightarrow \sim q$ is $\sim q \rightarrow \sim(\sim p)$, which we know is equivalent to $\sim q \rightarrow p$
(b) The contra-positive of $p \rightarrow (q \vee p)$ is $\sim(q \vee r) \rightarrow \sim p$ and since by De Morgan's Law, $\sim(q \vee r) \equiv \sim q \wedge \sim r$; hence, $p \rightarrow (q \vee p)$ is equivalent to $(\sim q \wedge \sim r) \rightarrow \sim p$.

In logical reasoning it is often necessary to prove a statement of the form "p if and only if q"; that is, $p \leftrightarrow q$. A truth table may be used to show that $p \leftrightarrow q$ is equivalent to $(p \rightarrow q) \wedge (q \rightarrow p)$; that is, $p \leftrightarrow q \equiv (p \rightarrow q) \wedge (q \rightarrow p)$.

$$p \rightarrow q \equiv q \vee \sim p$$
$$\&$$
$$q \rightarrow r \equiv r \vee \sim q$$

are equivalent to
$$(p \rightarrow q) \equiv (q \vee \sim p) \rightarrow r$$
$$\&$$
$$p \rightarrow (q \rightarrow r) \equiv$$
$$p \rightarrow (r \vee \sim q)$$

therefore
$$LHS = r \vee \sim(q \vee \sim p)$$
$$= r \vee \sim q \wedge p$$

and
$$RHS = (r \vee \sim q) \vee \sim p$$
$$= r \vee \sim q \vee \sim p$$

If you have significant debt than you should not by more cloths is equivalent to If you have money to buy new cloths, then you are not in significant debt.

If you have a National Merit Scholarship, then you can afford to go to university in the State of Florida or out of State is equivalent to If you cannot afford to go to university out of state and you cannot afford to go to university in the State of Florida, then you must not have a National Merit Scholarship.

EXERCISES

Critical Thinking

Given the logical statement, give the **converse**, **inverse**, and **contra-positive**.

2.5.1. "If John is blonde then Sandra is a brunette"

2.5.2. Given the logical statement "If Maria leaves for the moon then Liz will be going to Albuquerque"

Let p and q represent two statements where p is **false** and q is **true**. Determine the truth value of the following statements:

2.5.3. $p \rightarrow q$,
2.5.4. The inverse of $\sim p \rightarrow q$,
2.5.5. The converse of $\sim p$ only if q.

Let p and q represent two statements where p is **false** and q is **true**. Determine the truth value of the following statements:

2.5.6. $p \rightarrow q$
2.5.7. The inverse of $\sim p \rightarrow q$,
2.5.8. The converse of $\sim p$ only if q.

Show that the logical statements are **equivalent**.

2.5.9. $\sim(p \vee q) \vee (p \wedge \sim q)$ and $\sim q$
2.5.10. $p \underline{\vee} q$ and $(p \vee q) \wedge \sim (p \wedge q)$

Archival Note
*Categorical propositions are discussed in Aristotle's **Prior Analytics**. These types of propositions occur in categorical syllogisms.*

Syllogism

 From the Greek συλλογισμός meaning to conclude or infer

A discourse in which, certain things having been supposed, something different from the things supposed results of necessity because these things are so.

Aristotle

*Leonhard Euler was a Swiss mathematician and physicist working in such fields as mechanics, fluid dynamics, optics and astronomy; whereas in set theory, he introduce **Euler circles**, he also defined the mathematical constant **e**.*

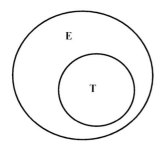

2.6   QUANTIFIED STATEMENT

There are four other commonly used forms of English phrases with which one should become familiar. These are the quantifying statements:

I.	All p are q	*Universal Affirmative*
II.	No p are q	*Universal Negative*
III.	Some p are not q	*Particular Affirmative*
IV.	Some p are q	*Particular Negative*

In order to see how to negate these forms, we must first consider exactly what they mean.

FORM I: Universal Affirmation

Form I: The ***universal affirmative***; this extreme "all p are q" is actually a conditional statement; for example in the statement "all teachers (are people who) give exams," if the **antecedent** is that "you are a teacher," then **consequence** is that "you give exams." In addition, as with any conditional statement, this implication only goes one way. That is, if you are known to give exams, this does not necessarily imply that you are in fact a teacher: for example, you may be a doctor giving a physical exam. Hence, the statement "all teachers give exams" is equivalent to "if you are a teacher, then you give exams." Thus, abbreviate "you are a teacher" as t and "people who give exams" as e, then mathematically, the statement "all teachers give exams" written as $t \rightarrow e$.

Another way to think of "all t are e" is in the context of containment (or as subsets). The group of teachers is contained in the large group of people who give exams. This idea can be illustrated using circles: a Swiss mathematician name Euler used them in the 1700s and for this reason these circles are sometimes called Euler circles. Whatever, if we let one circle represent the group of "teachers" (abbreviated **T**) and another circle represent the group of people who give "exams" (abbreviated **E**), then the relationship between these two circles or sets of people can be illustrated as shown to the left.

Example 2.6.1 "All *p* are *q*" to "If *p*, then *q*"

"All mothers are teachers" written in "if … then …" form is "if you are a mother, then you are a teacher."

Hence, "all *p* are *q*" is logically equivalent to $p \rightarrow q$.

- *A proposition* is a universal affirmative: *All S is P*
- *E proposition is a universal negative*: *No S is P*
- *I proposition* is a particular affirmative: *Some S is P*
- *O proposition* is a particular negative: *Some S is not P*

Universal Negatives:
➤ *No man is mortal.*
➤ *No publicity is bad*
➤ *No man is an island*

The above statements are equivalent to:
➤ *If you are a man, then you are not mortal*
➤ *If it gets you publicity, then it is not bad*
➤ *If you are a man, then you are not an island*

FORM II: Universal Negative

Form II: The **universal negative** is the extreme "none (no) *p* are *q*" is also a conditional; for example consider the statement "there are no good men" or "no man is good." Opinions aside, if the antecedent is that "you are a man," the consequence is that "you are not good." In addition, as with any conditional statement, this implication only goes one way. That is, if you are known to be bad (not good), this does not necessarily imply that you are in fact a man: because if you are bad, then you could have been a woman. Hence, the statement "there are no good men" is equivalent to "if you are a man, then your not good." Note: for all those who think this statement is false will have the opportunity to contradict me later, but for now this statement is assumed to be TRUE. Thus, if we let "you are a man" be abbreviate as *m* and "you are good" be abbreviated as *g*, then the statement "there are no good men" can be written mathematically as $m \rightarrow \sim g$.

Another way to think of "none (no) *m* are *g*" is in the context of non-containment (mutually exclusive). The group of men has nothing to do with the group of people who are good and the group of good people has nothing to do with the group of men. If we let one circle represent the group of "men" (abbreviated **M**) and another circle represent the group of people are "good" (abbreviated **G**), then the relationship between these two circles or sets of people can be illustrated as follows.

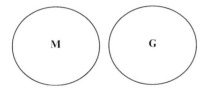

Example 2.6.2 "No *p* are *q*" to "If …, then …"

"No vegan eats eggs" written in "if … then…" form is "If you are a vegan, then you do not eat eggs" or "if you eat eggs, then you are not a vegan." Hence, "no *p* are *q*" is logically equivalent to $p \rightarrow \sim q$ or $q \rightarrow \sim p$.

FORM III: Negation of "All *p* are *q*"

Form III: This particular level of logic is not an extreme; in fact, "some *p* are not *q*" is the exact opposite of the extreme "all *p* are *q*"; for example, in the statement "some teachers do not give exams" is equivalent to the statement "Not all teachers give exams." Common sense aside, this statement leads to several consequences; that is, for you to be the one who make this statement true, then you must both be a teacher and not give exams; maybe you teach kindergarten? However, if you are not a teacher, you still may or may not give exams. Hence, the statement "Some teachers do not give exams" is not as easy to write any other way. Therefore, the best way to consider of "some are not" is in the context of partial-containment (overlapping sets). The group of teachers

*Using **Euler circles**, we can illustrate the universal negative as mutually exclusive events.*

Vegan

 *Coined by **Donald Watson** to distinguish those who abstain from all animal products including eggs and cheese and not just those who merely refuse to eat the meat from an animal.*

This second logic, then, I mean the worse one, the teach to talk unjustly, and prevail

*   **Aristophanes**
                     The Clouds (423 BCE)*

Particular Affirmative:
➢ *Some men are boys*
➢ *Sometimes too much is bad thing*
➢ *Some horses are white*

Is there any relationship among them?

∀
**is read as
"for all"**

is only partially contained in the group of people who give exams. If we let one circle represent the group of "teachers" (abbreviated **T**) and another circle represent the group of people who give "exams" (abbreviated **E**), then the relationship between these two circles or sets of people can be illustrated as follows.

"Some teachers do not give exams"

Example 2.6.3 "Some *p* are not *q*"

"Some automobiles are not cars" translates to "It is not the case that all automobiles are cars" or "it is not the case that if a vehicle is an automobile then it is a car." Hence, "some *p* are not *q*" is logically equivalent to $\sim(p \rightarrow q)$.

FORM IV: Negation of "No *p* are *q*"

Form IV: This particular level of logic is not an extreme; in fact "some *p* are *q*" is the exact opposite of the extreme "none (no) *p* are *q*"; for example the statement "some men are good" is equivalent to the statement "it is not that there no good men." Proper English aside, this statement can again lead to several consequences; that is if you are the one who make this statement true, then you must both be a man and you must be good; there are a few of you? However, if you are not a man, you still may or may not be good. Hence, the statement "some men are good" is not as easy to write any other way. Therefore, the best way to consider "some … are …" is again in the context of partial-containment (overlapping sets). The group of men is only partial contained in the group of people who are good. If we let one circle represent the group of "men" (abbreviated **M**) and another circle represent the group of people who are "good" (abbreviated **G**), then the relationship between these two circles or sets of people can be illustrated as follows.

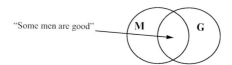

"Some men are good"

Example 2.6.4 "Some *p* are *q*"

"Some children are well behaved" translates to "It is not the case that there are no well-behaved children" or "It is not the case that if you are a child that you are not well behaved."

Hence, "some *p* are *q*" is logically equivalent to $\sim(p \rightarrow \sim q)$ or $\sim(q \rightarrow \sim p)$.

Let $p(x)$ be an ***open statement*** or ***predicate***; for example, let $p(x) =$ "if *x* is odd, then $x - 2 \geq 7$." Consider the truth value of this statement for $U = \{1, 2, 3, \dots\}$, then the open sentence $p(x)$ represents many statements, one for each $x \in U$.

$p(1) =$ "if 1 is odd, then $1 - 2 \geq 7$" is false, whereas
$p(2) =$ "if 2 is odd, then $2 - 2 \geq 7$" is vacuously true.

Definition 2.6.1 Universal Quantifier: $\forall$

The **universal quantifier** is "for all", denote by an upside-down A, $\forall$. The statement $\forall x \in U[p(x)]$ is true if and only if $p(x)$ is true for all $x \in U$.

For example, let the universal set be the set of integers, then "for all natural numbers n, n is greater than zero" can be translated as

$$\forall x (x > 0)$$

Example 2.6.5 "For all"

Given $U = \{0, 1, 2, 3, 4, 5, 6, 7, 8, 9\}$, determine the truth value for $\forall x (x^2 < 81)$.

Solution

$\forall x (x^2 < 81)$ translates to "for all digits, the digit squared is less than 81." However, the set of digits squared is $\{0, 1, 4, 9, 16, 25, 36, 49, 64, 81\}$ and hence the truth value of this is false; there exists a digit such that the digit square is not less than 81, but equal to 81.

Definition 2.6.2 Existential Quantifier: $\exists$

The **existential quantifier** is "**there exist**," denote by a backwards E, $\exists$. The statement $\exists x \in U[p(x)]$ is true if and only if there exist at least one $x \in U$ for which $p(x)$ is true.

For example, "there exists a integer such that this integer squared is less than 5," can be translated as

$$\exists x (x^2 < 5).$$

Example 2.6.6 "There exist"

Given $U = \{0, 1, 2, 3, 4, 5, 6, 7, 8, 9\}$, determine the truth value for $\exists x (x + 3 \geq 10)$

Solution

$\exists x (x + 3 \geq 10)$ translates to "there exists a digit such the value three more than the digit is greater than or equal to 10." Since the subset $A = \{7, 8, 9\}$ has three elements such that $x \in A \rightarrow x + 3 \geq 10$, there exists at least one value of x such that $x + 3 \geq 10$ is true, and thus $\exists x (x + 3 \geq 10)$ is true.

Example 2.6.7 "For all" and "There exist"

Let the universal set be the set of all college students and
    $p(x, y) = x$ is a friend of y
    $q(x, y) = x$ takes a class with y
Write the English sentence from the symbolic statement:
(a) $\forall x \forall y [p(x, y) \rightarrow q(x, y)]$
(b) $\forall x \exists y [p(x, y)]$

$\forall$
vs.
$\exists$

1. *The statement $P(x, y)$ holds for all x and for all y if and only if the statement $P(x, y)$ holds for all y and for all x.*
2. *Given there exists an x and there exists a y such that the statement holds true if and only if there exists a y and there exists an x such that the statement holds true.*
3. *Given there exists an x such that for all y the statement $P(x, y)$ holds true, if and only if for all y, there exists an x such that the statement $P(x, y)$ holds true.*

Solution

(a) For all college students, x and y, if x is a friend of y, then x takes a class with student y.

(b) For all college students, x, there exist a student y such that x is a friend of y.

Let $p(x,y)$ be an open statement regarding two variables x and y, the following are equivalent

(1) $\forall x \forall y\, p(x,y) \Leftrightarrow \forall y \forall x\, p(x,y)$

(2) $\exists x \exists y\, p(x,y) \Leftrightarrow \exists y \exists x\, p(x,y)$

(3) $\exists x \forall y\, p(x,y) \Leftrightarrow \forall y \exists x\, p(x,y)$

However, it should be noted that any other exchanges of $\forall$ and $\exists$ needs to be handled very carefully as they are unlikely to give equivalent statements for all cases.

EXERCISES

Critical Thinking

Write the given statements in symbolic form and illustrate with Euler circles.

2.6.1. All elephants are pink.

2.6.2. Some cars are Hondas

2.6.3. Some politicians are dishonest

2.6.4. Some people are not Democrats

2.6.5. No children are allow to drive

The following statements are from the writings of Lewis Carroll. Write each statement in "if … then …" form.

2.6.6. All my poultry are ducks.

2.6.7. All my sons are slim.

2.6.8. Opium-eaters have no self-control.

2.6.9. Donkeys do have not horns.

2.6.10. Some apples are not ripe.

2.6.11. No porcupines are talkative.

2.6.12. Some chickens are cats.

Translate the following in to complete English statements assuming the universal discourse is the set of all real numbers.

2.6.13. $\forall x, x^2 \geq 0$

2.6.14. $\forall x \exists y (y = 2x)$

2.6.15. $\forall x \forall y \left[(x = y) \rightarrow (x^2 = y^2) \right]$

2.6.16. $\exists x \exists y (xy = 2)$

2.7   NEGATING STATEMENTS

Symbolic logic may be usefully employed to solve an interesting problem; namely, that of forming an accurate and concise negation of an English statement. In order to correctly negate a statement, one must first translate it into symbolic form. To accomplish this task, we can use some of the algebra of statements discussed in previous sections. We should then be able to translate the symbolic negation into smooth and correct English. The principal rules that are involved in our task have negating English has been previously discussed. They are as follows:

Archival Note

Symbolic logic *was discussed in* ***Aristotle's Prior Analytics*** *as part of deductive reasoning*

Common: Negations

$\sim(\sim p) \equiv p$	Double negative, is the positive statement $\sim(p \vee q) \equiv \sim p \wedge \sim q$ Not at least one is equivalent to not having either; that is, not the first and not the second (De Morgan's Law)
$\sim(p \wedge q) \equiv \sim p \vee \sim q$	"Not both" is equivalent to not having at least one; that is, "not the first or not the second." (De Morgan's Law)
$\sim(p \rightarrow q) \equiv p \wedge \sim q$	When the first statement does not imply the second statement, the first statement can be true when (and) the second statement fails (that is, is false)

Note that **De Morgan's laws** state clearly the correct way to negate both the *disjunction* and the *conjunction* of two statements *p* and *q*. That is, De Morgan's laws indicate the method to use in negating English statements that involve the words "or" or "and" as the primary connective. Example 2.7.1 is concerned with the negation of conditional statements. With these rules one can accurately negate even very complex English statements. We shall illustrate their use in the following examples:

Example 2.7.1 ACC Championship

Consider the statement "**The ACC championship was won by UNC or Wake Forest.**" What is its negation?

Solution

Suppose that we wish to negate this statement. We could simply say that "it is not the case that the ACC championship was won by UNC or Wake Forest."

However, this style is stilted and it is still not clear exactly what is meant. To clarify matters let us first translate the original statement into symbolic form by letting

u: UNC won the ACC championship
w: Wake Forest won the ACC championship.

The original statement is $u \vee w$. By De Morgan's law, we have $\sim(u \vee w) \equiv \sim u \wedge \sim w$. Translating this into English we see that the negation of our original statement is

The ACC championship was not won by UNC and it was not won by Wake Forest.

Another way to express this is to say

The ACC championship was won by neither UNC nor Wake Forest.

This statement is concise and cannot be easily misunderstood. Also, note that *u* represents **U**NC and *w* represents **W**ake instead of the conventional *p* and *q*. Hence, statements can be represented by the standard lower case letters, *p*, *q* and *r*; or statements can be represented by distinct lower case letters that better represent the statement itself.

First symbolism in logic:

a = belongs to every
e = belongs to no
i = belongs to some
o = does not belong to some

Hence, categorical sentences may then be abbreviated as follows:

AaB = *A* belongs to every *B* (Every *B* is *A*)
AeB = *A* belongs to no *B* (No *B* is *A*)
AiB = *A* belongs to some *B* (Some *B* is *A*)
AoB = *A* does not belong to some *B* (Some *B* is not *A*)

Disjunction
vs.
Conjunctions

In Football, AAC stands for the Atlantic Coast Conference

UNC stands for the University of North Carolina

Wake Forest University is located in Winston-Salem, North Carolina

---

Example 2.7.2 Voter Approved

Let us now negate the statement "**The voters approved amendments one and three to the constitution.**"

Solution

This statement is the conjunction of the statements by letting

  o: The voters approved amendment one.

  t: The voters approved amendment three.

That is, the original statement symbolically is

$$o \wedge t$$

By **De Morgan's law** we have

$$\sim(o \wedge t) \equiv \sim o \vee \sim t$$

That is, the negation of the original statement is the English statement

The voters did not approve amendment one or they did not approve amendment three.

Another way to express this is to say

The voters failed to approve at least one of the two amendments offered.

Example 2.7.3 High School Pranks

Consider the conditional statement

If the fountain is turned on, then the students will put Jell-O in it.

Solution

Letting

  f: The fountain is turned on,

  j: The students put Jell-O in the fountain,

the original statement symbolically becomes $f \rightarrow j$.

If the cold weather does not break, then gas will become scarce and schools will close.

Example 2.7.4 Weather causes schools to close

Negate the statement "If the cold weather does not break, then gas will become scarce and school will be closed."

Solution

Letting

  c: The cold weather breaks,

  g: Gas will become scarce,

  s: School will close,

we have $\sim c \rightarrow (g \wedge s)$. Negating this symbolically we get

$$\sim[\sim c \rightarrow (g \wedge s)] \equiv \sim c \wedge \sim(g \wedge s) \equiv \sim c \wedge (\sim g \vee \sim s).$$

Solution—cont'd

Thus, the English negation of the original statement reads as follows:
"The cold will not break but gas will not become scarce or school will not close."

The negation of this statement (that is, when this statement is false) is when **the cold weather breaks and gas is not scarce and schools do not close**.

Common: Quantifying Statements
There are four other commonly used forms of English phrases with which one should become familiar. These are the **quantifying statements**:

I.	All *p* are *q*	Not "All *p* are *q*" is equivalent to "Some *p* are not *q*"
II.	No *p* are *q*	Not "No *p* are *q*" is equivalent to "Some *p* are *q*"
III.	Some *p* are *q*	Not "Some *p* are *q*" is equivalent to "No *p* are *q*"
IV.	Some *p* are not *q*	Not "Some *p* are not *q*" is equivalent to "All *p* are *q*"

In order to see how to negate these forms, we must first consider exactly what they mean.

Consider first **Form I**. This extreme "all *p* are *q*" is actually a conditional statement $p \to q$. The negation of Form I is, therefore,

$$\sim (p \to q) \equiv p \wedge \sim q.$$

That is, we may negate a statement of the form "All *p* are *q*" with a statement of the form "Some *p* are not *q*."

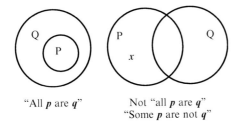

"All *p* are *q*" Not "all *p* are *q*" "Some *p* are not *q*"

Example 2.7.5 "All elections are honest"

Negate the statement "**All elections are honest**."

Solution

Letting
  e: The occurrence is an election,
  h: Occurrence is honest,
the original statement can be written as $e \to h$. Hence,

$$\sim (e \to h) \equiv e \wedge \sim h$$

That is, "**The occurrence is an election which is not honest**." Putting this thought into smoother English, we see that the negation of the statement "**All elections are honest**" is the statement "**Some elections are dishonest**."

Form II can be handled similarly. To say that "no *p* are *q*" is to say that "if one is a *p* then he is not *q*." Hence, Form II can be expressed symbolically as

$$p \to \sim q$$

and negated symbolically by $\sim (p \to \sim q) \equiv p \wedge q$.

"All B are A"

Negative of "All B are A"

"No B are A"

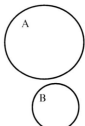

Negative of "No B are A"

All P are Q $\equiv p \to q$
Not all P are Q $\equiv p \wedge \sim q$
No P are Q $\equiv p \to \sim q$
Not the case that no P are Q $\equiv p \wedge q$

That is, we may negate a statement of the form "no *p* are *q*" with a statement of the form "Some *p* are *q*."

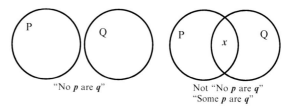

"No *p* are *q*"

Not "No *p* are *q*"
"Some *p* are *q*"

*When it is **not** the case that "**no fights are fixed**" then we can say that "**some fights are fixed**."*

Example 2.7.6 "No fights are fixed"

The negation of the statement "No fights are fixed" is the statement "Some fights are fixed." Here we see that to negate a **Form II** statement we use **Form III**. Hence, in order to negate **Form III**, we should use **Form II.** That is, the negation of a statement of the type "Some *p* are *q*" is a statement of the type "No *p* are *q*."

Negation: "Some fights are fixed"

*When is **not** the case that "**some people are democrats**" then we can say that **nobody is a democrat**.*

Example 2.7.7 "Some people are Democrats"

Negate the statement "Some people are Democrats." The desired negation is literally "No people are Democrats" or, in better English, "Nobody is a Democrat."

Example 2.7.8 Translations

In each of the following cases, express the given statement into symbolic form; negate the statement symbolically; and translate the negation into smooth, concise English:
(a) You will lose weight only if you stop stuffing yourself.
(b) A necessary condition for being a successful student is that you own a pair of grubby jeans.
(c) No mathematicians are sneaky.
(d) Cafeteria food is cheap and nourishing.

Solution

(a) Let

l: You lose weight,
s: You stop stuffing yourself.
Original statement: $\sim s \rightarrow \sim l \equiv l \rightarrow s$
Negation: $l \wedge \sim s$
Translation: You can lose weight and continue to stuff yourself.

(b) Let

s: You are a successful student,
g: You own a pair of grubby jeans.
Original statement: $\sim g \rightarrow \sim s \rightarrow s \rightarrow g$
Negation: $s \wedge \sim g$
Translation: You can be a successful student but not own a pair of grubby jeans.

Solution—cont'd

(c) Let

	m:	You are a mathematician,
	s:	You are sneaky.
Original statement:	$m \rightarrow \sim s$	
Negation:	$m \wedge s$	
Translation:	Some mathematicians are sneaky.	

(d) Let

	c:	Cafeteria food is cheap,
	n:	Cafeteria food is nourishing
Original statement:	$c \wedge n$	
Negation:	$\sim c \vee \sim n$	
Translation:	Cafeteria food is not cheap or else it is not nourishing.	

The following exercise is intended to extend the ideas presented in this section as well as to give the student practice in translating statements from English into symbols and vice versa. The ability to make such translations quickly is essential to the work to follow concerning the testing of arguments for validity.

Common: Universal/Existential Qualifiers

In addition, there are the **universal quantifier** and the **existential qualifier**, $\forall$ and $\exists$, respectively,

1. $\sim [\forall x \forall y \, p(x,y)] \Leftrightarrow \exists x \exists y [\sim p(x,y)]$
2. $\sim [\forall x \exists y \, p(x,y)] \Leftrightarrow \exists x \forall y [\sim p(x,y)]$
3. $\sim [\exists x \forall y \, p(x,y)] \Leftrightarrow \forall x \exists y [\sim p(x,y)]$
4. $\sim [\exists x \exists y \, p(x,y)] \Leftrightarrow \forall x \forall y [\sim p(x,y)]$

SYMBOLIC EQUVALENCES

$\sim \equiv$ *"not"*
$\forall \equiv$ *"for all"*
$\exists \equiv$ *"there exist"*
$\Leftrightarrow \equiv$ *"logical equvialent to"*

The above four equivalence, the first of which translates to "when it is not true that for all *x* and *y* the open statement $p(x,y)$ holds" is equivalent to "there exist an *x* and *y* for which $p(x,y)$ does not hold." The second statement translates to "when it is not true that for all *x*, there exist a *y* such that the open statement $p(x,y)$ holds" is equivalent to "there exist a *x* such that for all *y*, $p(x,y)$ does not hold."

EXERCISES

Critical Thinking

2.7.1. Consider the bi-conditional statement $p \leftrightarrow q$. Note that we have shown that $p \leftrightarrow q \equiv (p \rightarrow q) \wedge (q \rightarrow p)$. Use this information to symbolically negate $p \leftrightarrow q$.

2.7.2. Consider the exclusive disjunction:
   a. Show that $p \underline{\vee} q \equiv (p \vee q) \wedge \sim (p \wedge q)$.
   b. Use the above equivalence to symbolically negate $p \underline{\vee} q$.
   c. Verbally negate the statement "You can take calculus or finite mathematics but not both."

Basic Problem

2.7.3. Let

	d:	He drinks Singapore slings.
	p:	He sees pink elephants.
	g:	He has a good time at parties.

Using this notation put each of the following into symbolic form, negate each symbolically, and translate the negation into smooth English:
- **a.** If he drinks Singapore slings then he will see pink elephants.
- **b.** Drinking Singapore slings is necessary for having a good time at parties.
- **c.** He sees pink elephants only if he drinks Singapore slings.
- **d.** All people who see pink elephants have a good time at parties.
- **e.** Drinking Singapore slings is necessary and sufficient for having a good time at parties.
- **f.** In order not to see pink elephants, it is sufficient that one not drink Singapore slings.

2.7.4. Negate verbally each of the following:
- **a.** All politicians are devious.
- **b.** Some college presidents are devious.
- **c.** Some college presidents are politicians.
- **d.** All Southerners like grits.
- **e.** No Republicans voted for a Democrat.
- **f.** Some people did not vote.
- **g.** No Italians do not like spaghetti.

2.7.5. Let r: You can register early.

a: You are an athlete.

t: You are tall.

s: You are in style.

Using this notation put the following into symbolic form, negate each symbolically, and translate the negation into concise English:
- **a.** If you are an athlete or tall, then you can register early.
- **b.** If you are an athlete and tall, then you are in style.
- **c.** You are an athlete or tall.
- **d.** You are short but in style.
- **e.** You are neither tall nor in style but you can register early.
- **f.** For each integer, x, there exists an integer, y, such that $y = \sqrt{c}$

2.8   TESTING THE VALIDITY OF AN ARGUMENT

One of the most important applications of logic is to determine whether an argument is valid or fallacious (false). We begin our study of this topic with the following definitions:

Definition 2.8.1 Deductive Reasoning

Reasoning or **deductive reasoning** is a cognitive process using arguments to move from given statements or premises, which are true by assumption, to conclusions. The conclusions must be true when the premises are true.

Archival Note

Deductive reasoning *was largely advanced by the French philosopher and Mathematician* **Rene Descartes**.

Example 2.8.1 "All men are mortal"

Given "all men are mortal" and "Aristotle is a man," therefore we can deduce "Aristotle is mortal."

Deductive reasoning is often contrasted with inductive reasoning in that inductive reasoning is the process of reasoning in which the premises are an argument are believed to support the conclusion, how do not entail it; that is, they do not ensure it but is a generalization.

Example 2.8.2 "That which goes up, must come down"

Given the proposition, "**this object fell when dropped**," therefore one can infer "**all objects fall when dropped**."

Definition 2.8.2 Argument

An **argument** is an assertion that a given collection of statements $p_1, p_2, \ldots, p_n$ called **premises** yields another statement r called the **conclusion**.

We symbolize an argument as

$$\left.\begin{array}{c} p_1 \\ p_2 \\ \vdots \\ p_n \end{array}\right\} premises$$
$$\therefore r$$

Where the symbol $\therefore$ is read "**therefore**" and the p's represent the statement of the argument; the horizontal line simply separates these premises from the conclusion.

Definition 2.8.3 Valid/Fallacy

An argument is **valid** if the conclusion r is true whenever the conjunction of the premises $p_1, p_2, \ldots, p_n$ is true; that is, $p_1 \wedge p_2 \wedge \ldots \wedge p_n \to r$ is a tautology. Otherwise the argument is said to be a **fallacy**. In other words, an argument is valid whenever *all the premises are true, the conclusion is true*.

The validity of an argument can be checked by constructing a truth table. This procedure is illustrated by several examples.

Example 2.8.3 Law of Detachment

Test the validity of the following argument:

$$p \to q$$
$$\underline{p}$$
$$\therefore q$$

Solution

It suffices to show that $[(p \to q) \wedge p] \to q$ is a tautology. This means that if both premises, $p \to q$ and p, are true, then the conclusion q is true.

The truth table is

1	2	3	4	5
p	q	$p \to q$	$(p \to q) \wedge p$	$[(p \to q) \wedge p] \to q$
T	T	T	T	T
T	F	F	F	T
F	T	T	F	T
F	F	T	F	T

Hence, $[(p \to q) \wedge p] \to q$ is a **tautology**; and therefore if $p \to q$ is true and p is true, then the conclusion q is true. The above argument

$$p \to q$$
$$\underline{p}$$
$$\therefore q$$

Continued

Archival Note

*The word **argument** comes from the French meaning **a statement and reasoning in support of a proposition**. The word **premise** is the grounds or basis of the argument, that which comes before. The word **conclusion** is the deduction reached by reasoning.*

$$\therefore$$
is read
"therefore"

Valid
vs.
Fallacy

If you have a dime, you have ten cents.
You have a dime.
Therefore, you have ten cents.

This is rather redundant except the first two are premises and the third is a conclusion. The second statement has been detached from the condition set forth in the first statement.

Solution—cont'd

is called the **law of detachment.** In this form, the law of detachment is called **modus ponens**. Similarly, the argument

$$p \to q$$
$$\frac{\sim q}{\therefore \sim p}$$

using the contrapositive statement, the above argument is one form of the law of detachment called **modus tollens**.

Law of Detachment
&
Modus Ponents
&
Modus Tollens

Comparable to transitivity in equality:
If a=b and b=c, then a=c.

Archival Note
*The word **syllogism** comes from old French silogisme meaning **inference, conclusion, computation or calculation**.*

Example 2.8.4 Law of Syllogism

Show that the following argument is valid:

$$p \to q$$
$$q \to r$$
$$\frac{}{\therefore p \to r}$$

Solution

It suffices to show that $[(p \to q) \wedge (q \to r)] \to (p \to r)$ is a **tautology**. This can be shown by constructing its truth table.

1	2	3	4	5	6	7	8
p	q	r	$p \to q$	$q \to r$	$(p \to q) \wedge (q \to r)$	$p \to r$	$[(p \to q) \wedge (q \to r)] \to (p \to r)$
T	T	T	T	T	T	T	T
T	T	F	T	F	F	F	T
T	F	T	F	T	F	T	T
T	F	F	F	T	F	F	T
F	T	T	T	T	T	T	T
F	T	F	T	F	F	T	T
F	F	T	T	T	T	T	T
F	F	F	T	T	T	T	T

The above argument is called the **law of syllogism** (hypothetical syllogism) or the transitive property of implication.

Example 2.8.5 Fallacies in Arguments

Show that the following argument is fallacious:

$$p \to q$$
$$q$$
$$\frac{}{\therefore p}$$

Solution

If suffices to show that the statement $[(p \to q) \wedge q] \to p$ is not a tautology.
   The truth table is

1	2	3	4	5
p	q	$p \to q$	$(p \to q) \wedge q$	$[(p \to q) \wedge p] \to p$
T	T	T	T	T
T	F	F	F	T
F	T	T	T	F
F	F	T	F	T

Solution—cont'd

The third line tells us that the conclusion p is false when both the premise $p \to q$ and q is true. Hence, the argument is a fallacy or fallacious.

Example 2.8.6 Test the Validity

Test the validity of the argument
If it snows, then Maria will ski.

I did not snow.

Therefore, Maria will not ski.

Solution

We first translate the argument into symbolic form. Let p represent "it snows" and q represent "Maria will ski." Thus, the argument takes the form

$$p \to q$$
$$\sim p$$
$$\therefore \sim q$$

The truth table of the argument is given by

1	2	3	4	5	6	7
p	q	$p \to q$	$\sim p$	$(p \to q) \land \sim p$	$\sim q$	$[(p \to q) \land \sim p] \to \sim q$
T	T	T	F	F	F	T
T	F	F	F	F	T	T
F	T	T	T	T	F	F
F	F	T	T	T	T	T

In the third line of **Step 7** we observe that both the premise $p \to q$ and $\sim p$ are true, but the conclusion $\sim q$ is false. Thus, the argument is a fallacy.

Example 2.8.7 Test the Validity

Test the validity of the following argument:
If I study, then I will not fail statistics.
If I do not play tennis, then I will study.

I failed statistics.

Therefore, I played tennis.

Solution

Let p represent "**I study**," q represent "**I failed statistics**," and r represent "**I play tennis**." Thus, we can translate the argument into symbolic form:

$$p \to \sim q$$
$$\sim r \to p$$
$$q$$
$$\therefore r$$

To test the validity of this argument, we must show that whenever the premises $p \to \sim q$, $\sim r \to p$ and q are true, that the conclusion r must also be true.

2 = A NUMBER

1 = A NUMBER

2 = 1

If I win the lottery, then I will take you to dinner.
I took you to dinner.

...I win the lottery???

If I study, then I will not fail statistics.
If I do not play tennis, then I can study.
I failed statistics.

... I played tennis

	1	2	3	4	5	6	7
CASE	p	q	r	$\sim q$	$\sim r$	$p \to \sim q$	$\sim r \to q$
1	T	T	**T**	F	F	F	T
2	T	T	F	F	T	F	T
3	T	F	T	T	F	T	T
4	T	F	F	T	T	T	T
5	**F**	**T**	**T**	**F**	**F**	T	T
6	F	T	F	F	T	T	F
7	F	F	T	T	F	T	T
8	F	F	F	T	T	T	F

We observe, by crossing out all those that don't hold true, that the premises $p \to \sim q$, $\sim r \to p$ and q are true only in **Case 5**, and in that case the conclusion r is also true. Thus, the above argument is valid. Otherwise, we could have extended the table to include the conjunction of the premises and then finally the conditional statement $(p \to \sim q) \wedge (\sim r \to p) \wedge q \to r$; however, this becomes extremely tedious. Alternatively, we could have used equivalence and the law of detachment and law of syllogism.

Since $p \to \sim q \equiv q \to \sim p$ and $\sim r \to p \equiv \sim p \to r$, the contrapositive, the argument becomes

$$q \to \sim p$$
$$\sim p \to r$$
$$\frac{q}{\therefore r}$$

Hence, by the law of syllogism,

$$q \to \sim p$$
$$\frac{\sim p \to r}{\therefore q \to r}$$

yielding

$$q \to r$$
$$\frac{q}{\therefore r}$$

this is valid by the law of detachment or modus ponens. This type of argument will be extended to proof patterns later in this section.

If I fail statistics, then I did not study.
If I did not study, then I will play tennis.
I failed statistics.

∴ I played tennis

If I fail statistics, then I did not study and if I don't study, then I will play tennis.
∴ If I fail statistics, then I played tennis.

If I fail statistics, then I play tennis.
I failed statistics.

∴ I played tennis

> **Example 2.8.8 Test the Validity**
>
> Test the validity of the argument
> If Fred loves Maria, then Bill will leave town.
>
> Either Bill leaves town or Maria is divorced.
> Therefore, if Maria is divorced, then Fred does not love Maria.

> **Solution**
>
> Let p represent "Fred loves Maria," q represent "Bill will leave town," and r represent "Maria is divorced." Thus, the argument takes the form
>
> $$p \to q$$
> $$q \vee r$$
> $$\therefore r \to \sim p$$

Solution—cont'd

We now construct the truth table of the statements $p \to q$, $q \lor r$ and $r \to \sim p$

	1	2	3	4	5	6	7
CASE	p	q	r	$\sim p$	$p \to q$	$q \lor r$	$r \to \sim p$
1	T	T	T	F	T	T	F
2	T	T	F	F	T	T	T
3	T	F	T	F	F	T	F
4	T	F	F	F	F	F	T
5	F	T	T	T	T	T	T
6	F	T	F	T	T	T	T
7	F	F	T	T	T	T	T
8	F	F	F	T	T	F	T

Recall that an argument is valid if the conclusion is true whenever the premises are true. However, in Case 1 of the preceding truth table, the premises $p \to q$ and $q \lor r$ are both true, but the conclusion $r \to \sim p$ is false. Thus, the argument in Example 1.7.6 is a fallacy.

If an argument has two or three premises, we can always use the concept of a truth table to check its validity. However, when there are more than three premises, a truth-table analysis is quite awkward. An easier approach to check validity in such instances is by use of proof patterns. We have already seen two such patterns; namely, in Example 1.8.1: law of detachment

$$
\begin{array}{c}
p \to q \\
p \\
\hline
\therefore q
\end{array}
$$

and Example 1.8.4: law of syllogism

$$
\begin{array}{c}
p \to q \\
q \to r \\
\hline
\therefore p \to r
\end{array}
$$

The law of syllogism may be extended to more than two premises, all of which are conditionals. For example, we can write

$$
\begin{array}{c}
p \to q \\
q \to r \\
r \to s \\
s \to t \\
t \to u \\
\hline
\therefore p \to u
\end{array}
$$

We observe that in the preceding proof pattern, one just follows the arrows. The validity of most arguments can be tested using only the laws of detachment and syllogism. However, when only these proof patterns are used, it is often necessary to replace one or more of the premises with an equivalent statement; for example,

Statement	Equivalnce	Reason
$p \to q$	$\sim q \to \sim p$	Contrapositive
$\sim (p \land q)$	$\sim p \lor \sim q$	De Morgan's Law
$\sim (p \lor q)$	$\sim p \land \sim q$	De Morgan's Law
$\sim p \lor q$	$p \to q$	Disjunctive Syllogism

The last statement $\sim p \lor q$ is true when the implication is true since, $\sim p \lor p$ is a **tautology**; hence, if $p \to q$ is true, then

Detachment
 The action of detaching

 The condition of being detached

Syllogism
 Bring together, the premise and conclusion

 Deductive Reasoning

Contrapositive
vs.
Disjunction
vs.
De Morgan's Laws

$$\sim p \vee p \equiv \sim p \vee q.$$
$$\downarrow$$
$$q$$

Alternatively, $p \vee q$ is equivalent to $\sim p \rightarrow q$ since by double negatives, $p \vee q \equiv \sim (\sim p) \vee q$, which is equivalent to $\sim p \rightarrow q$ by implication.

We shall now give some examples on the use of proof patterns to prove the validity of certain arguments.

Example 2.8.9 Test the Validity

Prove the validity of the following argument using a proof pattern:

It is raining.

If it is cold, then it is not raining.

If it is not cold, then I cannot go skating.

Therefore, I cannot go skating.

Solution

Let p, q, and r represent the given statements:

p: It is raining

q: It is cold

r: I can go skating

In symbolic form the argument translates to

$$p$$
$$q \rightarrow \sim p$$
$$\underline{\sim q \rightarrow \sim r}$$

$$\therefore \sim r$$

	Statement	Reason
1.	p	Premise
2.	$q \rightarrow \sim p$	Premise
3.	$\sim q \rightarrow \sim r$	Premise
4.	$p \rightarrow \sim q$	Contrapositive of (2.)
5.	$p \rightarrow \sim r$	Syllogism using (3.) and (4.)
6.	$\sim r$	Detachment using (1.) and (5.)

Thus, the argument in Example 1.6.7 is valid because, when the premises p, $q \rightarrow \sim p$ and $\sim q \rightarrow \sim r$ are true, then the conclusion $\sim r$ is true.

Example 2.8.10 Test the Validity

Prove the validity of the following argument using a proof pattern:

If Jacob graduates, then he will go to Greece.

If he goes to Greece, then he will visit Athens.

If he does not visit Sparta, then he will not visit Athens.

Jacob did graduate.

Therefore, Jacob will visit Sparta.

Solution

Let p, q, r, and s represent the following statements:

p: Jacob graduates,

Solution—cont'd

q: He will go to Greece,

r: He will visit Athens,

s: He will visit Sparta.

Then, the above argument translates into

$$p \to q$$
$$q \to r$$
$$\sim s \to \sim r$$
$$\underline{p}$$
$$\therefore s$$

A proof pattern of the argument is constructed as follows:

Statement		Reason
1.	$p \to q$	Premise
2.	$q \to r$	Premise
3.	$\sim s \to \sim r$	Premise
4.	p	Premise
5.	$p \to r$	Syllogism using (1.) and (2.)
6.	$r \to s$	Contrapositive of (3.)
7.	$p \to s$	Syllogism using (5.) and (6.)
8.	s	Detachment using (4.) and (7.)

Thus, we conclude that the argument in Example 1.7.8 is valid because the conclusion s is true whenever the premises $p \to q$, $q \to r$, $\sim s \to \sim r$ and p are all true.

Premise

vs.

Syllogism

vs.

Contrapositive

Example 2.8.11 Test the Validity

Is the following argument valid?

Maria is a good dancer or Matthew is intelligent.

If Deb is a beautiful girl, then Maria is not a good dancer.

Matthew is not intelligent.

Therefore, Deb is a beautiful girl.

Solution

Let p, q, and r represent the given statements:

p: Maria is a good dancer

q: Matthew is intelligent

r: Deb is a beautiful girl

Then the argument translates to

$$p \vee q$$
$$r \to \sim p$$
$$\underline{\sim q}$$
$$\therefore r$$

Statement		Reason
1.	$p \vee q$	Premise
2.	$r \to \sim p$	Premise
3.	$\sim q$	Premise
4.	$\sim p \to q$	Disjunctive Syllogism using (1.) and (3.)
5.	$r \to q$	Syllogism using (2.) and (4.)
6.	$\sim q \to \sim r$	Contrapositive of (5.)
7.	$\sim r$	Detachment using (3.) and (7.)

Hence, r is false by definition of negation and therefore we see that the above argument is *not* valid; that is, this argument is invalid because the conclusion r is false when all the premises are true.

Example 2.8.12 Test the Validity

Is there a valid conclusion to the following argument?
   All men eat cake.
   I am a man.

Solution

Let *p* represent "people who are men" and *q* represent "people who eat cake."

Recall, this extreme in logic can be rewritten in terms of a conditional; for example, "all *p* are *q*" is equivalent to $p \to q$. Therefore, you could rewrite this extreme as a conditional and use the logic discussed previously.

Alternatively, you can use Euler circles. Just as above, a conclusion can be drawn in two situations. Let **P** be the set satisfying *p* and **Q** be the set satisfying *q*.

Then, the Example 2.44, the conditional argument is: "all *p* are *q*" and *p*; this situation can be illustrated as follows.

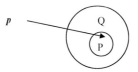

From this illustration, you can logically deduce *q*; that is, modus ponens. The premises can be written as

$$\begin{array}{c} p \to q \\ p \\ \hline \therefore ? \end{array}$$

which, by modus ponens, has *q* is the valid conclusion.

Archival Note
*Euler circles or diagrams are related to **Venn diagrams** and where used by **Leonhard Euler** to represents sets and their relationship.*

Example 2.8.13 Valid Argument

Is the following argument valid?
   All boys like bugs.
   Alexis does not like bugs.
   Therefore, Alexis is not a boy.

Solution

Let *p* represent "people who are boys" and *q* represent "people who like bugs."

The given argument is: "all *p* are *q*" and $\sim q$; this situation can be illustrated as follows.

From this illustration, you can logically deduce $\sim p$.

Hence, given the two premises, $p \to q$ and $\sim q$, the logical conclusion is $\sim p$. Note, the premise can be written in argument form are

$$\begin{array}{c} p \to q \\ \sim q \\ \hline \therefore \sim p \end{array}$$

which, by modus tollens, has $\sim p$ is the valid conclusion and therefore, this is a valid argument.

Example 2.8.14 Valid Conclusion

Is there a valid conclusion?
   All elephants are pink.
   Brownie is a not an elephant.

Solution

Let *p* represent "animals that are elephants" and *q* represent "animals that are pink."
   Consider the argument "all *p* are *q*" and ~*p*; this situation can be illustrated as follows.

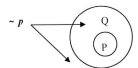

   From this illustration, you can see that there are two ways this can be situation can conclude, therefore a single conclusion cannot be logically deduced. Hence, there is no valid conclusion to this argument; any conclusion drawn would be invalid.

Example 2.8.15 Test the Validity

Is the following argument valid?
All gadgets are thingamajigs.

A wiper snap is a thingamajig.
Therefore, all wiper snaps are gadgets.

Thingamajig
   Something that is hard to classify or whose name is forgotten or unknown

Solution

Let *p* represent "things that are gadgets" and *q* represent "things that are thingamajigs."
   Given the argument "all *p* are *q*" and *q*; this situation can be illustrated as follows.

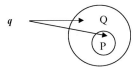

   From this illustration, you can see that there are still two ways this can be situation can be concluded, therefore a single conclusion cannot be logically deduced; any conclusion drawn under such premises would be invalid. A wiper snap may or may not be a gadget; therefore, this is an invalid argument.

Widgets

   A doodad or gadget

   An unnamed article considered in a hypothetical example

Jams

   Blocked or wedged

Example 2.8.16 Test the Validity

Is there a valid conclusion?
 No widgets are wedges.
 The doohickey is a widget.

Solution

Let p represent "widgets" and q represent "wedges."

 The statement "no p are q" is equivalent to $p \rightarrow \sim q$. Therefore, you could rewrite this extreme as a conditional and use the logic discussed previously.

 Alternatively, you can use **Euler** circles; just as before, a conclusion can be drawn in two situations. Let P be the set satisfying p and **Q** be the set satisfying q.

 Then, the case given in Example 2.45, "no p are q" and p can be illustrated as follows.

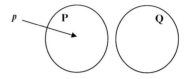

 From this illustration, you can logically deduce $\sim q$. Therefore, a valid conclusion is that the doohickey is not a wedge.

Example 2.8.17 Valid Conclusion

Is there a valid conclusion?
 No widgets are wedges.
 The jam is a wedge.

Solution

Let p represent "widgets" and q represent "wedges."

 Therefore, the argument "no p are q" and q can be illustrated as follows.

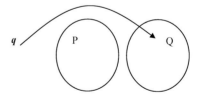

 From this illustration, you can logically deduce $\sim p$. Therefore, the jam is not a wedge.

Example 2.8.18 Valid Argument

Is the following argument valid?
No real man wears hoop skirts.

Dana does not wear hoop skirts.
Therefore, Dana is a realman.

Solution

Let *p* represent "people who are real men" and *q* represent "people who were hoop skirts."

Consider the argument "no *p* are *q*" and ~*p*; this argument can be illustrated as follows.

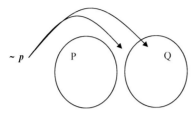

From this illustration, you can see that there are two ways this situation can be concluded, therefore a single conclusion cannot be logically deduced. Therefore, the conclusion "Dana is a real man" is an invalid argument. Not because Dana is not a real man, but because there is insufficient evident to prove that Dana is or is not a real man. Dana is a real man does not follow from the argument made.

Example 2.8.19 Valid Conclusion

Is there a valid conclusion?
    Some women are strong.
    Sam is a woman.

Solution

Let *p* represent "people who are women" and *q* represent "people who are strong."

Consider the argument: "some *p* are *q*" and *p*, which can be illustrated as follows.

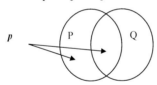

From this illustration, you can see that there are still two ways this can be situation can concluded, therefore a single conclusion cannot be logically deduced. That is, there is no valid conclusion. Sam may or may not be strong.

EXERCISES

Critical Thinking

2.8.1. Test the validity of the argument $\dfrac{\begin{matrix} p \to q \\ \sim q \end{matrix}}{\therefore \sim p}$

2.8.2. Show that the following argument is valid: $\dfrac{\begin{matrix} \sim p \to q \\ q \to \sim r \end{matrix}}{\therefore \sim p \to \sim r}$

2.8.3. Is the following argument valid? $\dfrac{\begin{matrix} \sim p \to \sim q \\ p \end{matrix}}{\therefore q}$

2.8.4. Test the validity of the following argument:
    If it stops raining, then Chris will play tennis.
    It did not stop raining.
    —————————————————————
    Therefore, Chris did not play tennis.

2.8.5. Determine the validity of the argument

If Diane invites Dennis, then John will attend her party.

Diane did not invite John.

Therefore, John attended the party.

2.8.6. Test the validity of the argument

$$\sim p \rightarrow q$$
$$r \rightarrow \sim p$$
$$q$$
$$\therefore r$$

2.8.7. Test the validity of the following argument:

If Linda does not study, then she will fail her course.

If Linda played tennis, then she did not study.

Linda passed her course.

Therefore, Linda did not play tennis.

2.8.8. Is the following argument valid?

If Chris marries Deb, then John joins the Navy.

Either Deb divorces Chris or John joins the Navy.

Therefore, if Deb is divorced, then John joins the Navy.

2.8.9. Using proof patterns prove or disprove the validity of the following argument:

It is hot.

If it is hot, then it is not raining.

If it is not raining, then Maria can go swimming.

Therefore, Maria did not go swimming.

2.8.10. Prove or disprove the validity of the argument

If Sue graduates in the top 10% of her class, then she will go to medical school.

If Sue goes to medical school, then she will specialize in heart disease.

If Sue did not specialize in heart disease, then she did not go to medical school.

Sue graduated in the top 10% of her class.

Therefore, Sue specialized in heart disease.

2.8.11. Show that the following argument is false:

If you like finite mathematics, then you will study.

Either you study or you will fail.

Therefore, if you failed, then you do not like finite mathematics.

2.8.12. Is the argument following argument valid?

$$p \rightarrow q$$
$$\sim q$$
$$\therefore \sim p$$

2.8.13. Illustrate Problem 2.8.8 using Euler circles.

2.8.14. Illustrate Problem 2.8.9 using Euler circles.

2.8.15. Illustrate Problem 2.8.10 using Euler circles.

2.9 APPLICATIONS OF LOGIC

Proof by Induction

The logic associated with conditional statements can be used to prove a property holds for an infinitely large set. For example, $2^n \geq n + 1$ for all natural numbers, let $P(n)$ be the statement, the property hold for the natural number n. Then if we can show that $P(n) \rightarrow P(n + 1)$ and $P(1)$ is true, then $P(n)$ holds for all natural numbers, by induction: $P(1)$ and $P(1) \rightarrow P(2)$ true, implies, by modus ponens, $P(2)$ is true, $P(2)$ and $P(2) \rightarrow P(3)$ true, implies, by modus ponens, $P(3)$ is true, so forth and so on for all n an element of the natural numbers. In general, proof by induction can be bounded below by $n = c$. That is, if the conditional statement in regards to sequentially defined equation "if it is true for $n = k$, then is it is true for $n = c$ greater than or equal to c."

It should be noted that Mathematical induction (proof by induction) is not a form of inductive reasoning, but rather an extended form of deductive reasoning. Proof by induction is a three-step procedure; two of the steps are the proving the conditional statement and one step to show for true for $n = 1$. "Show for true for $n = 1$" can be done first or last, clearly label each step.

STEPS: Proof by Induction

Step 1. Show for true for $n = 1$ or for $n = c$, where c is the first n.

Step 2. Assume the antecedent is true; that is assume true for $n = k$.

Step 3. Show true for $n = k + 1$; since the only conditions under which the conditional statement fails it when the antecedent is true and the conclusion is false.

If we can show that true for $n = k$, implies the equation it is true for $n = k + 1$; then we have shown that the conditional statement is always true.

Be sure to clearly state what you "need to show" (**NTS**); clearly define the left hand side (**LHS**) and the right hand side (**RHS**).

Example 2.9.1 Sum of Whole Numbers

Prove $1 + 2 + 3 + \cdots + n = \frac{n(n+1)}{2}$ for all natural numbers: $n \in N$, where $\in$ is read "an element of."

Proof

Step 1: Show true for $n = 1$

$$1 = \frac{1(1+1)}{2}$$

$$1 = 1$$

True

Step 2: Assume true for $n = k$; that is,

$$1 + 2 + 3 + \cdots + k = \frac{k(k+1)}{2}$$

Step 3: Show true for $n = k + 1$; that is, we NTS

$$1 + 2 + 3 + \cdots + k + (k+1) = \frac{(k+1)((k+1)+1)}{2}$$

or equivalently,

$$1 + 2 + 3 + \cdots + k + (k+1) = \frac{(k+1)(k+2)}{2}$$

Hence, $LHS = 1 + 2 + 3 \cdots + k + (k+1)$, which by assumption becomes

---

Archival Note

Recall, proof by inductions was introduced with great rigor by Aristotle. The basic idea is to get into an infinite loop.

First, show that the statement holds for $n = 1$.

Second, assume that the statement is true for $n = k$ in general.

Third, prove using the assumption stated in the second part that the statement holds true for $n = k + 1$.

$NTS \equiv "Need\,to\,show"$
$LHS \equiv "Left\,hand\,side"$
$RHS \equiv "Right\,hand\,side"$

Natural Number

The ordinary counting numbers:
1, 2, 3, …

$$= \underbrace{1 + 2 + 3 + \cdots + k}_{\frac{k(k+1)}{2}} + (k+1)$$

$$= \frac{k(k+1)}{2} + (k+1)$$

$$= \frac{k(k+1)}{2} + \frac{2(k+1)}{2}$$

$$= \frac{k(k+1) + 2(k+1)}{2}$$

$$= \frac{(k+1)(k+2)}{2} = RHS$$

Example 2.9.2 $n^2 + n$ **is a natural number given** $n \in N$

Prove $n^2 + n$ is even for all natural numbers: $n \in N$.

Solution

Note, an even number is a number of the form **2m** for some $m \in N$.

Proof by Induction

STEP 1:   Show true for **n = 1**

$$1^2 + 1 = 2 \times 1, \text{ and } 1 \in N$$

True: 2 is an even number.

STEP 2:   Assume true for **n = k**; that is,

$$k^2 + k = 2m; \text{ where } m \text{ is a natural number}$$

STEP 3:   Show true for **n = k + 1**; that is, we NTS

$$(k+1)^2 + (k+1) = 2j; \text{ where } j \text{ is a natural number}$$

Hence, $LHS = (k+1)^2 + (k+1)$, which by assumption becomes

$$= (k+1)^2 + (k+1)$$
$$= (k^2 + 2k + 1) + (k+1)$$
$$= \underbrace{(k^2 + k)}_{2m} + (2k+2)$$
$$= 2m + 2k + 2$$
$$= 2(m + k + 1)$$
$$= 2j$$

where $j = m + k + 1$ and hence the property holds for **n = k + 1**. Therefore, by induction, we have shown that $n^2 + n$ is even for all natural numbers **n**.

Boolean Algebra

The idea is based on zeros and ones, binary codes; let **0** represent a false statement and **1** represent a true statement. Let the operation, **+** be defined as the maximum of the given truth-values and the operation, **×** be defined as the minimum of the given truth-values. Recall the addition in general will represent "or" and multiplication in general will represent "and."

Using this algebra, consider the truth-values of $p \vee q \equiv Max(p, q)$ and $p \wedge q \equiv Min(p, q)$.

*George Boole was an English mathematician and philosopher inventing **Boolean logic**, the basis of modern digital computer logic.*

p	q	$p \vee q$	$p \wedge q$
1	1	**Max (1,1) = 1**	**Min (1,1) = 1**
1	0	**Max (1,0) = 1**	**Min (1,0) = 0**
0	1	**Max (0,1) = 1**	**Min (0,1) = 0**
0	0	**Max (0,0) = 0**	**Min (0,0) = 0**

Furthermore, if we consider the negation to be the (truth-value + 1) mod 2; or in layman's terms, the opposite truth-value, we have the following.

p	$\sim p$
1	**0**
0	**1**

Now for conditionals, it is easier to consider its disjunctive equivalent; but if you insist one just an algebraic rule, a conditional is the maximum of the opposite of the antecedent's truth-value and the consequence's truth-value; this is in essence the disjunctive equivalent.

p	q	$\sim p$	$p \to q \equiv \sim p \vee q$
1	1	0	**Max (0, 1) = 1**
1	0	0	**Max (0, 0) = 0**
0	1	1	**Max (1, 0) = 1**
0	0	1	**Max (1, 0) = 1**

Hence, if we interpret the **1**'s as true and the **0**'s as false, these truth-values create the same truth tables introduced previously.

Switching Circuits

The logic of compound statements is often utilized in the design of switching networks in electrical circuit theory. In this section we shall introduce some of the basic theory necessary for the construction of switching networks. We begin by defining what is meant by a switching network.

Definition 2.8.4 Switching Network

A **switching network** is a collection of wires and switches connecting two terminals, A and B. A switch may be either open, O, or closed, C. An open switch will not permit the current to flow while a closed switch will permit current to flow, (Figure 2.1).

We shall now proceed to develop the relationship between logic and the

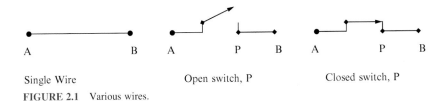

A   B   A   P   B   A   P   B

Single Wire   Open switch, P   Closed switch, P

FIGURE 2.1   Various wires.

design of simple switching networks. Given a switch, P, let *p* be a statement associated with this switch having the property that

p is true if and only if P is closed

or

Archival Note
Boolean logic *is the logical calculus of truth values developed in the 1840s*

Calculus
  From Latin **calculus** *"reckoning, account"*

Warning: **+ and ×**
When using **Boolean algebra**, remember we changed the meaning of **+** and **×**; these new definitions lead to similar but very different properties, some of which look alike and some of which look vastly different.

p is false if and only if P is closed.

Two switches, P and Q, may be connected in two fundamental ways: in series or parallel.

I SERIES

Figures 2.2 and 2.3 illustrate how the switches P and Q are connected in series.

Here, we observe that the current will flow from A to B if and only if both

A   P   Q   B

FIGURE 2.2   Closed switches.

A   P   Q   B

FIGURE 2.3   Open switches.

switches P and Q are closed; that is, if and only if *p ∧ q* is true. Also, current will not flow from A to B if one of the switches is open as shown by Figures 2.3 and 2.4.

That is, *p ∧ q* is false. The behavior of P and Q when they are connected in

A   P   Q   B

FIGURE 2.4   P closed, Q open.

series is summarized in the following table.

P	Q	Series Circuit		*p*	*q*	*p ∧ q*
C	C	C		T	T	T
C	O	O	OR	T	F	F
O	C	O		F	T	F
O	O	O		F	F	F

Thus, it is clear that *p ∧ q* is true only when both P and Q are closed.

II Parallel

When the switches P and Q are connected in *parallel* they appear as shown by Figures 2.5 and 2.6. These two figures illustrate two of the four possible switch positions.

From these figures it is clear that current will flow from A to B if and only if either P or Q is closed. That is, in logical terms, current will flow from A to B if

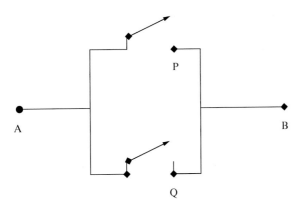

FIGURE 2.5   P and Q open.

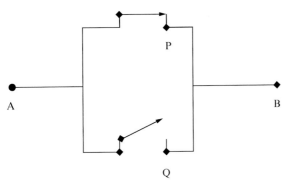

FIGURE 2.6   P closed, Q open.

and only if $p \lor q$ is true. Note also that in Figure 2.5 current will not flow from A to B since P and Q are open. That is, $p \lor q$ is false.

P	Q	Parallel Circuit		p	q	p ∨ q
C	C	C		T	T	T
C	O	C	OR	T	F	T
O	C	C		F	T	T
O	O	O		F	F	F

Hence, current will flow from A to B in three out of four possible combinations; namely, P open and Q open, P closed and Q open, P open and Q closed, or P closed and Q closed.

Example 2.9.3 Current flow from A to B?

Given the following network, when does current flow from A to B?

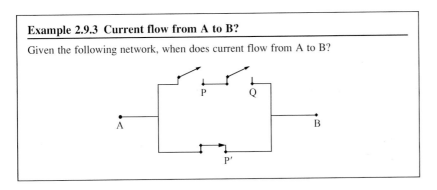

Solution

The above network corresponds to the logical statement, $[(p \wedge q) \vee r]$ because P and Q are in series $p \wedge q$, and R is in parallel with P and Q, $[(p \wedge q) \vee r]$. The question posed will be answered upon determining the combination of truth values for p, q, and r that will make the statement $[(p \wedge q) \vee r]$ true.

That is,

P	Q	R	P, Q in Series	(P and Q) in Parallel with R
C	C	C	C	C
C	C	O	C	C
C	O	C	O	C
C	O	O	O	O
O	C	C	O	C
O	C	O	O	O
O	O	C	O	C
O	O	O	O	O

OR

p	q	r	$p \wedge q$	$[(p \wedge q) \vee r]$
T	T	T	T	T
T	T	F	T	T
T	F	T	F	T
T	F	F	F	F
F	T	T	F	T
F	T	F	F	F
F	F	T	F	T
F	F	F	F	F

By observing the fifth column, we see that current will flow from A to B in five out of the eight cases; that is, if and only if

1. P closed, Q closed, R closed or
2. P closed, Q closed, R open or
3. P closed, Q open, R closed or
4. P open, Q closed, R closed or
5. P open, Q open, R closed

In the design aspects of switching networks it is sometimes necessary to require that one switch be open while another remains closed; that is, for the switches to be complementary.

Definition 2.8.5 Complementary

Two switches are said to be **complementary** if one switch is open and the other is closed, and vice versa. Thus, if one switch is P, the complementary switch will be labeled P′ (P prime).

The logical counterparts are

Switch	Statement
P	p
P′	$\sim p$

In the preceding sections we discussed the concept of logical equivalence of statements. Here, in the study of circuit theory, we can have electrical networks that are equivalent.

Definition 2.8.6 Equivalent Electrical Networks

Two electrical networks are said to be **equivalent** if they have the same electrical properties concerning the flow and non-flow of current.

This definition simply states that their corresponding statements are logically equivalent. We shall illustrate this concept in the following examples:

Example 2.9.4 Equivalent Electrical Network

Find a network equivalent to the one given in Figure 2.7.

Note that the dashed line indicates the path of the electrical flow. The current in this network will flow from A to B whenever the logical statement $(p \vee \sim p) \wedge (q \vee \sim q)$ circuits is true.

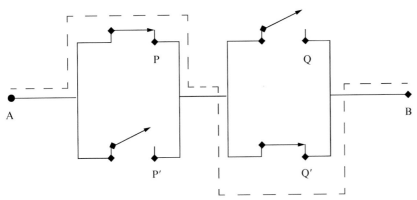

FIGURE 2.7 Network diagram of the electrical circuit $(p \vee \sim p) \wedge (q \vee \sim q)$.

Solution

A network equivalent to Figure 2.7 can be obtained as follows:

	Statement	Reason
1.	$(p \vee \sim p) \wedge (q \vee \sim q)$	Statement describing the circuit
2.	$\equiv \tau \wedge \tau$	Complement Law
3.	$\equiv \tau$	Identity Law

Thus, the above network can be designed equivalently by any tautology such as $p \vee \sim p$. That is,

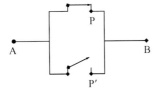

Example 2.9.5 Equivalent Electrical Network

Obtain a network equivalent to the network given in Figure 2.8

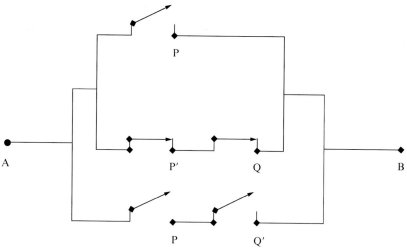

FIGURE 2.8   Network diagram of $[p \vee (\sim p \wedge q)] \vee (p \wedge \sim q)$.

Solution

The current in this network will flow from A to B whenever the logical statement, $[p \vee (\sim p \wedge q)] \vee (p \wedge \sim q)$ is true.

Note that although Figure 2.8 shows a pattern of open and closed switches, a number of different patterns that will allow current to flow from A to B are possible. The several patterns are illustrated in the following diagram:

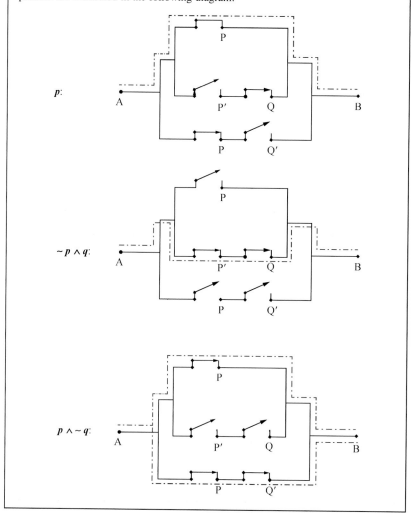

Solution—cont'd

Observe that the dashed path indicates the path of the electrical flow from A to B. Now, we proceed to find a statement equivalent to the logical statement already given. This statement is simply $p \vee q$ That is,

	Statement	Reason
1.	$[p \vee (\sim p \wedge q)] \vee (p \wedge \sim q)$	Statement describing the circuit
2.	$\equiv [(\sim p \vee p) \wedge (p \vee q)] \vee (p \wedge \sim q)$	Distributive Law (from the left)
3.	$\equiv [\tau \wedge (p \vee q)] \vee (p \wedge \sim q)$	Complement Law
4.	$\equiv (p \vee q) \vee (p \wedge \sim q)$	Identity Law
5.	$\equiv (p \wedge \sim q) \vee (p \vee q)$	Commutative Law
6.	$\equiv [(p \wedge \sim q) \vee p] \vee q$	Associative Law
7.	$\equiv [(p \wedge \sim q) \vee (p \wedge p)] \vee q$	Distribution Law (from the right)
8.	$\equiv [p \wedge (\sim q \vee p)] \vee q$	Distribution Law (in reverse)
9.	$\equiv (p \vee q) \wedge [(\sim q \vee p) \vee q]$	Distribution Law (from the right)
10.	$\equiv (p \vee q) \wedge [(p \vee \sim q) \vee q]$	Commutative Law
11.	$\equiv (p \vee q) \wedge [p \vee (\sim q \vee q)]$	Associative Law
12.	$\equiv (p \vee q) \wedge (p \vee \tau)$	Complement Law
13.	$\equiv (p \vee q) \wedge \tau$	Identity Law
14.	$\equiv p \vee q$	Identity Law

Thus, the complicated network shown by Figure 2.8 is equivalent to a network made up of two switches P and Q, connected in parallel; that is,

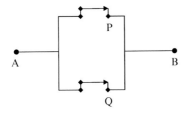

Example 2.9.6 Equivalent Electrical Networks

Draw a switching network corresponding to the compound statement

$$[(p \vee \sim q) \vee q \vee (\sim p \wedge q)] \wedge \sim q.$$

Also find an equivalent network simpler in design.

Solution

The switching network described by the preceding logical statement is

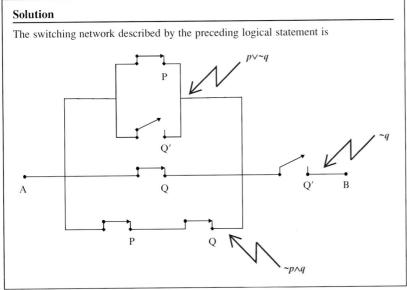

Continued

Solution—cont'd

To find a statement equivalent to this compound statement we proceed as follows:

	Statement	Reason
1.	$[(p \vee \sim q) \vee q \vee (\sim p \wedge q)] \wedge \sim q$	Statement describing the circuit
2.	$\equiv [p \vee (\sim q \vee q) \vee (\sim p \wedge q)] \wedge \sim q$	Associative Law
3.	$\equiv [(p \vee \tau) \vee (\sim p \wedge q)] \wedge \sim q$	Complement Law
4.	$\equiv [\tau \vee (\sim p \wedge q)] \wedge \sim q$	Identity Law
5.	$\equiv \tau \wedge \sim q$	Identity Law
6.	$\equiv \sim q$	Identity Law

Thus, the complicated network we have shown is equivalent to the simple network given by

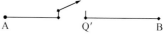

EXERCISES

2.9.1. Given the following network:

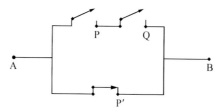

When does the current flow from A to 5?

2.9.2. In the network given in Problem 1 determine when the current does not flow from A to B.

In Problems 2.9.3-2.9.5 draw switching networks that will correspond to the given compound statement.

2.9.3. $(\sim p \vee q) \vee \sim q \vee (p \vee \sim q)$

2.9.4. $[(\sim p \vee q) \vee \sim q \vee (p \vee \sim q)] \wedge q$

2.9.5. $[(p \wedge r) \vee (q \wedge r)] \vee \sim q$

In Problems 2.160-2.163 draw a pair of switching networks to show each of the pairs of equivalent statements.

2.9.6. $\sim p \vee \sim p \equiv \sim p$

2.9.7. $[(q \vee \sim q) \wedge (\sim p \vee p)] \equiv q \vee \sim q$

2.9.8. $[\sim (p \vee q) \vee (\sim p \wedge p)] \equiv \sim p$

CRITICAL THINKING AND BASIC EXERCISE

Indicate which of the statements in Problems 2.1-2.18 are simple, compound, or non-statements. Also indicate the logical connective for each compound statement.

2.1. Play tennis.

2.2. The ocean is blue and the sun is yellow.

2.3. Stop drinking.

2.4. Peter is a good athlete but he smokes.

2.5. Do you think we can go steady?

2.6. Run for your life.

2.7. $\sqrt{69} = 8$

2.8. Sport cars are fun but they are dangerous.

2.9. Black is beautiful.

2.10. If butterflies have their wings, then they can fly.

2.11. Smoking cause cancer but we are not sure.

2.12. If the sun does not shine, then I will kiss you.

2.13. Why were you late?

2.14. Splendor in the grass but baby the rain must fall.

2.15. Yellow roses from Texas and oranges from Florida.

2.16. If football is violent, then soccer is fun.

2.17. Don't rock the boat and you will be successful.

2.18. Tarzan loves Jane but what happened to him?

2.19. Let p represent the statement "I love you" and let q be "the sun is shining." Give a verbal interpretation of the following statements:

(a) $\sim q$, (b) $\sim p$, (c) $p \vee q$, (d) $p \wedge q$, (e) $p \vee \sim q$, (f) $\sim p \vee \sim q$, (g) $\sim (\sim p) \vee q$.

2.20. Let p represent the statement "Jan is beautiful" and q be "Jan is smart." Express the following statements in symbolic form:

(a) Jan is beautiful and smart.

(b) Jan is not beautiful but she is smart.

(c) If Jan is beautiful, then she is smart.

(d) Jan is either smart or beautiful.

2.21. Write the negation of the following statements:

(a) Baby the rain must fall.

(b) Plato was a philosopher.

(c) Finite mathematics is an important subject.

(d) Cars are dangerous.

(e) All Hondas are cars

(f) No man is an island

(g) Some apples are purple

Determine the truth value of each of the compound statements in Problems 2.22-2.27 or give conditions that will make the statement true.

2.22. $3+3=5$ and $7+8=15$.

2.23. Sandy is beautiful or Sandy is smart.

2.24. Rome is in Italy but Tampa is in Florida.

2.25. It is not true that Sandy is smart.

2.26. It is true that $3+2=6$ and $4+4=8$ or John can add.

2.27. Football and tennis are fun or American history is motivating.

Construct truth tables for each of the statements in Problems 2.28-2.40.

2.28. $\sim p \wedge \sim q$

2.29. $\sim p \vee q$

2.30. $\sim (\sim p) \vee q$

2.31. $(p \wedge \sim q) \vee (\sim p \wedge q)$

2.32. $(p \vee q) \wedge r$

2.33. $(p \wedge \sim r) \vee (q \wedge \sim r)$

2.34. $\sim (p \vee r) \vee (\sim q \vee p)$

2.35. $p \rightarrow \sim q$

2.36. $\sim p \rightarrow \sim (\sim q)$

2.37. $p \rightarrow (\sim p \vee q)$

2.38. $[\sim (p \rightarrow q)] \wedge [\sim (q \wedge p)]$

2.39. $(p \rightarrow r) \rightarrow q$

2.40. $(p \vee \sim r) \rightarrow (\sim p \wedge q)$

Determine which of the logical statements in Problems 2.41-2.46 are tautologies, self-contradictions, or neither.

2.41. $\sim p \vee q$

2.42. $\sim p \wedge q$

2.43. $q \rightarrow \sim q$

2.44. $(p \wedge \sim p) \rightarrow p$

2.45. $q \leftrightarrow (p \vee q)$

2.46. $(p \wedge \sim q) \rightarrow (\sim p \rightarrow q)$

In Problems 2.47-2.52 determine whether the given statements are equivalent.

2.47. $q \rightarrow p; p \wedge \sim q$

2.48. $\sim (p \wedge q); \sim p \vee \sim q$

2.49. $\sim p \rightarrow \sim q; q \rightarrow p$

2.50. $\sim p \wedge \sim q; \sim p \wedge \sim q$

2.51. $p \wedge q; \sim p \rightarrow \sim (\sim q)$

2.52. $\sim (\sim p \wedge \sim q); p \vee q$

2.53. For each of the following statements write the converse, contrapositive, and inverse:

    (a) If Jan is smiling, then she is happy.

    (b) If Jan is smiling, then she is sick.

    (c) Jan smiles only when she is happy.

2.54. Show that the $(p \vee q) \vee r \equiv p \vee (q \vee r)$.

2.55. Verify that $(p \vee q) \wedge r \equiv (p \wedge q) \vee (p \wedge r)$.

2.56. Is $p \wedge p \equiv p$?

2.57. Show that $p \vee (p \wedge p) \equiv p$.

In Problems 2.58-2.61 determine whether the arguments are valid or fallacious.

2.58
$$\frac{\begin{matrix} p \wedge q \\ p \end{matrix}}{\therefore p}$$

2.60
$$\frac{\begin{matrix} p \\ \sim p \end{matrix}}{\therefore p \rightarrow \sim q}$$

2.59
$$\frac{\begin{matrix} p \rightarrow \sim q \\ r \rightarrow q \end{matrix}}{\therefore p \rightarrow q}$$

2.61
$$\frac{\begin{matrix} p \rightarrow \sim q \\ \sim q \end{matrix}}{\therefore p}$$

SUMMARY OF IMPORTANT CONCEPTS

Logic is derived from the Greek word logos, which means reason or discourse. A statement is a declarative sentence which is either true or false but not both. The truth value of a statement is the truth or falsity of the statement.

Definitions:

Logic symbolism: a convenient way to analyze logically compound statements formed by the connectives $\wedge, \vee, \sim, \rightarrow$ and $\leftrightarrow$ is by structuring their truth tables.

2.2.1. A **statement** is a declarative sentence which is either true or false, but not both. We shall denote statements symbolically by lower case letters $p, q, r \ldots$

2.2.2. The **truth value** of a statement is the truth or falsity of the statement. We shall denote true by **T** and false by **F**.

2.2.3. A **compound statement** is a statement composed of two or more statements connected by the logical connectives, "and," "or," "if then," "not," and "if and only if." A statement which is not compound is said to be a **simple statement**.

2.2.4. The **conjunction** of two statements p and q is the compound statement "p and q"; written symbolically, $p \wedge q$, where the symbol $\wedge$ is read "and" or "but."

2.2.5. The **disjunction** of two statements p and q is the compound statement "p or q"; written symbolically, $p \vee q$, where the symbol $\vee$ is read "or."

2.2.6. Given a statement p, the **negation** of p is denoted by $\sim p$. The truth value is obtained from the property: If p is true, then $\sim p$ is false and if p is false, then $\sim p$ is true. The symbol $\sim$ is read as not; that is, $\sim p$ is the statement "not p." Other common denotations of the negation of p are $\bar{p}$, $\sim p$ or even $\neg p$; however, we shall use the notation $\sim p$.

2.2.7. Statements of the form "if p then q" are called **conditional statements** and are denoted by $p \rightarrow q$. The statement p is called the **hypothesis (condition/antecedent)** while statement q is called the **conclusion (consequence)**. The truth value of $p \rightarrow q$ is given by the property: $p \rightarrow q$ is false only when p is true and q is false; otherwise $p \rightarrow q$ is true.

2.2.8. A statement of the form "p if and only if q" is called a **bi-conditional statement**, written in shorthand, you will often write "p iff q," symbolically, $p \leftrightarrow q$.

2.4.1. A compound statement, τ, is said to be a **tautology** or **logically true** if it is true for all possible truth values of its components.

2.4.2. A compound statement, ϕ, is said to be a **self-contradiction** or **logically false** if it is false for all possible truth values of its components.

2.4.3. A **paradox** is an apparently true statement or group of statements that leads to a contradiction.

2.4.4. Two statements r and s are said to be **logically equivalent** or simply **equivalent** if they have identical truth tables; that is, if $r \leftrightarrow s$ is a tautology. To symbolize two equivalent statements r and s, we write $r \equiv s$ or $r \Leftrightarrow s$.

Qualifiers are terms that indicate to what extent a property holds: *all are*, *none are*, *some are not* and *some are*. The universal qualifier is *for all*, $\forall$, and the existential qualifier is **there exist**, $\exists$.

2.6.1. The **universal quantifier** is "for all," denote by an upside-down A, $\forall$. The statement $\forall x \in U[p(x)]$ is true if and only if $p(x)$ is true for all $x \in U$.

2.6.2. The **existential quantifier** is "**there exist**," denote by a backwards E, $\exists$. The statement $\exists x \in U[p(x)]$ is true if and only if there exist at least one $x \in U$ for which $p(x)$ is true.

2.8.1. **Reasoning** or **deductive reasoning** is a cognitive process using arguments to move from given statements or premises, which are true by assumption, to conclusions. The conclusions must be true when the premises are true.

2.8.2. An **argument** is an assertion that a given collection of statements $p_1, p_2, \ldots, p_n$ called **premises** yields another statement r called the **conclusion**.

2.8.3. An argument is **valid** if the conclusion r is true whenever the conjunction of the premises $p_1, p_2, \ldots, p_n$ is true; that is, $p_1 \wedge p_2 \wedge \ldots \wedge p_n \rightarrow r$ is a tautology. Otherwise the argument is said to be a **fallacy**. In other words, an argument is valid whenever *all* the premises are true, the conclusion is true.

2.8.4. A **switching network** is a collection of wires and switches connecting two terminals, **A** and **B**. A switch may be either open, **O**, or closed, **C**. switch will not permit the current to flow while a closed switch will permit current to flow.

2.8.5. Two switches are said to be **complementary** if one switch is open and the other is closed, and vice versa. Thus, if one switch is **P**, the complementary switch will be labeled **P'** (**P** prime).

2.8.6. Two electrical networks are said to be **equivalent** if they have the same electrical properties concerning the flow and non-flow of current.

We have defined seven basic laws of algebra: *idempotent*, *associative*, *commutative*, *distributive*, *identity*, *complement*, and *De Morgan's* for statements under the equivalence relation. We have also considered various forms of the conditional statement.

Properties:

2.2.1. **Conjunction**: If p is true and q is true, then $p \wedge q$ is true; otherwise $p \wedge q$ is false.

2.2.2. **Disjunction (inclusive)**: When p is true or q is true or if both p and q are true, then $p \vee q$ is true; otherwise $p \vee q$ is false. Thus, the disjunction $p \vee q$ of the two statements p and q is false only when both p and q are false.

2.2.3. **Disjunction (inclusive)**: When p is true and q is false, or when p is false and q is true, then $p \underline{\vee} q$ is true; otherwise $p \underline{\vee} q$ is false. Thus, this alternative disjunction $p \underline{\vee} q$ of the two statements p and q is false only when both p and q are false and when both p and q are true.

2.2.4. **Negation**: When p is true, then $\sim p$ is false; if p is false, then $\sim p$ is true.

2.2.5. **Conditional**: The **conditional statement** $p \rightarrow q$ is false only when p is true and q is false; otherwise $p \rightarrow q$ is true.

2.2.6. **Bi-conditional**: If p and q are either both statements are true or both statements are false, then $p \leftrightarrow q$ is true; if p and q have opposite truth values, then $p \leftrightarrow q$ is false.

Rules:

2.4.1. **Idempotent**: $p \vee p \equiv p$ and $p \wedge p \equiv p$

2.4.2. **Associative**: $(p \vee q) \vee r \equiv p \vee (q \vee r)$ and $(p \wedge q) \wedge r \equiv p \wedge (q \wedge r)$

2.4.3. **Commutative**: $p \vee q \equiv q \vee p$ and $p \wedge q \equiv q \wedge p$

2.4.4. **Distributive**: $p \vee (q \wedge r) \equiv (p \vee q) \wedge (p \vee r)$ and $p \wedge (q \vee r) \equiv (p \wedge q) \vee (p \wedge r)$

2.4.5. **Identity**: If τ is a **tautology** and φ is any **contradiction**, then $p \vee \varphi \equiv p$, $p \wedge \tau \equiv p$, $p \wedge \varphi \equiv \varphi$ and $p \vee \tau \equiv \tau$

2.4.6. **Complement**: If τ is a **tautology** and φ is any **contradiction**, then $p \vee \sim p \equiv \tau$, $p \wedge \sim p \equiv \varphi$, $\sim(\sim p) \equiv p$, $\sim \tau \equiv \varphi$ and $\sim \varphi \equiv \tau$

2.4.7. **DeMorgan's Rule**: $\sim(p \vee q) \equiv \sim p \wedge \sim q$ and $\sim(p \wedge q) \equiv \sim p \vee \sim q$

Common: Conditional Statements

There are four common ways in which the **conditional statement** $p \to q$ can be expressed.

5. *p implies q* (Direct statement)
6. *p only if q*
7. *p* is *sufficient* for *q*
8. *q* is *necessary* for *p*

Common: Equivalent Statements

Analyzing **equivalent statements** we observe that

1. $p \to q$ is not equivalent to $q \to p$ because Columns 5 and 6 are **not identical**.
2. $p \to q$ is not equivalent to $\sim p \to \sim q$ because Columns 5 and 7 are **not identical**.
3. $p \to q$ is equivalent to $\sim q \to \sim p$ because Columns 5 and 8 are **identical**.
4. $q \to p$ is equivalent to $\sim p \to \sim q$ because Columns 6 and 7 are **identical**.

Common: Negations

$\sim(\sim p) \equiv p$	Double negative, is the positive statement
$\sim(p \lor q) \equiv \sim p \land \sim q$	Not at least one is equivalent to not having either; that is, not the first and not the second (De Morgan's Law)
$\sim(p \land q) \equiv \sim p \lor \sim q$	"Not both" is equivalent to not having at least one; that is, "not the first or not the second" (De Morgan's Law)
$\sim(p \to q) \equiv p \land \sim q$	When the first statement does not imply the second statement, the first statement can be true when (and) the second statement can fail (that is, is false)

Common: Quantifying Statements

There are four other commonly used forms of English phrases with which one should become familiar. These are the **quantifying statements**:

I.	All *p* are *q*	Not "All *p* are *q*" is equivalent to "Some *p* are not *q*"
II.	No *p* are *q*	Not "No *p* are *q*" is equivalent to "Some *p* are *q*"
III.	Some *p* are *q*	Not "Some *p* are *q*" is equivalent to "No *p* are *q*"
IV.	Some *p* are not *q*	Not "Some *p* are not *q*" is equivalent to "All *p* are *q*"

Common: Universal and Existential Qualifiers

In addition, there are the **universal quantifier** and the **existential qualifier**, $\forall$ and $\exists$, respectively,

1. $\sim[\forall x \forall y \, p(x,y)] \Leftrightarrow \exists x \exists y [\sim p(x,y)]$
2. $\sim[\forall x \exists y \, p(x,y)] \Leftrightarrow \exists x \forall y [\sim p(x,y)]$
3. $\sim[\exists x \forall y \, p(x,y)] \Leftrightarrow \forall x \exists y [\sim p(x,y)]$
4. $\sim[\exists x \exists y \, p(x,y)] \Leftrightarrow \forall x \forall y [\sim p(x,y)]$

Proof by Induction:

Step 1. Show for true for $n=1$ or for $n=c$, where c is the first n.

Step 2. Assume the antecedent is true; that is assume true for $n=k$.

Step 3. Show true for $n=k+1$; since the only conditions under which the conditional statement fails it when the antecedent is true and the conclusion is false.

We have studied several laws with respect to the validity of an argument:

Law of Detachment	$p \rightarrow q$ $\underline{p}$ $\therefore q$		
Law of syllogism	$p \rightarrow q$ $\underline{q \rightarrow r}$ $p \rightarrow r$		
Hypothetical syllogism	$p \rightarrow q$ $\underline{q \rightarrow r}$ $\therefore p \rightarrow r$		
Disjunctive syllogism	$p \vee q$ $\underline{\sim p}$ $\therefore q$	OR	$p \vee q$ $\underline{\sim q}$ $\therefore p$
Conjunctive Simplification	$\underline{p \wedge q}$ $\therefore p$	OR	$\underline{p \wedge q}$ $\therefore q$
Conjunctive Concatenation	p $\underline{q}$ $\therefore p \wedge q$		
Modus ponens	$p \rightarrow q$ $\underline{p}$ $\therefore q$	OR	all p are q $\underline{p}$ $\therefore q$
	$p \rightarrow \sim q$ $\underline{p}$ $\therefore \sim q$	OR	no p are q $\underline{p}$ $\therefore \sim q$
Modus tollens	$p \rightarrow q$ $\underline{\sim q}$ $\therefore \sim p$	OR	all p are q $\underline{\sim q}$ $\therefore \sim p$
	$p \rightarrow \sim q$ $\underline{q}$ $\therefore \sim p$	OR	no p are q $\underline{q}$ $\therefore \sim p$

Symbol (Abbr.)	Name/Origin	Meaning/ Read as	Description	Example/Description
p, q, r	Simple Statements	Abbreviation for a simple statement	Simple statements are abbreviated by single lowercase letters	Let s represent the simple statement: "The sun is shining."
$\vee$	Disjunction (Inclusive)	"Or"	At least one statement is true	"It is raining, or I would ride my bike."
$\underline{\vee}$	Disjunction (Exclusive)	"Either ... Or ..."	Exactly one statement is true	"You can have either ice cream or cake."
$\wedge$	Conjunction	"And"	Both statements are true	"I am a student and a full-time employee."
$\sim$	Negation	"Not"	The opposite of the statements in terms of validity	"I am not ugly."

Continued

Symbol (Abbr.)	Name/Origin	Meaning/ Read as	Description	Example/Description
$\rightarrow$	Implication/Conditional	"If ..., then..."	If the antecedent is true that implies that the consequence must be true	"If I win the lottery, then I will buy you dinner"
$\leftrightarrow$	Bi-conditional	"... if and only if ..."	The conditional holds in both directions	"I am a male if and only if I have a Y chromosome"
iff		"if and only if"	Abbreviation for "if and only if"	
$\forall$	Universal Qualifier	"for all ..."	An upside down A	For all real numbers x, $x^2 \geq 0$
$\exists$	Existential Qualifier	"there exist ..."	A backwards E	There exist $x \in Z$ such that $\sqrt{x}=2$
$\equiv$	Equivalent	"Is the same as saying"	Equality between compound statement	Related to the word: Equal
$\therefore$	Hebrew	"Therefore"	The abbreviation for "therefore"	Used when drawing conclusions
ϕ or c	Greek letter "Phi" or an abbreviation of contradiction	A contradiction	Statement that is never true	"This statement if false"
τ or t	Greek letter "Tau" or an abbreviation of **t**autology	A tautology	Statement that is always true	"A rose is a rose"
$\in$	Greek letter "Epsilon"	"An element of" or "is contained in"	Symbol illustrating containment	Let p represent a statement: $p \in S$
n	Abbreviation of **n**umber	"Number of simple statements"	Used when counting total number of possible situations	Given p,q,r: $n=3$

REVIEW TEST

Multiple-Choice:
  1. Which of the following is/are not statement(s)?
    i. Come here
    ii. That is a cute kitty
    iii. $2+2=5$
    A. i only
    B. i and ii only
    C. iii only
    D. all of them are statements
    E. none of them are statements
  2. Which of the following is/are not statement(s)?
    i. Watch your step
    ii. Susan will call you tonight.
    iii. $2+5=7$
    iv. $2 \times 5=7$
    A. i only
    B. ii and iii only
    C. iii only
    D. all of them are statements
    E. none of them are statements

True/False: Let *p* represents a true statement and *q* represent a false statement. Find the truth value of the following compound statements.

3. $\sim q \wedge (p \vee \sim q)$
4. $(p \vee q) \rightarrow (\sim p \wedge q)$

Construct a truth table and determine truth value under each condition.

5. $p \wedge \sim q$
6. $(\sim q \rightarrow p) \vee p$

Determine the value of *x* that makes the statement true.

7. $x + 5 = 7$, if $3 + 4 = 5$
8. $7 + 3 = 12$ if and only if $x + 5 = 8$

Let *p* represent "The sun is shining" and *q* represent "It is raining," write each of the following in a complete sentence

9. $\sim (p \wedge \sim q)$
10. $p \rightarrow \sim q$

Let *p* represent "The sun is shining" and *q* represent "It is raining," write each of the following in symbolic notation

11. If the sun is not shining, then it is raining.
12. The sun is shining and it is raining.

Construct a truth table and determine if the given statements are logically equivalent.

13. $p \rightarrow q, q \vee \sim p$
14. $p \rightarrow q, \sim p \rightarrow \sim q$

Give the negations of the following statements.

15. All students enroll in 12 credit hours per semester.
16. I will wash the dishes or I will wash the car.
17. If I offend you, then call the police.

Give a logically equivalent statement to the following statements.

18. I am broke or everything is not expensive
19. If you are traveling abroad, then you need an updated passport

Write the given statement in "if... then..." form.

20. Doing homework is necessary to pass this course.
21. All widgets are thingumabobs.

Write the contrapositive of the following statement.

22. If you are fifteen, then you are a minor.
23. If I win the lottery, then I will take you to dinner.

Write the inverse of the following statement.

24. If you call me, then I will not forget to bring drinks to the party.
25. If $x + 4 = 7$, then $x - 4 = 1$.

Determine what conclusion can be logically deduced given that:

26. (i) All engineers are intelligent, (ii) All intelligent people do their jobs well, and (iii) Some contractors do their job well.
27. (ii) No one who assigns work is well-liked, and (ii) All employers assign work.

Determine what conclusion is needed in order to make the argument valid, if any.

28.
$$\begin{array}{l} p \to q \\ \underline{p} \\ \therefore \end{array}$$

31.
$$\begin{array}{l} \sim p \to q \\ \underline{\sim q} \\ \therefore \end{array}$$

34.
$$\begin{array}{l} p \lor \sim q \\ \underline{\sim q} \\ \therefore \end{array}$$

29.
$$\begin{array}{l} p \to q \\ \underline{\sim q} \\ \therefore \end{array}$$

32.
$$\begin{array}{l} p \to \sim q \\ \underline{q} \\ \therefore \end{array}$$

35.
$$\begin{array}{l} \sim p \lor q \\ \underline{p} \\ \therefore \end{array}$$

30.
$$\begin{array}{l} p \to q \\ \underline{\sim p} \\ \therefore \end{array}$$

33.
$$\begin{array}{l} p \lor q \\ \underline{\sim q} \\ \therefore \end{array}$$

All of the following arguments have true conclusions but not all are valid. Determine if the argument if valid.

36. All cats are animals. All Maine coons are animals. Therefore, all Maine coons are cats.
37. All flowers have petals. An orchid is a flower. Therefore, an orchid has petals.
38. All birds have feathers. A dove has feathers. Therefore, a dove is a bird.

Determine if the argument if valid.

39. If it is a Washi, then it must be good. This machine is not a Washi. Therefore, it is not good.
40. Peter is at work or he is at home. Peter is not at work. Therefore, he is at home.
41. If you pass the final, then you will pass this course. You passed the course. Therefore, you passed the final.
42. If a man is not handsome, then he is funny. The man is handsome. Therefore, he is not funny.

Practice for the CLAST exam:

43. Select the negation of the statement "Fred is a farmer and George is a baker"
   A. Fred is not a farmer and George is not a baker
   B. Fred is not a farmer or George is not a baker
   C. Fred is not a farmer and George is a baker
   D. Fred is a farmer and George is not a baker
44. Select the negation of the statement "If you invite me, I will come to the party"
   A. If you invite me, I will not come to the party.
   B. If you do not invite me, I will not come to the party.
   C. You invited me and I did not go to the party.
   D. You invited me and I came to the party.
45. Select the statement that is logically equivalent to "It is not true that both Fred and George are good bowlers"
   A. If Fred is not a good bowler, George is a good bowler.
   B. Fred is a good bowler or George is a good bowler.
   C. Fred is not a good bowler and George is a good bowler.
   D. If George is not a good bowler, George is not a good bowler.
46. Select the statement that is *not* logically equivalent to "If the game is cancelled, Bob will go to the movies"
   A. Bob will go to the movies or the game is not cancelled.
   B. If Bob goes to the movies, then the game is cancelled
   C. If Bob does not go to the movies, then the game will not be cancelled
   D. The game is not cancelled or Bob will go to the movies.
47. Read the following valid argument and chose the symbolic form that illustrates the reasoning pattern: *Betty will wash the car or she will go to the store. She did not go to the store. Therefore, she will wash the car.*

   A.
$$\begin{array}{l} p \lor q \\ \underline{q} \\ \therefore p \end{array}$$

   B.
$$\begin{array}{l} p \lor q \\ \underline{p} \\ \therefore \sim q \end{array}$$

$$p \lor q$$
C. $\dfrac{\sim q}{\therefore \sim p}$

$$p \lor q$$
D. $\dfrac{\sim q}{\therefore p}$

48. Select the rule of logical equivalence which directly transforms first statement (i) into the second statement (ii).
    i. Not all men drive trucks.
    ii. Some men do not drive trucks.

    A. "All are not p" is equivalent to "none are p"
    B. "Not all are p" is equivalent to "all are not p"
    C. "Not all are p" is equivalent to "some are not p"
    D. "If not q then not p" is equivalent to "if p, then q"

49. Read the statements given and select a logical conclusion, if it exists.
    All pirates are treasure hunters. All treasure hunters seek gold and diamonds. Ted is seeking gold and diamonds.
    A. Ted is a pirate
    B. Ted is not a pirate
    C. Ted is a treasure hunter
    D. None of the above is warranted.

50. Read the statements given and select a logical conclusion, if it exists.
    If you study statistics, you will be successful. If you study engineering, you will be successful. You study statistics and not engineering.
    A. You will be successful.
    B. You will not be successful.
    C. You study chemistry as well.
    D. None of the above is warranted.

51. Select the statement that is logically equivalent to "It is not the case that I am happy and you are sad"
    A. I am not happy and you are happy.
    B. I am not happy or you are not sad.
    C. I am happy and you are not sad.
    D. I am happy and you are not happy.
    E. None of the above.

52. Select the statement that is *not* logically equivalent to "If you are wearing a suit, then you must wear dress shoes."
    A. You do not wear a suit or you must wear dress shoes.
    B. If you need not wear dress shoes, then you are not wearing a suit.
    C. You wear a suit and you do not wear dress shoes.
    D. You must wear dress shoes or you do not wear a suit.

REFERENCES

Church, A., 1956. Introduction to Mathematical Logic. Princeton University Press, Princeton, NJ.

Johnstone Jr., H.W., 1954. Elementary Deductive Logic. Crowell, New York.

Suppes, P., 1957. Introduction to Mathematical Logic. D. Van Nostrand, Princeton, NJ.

Tarski, A., 1946. Introduction to Logic, 2nd rev. ed. Oxford, New York.

Tsokos, C.P., 1978a. Mainstreams of Finite Mathematics with Application. Charles E. Merrill Publishing Company, A Bell & Howell Company.

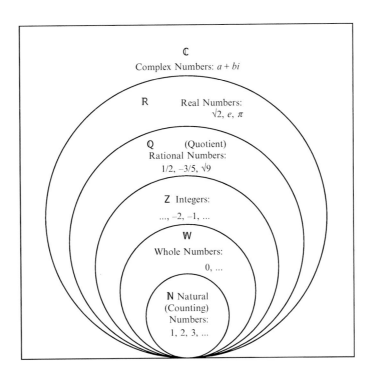

Number is the within of all things.

PYTHAGORAS

Chapter 3   Sets

The mathematical delving of Georg Cantor gave rise to the modern theory of the infinite. His imaginative and innovative ideas also led to new insights in the specialties of topology and geometry. In addition, he did valuable work in analysis and made a special study of rational and irrational numbers. The Cantor ternary set provides the basis for numerous counterexamples.

Cantor's parents immigrated to Russia from Denmark before his birth, and went from there to Germany when he was eleven. While his family was of Jewish descent, his mother was born a Roman Catholic and his father had converted to Protestantism. Cantor himself preferred Protestant Christianity, but he was intensely interested in medieval theology. This interest, combined with his studies in philosophy, physics, and mathematics, spurred him to focus on questions of continuity and the infinite. The direction of his studies was despite his father's urging him to study engineering, a much more financially rewarding occupation.

Most of Cantor's career was spent at the University of Halle, a small, second-rate school. The sensitive Cantor dearly hoped to achieve the distinction of a professorship at the University of Berlin, but he was thwarted in this desire by his contemporaries' objections to his concepts of the infinite. Cantor's mental health suffered greatly under the strain of continual attacks upon his ideas by colleagues, most notably Leopold Kronecker. These attacks eventually became personal rather than academic. At the age of thirty-nine, Cantor suffered the first of a series of nervous breakdowns. He never regained his self-confidence, and died in a mental institution.

Toward the end of his life, recognition for this innovator's achievements was being granted, and with the twentieth century, his work came to be accepted. Cantor is now admired for his brilliant thinking; Kronecker's objections are no longer taken seriously by most mathematicians. The mathematics of today is greatly indebted to Cantor's free spirit of a Set.

CANTOR SET

The set of points formed from the closed interval [0,1] by removing first the middle third of the interval, then the middle third of each remaining interval, and so on indefinitely, the intervals removed being open intervals.

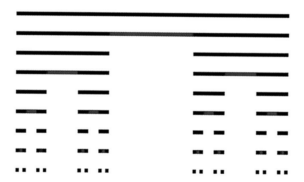

3.1 INTRODUCTION TO SET THEORY

The concept of set theory plays a major role in every branch of modern mathematics. Intuitively the term set means an arbitrary collection of objects with no restriction as to their nature or number, and no required relation between the objects. Georg Cantor devoted much of his life to developing the theory of sets.

Our aim in this chapter is to introduce some of the basic definitions and properties of the theory of sets. The usefulness of the concepts we study here will be illustrated in later chapters when we apply these concepts in practical analysis. In addition we shall discuss some basic techniques in counting that involve common problems of daily life.

Recall, we think faster than we speak, we speak faster than we write, therefore to think quickly and communicate these ideas, we must learn to abbreviate everything. A summary of modern symbolic logic and how this relates to the language of set theory is given throughout this chapter, as is a summary of the language used in set theory.

3.2 THE CONCEPT OF A SET

Let us begin by considering the thirty-eight presidents of the United States up to 1976. The members of this collection of individuals are all quite distinct, and all have met some basic requirements to qualify them as the president of the United States. This collection of individuals constitutes a set. Also, let us consider the fifty states that make up the United States of America. This collection is well defined and consists of states that are distinct from one another. This collection of states constitutes a set, and thus we have the following definition:

Definition 3.2.1 Set

A *set* is a collection of distinct objects; the objects that constitute the set are the **members** or **elements** of the set. We shall denote sets symbolically by capital letters A, B, C... and "an element of" by the Greek letter epsilon, $\in$. If an object "is not an element of," we shall write $\notin$.

For example, the positive integers greater than 5 and less than 13 constitute a set; namely, 6, 7, 8, 9, 10, 11, and 12 are the elements of the set. We commonly use braces **{ and }** as grouping symbol to enclose the members or elements of a set. Thus, we can write the preceding set as

$$\{6, 7, 8, 9, 10, 11, 12\}.$$

One need not associate numerical values to structure a set. Consider the American states that begin with the letter M. These states constitute a set

$$\{\text{Maine, Maryland, Massachusetts, Michigan,}$$
$$\text{Minnesota, Mississippi, Missouri, Montana}\}.$$

Here the set consists of eight elements, each element being a specific state. An important feature of a set is that it is well defined. In other words, it must be clear that a given object is either an element of the set or is not an element of the set.

We shall employ capital letters such as A, B, C... to denote sets and lowercase letters such as $a, b, c, \ldots$ to denote the elements (members) of the set. Thus, if we let $\mathbb{N}$ represent the set of all positive integers, 1, 2, 3, ...; that is, the *natural numbers* and hence we can write

$$\mathbb{N} = \{1, 2, 3, 4, \ldots\}.$$

We should also mention here that a set may consist of a *finite number* of elements or of *infinitely* many elements. For example, the set that consists of the eight states whose names begin with M is made up of a finite number of elements, whereas the set N consists of infinitely many positive integers; that is, the enumeration of elements in $\mathbb{N}$ continues in the natural manner.

To indicate that an element is a member of a given set, we use the symbol $\in$. Its negation $\notin$ denotes that the element is not a member of the set. For example, let us consider the set

$$A = \{a, b, c, d, e, f\}.$$

Here, $a \in A$; that is, a is an element of the set, while $g \notin A$, g is not an element of the set A.

In order to have a convenient way to describe certain sets we shall use the symbols > which means greater than and < which means less than. When the symbols are written in the form $\geq$ and $\leq$ it simply means greater than or equal to and less than or equal to, respectively. We shall also employ letters such as

$x, y, z, \ldots$ to represent elements of a set. To illustrate this, let C be the set of all natural numbers greater than 5 and less than 110, written in ***roster notation***:

$$C = \{6, 7, 8, 9, 10, 11, 12, \ldots, 108, 109\}$$

A more convenient way of writing the set C is in ***set-builder*** notation:

$$C = \{y \mid y \in N, 5 < y < 110\}$$

Here, the letter y represents an element of the set C. The vertical bar | after the y is read "such that." After the bar we place the conditions that must be satisfied to define the set C. Thus, we read the set C as follows: "C equals the set of all elements, y, such that y is an natural number greater than 5 and less than 110." That is, $C = \{y \mid y \in N, 5 < y < 110\}$.

Some common denotations of numerical sets:

Set	Name
➤ $\mathbb{N} = \{1, 2, 3, \ldots\}$	Natural (Counting Numbers)
➤ $\mathbb{W} = \{0, 1, 2, 3, \ldots\}$	Whole Numbers (Natural numbers including zero)
➤ $\mathbb{Z} = \{\ldots, -3, -2, -1, 0, 1, 2, 3, \ldots\}$	Integers (Positive/Negative Whole Numbers)
➤ $\mathbb{Q} = \left\{x \mid x = \dfrac{a}{b} \text{ where } a, b \in Z\right\}$	Rational (Quotients)
➤ $\mathbb{R} = \{x \mid -\infty < x < \infty\}$	Real Numbers
➤ $\mathbb{C} = \{a + bi \mid a, b \in R\}$	Complex Numbers where $i^2 = -1$

It should be noted that since it is not practical to write in bold, emphasize is often given to these sets by writing the capital letters with a double stroke as shown below. The relation between these sets is shown on the Chapter 2 title page.

Complex   Real   Rational   Integer   Whole   Natural

In set theory we are only interested in the collections of objects that are ***well-defined***; therefore we are only interested in sets—collections to which all objects are considered either to belong or not to belong. Non-sets or arbitrary collections not considered are collections described by opinions; for example the statement "dogs that are cute" does not form a well-defined set. "Bulldogs" is a well-defined set.

Example 3.2.1 First 100 Integers

Write the set that consists of all positive integers less than 100.

Solution

Let x represent the integers less than 100. Then we can write the desired set A as

$$A = \{x \mid x \in Z, x < 100\}.$$

Example 3.2.2 Roster Notation

List the elements of the set A in Example 3.2.1 in roster notation.

Solution

$$A = \{1, 2, 3, 4, 5, \ldots, 99\}$$

Example 3.2.3 Antibiotics

Consider the set of antibiotics

$A = \{$Penicillin, Streptomycin, Tetracycline, Erythromycin, Chlortetracycline$\}$

Determine which of the following statements are true and which are false.

(a) Erythromycin $\in A$

(b) Chlortetracycline $\notin A$

(c) Digitalis $\in A$

(d) Penicillin $\notin A$

Solution

Statement (a) is true because erythromycin is an element of the well-defined set A. Statement (b) is false because chlortetracycline is indeed an element of the set A. Statement (c) is false because digitalis is not an element of the set A. Statement (d) is false because penicillin is an element of the set A; that is, penicillin $\in A$.

Example 3.2.4 Subsets of the Integers

List the elements of the following sets:

(a) $A = \{x \,|\, x \text{ an integer}, -2 \leq x < 5\}$

(b) $B = \{y \,|\, y = 2n, n \text{ a positive integer}\}$

(c) $C = \{z \,|\, z \text{ a positive integer } z^2 \leq 36\}$

Solution

(a) $A = \{-2, -1, 0, 1, 2, 3, 4\}$

(b) Here $n \in N$, that is n simply represents the numbers $n = 1, 2, 3, \ldots$. Thus, we can write B as an infinite set of positive even integers:

$$B = \{2, 4, 6, \ldots\}$$

(c) The set C consists of all positive integers that when squared will be less than or equal to 36:

$$C = \{1, 2, 3, 4, 5, 6\}$$

Example 3.2.5 "Football"

Give a mathematical description of the set of all letters in the word *football*.

Solution

Let *x* represent a letter in the word *football*. Thus, the desired set in set-builder notation is written

$$A = \{x \mid x \text{ a letter in the word football}\}$$

The elements in the set can be listed in roster form:

$$A = \{f, o, t, b, a, l\}.$$

Note that although the word *football* contains two *o*'s and two *l*'s, we write them only once in structuring the set. In this context, the universal discourse is understood to be the letters of the alphabet; however, universal discourse is logistical jargon, in set theory, we shall say the universal set.

Definition 3.2.2 Universal Set

The **universal set** *U* is the largest set in a given context; that is, the **universal set** is the totality of the elements under consideration. Denoted by the capital letter, *U* normally written with the upper bars, $\overline{U}$, or a tail, *U*, as to be it distinguishable from the union symbol ∪.

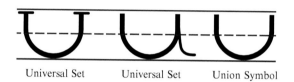

Universal Set   Universal Set   Union Symbol

This is the set from which all other sets will be taken, but it is important that you understand the given context. For example, if the set is defined in the context of the natural numbers, then the understood universe is $U = \{1, 2, 3, \ldots\}$ which the set of all counting numbers including large numbers like 100, 1000, and 1,000,000. However, if the set is defined in the context of the digits, then the understood universe is the limited set: $U = \{0, 1, 2, 3, 4, 5, 6, 7, 8, 9\}$ which does include the element zero by no integer larger than 9.

To help us better understand certain aspects of set theory we shall use circles and rectangles to denote sets. The diagram approach originated with an English logician named John Venn (1834-1923), and we refer to them as Venn diagrams. Similar to Euler circles in Logic, we can draw sets using **Venn diagrams**. The main difference begin, in Logic there are not always physical boundaries other than those represent by the statements themselves (the sets) and an implied universal discourse; whereas in Set Theory, there is the boundary of the contextual universal set. This universal set in a Venn diagram is emphasized in that it is represented by a rectangle which contains the primary set; for example, the universal set of digits can be illustrated as

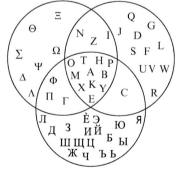

U

2	6	5
4		9
	8	
0		1
		3
	7	

Definition 3.2.3 Empty Set

This **empty set**, sometimes call the **null set** is the smallest set in any context; this set contains no elements. If we were to write it in "proper" set notation, it would be obvious that it has no elements, { } or the single symbol ∅.

$U = \{0, 1, 2, 3, 4, 5, 6, 7, 8, 9\}$
$U = \{x \,|\, x \text{ is a digit}\}$

| An empty set | The empty set | A contradicion | Null, Not, Nothing, Zilch | Zero |

Hence, it is possible for a set to have no elements, for example, the set

$$A = \{y \,|\, y \text{ a unit digit}, y < -1\}$$

has no elements because there is no digit that is less than or equal to −1. In a Venn diagram, letting a set be described by a circle, then the empty set would be represented by an empty circle.

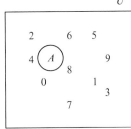

$A \subseteq U$

$A = \varnothing$

Example 3.2.6 Bounded Sets

List the elements of the set given by

$$B = \{z \,|\, z \text{ a positive integer, } z^2 = 25 \text{ and } 5 < z \leq 25\}.$$

Solution

Here, **B** is empty because there is no value of z that will satisfy both the condition (a) z is a positive integer, (b) $z^2 = 25$, and (c) that is greater than five and less than or equal to 25. The only positive number that when squared is 25 is 5; which is not greater than five.

There are many situations in which the elements of a given set are also the elements of another set. For example, let us consider the set **F** which consists of all female students of a finite mathematics course and the set **U** which consists of all students in our university. Here, all the members of the set **F** are also members of the set **U**. That is, all females students are university students; this idea of containment. Before we state the definition of a subset, let us consider another example. Define the sets of even digits **E** and the set of all digits **D** as

$$E = \{0, 2, 4, 6, 8\} \quad \text{and} \quad D = \{0, 1, 2, 3, 4, 5, 6, 7, 8, 9\}.$$

We note here that every element of the set **E** is also an element of the set **D**. We symbolize this situation by writing

$$E \subset D$$

This idea can be further illustrated in a Venn diagram. Assuming that the digits is the universal set

$U = \{0, 1, 2, 3, 4, 5, 6, 7, 8, 9\}$

$E = \{0, 2, 4, 6, 8\}$

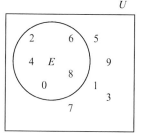

Definition 3.2.4 Subset

If every element of a set **A** is also a member of a set **B**, then we say that the set **A** is a **subset** of the set **B**. Logically, if $x \in A$, then $x \in B$.

The symbol $\subset$ in the statement $A \subset B$ is read "**A** is a subset of **B**" or "**A** is contained in **B**" in that this symbol looks similar to the letter C or c; however, not exactly. Recall, capital C usually represents a set and the lowercase c may represent an element or be an element, but a subset does not curve inward like a partially closed staple, but rather is straight out more like an unused staple. Moreover, this symbol indicates a **_proper subset_** in that the subset is "sub" or is "less than" the original set; this symbol is comparable to the symbol $<$ in algebra. Whereas if the subset is the original set, then we use the symbol $\subseteq$; this symbol is comparable to the symbol $\leq$ in algebra.

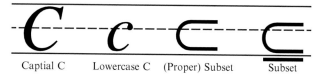

Captial C   Lowercase C   (Proper) Subset   Subset

It should be noted that we can compare two sets without the assumption that the universal set is known.

$A \subseteq B$

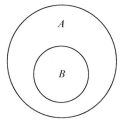

It is clear that every set **A** is a subset of itself; that is, $A \subseteq A$. Also, by definition the empty set, $\varnothing$, is a subset of every set **A**, $\varnothing \subset A$. In terms of the sets which constitute Complex numbers: the Natural numbers, Whole numbers, Integers, Rational numbers, Real number and Imaginary numbers, these sets are related as follows:

$$\mathbb{N} \subset \mathbb{W} \subset \mathbb{Z} \subset \mathbb{Q} \subset \mathbb{R} \subset \mathbb{C}$$

This relationship is such that if $x \in A$ and $A \subseteq B$, then $x \in B$, hence, the relationship of subsets in set theory is comparable to implication in logic; that is, $(x \in A \subseteq B) \rightarrow (x \in B)$.

implies -----→ ⊆

English   Logic   Set Theory

For example, for the set $C = \{1, 3, 4, 6\}$ we have the following subsets: $\varnothing$, $\{1\}$, $\{3\}$, $\{4\}$, $\{6\}$, $\{1,3\}$, $\{1,4\}$, $\{1,6\}$, $\{3,4\}$, $\{3,6\}$, $\{4,6\}$, $\{1,3,4\}$, $\{1,3,6\}$, $\{1,4,6\}$, $\{3,4,6\}$, and $\{1,3,4,6\}$. Thus, the set $C = \{1, 3, 4, 6\}$ possesses 16 possible subsets. In fact, there are 2^n possible subsets of a given set, where n is the number of elements that constitute the set. In the preceding example the set C consists of $n = 4$ elements and hence we have $2^4 = 16$ subsets. Note that the 2^n possible subsets of a given set including the empty set and the entire set.

Example 3.2.7 Subsets

Write all possible subsets of the set

$$A = \{x | x \text{ an integer}, 1 < x^2 \leq 16\}$$

Solution

Here, the set A in roster notation is written as

$$A = \{2, 3, 4\}$$

There are $2^3 = 8$ possible subsets of the set A; namely,

$$\varnothing, \{2\}, \{3\}, \{4\}, \{2, 3\}, \{2, 4\}, \{3, 4\}, \text{ and } \{2, 3, 4\}$$

If subset $\subseteq$ in set theory is comparable to implication or a conditional $\rightarrow$ in logic, then equality $=$ in set theory is comparable to biconditional in logic. That is, set equality, $A = B$ logically stated is: $x \in A$ if and only if $x \in B$.

Definition 3.2.5 Subset

Two sets A and B are said to be **equal** if they have the same elements and we write

$$A = B.$$

It is clear that if $A \subseteq B$ and $B \subseteq A$, then the two sets must be equal, $A = B$. Logically, $A \subseteq B$ translates to $x \in A \rightarrow x \in B$ and $B \subseteq A$ translates to $x \in B \rightarrow x \in A$; that is, $x \in A \leftrightarrow x \in B$.

Example 3.2.8 Temperatures

The daily temperatures during the first week in July, 1976 in Clearwater, Florida were observed to be

$$88, 92, 89, 89, 88, 88, 86$$

degrees Fahrenheit. Describe this set.

> **Solution**
>
> Putting these observations in a set form, we have
>
> $$C = \{86, 88, 89, 90, 92\}.$$
>
> Recall that all members of a set must be distinct. Therefore, although 88 and 89 were recorded twice, when we put them in set form we include the observations only once. Also, the order in which we list the elements of the set is irrelevant.
>
> Consider temperatures recorded for the same period in Sparkling City:
>
> $$90, 92, 89, 89, 88, 88, 86$$
>
> degrees Fahrenheit. Putting these observations in a set form, we have
>
> $$S = \{86, 88, 89, 90, 92\}$$
>
> Since both sets C and S contain the same recordings, $C = S$.

PROBLEMS

Critical Thinking

3.2.1. Indicate which of the following verbal descriptions a well-defined set:
   (a) The players of the Tampa Bay Buccaneers football team,
   (b) The collection of all good United States senators,
   (c) The states of the United States of America,
   (d) The collection of all secretaries who can type at least 75 words per minute,
   (e) The golf players who have won the United States Open Golf Tournament.

3.2.2. Let $B = \{-1, a, 2, b, c, d\}$. Indicate which of the following statements are correct:
   (a) $\varnothing \in B$,   (b) $a \notin B$,   (c) $d \in B$,   (d) $-1 \in B$

3.2.3. Consider the set $A = \{x | x \text{ a positive integer } 1 < x \leq 8\}$ Which of the following statements are true?
   (a) $A = \{2, 3, 4, 5, 6, 7\}$,   (b) $A = \{1, 2, 3, 4, 5, 6\}$
   (c) $4 \in A$,   (d) $1 \notin A$
   (e) $a \in A$,   (f) $7 \notin A$

3.2.4. Let $C = \{z | z \text{ is an even integer}, 1 \leq x < 13\}$. Indicate which of the following statements are true.
   (a) $C = \{1, 2, 4, 6, 8, 10, 12\}$
   (b) $C = \{2, 4, 6, 8, 10, 12\}$
   (c) $7 \notin C$,   (d) $1 \in C$,   (e) $8 \in C$

3.2.5. Consider the set of French philosophers during the Age of Reason, an intellectual movement of the 1700s.

$$P = \{\text{de Condorcet, Diderot, Helvetius, Rousseau, Voltaire}\}.$$

   Indicate which of the following are true or false:
   (a) de Condorcet $\in P$,   (b) Diderot $\in P$
   (c) Helvetius $\notin P$,   (d) Voltaire $\in P$

3.2.6. Consider the set of drugs useful in the treatment of certain cancers.

$$D = \{5 - \text{fluorourucil, methotrexate, cytoxin, Vincristine}\}.$$

   Indicate which of the following statements are true or false:
   (a) 7-fluorourucil $\in D$
   (b) cytoxin $\in D$
   (c) 5-fluorourucil $\notin D$
   (d) Vincristine $\in D$

3.2.7. Write the set which consists of all positive integers less than 15.

3.2.8. Write the set of all integers greater than -3 and less than or equal to 4.

3.2.9. Construct the set of all even integers greater than or equal to zero and less than 16.

3.2.10. Write the set of all integers greater than or equal to 8.

3.2.11. Construct the set of all integers less than zero.

3.2.12. Give a verbal description of the following sets:

(a) $A = \{a_1, a_2, a_3, a_4\}$

(b) $B = \{6, 9, 7, 8\}$

(c) $C = \{x | x$ is a positive integer, $1 < x \leq 10\}$

(d) $D = \{A, t, h, e, n, s\}$

3.2.13. Write a verbal description of the following sets:

(a) $A = \{-1, 1, -2, 2, -3, 3\}$

(b) $B = \{P, l, a, t, o\}$

(c) $C = \{y | y$ is a integer, $y^2 = 16\}$

(d) $D = \{O, l, y, m, p, i, a\}$

3.2.14. List explicitly, in roster notation, the elements of the following sets:

(a) $A = \{x | x$ is a positive integer, $x \leq 5\}$

(b) $B = \{y | y$ is an even positive integer, $4 \leq y^2 < 16\}$

3.2.15. List explicitly, in roster notation, the members of the following sets:

(a) $A = \{z | z$ is an integer, $z^2 = 64\}$

(b) $B = \{t | t = 2n - 1,$ n a positive integer$\}$

3.2.16. Give a mathematical description of the following:

(a) the set A of all the letters in the word "retina."

(b) the set B of all negative integers.

3.2.17. Write the elements of the set Z defined by

$$Z = \{z | z \text{ is an integer, } z = -4\}$$

3.2.18. Is the set E given by
$E = \{e | e$ is an integer, $e > 5$ and $e \leq 5\}$ empty? Why?

3.2.19. Write all possible subsets of the set $A = \{x | x$ is an integer, $-4 \leq x < 0\}$.

3.2.20. Consider the set of all the American states, the names of which begin with the letter N. How many subsets does this set have?

3.2.21. Construct a set that contains the sets $\{-2, 0\}$, $\{-1\}$, $\{0, 1, 3\}$ as subsets.

3.2.22. Write the set that contains the sets

{Matthew, Alee}, {George}, {Mark, Sean}, {Francis}

as subsets. How many possible subsets can we write?

3.2.23. Write all possible subsets of the set that consists of the names: Michael, Irene, Nick, Anastasia, and Chris.

3.3   THE ALGEBRA OF SETS

The basic set operators are union, intersection and complement; these operators in sets are comparable to the logical operators: disjunction, conjunction and negation. In fact the symbolism used in sets is comparable to the symbolism used in logic; similarly, set notation is a form of coding that abbreviates large amount of syntax into abridged versions that can be interpreted and manipulated in the mind more readily. Additional set operators include the set difference and symmetric difference.

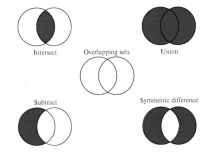

Remember: the speed at which these abbreviations can be interpreted depends on how often they are used.

Let us consider the sets $A = \{1, 3, 4, 6, 8\}$ and $B = \{1, 2, 3, 4, 5, 6\}$, we are interested in combining and comparing the elements of the two sets A and B into a new set, say C. Thus, to combine these two sets, we take them in union, $\cup$. Thus, we can write

$$C = A \cup B$$

which we read as "A union B."

Definition 3.3.1 Union

Let A and B be two sets. The set C if all elements that belong to either set A or B, or both is called the **union** of A and B. We denote this operation by

$$C = A \cup B.$$

Logically, $x \in C \leftrightarrow [(x \in A) \vee (x \in B)]$.

$$x \in A \cup B \text{ if and only if } (x \in A \text{ or } x \in B)$$

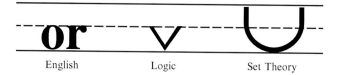

English   Logic   Set Theory

Example 3.3.1 Union

Construct the set $C = A \cup B$ of the sets:

$$A = \{1, 3, 4, 6, 8\} \quad \text{and} \quad B = \{1, 2, 3, 4, 5, 6\}$$

Solution

The elements that are in either A or B are given by

$$C = A \cup B = \{1, 3, 4, 6, 8\} \cup \{1, 2, 3, 4, 5, 6\}$$
$$= \{1, 2, 3, 4, 5, 6, 8\}$$

To further illustrate this set we can represent the previously defined sets A and B by circles, as shown in Figure 3.1.

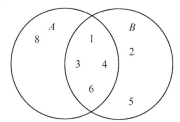

FIGURE 3.1   Circle diagram of set A in union with set B.

Example 3.3.2 Dissolved Oxygen

It has become increasingly evident that the world's most valuable natural resources—air and water—are being adversely affected by the activities of civilized man. Our water supply is threatened by the disposal of organic (and other) waste materials into rivers, lakes, and streams. This pollution has become a major concern of the scientific community, and various regulatory agencies have specified minimum levels for dissolved oxygen, DO, in natural bodies of water. These minimum levels of DO are significant in that if DO falls below a certain threshold value, the fish and other living organisms may die. Many organic compounds discharged into rivers, lakes, and streams are effectively degraded into inoffensive components by the action of bacteria in the water. As the organisms in a body of water degrade the pollution, they require oxygen, and hence use the dissolved oxygen that is available to them. Oxygen-consuming pollutants are

Example 3.3.2 Dissolved Oxygen—cont'd

measured in terms of the amount of oxygen required by the bacteria to stabilize them, called the biochemical oxygen demand, BOD. It is generally accepted that the BOD and DO, measured in parts per million, or ppm, are the primary indicators of water quality. Dissolved oxygen is obtained directly from the air by aeration and indirectly through the photosynthetic process of aquatic plants. In the River of No Return the average monthly recordings in ppm of DO and BOD for 1976 are

$$32, 28, 30, 33, 31, 32, 34, 33, 30, 31.28, 30$$

and

$$26.29, 27, 30, 29, 32, 31, 33, 32, 30, 31, 32,$$

respectively. Find the union of the recordings of the indicators of water quality.

Solution

Here, the elements of the two sets are

$$DO = \{28, 30, 31, 32, 33, 34\} \quad \text{and} \quad BOD = \{26, 27, 29, 30, 31, 32, 33\}.$$

Thus, the union of DO and BOD is $DO \cup BOD = \{26, 27, 28, 29, 30, 31, 32, 33, 34\}$ as shown by Figure 3.2.

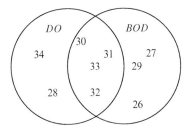

FIGURE 3.2 Circle diagram of Dissolved Oxygen and Biochemical Oxygen Demand.

Hence, the union of the two sets gives a distinct characterization of the numerical values of the primary indicator of water quality. This information is useful to the environmentalist in planning strategies for water quality control.

Here, the union of the two sets is more than a simple circle; the union of two sets is the combined circles as illustrated by the shaded region below.

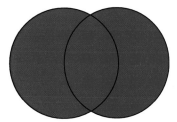

Example 3.3.3 Union with Integers

Find the union of the following sets:

$$A = \{x | x \text{ is an integer, } 1 < x \leq 6\}$$

$$B = \{y | y \text{ is a positive integer, } y < 5\}$$

$$C = \{z | z \text{ is an integer, } -3 < z \leq 2\}.$$

Solution

We begin by listing the elements of each of the sets in roster notation (Figure 3.3)

$$A=\{2,3,4,5,6\} \qquad B=\{1,2,3,4\} \qquad C=\{-2,-1,0,1,2\}$$

Thus, $A \cup B \cup C = \{-2,-1,0,1,2,3,4,5,6\}$

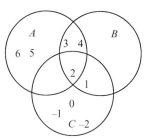

FIGURE 3.3 Three way circle diagram of the sets **A, B** and **C**.

Although we have defined the union of two sets and have illustrated its meaning with two and three sets, we are not restricted as to the number of sets.

Physician P_1 has a collection of five drugs—call them a,b,c,d,e—from which he selects to prescribe for his patients for a certain type of virus. Physician P_2, also selects from five drugs—a,b,d,f,g—for his patients with the same type of virus. A patient who has knowledge of the sets P_1 and P_2, is interested in just those drugs that are being used by both physicians. Here, we have the two sets

$$P_1=\{a,b,c,d,e\} \quad \text{and} \quad P_2=\{a,b,d,f,g\}$$

And we are interested in the elements common to both P_1 and P_2.

Definition 3.3.2 Intersection

Let A and B be two sets. The set C of all elements that belong to both A and B is called the **intersection** of the sets A and B. We denote this operation by

$$C=A \cap B.$$

Logically, $x \in C \leftrightarrow [(x \in A) \wedge (x \in B)]$.

$x \in A \cap B$ if and only if $(x \in A$ and $x \in B)$

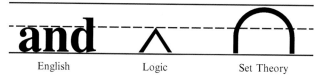

English   Logic   Set Theory

Here, the intersection of the two sets is less than a simple circle; the intersection of two sets is the overlapping circles as illustrated by the shaded region below.

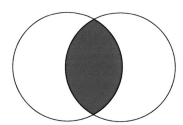

The intersection ∩ of the preceding sets P_1 and P_2 is given by

$$P_1 \cap P_2 = \{a, b, c, d, e\} \cap \{a, b, d, f, g\} = \{a, b, d\}$$

as shown by Figure 3.4.

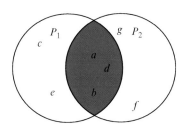

FIGURE 3.4 Circle graph of overlapping sets of drugs proscribed by each physicians.

Example 3.3.4 Humidity Readings _____

Find the intersection of the two sets that constitute the humidity recordings. The average monthly relative humidities as percentages for September to August, 1976 in the City of Tampa, Florida were

80, 78, 76, 68, 72, 74, 78, 82, 82, 78, 81, 82,

and the averages for St. Petersburg, Florida were

76, 68, 72, 74, 74, 76, 77, 78, 79, 80, 78, 80.

Solution _____

The two sets are

$$T = \{68, 72, 74, 76, 78, 80, 81, 82\}$$

and

$$P = \{68, 72, 74, 76, 77, 78, 79, 80\}.$$

Thus, the intersection of the two sets is $T \cap P = \{68, 72, 74, 76, 78, 80\}$, as shown by Figure 3.5.

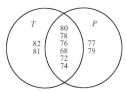

FIGURE 3.5 Circle diagram of humidity readings in Tampa and St. Petersburg Florida.

This means that there were six monthly readings during 1976 in which the average relative humidity in Tampa and St. Petersburg were the same.

Example 3.3.5 Intersection

Consider the sets given in Example 3.3.3; that is,

$$A = \{2, 3, 4, 5, 6\} \quad B = \{1, 2, 3, 4\} \quad C = \{-2, -1, 0, 1, 2\}$$

Find $A \cap B \cap C$.

Solution

$A \cap B \cap C = \{2\}$. That is, the only element that is common to all three sets is 2, as shown in Figure 3.6.

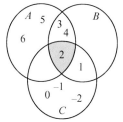

FIGURE 3.6 Three circle diagram of sets **A, B** and **C**.

Example 3.3.6 Intersections

Find (a) $A \cap B$, (b) $A \cap C$, (c) $A \cap B \cap C$, (d) $A \cap B \cap C$ and (e) $(A \cup B) \cap C$ given
$$A = \{2, 3, 4, 5, 6\} \quad B = \{1, 2, 3, 4\} \quad C = \{-2, -1, 0, 1, 2\}$$

Solution

(a) $A \cap B = \{5, 7\}$

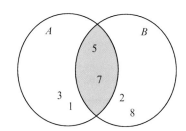

(b) $A \cap C = \{1, 3, 7\}$

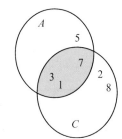

Solution—cont'd

(c) $A \cap B \cap C = \{7\}$

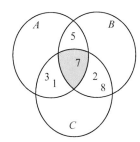

(d) $(A \cap B) \cup C = (\{1, 3, 5, 7\} \cup \{2, 5, 7, 8\}) \cap \{1, 2, 3, 7, 8\}$
$= \{5, 7\} \cup \{1, 2, 3, 7, 8\} = \{1, 2, 3, 5, 7, 8\}$

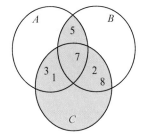

(e) $A \cup B = \{1, 2, 3, 5, 7, 8\}$

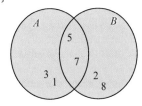

(f) $(A \cup B) \cap C = \{1, 2, 3, 5, 7, 8\} \cap \{1, 2, 3, 7, 8\} = \{1, 2, 3, 7, 8\}$

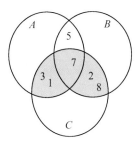

Sometimes it happens that the intersection of two sets A and B is empty; that is, they do not have any elements in common, and we say that the two sets are ***disjoint*** or ***mutually exclusive***.

Example 3.3.7 Intersection of D and E

Define the sets D and E as follows:

$$D = \{x \mid x \text{ an integer}, -3 \leq x < 1\}$$

and

$$E = \{y \mid y \text{ a positive integer}, 3 \leq y < 6\}.$$

Find the intersection of D and E.

Solution

The elements of the sets **D** and **E** are

$$D = \{-3, -2, -1, 0\} \quad \text{and} \quad E = \{3, 4, 5\}.$$

Thus, $D \cap E = \varnothing$ and the sets are disjoint, as shown by Figure 3.7.

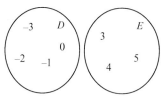

FIGURE 3.7 Circle diagram of Mutually Exclusive sets: **D** and **E**.

Another important definition in set theory is the **complement** of a set. That is, given a set. **A**, we are interested in considering a new set which consists of those elements which are not in **A**. We refer to this new set as the complement of **A** and denote it by **A**. For example, let **A** be the set of all students at your school who are majoring in modern dance. The **complement** of the set **A** will be all students in your school who are not majoring in modern dance. Thus, the union of **A** and its complement **A** constitute the largest possible set called the **universal set**, denoted by **U**. Recall, the universal set is the smallest possible set that is required so that every other set considered will be a subset of **U**. Now, using the universal set **U**, we can obtain a better understanding of the complement of a set by the following definition:

Definition 3.3.3 Complement

Let **A** be a subset of the universal set **U**, the **complement** of **A**, denoted by $\bar{A}$, is the set of all elements in **U** that are not in **A**. Other common denotation of the complement of **A** is A', or A^c; however, we shall use the notation $\bar{A}$. Logically, $x \in \bar{A}$ if and only if $x \in U$ and $x \notin A$.

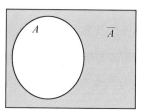

For example, the thirty-eight presidents of the United States (up to 1976) constitute the universal set, **U**. Let us now consider the set of presidents whose last name begins with the letter J; that is, the set **A** contains the names Thomas Jefferson, Andrew Jackson, Andrew Johnson, and Lyndon B. Johnson. It is clear that these four presidents constitute a subset of **U**. Thus, the complement of **A**, $\bar{A}$, is a subset of **U** which consists of the other thirty-four presidents.

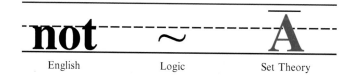

| not | ~ | A̅ |
| English | Logic | Set Theory |

Example 3.3.8 Integers

Let the universal set consist of the first ten even integers; that is,

$$U = \{2, 4, 6, 8, 10, 12, 14, 16, 18, 20\}.$$

Also, let

$$A = \{2, 8, 14, 20\} \quad \text{and} \quad B = \{4, 8, 14, 16\}$$

Find (a) the complement of A, and also B, (b) the complement of $A \cup B$, denoted $\overline{A \cup B}$.

Solution

(a) The complement of A is the set of all elements which are not in A.
That is,

$$\bar{A} = \{4, 6, 10, 12, 16, 18\},$$

as shown by Figure 3.8. Also,

$$\bar{B} = \{2, 6, 10, 12, 18, 20\},$$

which is displayed in Figure 3.9.

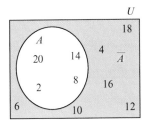

FIGURE 3.8   Venn diagram of set A within the Universe, U.

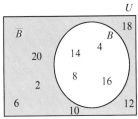

FIGURE 3.9   Venn diagram of set B within the Universe, U.

(b) $\overline{A \cup B}$ is the set of those elements which are not in $A \cup B$; that is, the set of all elements which are not in either A or B or both.

$$\overline{A \cup B} = \{6, 10, 12, 18\}$$

as shown by the Venn diagram, Figure 3.10.

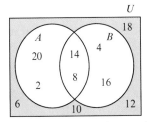

FIGURE 3.10   Venn diagram of the complement of the union of sets A and B; that is, what is not in A and is not in B.

Continued

Solution—cont'd

(c) The complement of $A \cap B$ is the set of elements which are not common to both A and B. That is,

$$A \cap B = \{8, 14\}$$

and

$$\overline{A \cap B} = \{2, 4, 6, 10, 12, 16, 18, 20\}$$

as shown by the Venn diagram, Figure 3.11.

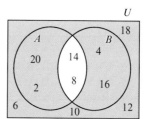

FIGURE 3.11   Venn diagram of the symmetric difference; that is, what is not in both set A and set B.

In terms of the sets which constitute Complex numbers: the Natural numbers, Whole numbers, Integers, Rational numbers, Real number and Imaginary numbers, we have the following relationships:

Set	Name
⟩ $\overline{Q}$	Irrational
⟩ $\overline{R}$	Imaginary numbers

In this section we have introduced two basic operations of sets; namely, the **union** $\cup$ and **intersection** $\cap$ of sets. These set operators are similar to their logistical counterparts. Logistically, the disjunctive "or" is translated as "at least one statement is true" and is denoted $\vee$ whereas in set theory union can be read as "or," translated as "in at least one set" and is denoted $\cup$. Logistically, the conjunction "and" is translated as "both statements are true" and is denoted by the inverted symbol $\wedge$ whereas in set theory the intersection can be read as "and," translated as "in both sets" and is denoted $\cap$. The symbolic language of these two dialects of mathematics are very similar, hence as in logic, there are various laws which hold true. We shall now state some properties of these operations and illustrate their meanings with Venn diagrams.

Let A, B, and C be subsets of the universal set U. Then we have

$$\{a, b\} \cup \{a, b\} = \{a, b\}$$
$$\{a, b\} \cap \{a, b\} = \{a, b\}$$

Rule 3.3.1 Idempotent

$$A \cup A = A \quad \text{and} \quad A \cap A = A$$

As in logic, **idempotent** describes the property of operations, as in mathematics and computer science, which yield the same result after the operation is applied multiple times.

Rule 3.3.2 Associative

$$(A \cup B) \cup C = A \cup (B \cup C) \quad \text{and} \quad (A \cap B) \cap C = A \cap (B \cap C)$$

As in logic and in algebra, **associative** describes the property of operations that enables statements to re-associate while yielding the same result.

Rule 3.3.3 Commutative

$$A \cup B = B \cup A \text{ and } A \cap B = B \cap A$$

$$\{a\} \cup \{b\} = \{b\} \cup \{a\}$$

$$\{a, c\} \cap \{b, c\} = \{b, c\} \cap \{a, c\}$$

As in logic and in algebra, **commutative** describes the property of operations that enables statements to move or commute while yielding the same result.

Rule 3.3.4 Distributive

$$A \cup (B \cap C) = (A \cup B) \cap (A \cup C)$$

and

$$A \cap (B \cup C) = (A \cap B) \cup (A \cap C)$$

As in logic, **distributive** describes the property of one operator to be expanded in a particular way which yield the same result; that is, an equivalent expression. The distributive law in set theory is more similar to logic. In logic, the operator for "and" is distributive over the operator for "or" and vice versa, whereas in set theory, this relation only works the same.

Rule 3.3.5 Identity

If U the **universal set** and $\varnothing$ is the **empty set**, then

$$A \cup \varnothing = A, \ A \cap U = A, \ A \cap \varnothing = \varnothing \text{ and } A \cup U = U$$

Recall, **identity** is the state or fact of remaining the same one or ones, as under varying aspects or conditions, to identify. Hence, the empty set in union identifies the original set A, but the intersection identifies the empty set. The universal set in union identifies the universal set, whereas the intersection identifies original set A.

Rule 3.3.6 Complement

If U the **universal set** and $\varnothing$ is the **empty set**, then

$$A \cup \bar{A} = U, A \cap \bar{A} = \varnothing, \ \overline{(\bar{A})} = A, \ \bar{U} = \varnothing \text{ and } \bar{\varnothing} = U$$

A **complement** is the part needed to make complete or perfect; in logic, this is the relationship between *true* and *false*, in set theory, an element is either in a given set or not in the given sets. In logic, either the statement is true or the statement is false and thus $p \vee \sim p$ is a tautology τ; in set theory, the set of elements that are either in a given set A or not in the given set, that is the complement of A, is in union the universal set. In logic, the statement cannot be both true and false; hence, $p \wedge \sim p$ is a contradiction φ; in set theory, an element cannot be both in a set A and not in the set A, hence the set of elements in both A and $\bar{A}$ is empty. Furthermore, as in English, a double negative is the positive statement; in logic, not "not p" then p and in set theory, not in $\bar{A}$, that is $\overline{(\bar{A})}$, is the set A. Finally, logically speaking, the negation of a tautology is a contradiction and the negation of a contradiction is a tautology. In set theory, the complement of the universal set is the empty set and the complement of the empty set is the universal set.

Rule 3.3.7 DeMorgan's Rule

$$\overline{(A \cup B)} = \bar{A} \cap \bar{B} \quad \text{and} \quad \overline{(A \cap B)} = \bar{A} \cup \bar{B}$$

De Morgan's Law illustrates the fact that "or" is the complement of "and" and vice versa. In both logic and set theory, "or" means at least one and "and" means both; hence, when *you do not have at least one*, $\overline{(A \cup B)}$, this is equivalent to you do not have either, or you are missing both $\bar{A} \cap \bar{B}$; *you do not have the first and you do not have the second*. Similarly, when *you do not have both*, $\overline{(A \cap B)}$, this is equivalent to you are missing at least one, $\bar{A} \cup \bar{B}$; *you do not have the first or you do not have the second*.

To further illustrate De Morgan's Law, draw the Venn diagram using lines for the various sets; in union. First, consider $\overline{(A \cup B)} = \bar{A} \cap \bar{B}$, the Venn diagram constructed using $\overline{(A \cup B)}$:

$$A \cup B \;\Rightarrow\; \overline{(A \cup B)}$$

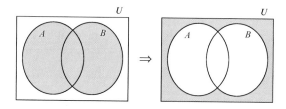

is equivalent to the Venn diagram constructed using $\bar{A} \cap \bar{B}$:

$$\bar{A} \cap \bar{B} \Rightarrow \bar{A} \cap \bar{B}$$

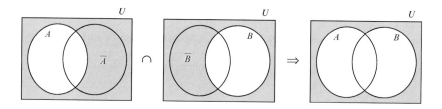

Second, consider $\overline{(A \cap B)} = \bar{A} \cup \bar{B}$; the Venn diagram constructed using $\overline{(A \cap B)}$

$$A \cap B \Rightarrow \overline{(A \cap B)}$$

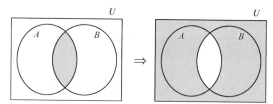

is equivalent to the Venn diagram constructed using $\bar{A} \cup \bar{B}$:

$$\bar{A} \cup \bar{B} \Rightarrow \bar{A} \cup \bar{B}$$

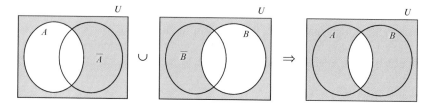

Example 3.3.9 Digits

Let $U = \{1, 2, 3, 4, 5, 6, 7, 8\}$, $A = \{1, 3, 5\}$, $B = \{5, 6, 7\}$, and $C = \{5, 7, 8\}$. Verify (a) the associative property for unions, (b) distribution property (intersection over union) and (c) De Morgan's Law for the complement of a union.

Solution

(a) Using Property 2 for unions:

$$A \cup B = \{1, 3, 5, 6, 7\} \Rightarrow (A \cup B) \cup C = \{1, 3, 5, 6, 7, 8\},$$

$$B \cup C = \{5, 6, 7, 8\} \Rightarrow A \cup (B \cup C) = \{1, 3, 5, 6, 7, 8\}$$

Thus, we have

$$(A \cup B) \cup C = A \cup (B \cup C)$$

(b) Using Property 4 for intersection over union:

$$A \cup (B \cap C) = (A \cup B) \cap (A \cup C)$$

$$B \cap C = \{5, 6, 7, 8\} \Rightarrow A \cup (B \cap C) = \{5\}$$

$$A \cap B = \{5\} \text{ and } A \cap C = \{5\} \Rightarrow (A \cup B) \cap (A \cup C) = \{5\}$$

Thus, we have

$$A \cup (B \cap C) = (A \cup B) \cap (A \cup C)$$

(c) Using Property 7 for the complement of a union:

$$\overline{(A \cup B)} = \bar{A} \cap \bar{B}$$

$$A \cup B = \{1, 3, 5, 6, 7\} \Rightarrow \overline{A \cup B} = \{2, 4, 8\}$$

$$\bar{A} = \{2, 4, 6, 7, 8\} \text{ and } \bar{B} = \{1, 2, 3, 4, 8\} \Rightarrow \bar{A} \cap \bar{B} = \{2, 4, 8\}$$

Thus, we have

$$\overline{(A \cup B)} = \bar{A} \cap \bar{B}$$

Verification of the remaining properties is left to the student.

Additional set operators include the symmetric difference and the set difference.

Definition 3.3.4 Set Difference

Let A and B be two sets. The set C of all elements that belong to A but not B is called the **set difference** of the sets A and B. We denote this operation by

$$C = A - B \text{ or } C = A \backslash B.$$

We shall use the notation $C = A - B$. Logically, $x \in C \leftrightarrow [(x \in A) \wedge (x \notin B)]$; that is, $x \in C$ if and only if $x \in A$ and $x \notin B$.

Hence, alternatively $A - B = A \cap \bar{B}$.

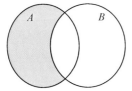

Example 3.3.10 Digits

Let the universal set consist of the first ten even integers; that is,

$$U = \{0, 1, 2, 3, 4, 5, 6, 7, 8, 9\}.$$

Also, let

$$A = \{2, 5, 7, 9\} \quad \text{and} \quad B = \{1, 2, 4, 7, 8\}$$

Find the set differences: (a) $A - B$ and (b) $B - A$.

Solution

(a) $A - B = \{5, 9\}$ and (b) $B - A = \{1, 4, 8\}$

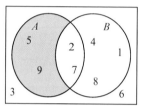

Definition 3.3.5 Symmetric Difference

Let A and B be two sets. The set C of all elements that belong to either A or B is called the **symmetric difference** of the sets A and B. We denote this operation by

$$C = A \Delta B \quad \text{or} \quad C = A \ominus B.$$

Logically, $x \in C \leftrightarrow \left[(x \in A) \underline{\vee} (x \in B) \right]$; that is, $x \in C$ if and only if $x \in A$ and $x \notin B$ or $x \notin A$ but $x \in B$. We shall use the Greek letter delta, Δ, to denote the symmetric difference.

Alternatively,

$$A \Delta B = (A \cup B) - (A \cap B)$$

or

$$A \Delta B = (A - B) \cup (B - A)$$

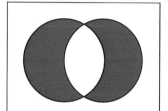

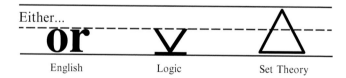

Either...

| English | Logic | Set Theory |

Example 3.3.11 Symmetric Difference

Using the sets defined in Example 3.3.10, find the symmetric differences $A \Delta B$.

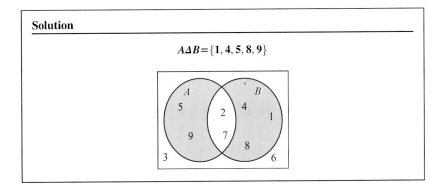

Solution

$$A \triangle B = \{1, 4, 5, 8, 9\}$$

PROBLEMS

Critical Thinking

3.3.1. Let $A = \{2, 4, 6, 8\}$, $B = \{6, 8, 10, 12\}$ and $C = \{8, 10, 12, 14\}$. Find
   a. $A \cup B$ e. $A \cap B$
   b. $A \cup C$ f. $A \cap C$
   c. $B \cup C$ g. $B \cap C$
   d. $A \cup B \cup C$ h. $A \cap B \cap C$

3.3.2. Illustrate the solution to Problem 3.3.1 by using Venn diagrams.

3.3.3. Given the following Venn diagram, list the elements of the three sets A, B, and C.

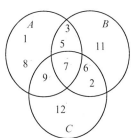

3.3.4. Using the sets A, B, and C obtained in Problem 3.3.3, compute
   a. $A \cup B$
   b. $A \cap B$
   c. $A \cup (B \cap C)$
   d. $A \cap (B \cup C)$
   e. $(A \cup B) \cap C$
   f. $A \cup \varnothing$
   g. $B \cap \varnothing$
   h. $(A \cap \varnothing) \cup C$
   i. $(A \cup \varnothing) \cup B$

3.3.5. Draw the Venn diagram for Parts (c), (e), (g), and (i) of Problem 3.3.4.

3.3.6. Draw the Venn diagram for Parts (b), (d), (f), and (h) of Problem 3.3.4.

3.3.7. Are the sets A and B, A and C, and B and C defined in Problem 1 disjoint sets?

3.3.8. Let $A = \{x \mid x \text{ an integer}, -1 < x \leq 4\}$, $B = \{y \mid y \text{ an integer}, 0 \leq y \leq 6\}$, and $C = \{z \mid z \text{ a positive integer}, z \leq 8\}$. Find
   a. $A \cup B$
   b. $A \cap B$
   c. $(A \cup B) \cap (A \cap B)$

     d. $(A \cap C) \cup \emptyset$

     e. $(A \cup \emptyset) \cap (B \cap \emptyset)$

     f. $(A \cup B) \cup C$

3.3.9. Draw Venn diagrams illustrating the solution to Problem 3.3.8.

3.3.10. Let the universal set U and the subsets A, B, and C be defined as follows: $U = \{a, b, c, d, e, f, g, h, i, j, k, l\}$, $A = \{a, b, f\}$, $B = \{c, d, g, h, l\}$, and $C = \{a, b, k, l\}$. Find the following sets:

     a. $A \cup B \cup C$

     b. $A \cap C$

     c. $A \cap B \cap C$

3.3.11. Using the sets defined in Problem 3.3.10, find

     a. $A \cup B \cup C$

     b. $A \cap C$

     c. $A \cap B \cap C$

     d. $\bar{A} \cap \bar{B} \cap \bar{C}$

3.3.12. Find the following sets using the sets defined in Problem 3.3.10:

     a. $\overline{A \cap C}$

     b. $\overline{U \cup \emptyset}$

     c. $(A \cap B) \cup (\bar{A} \cap C)$

3.3.13. Draw the Venn diagrams for the results in Problem 3.3.10.

3.3.14. Using Problem 3.3.11, illustrate the results using Venn diagrams.

3.3.15. Show the behavior of the results of Problem 3.3.12 employing Venn diagrams.

3.3.16. Let A, B, and C be subsets of the universal set U. Draw Venn diagrams for the following sets and shade the indicated areas:

     a. $(A \cup B) \cap (A \cap C)$

     b. $(A \cup \bar{B}) \cap (C \cup \bar{B})$

     c. $(A \cup \bar{B}) \cap (A \cup \bar{B})$

     d. $(A \cup \bar{B}) \cap (\bar{A} \cup C)$

3.3.17. Using Venn diagrams, prove that the following identities are true:

     a. $\bar{A} \cap \bar{B} = \overline{A \cup B}$

     b. $\bar{A} \cup \bar{B} = \overline{A \cap B}$

3.3.18. Let the universal set U be defined by $U = \{x | x$ a positive integer, $x \le 12\}$, and subsets $A = \{y | y$ an integer, $0 \le y < 6\}$, $B = \{z | z$ an integer, $4 \le z < 9\}$, $C = \{t | t$ an integer, $1 < t \le 3$ or $6 < t \le 10\}$. Using these sets verify (a) the idempotent law, (b) the commutative law, and (c) the associative law.

3.3.19. Using the universal set U and subsets A, B, and C defined in Problem 3.3.18, show that the following laws are valid: (a) the distributive law, (b) the identity law, (c) the complement law, and (d) De Morgan's Law.

3.3.20. Illustrate the solutions to Problem 3.3.18 using Venn diagrams.

3.3.21. Using Venn diagrams, show the results of Problem 3.3.18.

3.3.22. Let U be the universal set and A, B, and C be subsets of U. Prove or disprove each of the following statements:

     a. $(A \cup \bar{B}) \cup \bar{C} = A \cup \overline{(B \cup C)}$

     b. $A \cap (B \cup \bar{C}) = (A \cap B) \cup \overline{(A \cap C)}$

     c. $(A \cup \bar{B}) \cup (A \cup \bar{C}) = A \cup \overline{(B \cap C)}$

     d. $(A \cup \bar{B}) \cap (A \cup \bar{C}) = A \cup \overline{(B \cup C)}$

3.3.23. Determine whether the following statements are true or false:

     a. $\emptyset \in U$

     b. $\emptyset \subset U$

     c. $\emptyset \cup U = U$

     d. $\{a\} \in (a, b.c)$

     e. $\{1, 2, 3\} \subset (1, 2, 3, 4, 5, 6)$

     f. $\emptyset \subseteq \emptyset$

3.3.24. Using the sets defined in Problem 3.3.10, find

     a. $A - B$

     b. $B - A$

     c. $A \, \Delta \, B$

3.3.25. Using the sets defined in Problem 3.3.18, find

    a. $A - B$

    b. $B - A$

    c. $A \, \Delta \, B$

CRITICAL THINKING AND BASIC EXERCISE

3.1. Write the set of all positive integers less than 10.

3.2. Write in a convenient mathematical form the set of all positive integers less than 201.

3.3. List explicitly the elements of the following sets:

    a. $A = \{x | x$ a negative integer, $-9 \le x\}$.

    b. $B = \{y | y$ an integer, $-3 < y < 5\}$.

3.4. List the elements of the following sets:

    a. $A = \{z | z$ a negative integer, $1 < z\}$.

    b. $B = \{t | t$ an integer, $-3 < t \le 5\}$.

3.5. Give a mathematical description of the set of all letters in the word Mississippi. Also list all the elements of the set.

3.6. List the elements of the set $D = \{d | d$ a positive integer, $d^3 = 27\}$.

3.7. Let the set A be given by $A = \{a, b, c, d\}$. Write all possible subsets of the set A.

3.8. Let $B = \{x | x$ an even integer, $4 < x \le 8\}$. Write all possible subsets of the set B.

3.9. If a set consists of six elements, how many possible subsets can we write from this set?

3.10. Let $A = \{2, 3, 4, 5, 6\}$ and $B = \{x | x$ an integer, $1 < x^2 \le 36\}$. Are the sets A and B equal? Why?

3.11. Construct a set of all the presidents of the United States whose last names begin with (a) A and (b) T.

3.12. Construct a set which contains the sets $\{0, 1\}$, $\{-1, 1, 3\}$, $\{a, -a\}$ as subsets.

3.13. A partial listing of the subsets of a given set are $\{Jan, Chris\}$, $\{Pete\}$, $\{Alex, Liz\}$. Write the complete set. How many possible subsets can we write from this set?

3.14. Let $A = \{2, 4, 5, 8, 10\}$ and $B = \{1, 2, 3, 4, 5\}$. Construct the sets (a) $A \cup B$ and (b) $A \cap B$.

3.15. In Problem 3.3, compute (a) $n(A \cup B)$ and (b) $n(A \cap B)$.

3.16. Construct a Venn diagram and shade in the appropriate portions for Problem 3.14.

3.17. Let

    i. $A = \{x | x$ an even integer, $2 \le x \le 12\}$,

    ii. $B = \{y | y$ an integer, $1 < y \le 7\}$, and

    iii. $C = \{-1, 0, 2, 4, 6\}$.

        Find

    a. $A \cup B$

    b. $A \cup C$

    c. $B \cup C$

    d. $A \cup B \cup C$

3.18. In Problem 3.3, find

    a. $A \cap B$

    b. $A \cap C$

    c. $B \cap C$

    d. $A \cap B \cap C$

3.19. In Problem 3.3, compute

    a. $n(A \cup B)$

    b. $n(A \cup C)$

    c. $n(B \cup C)$

    d. $n(A \cap B)$

    e. $n(A \cap C)$

    f. $n(B \cap C)$

    g. $n(A \cap B \cap C)$

3.20. Construct a Venn diagram indicating the relative portions of Problem 3.3.

3.21. Using the sets defined in Problem 3.17, find

    a. $(A \cap B) \cup C$

    b. $(A \cap C) \cup B$

    c. $(A \cup B) \cap (A \cup C)$

3.22. Write the appropriate Venn diagrams to Problem 3.17.

3.23. Let the universal set U be given by

$$U = \{a_1, b_1, a_2, b_2, c_1, d_1, c_2, d_2\}.$$

    Also, let

$$A = \{a_1, b_2, c_1, d_2\} \quad \text{and} \quad B = \{a_2, b_1, b_2, c_2, d_1\}$$

    Find

    a. $\bar{A}$

    b. $\bar{B} \cap A$

    c. $\overline{A \cap B}$

    d. $\bar{A} \cap \bar{B}$

    e. $(A \cup B) \cap \bar{A}$

    f. $(A \cup B) \cup \overline{(A \cup B)}$

3.24. Construct a sequence of Venn diagrams to illustrate the solution to Problem 3.23.

3.25. Using the following sets, verify (a) the idempotent law, (b) commutative law, and (c) associative law.

    i. $U = \{x | x \text{ an integer} -3 \leq x \leq 10\}$,

    ii. $A = \{1, 2, 3, 4, 5\}$,

    iii. $B = \{x | x \text{ an integer}, 2 \leq x \leq 7\}$,

    iv. $C = \{y | y \text{ an integer}, 0 < y^2 \leq 9\}$.

3.26. Verify (a) the distributive law, (b) identity law, and (c) complement law using the sets defined in Problem 3.25.

3.27. Using the sets defined in Problem 3.25, compute:

    a. $n(A \cup \bar{A})$, **b.** $n(A \cap \varnothing)$, **c.** $n(A \cup B \cup C)$, **d.** $n(A \cap B \cap C)$

3.28. For the sets given in Problem 25, compute:

    a. $n(\bar{A} \cap B)$, **b.** $n(\bar{A} \cup \bar{B})$, **c.** $n[(A \cup B) \cap C]$, **d.** $n[\bar{A} \cap (A \cup C)]$

3.29. A survey of 1200 people who subscribe to at least one of the "financial assistance newsletters," More Dollars and Tomorrow's Dollar, revealed the following information: 400 subscribe to both More Dollars and Tomorrow's Dollar, 600 subscribe to only More Dollars. Find how many of the subscribers questioned subscribe only to Tomorrow's Dollar.

3.30. Let $A = \{a_1, a_2, a_3\}$, $B = \{b_1, b_2\}$ Find the product set $A \times B$ and $n(A \times B)$.

3.31. Let $A = \{1, 3, 5\}$, $B = \{x, y\}$ and $C = \{c_1, c_2\}$. Find the Cartesian product $A \times B \times C$ and $n(A \times B \times C)$.

3.32. Construct a tree diagram to determine the number of ordered pairs that constitute the product set $A \times B$ of the sets given in Problem 3.30.

3.33. Obtain the product set $A \times B \times C$ using the tree diagram approach of the sets given in Problem 3.32. How does your answer compare with that obtained in Problem 3.30?

3.34. From the Venn diagram, determine $n(A \times B \times C)$.

SUMMARY OF IMPORTANT CONCEPTS

Definitions:

3.2.1. A **set** is a collection of distinct objects; the objects that constitute the set are the **members** or **elements** of the set. We shall denote sets symbolically by capital letters A, B, C... and "an element of" by the Greek letter epsilon, $\in$. If an object "is not an element of," we shall write $\notin$.

3.2.2. The **universal set** U is the largest set in a given context; that is, the **universal set** is the totality of the elements under consideration. Denoted by the capital letter, U normally written with the upper bars, U, or a tail, U, as to be it distinguishable from the union symbol $\cup$.

3.2.3. This **empty set**, sometimes call the **null set** is the smallest set in any context; this set contains no elements. If we were to write it in "proper" set notation, it would be obvious that it has no elements, { } or the single symbol $\varnothing$.

3.2.4. If every element of a set A is also a member of a set B, then we say that the set A is a **subset** of the set B. Logically, if $x \in A$, then $x \in B$.

3.2.5. Two sets A and B are said to be **equal** if they contain the same elements and we write $A = B$. It is clear that if $A \subseteq B$ and $B \subseteq A$, then the two sets must be equal, $A = B$. Logically, $A \subseteq B$ translates to $x \in A \rightarrow x \in B$ and $B \subseteq A$ translates to $x \in B \rightarrow x \in A$; that is, $x \in A \leftrightarrow x \in B$.

3.3.1. Let A and B be two sets. The set C if all elements that belong to either set A or B, or both is called the **union** of A and B. We denote this operation by $C = A \cup B$. Logically, $x \in C \leftrightarrow [(x \in A) \vee (x \in B)]$. $x \in A \cup B$ if and only if ($x \in A$ or $x \in B$).

3.3.2. Let A and B be two sets. The set C of all elements that belong to both A and B is called the **intersection** of the sets A and B. We denote this operation by $C = A \cap B$. Logically, $x \in C \leftrightarrow [(x \in A) \wedge (x \in B)]$. $x \in A \cap B$ if and only if ($x \in A$ and $x \in B$).

3.3.3. Let A be a subset of the universal set U, the **complement** of A, denoted by $\bar{A}$, is the set of all elements in U that are not in A. Other common denotation of the complement of A is A', or A^c; however, we shall use the notation $\bar{A}$. Logically, $x \in \bar{A}$ if and only if $x \in U$ and $x \notin A$.

3.3.4. Let A and B be two sets. The set C of all elements that belong to A but not B is called the **set difference** of the sets A and B. We denote this operation by $C = A - B$ or $C = A \backslash B$. We shall use the notation $C = A - B$. Logically, $x \in C \leftrightarrow [(x \in A) \wedge (x \notin B)]$; that is, $x \in C$ if and only if $x \in A$ and $x \notin B$. Hence, alternatively $A - B = A \cap \bar{B}$.

3.3.5. Let A and B be two sets. The set C of all elements that belong to either A or B is called the **symmetric difference** of the sets A and B. We denote this operation by $C = A \Delta B$ or $C = A \ominus B$. Logically, $x \in C \leftrightarrow [(x \in A) \underline{\vee} (x \in B)]$; that is, $x \in C$ if and only if $x \in A$ and $x \notin B$ or $x \notin A$ but $x \in B$. We shall use the Greek letter delta, Δ, to denote the symmetric difference. Alternatively, $A \Delta B = (A \cup B) - (A \cap B)$ or $A \Delta B = (A - B) \cup (B - A)$.

Rules

3.3.1. **Idempotent:** $A \cup A = A$ and $A \cap A = A$

3.3.2. **Associative:** $(A \cup B) \cup C = A \cup (B \cup C)$ and $(A \cap B) \cap C = A \cap (B \cap C)$

3.3.3. **Commutative:** $A \cup B = B \cup A$ and $A \cap B = B \cap A$

3.3.4. **Distributive:** $A \cup (B \cap C) = (A \cup B) \cap (A \cup C)$ and $A \cap (B \cup C) = (A \cap B) \cup (A \cap C)$

3.3.5. **Identity:** If U the **universal set** and $\varnothing$ is the **empty set**, then $A \cup \varnothing = A$, $A \cap U = A$, $A \cap \varnothing = \varnothing$ and $A \cup U = U$

3.3.6. **Complement:** If U the **universal set** and $\varnothing$ is the **empty set**, then $A \cup \bar{A} = U$, $A \cap \bar{A} = \varnothing$, $(\bar{A}) = A$, $\bar{U} = \varnothing$ and $\bar{\varnothing} = U$

3.3.7. **De Morgan's Rule:** $\overline{(A \cup B)} = \bar{A} \cap \bar{B}$ and $\overline{(A \cap B)} = \bar{A} \cup \bar{B}$

Symbol (abbr.)	Name	Meaning/Read as	Description	Example/Description
A, B, C	Sets	Abbreviation for a sets	Sets are abbreviated by one capital letter	Let S represent the set of statements
$\{\ldots\}$	Braces	"The set"	Another way to indicated a set	$D = \{1, 2, 3, 4, 5, 6\}$—in this case, D for a Die
$\vert$	Bar	"Such that"	A partition used in conjunction with $\{\ldots\}$	$E = \{x \vert x \text{ is even integer}\}$
$\in$	Greek letter "Epsilon"	"An element of"	Symbol illustrating containment	$2 \in D$
$\notin$	"Epsilon" crossed off	"Not an element of"	Symbol illustrating non-containment	$5 \notin E$
U	Capital letter U	Universal Set	The largest set in a given context	$U = \{x \vert x \text{ is a digit}\}$ $U = \{0, 1, 2, \ldots, 9\}$
$\varnothing$	Null Set	Empty Set	Set containing nothing—smallest set	Sometimes denoted $\varnothing = \{\}$
$\cup$	Union	"Or" (Inclusive)	Elements are in at least one of the sets	$A \cup B = \{x \vert x \in A \vee x \in B\}$
$\cap$	Intersection	"And"	Elements are in both sets	$A \cap B = \{x \vert x \in A \wedge x \in B\}$
$\bar{A}$ or A'	Complement	"Not"	The opposite set as compared to the U	$A' = \{x \vert x \notin A\}$
$-$	Minus sign	"Set difference"	The minus sign indicating difference	$A - B = A \cap \bar{B}$
Δ or $\ominus$	Greek letter "Delta" or a emphasized	"Symmetric difference"	An emphasized minus sign indicating a type of difference	$A \Delta B = (A \cap \bar{B}) \cup (\bar{A} \cap B)$
$\subset$ or $\subseteq$	Subset	"Is contained in"	An element in the first set is implied to be in the second set	$D \subset U$
$\not\subset$	Not a Subset	"Not entirely contained in"	Implies at most partial containment	$\{1, 2, 3\} \not\subset \{1, 3, 5\}$
$=$	Equal	"Is the same set as"	Equality between sets	$\{1, 2, 3\} = \{3, 2, 1\}$
$\cong$ or $\sim$	Equivalent	"Is the same size set as"	Related to the word: equal	$\{1, 2, 3\} \cong \{x, y, z\}$
n	Abbreviation of "number"	"Number of elements in"	Used when counting elements in a set	$n(D) = 6$
$_nP_r$	Abbreviation of Permutation	"n objects permuted r at a time"		$_3P_2 = 3 \times 2 = 6$
$_nC_r$	Abbreviation of Combination	"n objects combined r at a time"		$_3C_2 = \dfrac{3 \times 2}{2 \times 1} = 3$
$\sum$	Greek letter "Sigma"	"the sum of"	Summation Notation	$\displaystyle\sum_{k=1}^{3} k = 1 + 2 + 3$

REVIEW TEST

1. List the elements of $\{x | x \text{ is a counting number less than } 6\}$.
2. Complete the blank with the correct symbol so that the resulting statement is true.
   A. $\{0\}$ ____ $\{-1, 0, 1, 2, 3\}$
   B. 0 ____ $\{-1, 0, 1, 2, 3\}$
3. If $A = \{a, b, c\}$, then list all possible subsets of A.
4. Which of the following are true statements?
   A. $\varnothing \subseteq 1,2,3$
   B. $\varnothing \cap 1,2,3 = \varnothing$
   C. $\varnothing \in 1,2,3$
   D. $\varnothing \cup 1,2,3 = \varnothing$
5. Set A is a subset of B $(A \subseteq B)$ provided that:
   A. Every element of A is an element of B
   B. Every element of B is an element of A
   C. At least one element of A is in B
   D. At least one element of A is not in B
   E. None of the above
6. Given a set A and its complement A', simplify the following:
   A. $A \cap A'$
   B. $A \cup A'$
7. Find the union of sets $\{a, e, i, o, u\}$ and $\{s, e, t\}$.
8. Find the intersection of sets $\{a, e, i, o, u\}$ and $\{s, e, t\}$.
9. Let $X = \{1, 3, 5, 7\}$, $Y = \{2, 4, 7, 9\}$, $Z = \{1, 3, 8, 9\}$ and the universal set is the digits, $U = \{0, 1, ..., 9\}$. Determine the following sets:
   A. X'
   B. $X' \cup Y'$
   C. U'
   D. $(X' \cap Z)'$
   E. $(X \cap Y) \cap Z$
10. Determine the number of subsets of $\{1, 2, 3, 4\}$.
11. Describe the shaded portion of the Venn diagram below symbolically.

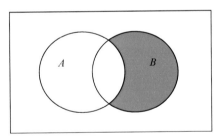

12. Draw the Venn diagram of the set $A \cap (B \cup C)$.
13. Draw the Venn diagram of the set $A' \cap (B \cup C)$.
14. What is the maximum number of elements in any subset of $S = \{2, 4, 6, 8\}$?
15. In the figure below, A represents $\{2, 4, 6, 8, 10, 12\}$, B represents $\{3, 6, 9, 12, 15\}$ and C represents $\{2, 6, 10, 11, 16\}$. What set is represented by the shaded region?

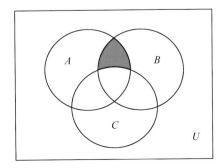

16. A survey of book club produced the following results:

15 like novels	9 like novels and plays
13 like plays	9 like novels and biographies
14 like biographies	8 like biographies and plays
7 like all three	10 don't like any of the three

   A. How many people were there all together?

   B. How many people like biographies only?

   C. How many people like exactly one type of literature?

   D. How many people do not like biographies?

   E. How many people like anything but plays?

17. Let $n(A)=23$, $n(B)=17$ and $n(A \cup B)=35$. Find $n(A \cap B)$.

18. Given $A=\{1, 3, 5, 7, ...\}$ and $U=\{0, 1, 2, 3, ...\}$, then find A'.

Practice for the CLAST exam:

19. Sets A, B and C are related as shown in the figure below where U is the universal set.

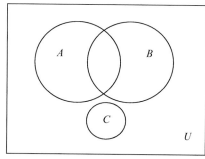

   Which of the following statements is/are true assuming none of the shown regions is empty?

   i. Any element which is a member of U is also a member of set A or B.

   ii. Any element which is a member of C is not a member of set A or B.

   iii. No element is a member of all three sets.

     A. i only

     B. ii only

     C. iii only

     D. ii and iii

20. Sets X and Y are related as shown in the figure below where U is the universal set. Which of the following statements is/are true assuming none of the shown regions is empty?

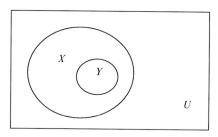

   A. $U \subseteq X$

   B. $Y \subseteq X$

   C. $X \cap Y = X$

   D. $X \subseteq Y$

21. Sets A, B and C are related as shown in the figure below where U is the universal set. Which of the following statements is/are true assuming none of the shown regions is empty?

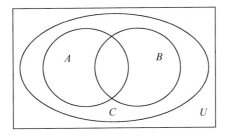

    A. $C \subseteq A \cup B$

    B. $U \subseteq C$

    C. $A \cup B \subseteq C$

    D. $A \subseteq B$

22. Sets A, B and C are related as shown in the figure below where U is the universal set. Which of the following statements is/are true assuming none of the shown regions is empty?

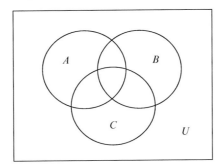

    A. Any element which is a member of U is also a member of set A.

    B. Any element which is a member of C is not a member of set U.

    C. No element is a member of all three sets

    D. None of the above statements is true.

REFERENCES

Breuer, J., 1958. Introduction to the Theory of Sets. Prentice-Hall, Englewood Cliffs, NJ.

Dinkines, F., 1964. Elementary Theory of Sets. Appleton-Century-Crofts, New York.

Fraenkel, A.A., 1953. Abstract Set Theory. North-Holland Publishing Company, Amsterdam.

Tsokos, C.P., 1978b. Mainstreams of Finite Mathematics with Application. Charles E. Merrill Publishing Company, A Bell & Howell Company.

Counting

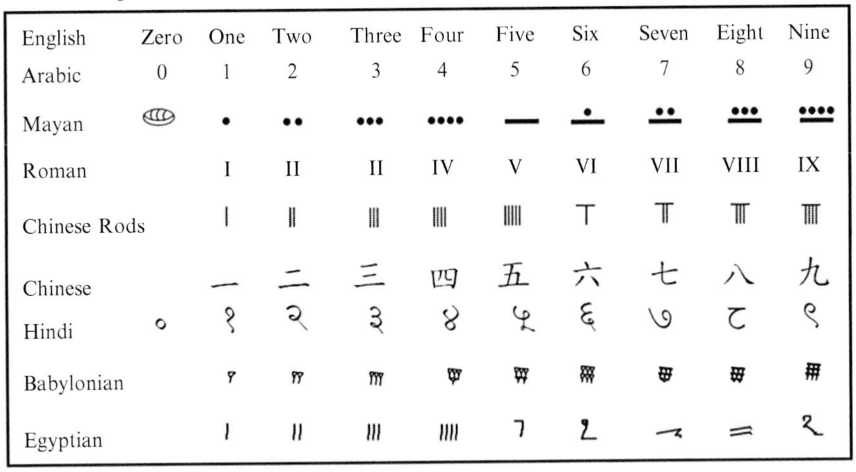

English	Zero	One	Two	Three	Four	Five	Six	Seven	Eight	Nine
Arabic	0	1	2	3	4	5	6	7	8	9
Mayan	〇	•	••	•••	••••	—	•̱	••̱	•••̱	••••̱
Roman		I	II	II	IV	V	VI	VII	VIII	IX
Chinese Rods		I	II	III	IIII	IIIII	丁	丌	丌	丌
Chinese		一	二	三	四	五	六	七	八	九
Hindi	०	१	२	३	४	५	६	७	८	९
Babylonian		𒁹	𒈫	𒐗	𒐘	𒐙	𒐚	𒐛	𒐜	𒐝
Egyptian		I	II	III	IIII	�	ꞏꞏ	⌐	=	ꝅ

Mathematics is the language with which God has written the Universe.

GALILEO GALILEI

Chapter 4 — Counting Techniques

4.1 INTRODUCTION TO COUNTING

Counting began with shepherds who would tie a knot in a piece of string, one for each sheep they send out to pasture. As the sheep returned, a knot was untied and if any knots remain, this indicated that there were lost sheep. Hence, the idea of one, two, three… that is, the natural numbers was created. However, **enumeration** is only one way of counting; additional techniques include the **multiplication principle**, **factorials**, **permutations** and **combination**, among others.

Below is an outline of the chapter and the developed formulas. Note, in sets we are only interested in the collections of objects that are well defined; and now we only are interested in how many elements are in various sets. But most of the notations are the same. There are a few new ones.

Definition 4.1.1 Enumeration

Enumeration is the act of enumerating, the mention separately as if in counting; name one by one as in a list. That is, to ascertain the number of items or elements belonging to an outlined set.

Pagination

  The enumeration of pages or leaves of a book marked to indicate their sequence

---

Example 4.1.1 Pagination

As student turns in a lengthy report which has a 25 page minimum requirement; to help assess the number of pages turned in by each student, the professor requires that the pages be numbered.

Definition 4.1.2 Countable (Finite)

A set is **countable** if there is an enumeration; that is, a set S having a **finite** number of elements which form a one-to-one correspondence with a subset the natural numbers, N:

$$f : N \to S$$

Example 4.1.2 Natural Numbers

The natural numbers are enumerable by the identity function $f(n) = n$, where $n \in N$ the set of natural numbers; $f : N \to N$.

Example 4.1.3 Even Numbers

The natural numbers are countable using the function $f(n) = 2n$ where $n \in N$ the set of natural numbers; $f : N \to E$ where E is the set of even numbers.

Example 4.1.4 Integers

The set of integers, $Z = \{\ldots -3, -2, -1, 0, 1, 2, 3, \ldots,$ are countable using the function

$$f(n) = \begin{cases} \dfrac{n}{2} & n \in E \\ -\dfrac{1+n}{2} & n \in \bar{E} \end{cases} \quad \text{where } n \in N = E \cup \bar{E} \text{ the set of natural numbers;}$$

$f : N \to Z$ where E is the set of even numbers and $\bar{E}$ is the set of odd numbers.

These are examples of counting, but how might we count larger values.

Definition 4.1.3 Countably Infinite

A set is **countably infinite** if there is an enumeration; that is, a set S having an infinite number of elements which form a one-to-one correspondence with the entire set of the natural numbers, N:

$$f : N \to S.$$

PROBLEMS

4.1.1. Determine if the following sets are countable (finite) or countably infinite or neither.
   a. Real Numbers
   b. Rational Numbers
   c. English Alphabet
   d. Whole Numbers

4.1.2. Determine if the following sets are countable (finite) or countably infinite or neither.
   a. $\{1,2,3,4,5,6\}$
   b. $\{x \mid 0 \leq x \leq 1\}$
   c. $\{1.1, 1.11, 1.111, \ldots\}$
   d. $\{x \mid x$ is an integer and $0 < x \leq 10\}$

4.1.3. Determine if the following sets are countable (finite) or countably infinite or neither.
   a. Deck of standard playing cards
   b. Irrational Numbers
   c. Set of Digits
   d. Integers

4.1.4. Outline a function to show that the set of positive odd integers is countable.

4.1.5. Outline a function to show that the set of positive even integers starting with four is countable.

4.1.6. Outline a function to show that the set of integer is countable.

4.1.7. Outline a function to show that the set of square is countable.

4.1.8. Outline a function to show that the set $\left\{1, \frac{1}{2}, \frac{1}{3}, \frac{1}{4}\right\}$ is countable.

4.1.9. Outline a function to show that the set $\left\{1, \frac{1}{3}, \frac{1}{5}, \frac{1}{7}\right\}$ is countable.

4.1.10. Outline a function to show that the set $\{7, 10, 13, 16, \ldots\}$ is countable.

4.1.11. Outline a function to show that the set $\left\{1, \frac{1}{2}, \frac{1}{4}, \frac{1}{8}, \ldots\right\}$ is countable.

4.1.12. Outline a function to show that the set $\{1, 8, 27, 64, \ldots\}$ is countable.

4.1.13. Outline a function to show that the set $\{1, 0.1, 0.01, 0.001, 0.0001, \ldots\}$ is countable.

4.1.14. Outline a function to show that the set $\{1, 0.1, 0.01, 0.001, 0.0001, \ldots\}$ is countable.

4.1.15. TRUE/FALSE: If you remove a countable number of elements from a countable set, there are a countable number of elements left.

4.1.16. TRUE/FALSE: If you remove a finite number of elements from a countable set, there are a countable number of elements left.

4.1.17. TRUE/FALSE: The union of two countable sets is countable.

4.1.18. TRUE/FALSE: The intersection of two countable sets is countable.

4.1.19. TRUE/FALSE: You can remove a countably infinite set from a countably finite set.

4.1.20. TRUE/FALSE: You can remove a countably finite set from a countably infinite set.

4.1.21. TRUE/FALSE: The union of two countable infinite set form an uncountable set.

4.2 MULTIPLICATION PRINCIPLE

We have seen that the word "and" is interpreted as "multiply" when performing more than one event. Recall from Chapter 1, in logic, when we had more than one statement, each having two possible truth-values, we multiplied: $2 \times 2 \times \cdots \times 2$, depending on the number of statements. In sets, when we counted the number of possible subsets for a given set, each element was taken to be in the new set or it was not taken and so we multiplied: $2 \times 2 \times \cdots \times 2$, depending this time on the number of elements in the original set. This idea is referred to as the multiplication principle.

And similarly, if two dice are tossed, assuming order matters (the first die is red and the second die is blue) there are 6×6 different possible outcomes.

Multiplication Principle or Sequential Counting Principle

So this idea of "and" means "multiply" is between a multiple number of events or tasks; to be done in sequence, succession or in order and can be extended to individual events which can be taken several ways. Unfortunately, this idea is not as easy to formulate. But let's try:

Rule 1
Multiplication Principle

Definition 4.2.1 Multiplication Principle

If there are n tasks to be done or events to perform, and each task can be done in $m_1, m_2, \ldots, m_n$, then the number of ways to perform all given tasks is $m_1 \times m_2 \times \cdots \times m_n$.

For example, given the events: a coin is tossed and (then) a die is tossed, there are 2×6 different outcomes. If we let the collection of possible outcomes be called our **sample space**; then $S = \{H1, H2, H3, H4, H5, H6, T1, T2, T3, T4, T5, T6\}$. This set may also be referred to as the universal set in this given situation.

Example 4.2.1 Entrée and Dessert

Given there are five entrees and three desserts, how many different combinations of one entree and one dessert be selected?

Solution

$$m_1 = 5, m_2 = 3 \Rightarrow n = 5 \times 3 = 15.$$

There are fifteen different selection of an entree and dessert.

Example 4.2.2 Outfits

A woman is selecting an outfit from two pant, three shirts and two pair of shoes. How many different outfits can be created?

Solution

$$m_1 = 2, m_2 = 3, m_2 = 2 \Rightarrow n = 2 \times 3 \times 2 = 12$$

There are twelve different outfits that can be created.

(1) Pant 1, Shirt 1 and Shoe 1
(2) Pant 1, Shirt 1 and Shoe 2
(3) Pant 1, Shirt 2 and Shoe 1
(4) Pant 1, Shirt 2 and Shoe 2

Solution—cont'd

(5) Pant 1, Shirt 3 and Shoe 1
(6) Pant 1, Shirt 3 and Shoe 2
(7) Pant 2, Shirt 1 and Shoe 1
(8) Pant 2, Shirt 1 and Shoe 2
(9) Pant 2, Shirt 2 and Shoe 1
(10) Pant 2, Shirt 2 and Shoe 2
(11) Pant 2, Shirt 3 and Shoe 1
(12) Pant 2, Shirt 3 and Shoe 2

Definition 4.2.2 Multiple Selection

Suppose that k objects are selected from a set of N objects one at a time **with replacement**. Then the sample space for this **multiple selection** consists of N^k outcomes.

Example 4.2.3 Yahtzee

In a game of dice, how many different ways can six dice land on the first toss of six dice in a game of Yahtzee.

Solution

$$N=6, k=6 \Rightarrow N^k = 6^6 = 46,656$$

Another way to illustrate these events in using what is call a tree diagram. A **tree diagram** is a diagram of the possible events that looks like the branches of a tree; a "tree" diagram, but we normally draw the branching from left to right and not bottom to top and each new branching represents the next event.

Rule 2
Multiple Selections

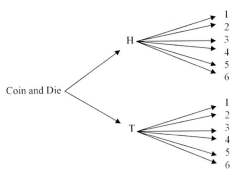

This idea is a very powerful one, in general when performing more than one event, "and" means "multiply" and order is implied; the tasks or events are done in sequence. Furthermore, repetition is allowed or it is not relevant. For example, when tossing two distinct die, repetition of numbers is allowed, but when a coin was tossed and then a die, repetition is not an issue; these events are independent of each other.

Definition 4.2.3 Tree Diagram

A **tree diagram** is a network diagram used to find and analyze the various outcomes in a multi-stage sampling. Starting at a point, create branches to the first set of possible outcomes. From each of the secondary points created above, create additional branches to the second set of outcomes. Repeat as needed.

If the events do depend on each other, "and" still means "multiply" but you have to be careful to compensate for the dependent event.

For example, when two cards are drawn without replacement, the first card can be chosen out of the entire deck of 52 cards, but the second card depends on the fact that there is now one less card in the deck and must be chosen out of the remaining 51 cards; therefore, there are **52×51** ways to choose two cards in succession without replacement. If the card were to be replaced, these events would become independent from each other; that is, both the first and the second card can be chosen out of the entire deck of 52 cards yielding **52×52** ways to choose two cards in succession with replacement—but "and" still means "multiply."

However, replacement seems to plays a very important role. It does, so much so that we will recognize this specific type of counting as permutations. If we take it a step further and removed the order, we recognize this specific type of counting as combinations. But no matter: "and" means "multiply" and order is implied.

Example 4.2.4 Three Fair Coins

Three fair coins are tossed. Count the number of possible outcomes and construct a tree diagram illustrating the various outcomes.

Solution

$$N=2, k=3 \Rightarrow N^k = 2^3 = 8$$

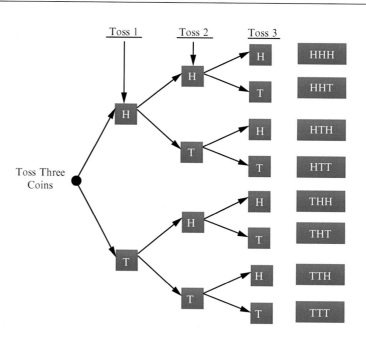

PROBLEMS

4.2.1. How many three course meals consisting of an appetizer, an entree, and a dessert are possible if there are:
  a. 3 appetizers, 4 entrees and 2 desserts
  b. 4 appetizers, 5 entrees and 2 desserts
  c. 5 appetizers, 10 entrees and 4 desserts
  d. 3 appetizers, 8 entrees and 4 desserts

4.2.2. How many outfits consisting of a shirt, a pant, and a jacket are possible if there are:
  a. 2 pants, 5 shirts and 2 jackets
  b. 3 pants, 2 shirts and 2 jackets
  c. 4 pants, 2 shirts and 3 jackets
  d. 2 pants, 3 shirts and 4 jackets

4.2.3. Find the number of possible choices when you choose one item from each category.
  a. 4 chairs, 6 colors
  b. 2 pencils, 3 binders
  c. 3 cone sizes, 7 flavors of ice cream
  d. 8 men, 5 women

4.2.4. Find the number of possible choices when you choose one item from each category.
  a. 3 pairs of shoes, 7 sweaters, 3 pants
  b. 5 types of pasta, 2 meats and 3 cheese toppings
  c. 2 desserts, 2 drinks, 5 vegetables
  d. 7 socks, 12 hats, 3 pairs of shoes

4.2.5. Find the number of possible choices when you choose one item from each category.
  a. 2 car type, 3 makes, and 5 colors
  b. 3 fruits, 4 beverages and 2 snacks

4.2.6. Find the number of possible choices when you choose one item from each category.
  a. 4 pairs of shoes, 9 socks, 6 shirts, 2 pairs of pants, 3 hats
  b. 5 scarves, 12 socks, 5 pairs of pants, 2 pairs of sunglasses, 7 pairs of shoes

4.2.7. How many three-digit codes using digits can be created with repetition of digits?

4.2.8. How many three-digit numbers exist with repetition of digits? Note: There are 10 digits: 0, 1, 2, ..., 9 and to be a four digit number, the first digit cannot be zero.

4.2.9. Draw the associated tree diagram for three coins are tossed.

4.2.10. Draw the associated tree diagram for two dice are tossed.

4.2.11. Draw the associated tree diagram for a die is tossed and then a coin is tossed.

4.3 PERMUTATIONS AND COMBINATIONS

We will go back to calling elements, objects. We will now study how many ways to order n objects.

Definition 4.3.1 Factorials

Suppose that n objects are selected at random one at a time without replacement. Then the sample space consists of $n!=n\times(n-1)\times\cdots\times2\times1$ outcomes. Here, $n!$ is said "n **factorial**." $n!$ counts the number of ways to order n objects. Note: $0!=1$, there is one way to order nothing—do nothing.

Example 4.3.1 Books on a Shelf

How many whys can an obsessive compulsive individual order five books on a shelf?

$$0! = 1$$
$$1! = 1$$
$$2! = 1 \cdot 2 = 2$$
$$3! = 1 \cdot 2 \cdot 3 = 6$$
$$4! = 1 \cdot 2 \cdot 3 \cdot 4 = 24$$
$$5! = 1 \cdot 2 \cdot 3 \cdot 4 \cdot 5 = 120$$
$$6! = 1 \cdot 2 \cdot 3 \cdot 4 \cdot 5 \cdot 6 = 720$$

Solution

Since there are five books, the first book can be selected one of five ways, 5, and the second book can therefore be selected from the four remaining books, so forth and so on, until we have

$$5! = 5 \times 4 \times 3 \times 2 \times 1 = 120.$$

Therefore, there are 120 ways an individual can order five books.

Given $n(A) = n$ and $r \leq n$; if r objects are taken from A in succession without replacement, then this event is called a **permutation**; order is implied and repetition is not allowed; permute meaning to interchange.

Unfortunately the idea is not easy to formulate. There are n ways to choose the first object and there are $(n-1)$ ways to choose the second event; if we continue like this, our counting is off by one! Therefore when we let the number of permutation be abbreviated by P; then the number of permutations of n, chosen r at a time is written mathematically:

$$P(n,r) = n \times (n-1) \times \cdots \times (n-r+1).$$

But the "$\cdots$" are vague.

So, let's consider an alternative notation: what happens when $r = n$; that is how many permutations consist of all n objects? We have, $P(n,n) = n \times (n-1) \times \cdots \times (2) \times (1)$, where one is a nice number to end with, it looks like we are counting backwards down to one; in fact, that is what we are doing! Using this notation, $P(3,3) = 3 \times (2) \times (1) = 6$. In general, we will call this idea a factorial and will denote this using the abbreviation "!", the exclamation mark. Therefore n factorial is $n! = n \times (n-1) \times \cdots \times (2) \times (1)$.

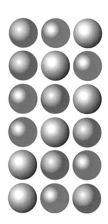

But in general, $P(n,r) = n \times (n-1) \times \cdots \times (n-r+1)$ where $0 < r < n$, $n!$ has too many objects, in fact $n! = n \times (n-1) \times \cdots \times (n-r+1) \times (n-r) \times \cdots \times (2) \times (1)$ has exactly $(n-r)! = (n-r) \times (n-r-1) \times \cdots \times (2) \times (1)$ times too many objects. And the opposite of multiply (times) is divided (division), we have $P(n,r) = \dfrac{n!}{(n-r)!}$; where $n!$ orders all n objects and dividing by $(n-r)!$ removes the $(n-r)$ objects not wanted.

Rule 3
Permutations

Definition 4.3.2 Permutations

Suppose that k objects are drawn without replacement, with order being preserved, from a collection of n distinct objects. Then the sample space of **permutation** consists of

$$P(n,k) = P_n^k = {}_nP_k = \frac{n!}{(n-k)!}$$

sample points; that is, the number of ways of ordering n distinct objects taken k at a time; permutations

Example 4.3.2 Permute 2 out of 3

Consider the set $A = \{x, y, z\}$, how many ways are there to permute two out of the three letters?

Solution

There are six ways to permute these three objects two at time:

$$P(3,2)=\underbrace{3\times 2}_{\substack{idea\\MP}}=\underbrace{\frac{3!}{1!}=\frac{3\times 2\times 1}{1}}_{formula}=6;$$

namely, $\{xy,yx,yz,zy,xz,zx\}$.

Note there is an issue here, the formula $P(n,r)=\dfrac{n!}{(n-r)!}$ is true when $0<r<n$, but when $r=n$, $P(n,n)=n!$, what would we have to define $0!$ equal to so as to maintain this formulation: $P(n,n)=\dfrac{n!}{0!}=n!$? What would we have to define $0!$ equal to so as to maintain this formulation: $P(n,0)=\dfrac{n!}{n!}=1$ and therefore $\dfrac{0!}{0!}=1$? Well, there is only one what to take nothing; even if you have nothing, there is still one way to take nothing; this is the subset the empty set: $\varnothing$. Hence we define $0!=1$; this will allow the formula to work $\forall r, r=1,2,...,n$.

In many respects, the idea is easier to remember than the formula: start with n and count down the r objects you are choosing, and then "and" means "multiply."

Now consider the case when order does not matter. Given $n(A)=n$ and $r\leq n$; if r objects are taken from A in without replacement but the order in which the objects may be permuted is not considered, then this event is called a **combination**; order does not matter, it is not implied however, repetition is still not allowed; combine meaning to join.

And recall it was the order which we multiply that dictated the ordering; therefore it stands to reason that division would remove the dictation between different ordering. Using this idea, if we want to count the number of combinations instead of the permutation; there are $P(n,r)$ ways that r objects out of the n objects can be ordered and there are $r!$ ways that the r objects could have been ordered. Therefore, if we let C denote the combinations of n chosen r at a time is written mathematically:

$$C(n,r)=\frac{P(n,r)}{r!}=\frac{n!}{r!(n-r)!}.$$

Here again, $0!=1$.

In many respect, the idea is easier to remember than the formula: start with n and count down the r objects you are choosing; and then count up the order on the r objects chosen. Here "and" means "multiply" with order and "divide" removes order.

Combinations are permutations removing order:

ABC	ABD	ACD	BCD
ACB	ADB	ADC	BDC
BAC	BAD	CAD	CBD
BCA	BDA	CDA	CDB
CAB	DAB	DAC	DBC
CBA	DBA	DCA	DCB

Rule 4
Combinations

Definition 4.3.3 Combinations

Suppose that k objects are drawn without replacement and without regard to order from a collection of n distinct objects. Then the sample space of **combinations** consists of

$$C(n,k)=\binom{n}{k}={}_nC_k=\frac{n!}{k!(n-k)!}$$

sample points; combinations in set theory.

Example 4.3.3 Combine 2 out of 3

For example, consider the set $A = \{x, y, z\}$, there are three ways to combine these three objects two at time:

$$C(3, 2) = \frac{3 \times 2}{1 \times 2} = \underbrace{\frac{3!}{2!1!}}_{\substack{idea \\ MP}} = \underbrace{\frac{3 \times 2 \times 1}{2 \times 1 \times 1}}_{formula} = 3;$$

namely, $\{xy, yz, xz\}$.

Know this: for each combination, there are many permutations.

We have seen that factorials order all, but what if objects were allowed to repeat? How would you count the distinct orderings?

Consider the number of distinct orderings of the letters in the word ***Mississippi***. As there are repetition of letters, switching around the various *s*'s does not change the word as it is not distinct. Only if letters are swapped with different letters does this create a new distinct word.

Rule 5
Partial Orderings

Definition 4.3.4 Partial Ordering

The distinct ordering of n objects where each of the k distinguishable objects $i = 1, 2, \ldots, k$ occurs n_i times where $n_1 + n_2 + \cdots + n_k = n$. Just as before factorials order all: $n!$ and division removes the individual orderings; hence, the number of distinct **partial orderings** is given by:

$$\frac{n!}{n_1! \times n_2! \times \cdots \times n_k!}$$

Example 4.3.4 "Noon"

How many distinct words can be created using the letter in the word "noon"?

Solution

$$n = 4, n_1 = n_2 = 2 \Rightarrow \frac{4!}{2!2!} = \frac{4 \times 3 \times 2 \times 1}{2 \times 1 \times 2 \times 1} = 6$$

There are six distinct arrangements of the letters "noon":
noon, nono, nnoo, onno, onon, oonn.

What if there is overlap; that is, if $A \cap B \neq \varnothing$, then the elements in $A \cap B$, have been counted twice—once as elements in A and once as elements in B, therefore subtracting out the one additional counting of $A \cap B$ will correctly yield the number of elements that belong to A or B: $A \cup B$.

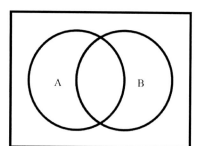

Property 4.3.1 Overlapping Events

The number of elements in two overlapping sets is the number of elements minus the number of elements the sets have in common as they have been added twice.

$$n(A \cup B) = n(A) + n(B) - n(A \cap B)$$

Example 4.3.5 A Fair Die

Consider the experiment of tossing a fair die and observing the number that appears on top. What is the number of ways one can observe an odd number or prime number?

Solution

Here the universe consists of six possible outcomes: $U=\{1, 2, 3, 4, 5, 6\}$.

Let A be the event that an odd number occurs and B the event that a prime number occurs:

$$A=\{1, 3, 5\} \quad \text{and} \quad B=\{2, 3, 5\}$$

Then, $A \cap B=\{3, 5\}$ and

$$n(A \cup B)=n(A)+n(B)-n(A \cap B)$$
$$=3+3-2=5$$

Thus, there are 5 ways the number that appears on a single toss of a die will be an odd number or a prime number.

Example 4.3.6 Three Fair Coins

Consider the experiment of tossing a coin three times. (See Example 4.2.4.) How many ways can two or more heads appearing consecutively or of all the tosses being the same?

Solution

Here the sample space consists of eight possible outcomes: $S=\{HHH, HHT, HTH, THH, HTT, THT, TTH, TTT\}$, where H and T stand for heads and tails, respectively and HHH represents the triplet (H,H,H), HHT represents the triplet (H,H,T), etc.

Let A be the event that two or more heads appear consecutively and B the event that all tosses are the same; that is,

$$A=\{HHH, HHT, THH\} \quad \text{and} \quad B=\{HHH, TTT\}.$$

Thus, $A \cap B=\{HHH\}$ and the desired probability is given by

$$n(A \cup B)=n(A)+n(B)-n(A \cap B)$$
$$=3+2-1=4$$

That is, there are four out of eight ways we obtain 3 heads and 2 tails. Now consider events which are mutually exclusive.

Definition 4.3.5 Mutually Exclusive

Two sets, A and B, in a given universal set U are said to be mutually exclusive if they do not have any elements in common and we denote it by

$$A \cap B=\varnothing.$$

When we consider a single universe with overlapping sets, to count the total number of elements that belong to A or B; simply adding the number of element belong to A with the number of element belong to B is a bit much; however, if $A \cap B=\varnothing$, then $n(A \cup B)=n(A)+n(B)$ as there is no overlapping regions, $n(A \cap B)=0$, and therefore we can easily determine the number of elements that belong to A or B: $A \cup B$.

Property 4.3.2 Mutually Exclusive Events

The number of elements in two mutually exclusive (non-overlapping) sets is the number of elements.

$$n(A \cup B) = n(A) + n(B)$$

Example 4.3.7 One Missing Sock

Consider the experiment of drawing a sock from a drawer containing six white, four blue and three red socks. The number of elements in the event a white or red sock is draw is

$$n(W \cup R) = 6 + 3 = 9$$

PROBLEMS

4.3.1. Evaluate the following factorials:
  A. 0!
  B. 3!
  C. 6!
  D. 10!

4.3.2. Write 10! in terms of 9!

4.3.3. Write 100! in terms of 99!

4.3.4. How many ways can an individual with OCD place four books on a shelf?

4.3.5. How many ways can eight horses cross a finish line?

4.3.6. How many distinct orderings of the digits $\{1, 2, 3, 4\}$ are there?

4.3.7. $_8P_2$

4.3.8. $_{10}P_5$

4.3.9. $_8P_7$

4.3.10. $_9P_3$

4.3.11. P(11,3)

4.3.12. P(10,1)

4.3.13. P(100,1)

4.3.14. P(10,3)

4.3.15. P(10,7)

4.3.16. P(5,3)

4.3.17. P(5,2)

4.3.18. How many three-digit codes using digits can be created without repetition of digits?

4.3.19. How many permutations of three letters are possible from the letters P, E, T, and S?

4.3.20. In how many ways can Alex, Becky, Cindy, and David stand in line?

4.3.21. How many three-digit numbers can you make by arranging the numbers 1, 2, and 3?

4.3.22. Write the list of all permutations of the set $A = \{1, 2, 3, 4\}$ chosen **3** at a time. Note: there are twenty-four possible permutations.

4.3.23. Write the list of all permutations of the set $A = \{1, 2, 3, 4\}$ chosen **2** at a time. Note: this time there are only twelve possible permutations.

4.3.24. Without writing the list of all permutations of the set $A = \{1, 2, 3, 4\}$ chosen **4** at a time, compute the number of such permutations.

4.3.25. Without writing the list of all permutations of the set $A = \{1, 2, 3, 4\}$ chosen **3** at a time, compute the number of such permutations.

4.3.26. Without writing the list of all permutations of the set $A = \{1, 2, 3, 4, 5\}$ chosen **4** at a time, compute the number of such permutations.

4.3.27. Without writing the list of all permutations of the set $A = \{1, 2, 3, 4, 5, 6\}$ chosen **4** at a time, compute the number of such permutations.

4.3.28. Write the list of all combination of the set $A = \{1, 2, 3, 4\}$ chosen **3** at a time. Note: there are only four possible combinations.

4.3.29. Write the list of all combination of the set $A = \{1, 2, 3, 4\}$ chosen **2** at a time. Note: this time there are only six possible combinations.

4.3.30. Without writing the list of all combination of the set $A = \{1, 2, 3, 4\}$ chosen **4** at a time, compute the number of such combinations.

4.3.31. Without writing the list of all combination of the set $A = \{1, 2, 3, 4, 5\}$ chosen **3** at a time, compute the number of such combinations.

4.3.32. Without writing the list of all combination of the set $A = \{1, 2, 3, 4, 5\}$ chosen **4** at a time, compute the number of such combinations.

4.3.33. Without writing the list of all combination of the set $A = \{1, 2, 3, 4, 5, 6\}$ chosen **4** at a time, compute the number of such combinations.

4.3.34. Write out all subsets of the set $A = \{a, b, c\}$. Hint: there are a total of eight combinations including the empty set as $C(3, 0) = 1$.

4.3.35. How many distinct orderings of the word "happy" are there?

4.3.36. How many distinct orderings of your first name are there?

4.3.37. How many distinct orderings of the word "automatic" are there?

4.3.38. How many distinct orderings of the word "statistics" are there?

4.3.39. How many distinct orderings of the word "Chris" are there?

4.3.40. If $n(A \cap B) = 17$, $n(A) = 47$ and $n(B) = 74$, then $n(A \cup B) = $ _____

4.3.41. If $n(A \cap B) = 17$, $n(A) = 17$ and $n(B) = 34$, then $n(A \cup B) = $ _____

4.3.42. If a committee of three is to be chosen from a group of ten people: six men and four women. In how many different ways can this committee be chosen if the committee must consist of two men and one women

4.3.43. If a committee of three is to be chosen from a group of ten people: six men and four women. In how many different ways can this committee be chosen if the committee must consist of three men and no women

4.3.44. If a committee of three is to be chosen from a group of ten people: six men and four women. In how many different ways can this committee be chosen if the committee must consist of all men

4.3.45. If a committee of three is to be chosen from a group of ten people: six men and four women. In how many different ways can this committee be chosen if the committee must consist of all women

4.4 OTHER COUNTING TECHNIQUES

In general, when we consider the event with exactly two outcomes. Given two events, F as the "first" outcome and $S = F'$ as the "second" outcome; or S for "success" and F for "failure." Either way, there are two—"bi" numbers or outcomes "nomials," that is binomials.

So, either S or F happens. So, I pose a new question, what happens if flip this coin fifty times? Ok, for argument sake, let's flip this hypothetical coin three times.

That is (S or F) and (S or F) and (S or F); recall "or" means "add" and "and" means "multiply"; therefore, (S or F) and (S or F) and (S or F) = $(S + F)^3$. Moreover, if you foil this out like algebra, it will list all the possible sequential outcomes.

$$(S + F)^3 = (S + F)(SS + SF + FS + FF)$$
$$= SSS + SSF + SFS + SFF$$
$$+ FSS + FSF + FFS + FFF$$

Neat, huh...

And if you incorporate more algebraic abbreviations, it will count the number of permutation for each combination—that is $(S + F)^3 = S^3 + 3S^2F + 3SF^2 + F^3$ can be interpreted as there is one distinct permutations of the repetitive combination of three S and no F's, there are three distinct permutations of the repetitive combination of letter with two S's and one F, and three distinct permutations of the repetitive combination of letter with one S and two F's and finally, there is one distinct permutations of the repetitive combination of three F and no S's.

Note: $\binom{3}{3} = 1$, $\binom{3}{2} = 3$, $\binom{3}{1} = 3$, and $\binom{3}{0} = 1$, this is very interesting.

In general than, if you choose **S** or **F**, over and over again **n** times, we get

$$(S+F)^n = \binom{n}{n}S^n F^0 + \binom{n}{n-1}S^{n-1}F + \binom{n}{n-2}S^{n-2}F^2 + \cdots + \binom{n}{0}S^0 F^n$$

sometimes referred to as the **binomial expansion**.

Therefore, if you want to know how many times you will have **r** "successes" in **n** "trials"; the rest being "failures," this can easily be counting using combinations is the number of successes.

$(a + b)^0 = \qquad\qquad 1$

$(a + b)^1 = \qquad\qquad a + b$

$(a + b)^2 = \qquad\qquad a^2 + 2ab + b^2$

$(a + b)^3 = \qquad\qquad a^3 + 3a^2b + 3ab^2 + b^3$

$(a + b)^4 = \qquad a^4 + 4a^3b + 6a^2b^2 + 4ab^3 + b^4$

$(a + b)^5 = a^5 + 5a^4b + 10a^3b^2 + 10a^2b^3 + 5ab^4 + b^5$

Rule 4.4.1 Binomial Coefficients

In the **binomial expansion** of $(a+b)^n$, the coefficient on the term with a^r and b^{n-r}, the **binomial coefficient**, is given by

$$C(n,r) = \binom{n}{r}.$$

Example 4.4.1 $(a+b)\cdot(a+b)^n$

If you choose either **a** or **b** four times. How many ways can you distinct permute a combination with three **a**'s and one **b**.

Solution

Since $n=4$ and $r=3$, there are $\binom{4}{3} = \underbrace{\frac{4\times3\times2}{3\times2\times1}}_{idea} = 4$; namely, $\{aaab, aaba, abaa, baaa\}$

Another interesting way to generate these counters is by building Pascal's triangle. Pascal's triangle is infinite numerical table that forms a triangle in shape; this triangle is such that the nth row of the triangle corresponds to the coefficients in the binomial expansion of any binomial, say $(a+b)^n$.

0th				1				There is only one way to take nothing
1st			1		1			Either a or $b = a$ or $b = (a+b)$
2nd		1		2		1		aa, ab or $ba, bb = (a+b)^2$
3rd	1		3		3		1	This will be a homework assignment
4th	1	4		6		4	1	This will be a homework assignment

Pascal's triangle is continued by adding the two numbers above to the left and to the right of where the new number will be placed. If not number exist above to the left or to the right, then the sum will stay one—add zero.

Written mathematically:

$$\binom{n}{r} = \binom{n-1}{r-1} + \binom{n-1}{r}; \text{ for } r < n \text{ and } n > 1$$

For example the 5th row would be, $0+1=1, 1+4=5, 4+6=10, 6+4=10, 4+1=5, 1+0=1$: 1, 5, 10, 10, 5, 1.

Note: each row is symmetric; written mathematically:

$$\binom{n}{r} = \binom{n}{n-r}$$

Alternative Counting Method

If you asked to find the binomial expansion, an alternative to using the binomial expansion (combinations) or Pascal's triangle is the rule:

Rule 4.4.2 Next Coefficient

Start with a coefficient of one. The **next coefficient** is the current coefficient times the power of the first entry, then divided by the power of the second entry plus one.

You just have to remember for the first term, the first coefficient is 1, the first power is n and the second power is 0; and for each consecutive term the first power goes down one and the second power goes up one.

Start with, $(a+b)^n = 1 \times a^n \times b^0 + \ldots$

$$\frac{Coefficient \times First\, Power}{Second\, Power + 1} = \frac{1 \times n}{0+1} = n$$

$$(a+b)^n = a^n + n \times a^{n-1} \times b^1 + \cdots$$

So forth and so on.

Example 4.4.2 $(x+y)^3$

Expand $(x+y)^3$ using the binomial expansion.

Solution

$$(x+y)^3 = 1x^3y^0 + \underset{\left(\frac{1\times3}{0+1}\right)}{3}\, x^2y^1 + \underset{\left(\frac{3\times2}{1+1}\right)}{3}\, x^1y^2 + \underset{\left(\frac{3\times1}{2+1}\right)}{1}\, x^0y^3$$

PROBLEMS

4.4.1. Write the binomial expansion of $(a+b)^2$.
4.4.2. Write the binomial expansion of $(x+y)^3$.
4.4.3. Write the binomial expansion of $(p+q)^4$.
4.4.4. Write the binomial expansion of $(x+3y)^3$.
4.4.5. Write the binomial expansion of $(x-y)^4$.
4.4.6. Write the binomial expansion of $(2x+7y)^3$.
4.4.7. Write the term in the binomial expansion of $(x+y)^5$ which has x^4 in it.
4.4.8. Write the term in the binomial expansion of $(x^2+y)^3$ which has x^6 in it.
4.4.9. Write the term in the binomial expansion of $(2x-y)^6$ which has x^4 in it.
4.4.10. Write the term in the binomial expansion of $(x+3y)^6$ which has no x term in it.
4.4.11. Write the term in the binomial expansion of $(x+5y)^4$ which has only x term in it.
4.4.12. Construct the first 6 rows of Pascal's triangle.
4.4.13. Explain how one row in Pascal's triangle is generated from the previous row.

CRITICAL THINKING AND BASIC EXERCISE

Multiplication Principle:
How many outfits consisting of a shirt, a pant, and a jacket are possible if there are:

4.1. Two shirts, three pants and four jackets to be chosen from.
4.2. Three shirts, three pants and two jackets to be chosen from.
4.3. Four shirts, two pants and two jackets to be chosen from.

Find the number of possible choices when you choose one item from each category.

4.4. 6 cars, 4 colors

4.5. 5 pens, 2 notebooks

4.6. 6 sizes, 10 flavors

4.7. 18 boys, 15 girls

4.8. 3 pairs of shoes, 7 sweaters, 4 shorts

4.9. 5 flavors, 3 toppings

4.10. 2 desserts, 2 drinks, 5 vegetables

4.11. 7 socks, 12 hats, 3 pairs of shoes

4.12. 12 flavors of ice cream, 3 cones

4.13. 4 shorts, 5 pairs of shoes, 8 sweaters

4.14. 5 colors, 4 styles

4.15. 15 video games, 7 CDs

4.16. 5 vegetables, 4 desserts, 3 drinks

4.17. 6 pairs of shoes, 12 socks, 3 shirts, 2 pairs of pants, 3 hats

4.18. 8 scarves, 10 socks, 3 pairs of pants, 3 pairs of sunglasses, 14 pairs of shoes

4.19. How many four-digit codes using digits can be created with repetition of digits?

4.20. How many four digit numbers exist with repetition of digits? Note: There are 10 digits: 0, 1, 2, …, 9 and to be a four digit number, the first digit cannot be zero.

Draw the associated tree diagrams

4.21. Two coins are tossed

4.22. A coin is tossed and then a die is tossed

4.23. A die is tossed and then a coin is tossed

4.24. Three distinct coins are tossed

Factorials and Permutations

4.25. Evaluate 0!

4.26. Evaluate 13!

4.27. Evaluate 5!

4.28. Evaluate 9!

4.29. Write 10! in terms of 8!

4.30. Write 100! in terms of 95!

4.31. Write n! in terms of $(n-1)$!

4.32. Write n! in terms of $(n-2)$!

4.33. How many ways can an individual with OCD place five books on a shelf?

4.34. How many ways can five horses cross a finish line?

4.35. How many distinct orderings of the digits are there?
      Note: There are 10 digits: 0, 1, 2, …, 9.

4.36. $_{12}P_7$ 4.40. $_9P_5$ 4.44. P(100.3)

4.37. $_5P_2$ 4.41. P(15,3)   4.45. P(10,10)

4.38. $_{12}P_5$ 4.42. P(20,1)   4.46. P(10,0)

4.39. $_{10}P_7$ 4.43. P(100,1) 4.47. P(5,0)

4.48. How many ways can an individual with OCD place three out of five books on a shelf?

4.49. How many ways can three out of eight horses cross a finish line?

4.50. How many four-digit codes using digits can be created without repetition of digits?

4.51. How many permutations of three letters are possible from the letters I, P, O, and C?

4.52. In how many ways can Stephanie, Katherine, Sierra, and Jacob stand in line?

4.53. How many three-digit numbers can you make by arranging the numbers 9, 1, and 2?

4.54. Write the list of all permutations of the set $A=\{a, b, c, d\}$ chosen **3** at a time. Note: there are twenty-four possible permutations.

4.55. Write the list of all permutations of the set $A = \{a, b, c, d\}$ chosen **2** at a time. Note: this time there are only twelve possible permutations.

4.56. Without writing the list of all permutations of the set $A = \{a, b, c, d\}$ chosen **4** at a time, compute the number of such permutations.

4.57. Without writing the list of all permutations of the set $A = \{a, b, c, d, e\}$ chosen **3** at a time, compute the number of such permutations.

4.58. Without writing the list of all permutations of the set $A = \{a, b, c, d, e\}$ chosen **4** at a time, compute the number of such permutations.

4.59. Without writing the list of all permutations of the set $A = \{a, b, c, d, e, f\}$ chosen **4** at a time, compute the number of such permutations.

4.60. Write out all permutations of the set $A = \{1, 2, 3\}$. Hint: there are a total of sixteen permutations including the empty set as $P(3, 0) = 1$.

Combinations

4.61. Write the list of all combination of the set $A = \{a, b, c, d\}$ chosen **3** at a time. Note: there are only four possible combinations.

4.62. Write the list of all combination of the set $A = \{a, b, c, d\}$ chosen **2** at a time. Note: this time there are only six possible combinations.

4.63. Without writing the list of all combination of the set $A = \{a, b, c, d\}$ chosen **4** at a time, compute the number of such combinations.

4.64. Without writing the list of all combination of the set $A = \{a, b, c, d, e\}$ chosen **3** at a time, compute the number of such combinations.

4.65. Without writing the list of all combination of the set $A = \{a, b, c, d, e\}$ chosen **4** at a time, compute the number of such combinations.

4.66. Without writing the list of all combination of the set $A = \{a, b, c, d, e, f\}$ chosen **4** at a time, compute the number of such combinations.

4.67. Write out all combination of the set $A = \{1, 2, 3\}$. Hint: there are a total of eight combinations including the empty set as $C(3, 0) = 1$.

Distinct Orderings

4.68. How many distinct orderings of the word "Mississippi" are there?

4.69. How many distinct orderings of your last name are there?

4.70. How many distinct orderings of the word "hippopotamus" are there?

4.71. How many distinct orderings of the word "racecar" are there?

4.72. How many distinct orderings of the word "Rebecca" are there?

Overlapping Events

4.73. If $n(A \cap B) \neq 0$, then $A \cap B \neq \varnothing$ and a formula for $n(A \cap B) =$ _____

4.74. Write the formula in problem one of this section twenty times neatly as if you were writing a spelling list.

4.75. If $n(A \cap B) = 17$, $n(A) = 47$ and $n(B) = 74$, then $n(A \cup B) =$ _____

4.76. If $n(A \cap B) = 17$, $n(A) = 17$ and $n(B) = 34$, then $n(A \cup B) =$ _____

4.77. What consequential set relation must hold true between the sets A and B in problem five of the section.

Mutually Exclusive Events

4.78. If $A \cap B = \varnothing$, then $n(A \cap B) =$ _____

4.79. If $n(A \cap B) = 0$, then $A \cap B = \varnothing$ and $n(A) + n(B) =$ _____

4.80. If A and B are mutually exclusive, $n(A) = 47$ and $n(B) = 74$, then $n(A \cup B) =$ _____

4.81. If A and B are mutually exclusive, $n(A) = 17$ and $n(B) = 34$, then $n(A \cup B) =$ _____

Binomials

4.82. Write the binomial expansion of $(a + b)^3$.

4.83. Write the binomial expansion of $(x + y)^4$.

4.84. Write the binomial expansion of $(x + y)^5$.

4.85. Write the binomial expansion of $(x + 2y)^3$.

4.86. Write the binomial expansion of $(x-y)^3$.

4.87. Write the binomial expansion of $(3x+5y)^3$.

4.88. Write the term in the binomial expansion of $(x+y)^6$ which has x^4 in it.

4.89. Write the term in the binomial expansion of $(x^2+y)^3$ which has x^4 in it.

4.90. Write the term in the binomial expansion of $(2x+y)^6$ which has x^4 in it.

Random Counting

4.91. $5!$ **4.93.** $4!-3!$ **4.95.** $P(6,2)$

4.92. $(5-3)!$ **4.94.** $C(7,3)$

4.96. How many different ways (assuming no ties) can eight horses cross a finish line?

4.97. How many different ways can an obsessive-compulsive person arrange six different books on a shelf?

4.98. How many different ways can someone choose four books out of six books to be placed in a backpack?

4.99. How many different ways can a committee of four be chosen from a group of ten people?

For the next set of problems, if a committee of four is to be chosen from a group of ten people: six men and four women. In how many different ways can this committee be chosen if the following must hold?

4.100. The committee consist of two men and two women

4.101. The committee consist of three men and one women

4.102. The committee consist of all men

4.103. The committee consist of all women

4.104. Any four members

4.105. A president, a vice-president, a secretary and a treasurer

For the next set of problems, if a committee of five is to be chosen from a group of fifteen people of which six men. In how many different ways can this committee be chosen if the following must hold?

4.106. The committee consist of exactly two women

4.107. The committee consist of exactly three men

4.108. The committee consist of all men

4.109. The committee consist of all women

4.110. Any four members

4.111. A president, a vice-president, a secretary and a treasurer and a minuteman.

4.112. How many four-digit natural numbers can be made from the set of digits: $D=\{0,1,2,3,4,5,6,7,8,9\}$. Note to be a four-digit number, the first digit cannot be zero.

4.113. How many nine-digit S.S.N.(s) can be made from the set of digits: $D=\{0,1,2,3,4,5,6,7,8,9\}$.

4.114. How many passwords of length five are there if the characters can be a natural numbers can be made from the set of digits or a letter from the English alphabet. Repetition allowed.

4.115. How many passwords of length five are there if the characters can be a natural numbers can be made from the set of digits, or a letter form the English alphabet where the letters are consider "case sensitive" and repetition allowed.

4.116. How many passwords of length four are there if the characters can only be taken from the set of digits: $D=\{0,1,2,3,4,5,6,7,8,9\}$ where repetition is not allowed.

4.117. How many passwords of length four for less are there if the characters can only be letters form the English alphabet and repetition is not allowed.

4.118. How many passwords of length seven are there if the characters can only be taken from the set of digits: $D=\{0,1,2,3,4,5,6,7,8,9\}$ and repetition is allowed.

4.119. How many passwords of length seven are there if the characters can only be taken from the set of digits: $D=\{0,1,2,3,4,5,6,7,8,9\}$ and repetition is not allowed.

4.120. On the multiple choose final, there are 55 question each having four possible answers. In how many ways could you "Christmas tree" then test; that is how many ways can the test be completed? Do not compute; leave in exponential form.

4.121. In how many ways can five cards be dealt from a standard 52-card deck?

4.122. In how many ways can four workers be selected for a group of eight applicants, assuming all four jobs are considered to be the same?

4.123. In how many ways can four workers be selected for a group of eight applicants, assuming all four jobs are considered to be different?

4.124. How many three-letter initials can be formed from the English alphabet.

4.125. How many even four digits counting numbers are there?

4.126. How many distinguishable arrangements of the letters in the word "bubble" are there?

4.127. A coin is tossed five times, how many sequences of heads and tails are there?

4.128. From a group of twelve men and eleven women, two men and two women will be selected to attend a conference. How many possible ways can the selection be made?

4.129. How many two-element subsets can be taken from a larger set containing six elements?

4.130. How many ordered triplets can result in the tossing of three dice?

4.131. From a list of five students who have been approved by the dean of your school we need to select three to represent the student body to the academic, finance, and athletic committees. In how many ways can we select the three representatives?

4.132. Compute

a. $_6P_2$, b. $_nP_2$, c. $_nP_{n-k}$, d. $_aP_b$

4.133. Compute

a. $\binom{14}{3}$

b. $\binom{n}{n-1}$

c. $\binom{n-2}{n-3}$

4.134. In how many ways can we arrange two candidates for public office and three journalists in a straight line in front of a television camera?

4.135. How many distinct permutations can we form from the letters of each of the following words:

a. *Plato*, b. *Demetri*, c. *Mississippi*.

4.136. A finite mathematics class in your school consists of 16 women and 10 men. The teacher wants to choose a committee of five students to represent the class in a school contest. In how many ways can the teacher choose this committee?

4.137. Maria is interested in having a dinner party with four men and three women from the list of her close friends. There are a total of eight men and 10 women. In how many ways can she select her dinner guests?

4.138. Find the number of different permutations of the letters in each of the following words:

a. *Massachusetts*, b. *hippopotamus*, c. *Ohoopee*.

4.139. Determine the number of different permutations of the letters in each of the following words:

a. *Okeechobee*, b. *appendicitis*, c. *Chattahoochee*

PROJECTS

Project I: Experiment with Coins

On a single sheet of paper

1. Create a tree diagram for the event: Three coins are tossed and generate the implied universe; write in roster notation. Hint: there are eight elements.

2. Use the multiplication principle to count the total number of outcomes associated with this event.

Project II: Experimenting with Dice

On a single sheet of paper

1. State the universe for the event: a single die is tossed and the number uppermost is observed.

2. Create a box chart illustrating the event: A blue die and a red die are tossed and the uppermost numbers are observed, first on the red and then on the blue.

red\blue	1	2	3	4	5	6
1						
2						
3						
4						
5						
6						

3. Use the multiplication principle to count the total number of outcomes associated with this event.
4. Create another box chart illustrating the event: A blue die and a red die are tossed and the uppermost sums are observed.
5. How many different outcomes are possible. Explain the difference in the count.

Project III: Combinations

On a single sheet of paper

1. Illustrate the formula or idea for counting the different combinations (no order, no repetition) of n objects chosen r at a time.
2. Use this formula for your favorite three-element set to compute the number of various combinations; that is, compute: $C(3,0)$, $C(3,1)$, $C(3,2)$ and $C(3,3)$.
3. List all combinations of your favorite three-element set. Hint: there are eight including the empty set.

Project IV: Permutations

On a single sheet of paper

1. Illustrate the formula or idea for counting the different permutations (order, no repetition) of n objects chosen r at a time.
2. Use this formula for your favorite three-element set to compute the number of various permutations; that is, compute: $P(3,0)$, $P(3,1)$, $P(3,2)$ and $P(3,3)$.

List all permutations of your favorite three-element set. Hint: there are sixteen including the empty set.

SUMMARY OF IMPORTANT CONCEPTS

Definitions:

4.1.1. **Enumeration** is the act of enumerating, the mention separately as if in counting; name one by one as in a list. That is, to ascertain the number of items or elements belonging to an outlined set.

4.1.2. A set is **countable** if there is an enumeration; that is, a set S having a finite number of elements which form a one-to-one correspondence with the natural numbers, $N: f : N \to S$.

4.1.3. A set is **countably infinite** if there exist an enumeration; that is, a set S having an infinite number of elements which form a one-to-one correspondence with the entire set of the natural numbers, N:

$$f : N \to S.$$

4.2.1. If there are n tasks to be done or events to perform, and each task can be done in $m_1, m_2, \ldots, m_n$, then the number of ways to perform all given tasks is $m_1 \times m_2 \times \cdots \times m_n$.

4.2.2. Suppose that k objects are selected from a set of N objects one at a time **with replacement**. Then the sample space for this **multiple selection** consists of N^k outcomes.

4.2.3. A **tree diagram** is a network diagram used to find and analyze the various outcomes in a multi-stage sampling. Starting at a point, create branches to the first set of possible outcomes. From each of the secondary points created above, create additional branches to the second set of outcomes. Repeat as needed.

4.3.1. Suppose that n objects are selected at random one at a time without replacement. Then the sample space consists of $n!=n\times(n-1)\times\cdots\times2\times1$ outcomes. Here, $n!$ is said "n **factorial**." $n!$ counts the number of ways to order n objects. Note: $0!=1$, there is one way to order nothing—do nothing.

4.3.2. Suppose that k objects are drawn without replacement, with order being preserved, from a collection of n distinct objects. Then the sample space of **permutation** consists of $P(n,k)=P_n^k=\,_nP_k=\frac{n!}{(n-k)!}$ sample points; that is, the number of ways of ordering n distinct objects taken k at a time; permutations.

4.3.3. Suppose that k objects are drawn without replacement and without regard to order from a collection of n distinct objects. Then the sample space of **combinations** consists of $C(n,k)=\binom{n}{k}=\,_nC_k=\frac{n!}{k!(n-k)!}$ sample points; combinations in set theory.

4.3.4. The distinct ordering of n objects where each of the k distinguishable objects $i=1,2,\dots,k$ occurs n_i times where $n_1+n_2+\cdots+n_k=n$. Just as before factorials order all: $n!$ and division removes the individual orderings; hence, the number of distinct **partial orderings** is given by: $\dfrac{n!}{n_1!\times n_2!\times\cdots\times n_k!}$.

4.3.5. Two sets, A and B, in a given universal set U are said to be mutually exclusive if they do not have any elements in common and we denote it by $A\cap B=\varnothing$.

Properties:

4.3.1. **Overlapping Events:** The number of elements in two overlapping sets is the number of elements minus the number of elements the sets have in common as they have been added twice. $n(A\cup B)=n(A)+n(B)-n(A\cap B)$

4.3.2. **Mutually Exclusive Events:** The number of elements in two mutually exclusive (non-overlapping) sets is the number of elements. $n(A\cup B)=n(A)+n(B)$

Rules:

4.4.1. In the **binomial expansion** of $(a+b)^n$, the coefficient on the term with a^r and b^{n-r}, the **binomial coefficient**, are given by $C(n,r)=\binom{n}{r}$.

4.4.2. Start with a coefficient of one. The **next coefficient** is the current coefficient times the power of the first entry, then divided by the power of the second entry plus one. You just have to remember for the first term, the first coefficient is 1, the first power is n and the second power is 0; and for each consecutive term the first power goes down one and the second power goes up one. Start with, $(a+b)^n=1\times a^n\times b^0+\dots$, then $\dfrac{Coefficient\,First\,Power}{Second\,Power+1}=\dfrac{1\times n}{0+1}=n$; thus, $(a+b)^n=a^n+n\times a^{n-1}\times b^1+\dots$ So forth and so on.

Below is a list of every idea that might be symbolically abbreviated when counting; that is the counting.

Symbol (Abbr.)	Name	Meaning/Read as	Description	Example/ Description
$\forall$	For all	"For all"		$x>0,\ \forall x\in N$
$n!$	n "Factorial"	Ways to order all n objects		$5!=5\times4\times3\times2\times1=120$
MP	Multiplication Principle		"And" means "Multiply"	
$P(n,r)$	Permutation	Number of ways to permute n objects r at a time		$P(3,3)=6$
$C(n,r)$ or $\binom{n}{r}$	Combination; be careful, this is not a fraction!	Number of ways to combine n objects r at a time		$C(3,3)=1$ $\binom{3}{2}=3$

Chart on the various ways to count:

	Repetition Allowed	Repetition Not Allowed
Order	Multiplication Principle $m_1 \times m_2 \times \cdots \times m_n$	Permutations $P(n,r) = \frac{n!}{(n-r)!}$
Partial or No Order	Distinct Orderings (Partial) $\frac{n!}{n_1! \times n_2! \times \cdots \times n_k!}$	Combination (No Order) $C(n,r) = \frac{n!}{r!(n-r)!}$

REVIEW TEST

1. Determine the number of squares of any size in the figure below.

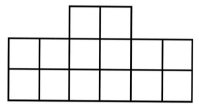

2. Draw the tree diagram representing the experiment of tossing a six sided die and then a coin.
3. Evaluate the following expressions:
   A. **5!**
   B. **$(5-3)!$**
   C. **$5! - 3!$**
   D. **$C(5,3)$**
   E. **$P(5,3)$**
4. How many five member committees can be selected from a club which has 13 members?
5. A restaurant offers three choices of salad, four choices of entree, and three choices of dessert. How many complete meal could you order if:
   A. One of the desserts is milk based and you cannot eat it.
   B. There is only one entree you will eat, the one with chicken.
6. How many four-digit codes can be formed from the digits 0, 1, 2, 3, 4 and 5 if repetition is not allowed?
7. How many four-digit codes can be formed from the digits 0, 1, 2, 3, 4 and 5 if repetition is allowed?
8. How many four-digit numbers can be formed from the digits 0, 1, 2, 3, 4 and 5 if repetition is allowed? Hint: a four-digit number cannot start with 0.
9. How many ways can eight horses competing in a race cross the finish line if ties are not allowed?
10. An exam contains 10 true/false questions. How many different ways can the exam be completed?
11. How many ways can three students be selected from a group of seven if order does not matter?
12. A box contains 6 apples and 4 oranges. How many ways can two apples and two oranges be selected?
13. How many ways can 5 cards be dealt from a standard deck of 52 cards if order does not matter?
14. How many different sequences of heads and tails exist when a coin is tossed five times?
15. How many different three letter initials are there using the English alphabet?
16. How many even four-digit counting numbers are there?
17. How many distinct arrangements are there of the letters in the word **happy**?
18. Study the examples illustrating the various ways to combine two out of **n** objects:
   $n = 3$, $A = \{a,b,c\}$ with three combinations: **ab**, **ac** and **bc**.
   $n = 4$, $A = \{a,b,c,d\}$ with six combinations: **ab**, **ac**, **ad**, **bc**, **bd** and **cd**.
   $n = 5$, $A = \{a,b,c,d,e\}$ with ten combinations: **ab**, **ac**, **ad**, **ae**, **bc**, **bd**, **be**, **cd**, **ce** and **de**.

How many combinations can be formed taking two out of six objects?

A. 15

B. 18

C. 21

D. 30

19. A soft drink company decides to test five different recipes for a new soda in a pair-wise comparison. How many different pairs will result from selecting two recipes at a time?

A. 5

B. 6

C. 10

D. 20

20. From a group of three men and two women, how many ways can one man and one woman be selected to be on a committee?

A. 5

B. 6

C. 3

D. 2

REFERENCES

Hall, M., 1967. Combinational Theory. Blaisdell, Waltham, MA.

Tsokos, C.P., 1972a. Probability Distributions: An Introduction to Probability Theory with Applications. Wadsworth, Belmont, CA.

Tsokos, C.P., 1978c. Mainstreams of Finite Mathematics with Application. Charles E. Merrill Publishing Company, A Bell & Howell Company.

We see … *that the theory of probabilities is at bottom only common sense reduced to calculations; it makes us appreciate with exactitude what reasonable minds feel by a sort of instinct, often without being able to account for it. … it is remarkable that this science, which originated in the consideration of games of chance, should have become the most important object of human knowledge*

PIERRE S. DE LAPLACE

Chapter 5	Basic Probability

Blaise Pascal was born in the French province of Auvergne in 1623 and showed phenomenal ability in mathematics at a very early age. At fourteen he was admitted to the weekly meetings of Roberval, Mersenne, Mydorge, and other French geometricians from which, ultimately, the French Academy sprung. At sixteen Pascal wrote an essay on conic sections; and in 1641, at the age of eighteen, he constructed the first arithmetical machine, an instrument that he further improved eight years later.

In 1650, suffering from ill health, Pascal decided to abandon mathematics and science and devote himself to religious contemplation. Three years later, however, he returned briefly to mathematical research. At this time he wrote Traite du Triangle Arithmetique[1], *conducted several experiments on fluid pressure, and in correspondence with Fermat assisted in laying the foundation of probability theory.*

Only once again, in 1658, did Pascal return to mathematics. While suffering with a toothache some geometrical ideas occurred to him, and his teeth suddenly ceased to ache. Regarding this as a sign of Divine Will, he obediently applied himself toward developing these ideas, producing in eight days a fairly full account of the geometry of the cycloid curve and solving some problems that subsequently, when issued as challenge problems, baffled other mathematicians.

Pascal's famous Provinciales *and* Pensèes, *which are read today as models of early French literature, were written toward the close of his brief life. He died in Paris in 1662 at the age of thirty-nine. His father Etienne Pascal, (588-1640) was also an able mathematician; it is for the father that the limaçon of Pascal is named.*

5.1   INTRODUCTION TO PROBABILITY

Webster defines *probability* as "the quality or state of being probable," "likelihood," or "the number of times something will probably occur over the range of possible occurrences expressed as a ratio"[2]. The idea of probability originated with *games of chance* in the seventeenth century. The earliest writings in the area were the result of the collaboration between the eminent mathematicians Blaise Pascal and Pierre Fermat, and a gambler, Chevalier de Mere. There seemed to them to be contradictions between mathematical calculations and the events of actual games of chance involving throwing dice, tossing a coin, spinning a roulette wheel, or playing cards. For example, in repeated throws of a die, it was observed that each face of the die, 1 to 6, turned up approximately equally often; that is, each number appeared with a frequency of **1/6**. However, if two dice were rolled, the sum of numbers showing on the two dice; that is, 2, 3, 4, and so on to 12, did not appear equally often. Similar experiments were conducted, the outcomes spurring interest in the development of some basic rules. Probability was developed primarily for application to games of chance until the eighteenth century.

In recent years, probability has emerged as one of the most important disciplines. It has been successfully applied with increasing frequency in areas such as statistics, engineering, operations research, physics, medicine, business, economics, accounting, education, sociology, physiology, agriculture, meteorology, linguistics, and political science. In every profession, probabilistic questions are encountered. For example,

---

1. *Translated* Treatise on the Arithmetic Triangle.
2. Webster's New *World Dictionary of the American Language, 2nd* college ed', s.v. "Probability".

In medicine	What is the probability that a particular drug will be effective on a given patient?
In meteorology	What is the probability that it will rain in a certain city in Florida on a specific date of the year?
In engineering	What is the probability that the radio we designed will be operable?
In statistics	What is the probability that the number of automobile accidents during this year's Christmas holidays will exceed the previous year's number of accidents?
In sociology	What is the probability that a child in an urban ghetto will fail to complete his high school education?
In education	What is the probability that an individual's score on an intelligence test will show significant improvement following a refresher course in verbal skills?
In sports	What is the probability that a certain swimmer can break the world record in the butterfly event?
In operations	What is the probability that a system (or component) research will perform its purpose adequately for the period of time intended under the operating conditions encountered?
In chemistry	What is the probability that an acid solution made by a specific method will satisfactorily etch a tray?
In business	What is the probability that the fuel supply of a certain airport will be inadequate on any given day?
In physiology	What is the probability that an experimental animal will convulse upon administration of a certain pharmacological agent?

Our aim in this chapter is to introduce the basic concepts of probability and to illustrate the usefulness of the discipline.

5.2 DEFINITION OF PROBABILITY

The development of the concept of probability can be studied by familiarizing ourselves with the various definitions of the term. We shall begin by introducing these definitions.

Definition 5.2.1 Personal Probability

Personal probability is a judgment about the probability of an event happening in a particular manner, and is often based on personal feelings or beliefs rather than on a mathematical analysis of the situation.

For example, if a person says, "I'm almost certain that we will land on Mars," the high probability that he or she has assigned to this possible happening is not the result of an exact method of determination but rather the expression of a belief. As a second example, a person might say concerning the occurrence of earthquakes, "It is probable that on a specific date there will be an earthquake on the island of Rhodes." Again, this individual is assigning a certain probability to the happening in an intuitive manner—through feelings or beliefs.

To obtain probabilities from a set, let us begin by denoting a certain happening or event by the capital letter A. The happening could be the result of a certain experiment that might be quantitative or non-quantitative in nature. We are interested in determining the probability of the occurrence of A, which we shall denote by $\mathbf{Pr}(A)$ or $P(A)$. We shall assume that there are n outcomes, and each outcome is equally likely to possess characteristic A.

Definition 5.2.2 Classical Probability

The classical definition of the **probability** of the event A is equal to the ratio of the number of ways in which event A may occur in a particular situation, denoted by n_A, and the total number of possible outcomes in a given situation, denoted by n. That is,

$$\mathrm{Pr}(A) = \frac{\textit{number of elements which attain attribute } A}{\textit{total number of possible outcomes}} = \frac{n_A}{n}$$

We shall approach this definition more precisely using some of the basic concepts of set theory that we have studied previously. First, however, let us give some examples to illustrate how we obtain probabilities using the classical definition of probability.

Example 5.2.1 Tossing A Fair Coin

Consider the situation of tossing a fair coin. There are two possible outcomes to this situation, each of which is equally likely to occur. The probability of obtaining a head, H, is given by

$$\mathrm{Pr}(H) = \frac{n_H}{n} = \frac{1}{2}.$$

Example 5.2.2 Tossing a Fair Die

Consider the situation of tossing a fair die once; we are interested in obtaining the probability that an odd number will occur. Here, there are six (6) possible outcomes, $n = 6$, each of which is equally likely to occur, and three of which result in an odd number, $n_O = 3$. Thus, the probability of an odd number (O) is

$$\mathrm{Pr}(O) = \frac{n_O}{n} = \frac{3}{6} = \frac{1}{2}.$$

Example 5.2.3 Drawing a Card

Consider the situation of drawing a card from an ordinary deck of 52 cards. Here there are 52 possible outcomes, $n = 52$, and we assume that each card has the same chance of being selected. Suppose that we are interested in obtaining the following probabilities:
(a) The chosen card is a heart, H.
(b) The chosen card is a picture card, P.
(c) The chosen card is an ace, A.

Example 5.2.3 Drawing a Card—cont'd

For (a), since 13 of the 52 cards are hearts, we have $n_H = 13$ and

$$\mathbf{Pr}(H) = \frac{n_H}{n} = \frac{13}{52} = \frac{1}{4}.$$

For (b), since there are 12 picture cards in an ordinary deck of cards, we have $n_p = 12$ and

$$\mathbf{Pr}(P) = \frac{n_p}{n} = \frac{12}{52} = \frac{3}{13}.$$

For (c), we have $n_A = 4$, since there are four aces in the deck of 52 cards, and

$$\mathbf{Pr}(A) = \frac{n_A}{n} = \frac{4}{52} = \frac{1}{13}.$$

In the preceding examples we use the term "fair"—fair coin, fair die, etc., so that the classical definition of probability will be applicable. For example, if the die is *not* fair, or in other words, the appearance of each of the six faces is not equally likely, the assumptions of the definition have not been met and if used, the results will not be correct.

Finally, we should mention that the classical definition of probability is also called the a *priori definition of probability*. That is, prior to and without actually performing the experiment, we can obtain probabilities of various happenings or attributes in a given situation.

To give a rigorous presentation of the **relative-frequency** definition of probability, we need some advanced concepts of mathematics which are beyond the scope of this book. However, we can give a workable description of the definition.

Let n be the total number of identical trials in a given problem and n_B the number of occurrences of attribute or characteristic B. Here, *identical* means that each trial in a given situation is performed under identical conditions. The probability of B happening, $\mathbf{Pr}(B)$, is given by

$$\mathbf{Pr}(B) = \frac{n_B}{n}, \text{ as } n \text{ becomes very large}$$

"As n becomes very large" means that the accuracy of measuring the probability of the happening B becomes greater as the number of trials becomes very large. For example, in tossing a coin (which need not be fair, as was required by the classical definition of probability), we are interested in obtaining the probability of getting a head, which will be

$$\mathbf{Pr}(H) = \frac{n_H}{n}.$$

This expression does not give us a specific answer, but it does tell us an approximate value for a particular number of trials, and that this approximation will improve as n increases. That is, if we toss a coin 100 times ($n = 100$) and count the number of heads that have occurred, say $n_H = 60$, the ratio of $\frac{n_H}{n} = \frac{60}{100} = \frac{3}{5}$. If we had employed the classical definition we would have estimated the $\mathbf{Pr}(H)$ to be equal to $\frac{1}{2}$. It is of interest to mention that in tossing a fair coin, the probability of obtaining a head, H, approaches that of obtaining a tail, T. That is,

$$\mathbf{Pr}(H) = \frac{1}{2} = \mathbf{Pr}(T), \text{ as } n \text{ becomes large (say 2000, 3000, etc.).}$$

Note that the relative-frequency definition of probability is also called *a posteriori* because the desired probabilities are calculated *after* the necessary information has been collected; whereas in the classical definition the probabilities are calculated *before* conducting the experiment. We have been using the word *experiment* loosely, but in the next section the term will be carefully defined.

The **axiomatic** definition of probability originated with a Russian scientist, *A. N. Kolmogorov*[3]. It is the most important definition of probability because it eliminates most of the difficulties we encounter using the other definitions. Before stating the axiomatic definition, we shall discuss some additional concepts in Section 5.3.

EXERCISES

The following exercise will be helpful in understanding the classical and relative-frequency definitions of probability.

5.2.1. A card is selected from an ordinary deck of 52 cards.
- **a.** What is the probability that the selected card is a picture card and a heart?
- **b.** What is the probability that the selected card is an ace of hearts?
- **c.** What is the probability that the selected card is a king?

5.2.2. An urn contains 25 balls out of which ten are red and 15 black. A ball is drawn from the urn.
- **a.** What is the probability that the selected ball is black?
- **b.** What is the probability that the drawn ball is red?
- **c.** What is the probability that the chosen ball is yellow?

5.2.3. A doctor purchases a certain drug from three companies: *A*, *B*, and *C*. If equal amounts of drug are purchased from each company, what is the probability that the drug administered to a specific patient was purchased from Company *C*?

5.2.4. A certain cartridge tape was purchased from a shelf containing 10 such tapes. If we were told that 10% of the tapes were defective, what is the probability that the purchased cartridge tape is non-defective?

5.2.5. A fair die is rolled once.
- **a.** What is the probability that an one will turn up?
- **b.** What is the probability that an even number will turn up?

5.3   THE SAMPLE SPACE

The sample space is a set of all possible outcomes in given experiment.

> **Definition 5.3.1   Experiment**
>
> An **experiment** is the method by which an observation or a measurement is obtained.

Some examples of experiments are

1. making a measurement of the day's highest temperature,
2. rolling a die and observing the number that appears on the top,
3. recording the effectiveness of a certain drug administered to a patient in a hospital,
4. recording the weight of a newborn baby,
5. recording a grade on a certain examination, and
6. ascertaining the preference of a voter before an election.

Thus, by an experiment we mean the following of a specific procedure at the completion of which we observe certain results.

> **Definition 5.3.2   Sample Point**
>
> A possible outcome of an experiment is called a **sample point**; the actual observations of an experiment constitute a **sample**.

---

3. In probability theory, Kolmogorov's specialty, he established new directions in research, formulated two important systems of equations in partial derivation, and worked on the problems of random stationary processes. His books on this subject include *Foundations of the Theory of Probability* and *Analytical Methods of Probability, Theory*.

Example 5.3.1 Tossing a Fair Coin ×5

Conducting the experiment of tossing a coin five times and recording the outcome each time will constitute a sample. The outcome of each individual toss is a sample point.

$$\{HHHHH\}, \{HHHHT\}, etc.$$

Definition 5.3.3 Population

A group of individual persons, objects, or items from which samples are taken for statistical measurement constitutes a **population**.

Definition 5.3.4 Event

An **event** is a set of sample points.

We shall denote events by capital letters and sample points by lowercase.

Example 5.3.2 Tossing a Fair Coin ×2

Consider the experiment of flipping a coin twice. Some of the possible events are as follows:
(a) Event *A*: You observe two heads.
(b) Event *B*: You observe two tails.
(c) Event *C*: You observe one head and one tail.

Definition 5.3.5 Simple Event

An event that consists of only one sample point is called a **simple** or **elementary event**.

Example 5.3.3 Two Sample Points

Events *A* and *B* in Example 5.3.2 are simple or elementary events. Event *C* is not, because it can be thought of as consisting of two sample points; namely, *HT* and *TH*.

Definition 5.3.6 Compound Event

An event that consists of more than one sample point is called a **compound event**.

Example 5.3.4 Rolling a Fair Die

Consider the experiment of rolling a fair die and the event an odd number is observed.
    There are six possible outcomes; namely,
    Event *A*: a 1 is observed,
    Event *B*: a 2 is observed,
    Event *C*: a 3 is observed,
    Event *D*: a 4 is observed,

Continued

Example 5.3.4 Rolling a Fair Die—cont'd

Event *E*: a 5 is observed, and

Event *F*: a 6 is observed.

Each of the preceding events is an elementary event. With events *A*, *C*, or *E* we observe an odd number; that is, we observe a 1, 3, or 5. Thus, this event of an odd number is a compound event.

In view of the preceding definitions, we can say that an event that cannot be decomposed is a simple or elementary event. Similarly an event that can be decomposed into two or more simple events is a compound event. It is clear that in Example 5.3.4 events *A*, *B*, *C*, *D*, *E*, and *F* cannot be decomposed.

However, the event of observing an odd number is decomposable.

Definition 5.3.7 Sample Space

A set of all possible outcomes of an experiment is called a **sample space**. We shall denote the sample space by **S**.

Hence, that the sample space is simply the set of all possible sample points of a given experiment.

Example 5.3.5 Tossing a Fair Coin

Consider the experiment of tossing a fair coin. The sample space *S* consists of the two possible outcomes

H = a head was observed and *T = a tail was observed*.

Note that outcomes *H* and *T* are sample points. Using set theory we symbolize the sample space by $S = \{H, T\}$.

Example 5.3.6 Tossing a Fair Coin ×2

Consider the experiment of tossing a coin twice. The possible outcomes are as follows:

Event $A = \{(H, H)\} = $ *head in the first toss, head in the second toss*,

Event $B = \{(H, T)\} = $ *head in the first toss, tail in the second toss*,

Event $C = \{(T, H)\} = $ *tail in the first toss, head in the second toss*, and

Event $D = \{(T, T)\} = $ *tail in the first toss, tail in the second toss*.

Thus, the sample space *S* consists of the simple events (sample points) *A*, *B*, *C*, and *D*. Again, using set notation we can write the sample space as follows:

$$S = \{(H, H), (H, T), (T, H), (T, T)\}$$

or

$$S = \{A, B, C, D\}.$$

The sample space may be viewed symbolically in terms of the Venn diagram. The sample space of Example 5.3.6 is shown by Figure 5.1.

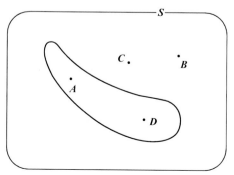

FIGURE 5.1   Venn diagram for Example 5.3.6.

Having constructed the sample space, we can formulate additional events by grouping the necessary sample points. That is, in Example 5.3.6 the event of obtaining either two heads or two tails will consist of the simple events A and D. This event is represented on the Venn diagram by circling the sample points A and D, as shown by the enclosed‘ portion of S in Figure 5.1.

Definition 5.3.8 Mutually Exclusive

Two events, A and B, of the sample space S are said to be **mutually exclusive** if $A \cap B = \varnothing$.

That is, events A and B do not have any elements in common.

Example 5.3.7 Mutually Exclusive Events

In Example 5.3.6, the events A and D are mutually exclusive; the definition of mutually exclusive events can similarly be extended to a larger number of events. For example, the events A, B, and C are mutually exclusive if $A \cap B = \varnothing$, $A \cap C = \varnothing$, $B \cap C = \varnothing$, and

$$A \cap B \cap C = \varnothing.$$

EXERCISES

5.3.1. Make a list of five examples that illustrate the meaning of an experiment.

5.3.2. Which of the following describe an experiment:
   a. recording a test grade,
   b. tossing a coin and observing the face which appears,
   c. making a measurement of the daily rainfall,
   d. recording the rushing yards of a running back in a football game?

5.3.3. Consider the experiment of flipping a coin three times. Make a list of all possible sample points.

5.3.4. Consider the experiment of tossing a coin and a die. Make a list of the 12 possible sample points.

5.3.5. Do the sample points which you have listed in Problem 5.3.3 constitute simple or compound events?

5.3.6. Distinguish which of the observations you have listed in Problem 5.3.4 are elementary and which are compound events.

5.3.7. Consider the experiment of tossing three coins and recording the number of heads observed, Define the sample space of the experiment.

5.3.8. Construct the sample spaces of Problems 5.3.3 and 5.3.7.

5.3.9. The sample space S consists of five elements:

$$S = \{s_1, s_2, s_3, s_4, s_5\}.$$

Let A_1 and A_2 be subsets of S: $A_1 = \{s_1, s_3, s_5\}$ and $A_2 = \{s_2, s_4\}$ Draw a Venn diagram to illustrate symbolically S, A_1, and A_2.

5.4 COMPUTING PROBABILITIES OF EVENTS

We have defined the sample space S to be the set of all possible outcomes of an experiment. Now we shall be concerned with calculating probabilities of different events or a collection of events defined on the sample space. It is of interest at this point to reiterate the fact that if the sample space consists of n sample points, then we can obtain probabilities of 2^n possible events in S. That is, there are 2^n possible subsets of a set that consists of n elements.

We proceed by stating the *axiomatic definition of probability.*

Definition 5.4.1 Axiomatic Probability

Let $A_1, A_2, \ldots, A_n$ be events defined on the sample space S. The probability of each event, denoted by $\mathbf{Pr}(A_i)$, $i = 1, 2, \ldots, n$, are numbers that satisfy the following **axioms**:

1. $\mathbf{0 \leq Pr}(A_i)$
2. $\mathbf{Pr}(S) = 1$
3. $\mathbf{Pr}(A_1 \cup A_2 \cup \ldots \cup A_n) = \mathbf{Pr}(A_1) + P(A_2) + \ldots + \mathbf{Pr}(A_n)$,   when $A_1, A_2, \ldots, A_n$ are mutually exclusive events, $\mathbf{Pr}(A_i)$.

Axiom 1 simply says that the probability of every event defined on the sample space is greater than or equal to zero. If the sample space has n points, the empty event on S, the probability of which will be equal to zero, is the **impossible event**, that is, an event containing no sample points. You recall that we denote such an event by $\varnothing$, It is clear why the probability of $\varnothing$ is zero,

$$\mathbf{Pr}(\varnothing) = \frac{n_\varnothing}{n} = \frac{0}{n} = \mathbf{0}.$$

Axiom 2 says that the probability of the set S, the sample space, is one. That is,

$$\mathbf{Pr}(S) = \frac{n_s}{n} = \frac{n}{n} = \mathbf{1}.$$

We should also mention here that if we determine the probability of every event on the sample space S, then we say that S is a **probability space**. Furthermore, if we sum the probabilities of every possible simple event on S, the sum will be equal to one. These facts will be illustrated in detail with the examples to follow.

Axiom 3 says that the probability of the union of a sequence of events defined on S is equal to the sum of their probabilities, provided that the sequence of events is mutually exclusive. You recall that two events, A_1 and A_2, of the sample space S are said to be mutually exclusive if $A_1 \cap A_2 = \varnothing$. In order to meet the condition of the third axiom, every possible pair in the sequence must be mutually exclusive. (In a later section we shall discuss techniques to allow us to calculate probabilities of events that are not mutually exclusive.)

We would like to emphasize here that the axiomatic approach to probability in no way assigns a numerical value to the probability of any event. It is used primarily to manipulate probabilities of compound events so that we may express them in terms of probabilities of simple events, which in turn may be evaluated by one of the other definitions of probability already mentioned. The following example illustrates this concept:

Example 5.4.1 Tossing a Fair Die ×2

Consider the experiment of tossing two fair dice once in the interest of obtaining the following four probabilities:

1. The probability that the sum showing on the two dice is equal to four,
2. The probability that the sum showing on the two dice is less than five,
3. The probability that the absolute value of the difference on the dice is equal to three, and
4. The probability that the absolute value of the difference on the dice is greater than or equal to four.

Here, the sample space consists of 36 sample points (simple events). Let us represent each sample point by an ordered pair in which the first and second numbers refer to the observed values on the first and second die, respectively. Thus, the sample space S contains the following sample points:

$$\begin{aligned}
S = \{ & (1,1)\ (1,2)\ (1,3)\ (1,4)\ (1,5)\ (1,6) \\
& (2,1)\ (2,2)\ (2,3)\ (2,4)\ (2,5)\ (2,6) \\
& (3,1)\ (3,2)\ (3,3)\ (3,4)\ (3,5)\ (3,6) \\
& (4,1)\ (4,2)\ (4,3)\ (4,4)\ (4,5)\ (4,6) \\
& (5,1)\ (5,2)\ (5,3)\ (5,4)\ (5,5)\ (5,6) \\
& (6,1)\ (6,2)\ (6,3)\ (6,4)\ (6,5)\ (6,6) \}.
\end{aligned}$$

Example 5.4.1 Tossing a Fair Die ×2—cont'd

Before we proceed to answer the posed questions, we shall illustrate the meaning of Axioms 1 and 2. Each of the 36 simple events (sample points) has an equal chance of occurring. According to the classical definition, the probability of each simple event A_i, is $\frac{1}{36}$; that is,

$$\Pr(A_i) = \frac{1}{36}, \ i = 1, 2, \dots, 36.$$

Thus, the probability of each simple event, $\Pr(A_i)$, is greater than zero, and it will be equal to zero for any other event which is not in $S-$ Axiom 1. Now, from Axiom 2 we see that

$$\begin{aligned}
\Pr(S) &= \Pr\{A_1 \cup A_2 \cup \dots \cup A_{36}\} \\
&= \Pr(A_1) + \Pr(A_2) + \dots + \Pr(A_{36}) \\
&= \Pr\{(1, 1)\} + \Pr\{(1, 2)\} + \Pr\{(1, 3)\} + \dots + \Pr\{(6, 6)\} \\
&= \frac{1}{36} + \frac{1}{36} + \frac{1}{36} + \dots + \frac{1}{36} \\
&= 1.
\end{aligned}$$

To answer the first part of Example 5.4.1; that is, the event, B_1, that the sum of the faces is equal to four, we find only three sample points which yield a sum of four: (3,1), (2,2), and (1,3). Therefore

$$B_1 = \{(3, 1), (2, 2), (1, 3)\}.$$

The probability of this event, $\Pr(B_1)$, can be obtained by applying Axiom 3; that is, by adding the probabilities of each of the simple events contained in B_1, Hence,

$$\begin{aligned}
\Pr(B_1) &= \Pr\{(3, 1)\} + \Pr\{(2, 2)\} + \Pr\{(1, 3)\} \\
&= \frac{1}{36} + \frac{1}{36} + \frac{1}{36} = \frac{3}{36} = \frac{1}{12}.
\end{aligned}$$

To answer the second part, we refer to the sample space of the problem and we see that there are six sample points constituting the compound event B_2; namely, (1,1), (1,2), (2,1), (3,1), (2,2), and (1,3). That is,

$$B_2 = \{(1, 1), (1, 2), (2, 1), (3, 1), (2, 2), (1, 3)\}.$$

Therefore,

$$\begin{aligned}
\Pr(B_2) &= \Pr\{(1, 1)\} + \Pr\{(1, 2)\} + \Pr\{(2, 1)\} + \Pr\{(3, 1)\} + \Pr\{(2, 2)\} + \Pr\{(1, 3)\} \\
&= \frac{1}{36} + \frac{1}{36} + \frac{1}{36} + \frac{1}{36} + \frac{1}{36} + \frac{1}{36} = \frac{6}{36} = \frac{1}{6}.
\end{aligned}$$

For the third part, there are six sample points in the sample space where the absolute value of the difference on the dice is equal to three; namely, (1,4), (4,1), (2,5), (5,2), (6,3), and (3,6). Let B_3 be the compound event which consists of the six sample points; that is,

$$B_3 = \{(1, 4), (4, 1), (2, 5), (5, 2), (3, 6), (6, 3)\}.$$

Thus, using Axiom 3 we have

$$\begin{aligned}
\Pr(B_3) &= \Pr\{(1, 4)\} + \Pr\{(4, 1)\} + \Pr\{(2, 5)\} + \Pr\{(5, 2)\} + \Pr\{(3, 6)\} + \Pr\{(6, 3)\} \\
&= \frac{1}{36} + \frac{1}{36} + \frac{1}{36} + \frac{1}{36} + \frac{1}{36} + \frac{1}{36} = \frac{6}{36} = \frac{1}{6}.
\end{aligned}$$

To answer the fourth part, we refer to the sample space of the problem and see that there are six sample points where the absolute value of the difference on the dice is greater than or equal to four; namely, (1,5), (5,1), (2,6), (6,2), (1,6), and (6,1). Let B_4 represent the compound event that consists of the six sample points; that is,

$$B_4 = \{(1, 5), (5, 1), (2, 6), (6, 2), (1, 6), (6, 1)\}.$$

Continued

> **Example 5.4.1 Tossing a Fair Die ×2—cont'd**
>
> Therefore, applying Axiom 3, we can write
>
> $$\mathbf{Pr}(B_4) = \mathbf{Pr}\{(1, 5)\} + \mathbf{Pr}\{(5, 1)\} + \mathbf{Pr}\{(2, 6)\} + \mathbf{Pr}\{(6, 2)\} + \mathbf{Pr}\{(1, 6)\} + \mathbf{Pr}\{(6, 1)\}$$
> $$= \frac{1}{36} + \frac{1}{36} + \frac{1}{36} + \frac{1}{36} + \frac{1}{36} + \frac{1}{36} = \frac{6}{36} = \frac{1}{6}.$$
>
> This means that the probability that the absolute value of the difference on the dice is greater than or equal to four is $\frac{1}{6}$.

EXERCISES

5.4.1. The sample space of the experiment of tossing a coin and a die was obtained in 5.3 Exercise, Problem 5.3.4. Calculate the probability of the following events:

    a. the event of a tail on the coin and an even number on the die,

    b. the event of a head and an odd number, and

    c. the event of a tail and an even number or a head and an odd number.

5.4.2. Refer to 5.3 Exercises, Problem 5.3.3 and obtain the probability of the following events:

    a. the event of two heads,   **c.** the event of two or more tails, and

    b. the event of three heads,   **d.** the event of two heads or two tails.

5.4.3. The results of a certain experiment are shown in the accompanying diagram.

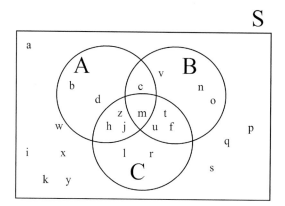

5.4.4. Calculate the probability of the following events assuming all sample points are equally likely:

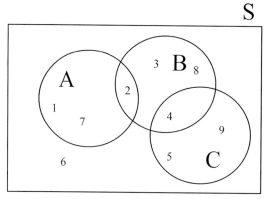

    (a) **Pr(A)**, (b) **Pr(B)**, (c) **Pr(A)** (d) **Pr(A ∪ C)**.

5.4.5. Three horses, H_1, H_2, and H_3, participate in a race. Suppose all possible results are equally likely.

    a. What is the probability that horse H_1 wins?

    b. What is the probability that horse H_3 wins?

    c. What is the probability that horse H_1 "shows" (comes in first or second)?

5.4.6. On a bargain table in a shoe store are many shoes of a single size and style, not matched in pairs. If the number of left shoes equals the number of right shoes,

a. What is the probability of picking two shoes and getting a matched pair (left and right)?

b. What is the probability of obtaining a matched pair when three shoes are picked?

5.4.7. Two people play a game involving coins. Each tosses a coin. If they match, **A** wins; if not, **B** wins. What is the probability that **A** wins?

5.4.8. One way of eliminating one of three contest winners is to have all three toss coins. In "odd man out," the one whose coin does not match the other two loses. However, if all three outcomes of the coins match, no one loses. What is the probability that no one is eliminated in the first toss?

5.4.9. An experiment consists of tossing two fair dice once. What is the probability that the sum showing on the two dice is equal to seven?

5.4.10. In Problem 5.4.9, calculate the probability that the sum showing on the dice is greater than five.

5.4.11. In Problem 5.4.9, what is the probability that the absolute value of the difference on the dice is equal to six?

5.5 LAWS OF PROBABILITY

Consider the experiment of drawing a single card from a well-shuffled deck. We might be interested in obtaining the probability of the event that the card drawn is a diamond or a picture card. Here, the sample space consists of 52 sample points corresponding to the 52 possible cards that might be drawn. Let A_1 represent the occurrence of a diamond in a single draw and A_2 the occurrence of a picture card in a single draw. The question we have posed requires us to find the

$$\mathbf{Pr}(A_1 \cup A_2)$$

However, since the events A_1, and A_2 are not mutually exclusive, we cannot apply the axiomatic definition of probability to obtain the desired result. Clearly, the reason the events are not mutually exclusive is the fact that A_1 can be a picture diamond.

Many other problems involve events that are not mutually exclusive. To allow us to overcome this difficulty we state the following basic laws:

Property 5.4.1 Complement

If A is any event of the sample space S, then

$$\mathbf{Pr}(\bar{A}) = 1 - \mathbf{Pr}(A),$$

where $\bar{A}$ denotes the **complement** of event A.

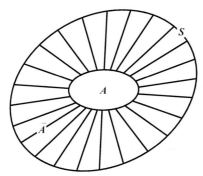

FIGURE 5.2 Venn diagram of Theorem 1.

From Figure 5.2, we can write the following equation:

$$S = A \cup \bar{A};$$

that is, the sample space is the union of the subset A and its complement, $\bar{A}$. Since A and $\bar{A}$ are mutually exclusive, we can apply Axiom 3 to the preceding equation; that is,

$$\mathbf{Pr}(S) = \mathbf{Pr}(A \cup \bar{A}) = \mathbf{Pr}(A) + \mathbf{Pr}(\bar{A}).$$

From Axiom 2 we know that $\mathbf{Pr}(S) = \mathbf{1}$, thus

$$\mathbf{1} = \mathbf{Pr}(A) + \mathbf{Pr}(\bar{A}) \quad or \quad \mathbf{Pr}(\bar{A}) = \mathbf{1} - \mathbf{Pr}(A).$$

Example 5.5.1 Contents of an Urn

Consider an urn that contains 15 red and 10 black balls

$$R = 15$$
$$B = 10$$
$$Total = 25$$

What is the probability that a ball drawn will be black? Thus,

$$\mathbf{Pr}(B) = \frac{n_B}{n} = \frac{\mathbf{10}}{\mathbf{25}} = \frac{\mathbf{2}}{\mathbf{5}}$$

and the probability that the ball drawn is not black is

$$\mathbf{Pr}(\bar{B}) = \mathbf{1} - \mathbf{Pr}(B) = \mathbf{1} - \frac{\mathbf{2}}{\mathbf{5}} = \frac{\mathbf{3}}{\mathbf{5}} = \mathbf{Pr}(R),$$

due to Property 5.4.1.

The following results give a more rigorous reasoning concerning the probability of the empty set (impossible event), $\varnothing$.

Property 5.4.2 Empty Set

If $\varnothing$ is the impossible event (**empty set**) of the sample space S, then

$$\mathbf{Pr}(\varnothing) = \mathbf{0}.$$

We know that $S = S \cup \varnothing$ and that S and $\varnothing$ are mutually exclusive since they do not have any elements in common. Thus, applying Axiom 3 we can write

$$\mathbf{Pr}(S) = \mathbf{Pr}(S \cup \varnothing) = \mathbf{Pr}(S) + \mathbf{Pr}(\varnothing).$$

But we know from Axiom 2 that $\mathbf{Pr}(S) = \mathbf{1}$, Therefore

$$\mathbf{1} = \mathbf{1} + \mathbf{Pr}(\varnothing) \quad or \quad \mathbf{Pr}(\varnothing) = \mathbf{0},$$

Example 5.5.2 Tossing Two Fair Dice

In the experiment consisting of tossing two dice and observing the sum uppermost, what is the probability that the sum showing on the two dice is equal to 13?

Solution

Here, the event in question is impossible (empty) because one can never obtain a sum of 13 in tossing a pair of fair dice. Thus, the desired probability is zero.

Property 5.4.3 Union
_____

If A and B are any two events of the sample space S, then the probability of their union is given by

$$\Pr(A \cup B) = \Pr(A) + \Pr(B) - \Pr(A \cap B).$$

This property says that if the events A and B are not mutually exclusive, then the probability of the union of these events is equal to the sum of the probabilities of each of the events minus the probability of the intersection of the events. The proof of this statement can be reasoned as follows: The sum, $\Pr(A) + \Pr(B)$, contains the sum of the probabilities of all sample points in $A \cup B$ but includes a double counting of the probabilities of all points in the intersection, $A \cap B$. Subtracting the $\Pr(A \cap B)$ gives the correct result. That is, when we calculated the $\Pr(B)$ and the $\Pr(A)$ we used the common points of A and B twice therefore, we must subtract these once to obtain the correct result.

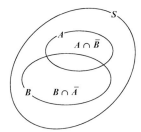

FIGURE 5.3   Venn diagram for Example 5.5.3.

We can also show this analytically. From Figure 5.3, we see that we can write $A \cup B = (A \cap \bar{B}) \cup B$. Note that the events $A \cap \bar{B}$ and B are mutually exclusive events. Thus, applying Axiom 3 we have

$$\Pr(A \cup B) = \Pr(A \cap \bar{B}) + \Pr(B).$$

However, we can write event A as a pair of mutually exclusive events; that is,

$$A = (A \cap B) \cup (A \cap \bar{B})$$

and thus,

$$\Pr(A) = \Pr(A \cap B) + \Pr(A \cap \bar{B}) \text{ or } \Pr(A \cap \bar{B}) = \Pr(A) - \Pr(A \cap B)$$

by Axiom 3, since the events $A \cap B$ and $A \cap \bar{B}$ are mutually exclusive events. Substituting the preceding expression, $\Pr(A \cap \bar{B})$, into the expression for $\Pr(A \cup B)$ we have

$$\Pr(A \cup B) = \Pr(A) + \Pr(B) - \Pr(A \cap B).$$

This expression is called the **additive law of probability**.

If the events A and B are mutually exclusive, then $\Pr(A \cap B) = 0$ because A and B do not have any sample points in common. We then write

$$\Pr(A \cup B) = \Pr(A) + \Pr(B).$$

Thus, we compute the $\Pr(A \cup B)$ when events A and B are not mutually exclusive and when the outcomes are equally likely as follows:

$$\Pr(A \cup B) = \frac{n_A}{n} + \frac{n_B}{n} - \frac{n_{A \cap B}}{n}.$$

Example 5.5.3 Tossing a Fair Die
_____

Consider the experiment of tossing a fair die and observing the number that appears on top. What is the probability of an odd number or prime number? Here the sample space consists of six possible outcomes:

$$S = \{1, 2, 3, 4, 5, 6\}.$$

Continued

Example 5.5.3 Tossing a Fair Die—cont'd

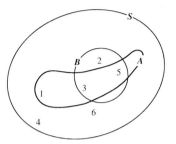

FIGURE 5.4 Venn diagram for Example 5.5.3.

Let A be the event that an odd number occurs and B the event that a prime number occurs:

$$A = \{1, 3, 5\} \quad \text{and} \quad B = \{2, 3, 5\}$$

Then, $A \cap B = \{3, 5\}$ and

$$\mathbf{Pr}(A \cup B) = \mathbf{Pr}(A) + \mathbf{Pr}(B) - \mathbf{Pr}(A \cap B)$$
$$= \frac{n_A}{n} + \frac{n_B}{n} - \frac{n_{A \cap B}}{n}$$
$$= \frac{3}{6} + \frac{3}{6} - \frac{2}{6} = \frac{4}{6} = \frac{2}{3}$$

Thus, there is a 2/3 chance that the number that appears on a single toss of a die will be an odd number or a prime number.

Example 5.5.4 Tossing a Fair Coin ×3

Consider the experiment of tossing a coin three times. What is the probability of two or more heads appearing consecutively or of all the tosses being the same? Here the sample space consists of eight possible outcomes

$$S = \{(H, H, H), (H, H, T), (H, T, H), (H, T, T), (T, H, H),$$
$$(T, H, T), (T, T, H), (T, T, T)\},$$

where H and T stand for heads and tails, respectively, Let A be the event that two or more heads appear consecutively, and B the event that all tosses are the same; that is,

$$A = \{(H, H, H), (H, H, T), (T, H, H)\} \text{ and } B = \{(H, H, H), (T, T, T)\}.$$

Thus, $A \cap B = \{(H, H, H)\}$ and the desired probability is given by

$$\mathbf{Pr}(A \cup B) = \mathbf{Pr}(A) + \mathbf{Pr}(B) - \mathbf{Pr}(A \cap B)$$
$$= \frac{3}{8} + \frac{2}{8} - \frac{1}{8} = \frac{4}{8} = \frac{1}{2}.$$

We shall now reconsider the problem we posed in introducing these basic laws of probability.

Example 5.5.5 Drawing a Card

Consider the experiment of drawing a card from a well-shuffled deck of 52 cards. We would like to find the probability of the following events:
1. The event of drawing a diamond or a picture card (including a picture diamond); and
2. The event of not obtaining a diamond or a picture card; that is, of obtaining; any card other than a picture card or a diamond.

Example 5.5.5 Drawing a Card—cont'd

The sample space consists of 52 sample points corresponding to the 52 possible cards that might be drawn. Let A be the event that the card drawn is a diamond and B the event that the card drawn is a picture card. It is clear that events A and B are not mutually exclusive. For the first part, we must apply Theorem 3:

$$\Pr(A \cup B) = \Pr(A) + \Pr(B) - \Pr(A \cap B)$$

There are 13 diamonds in an ordinary deck of 52 cards; thus, using the classical definition of probability, we have

$$\Pr(A) = \frac{n_A}{n} = \frac{13}{52} = \frac{1}{4}.$$

Similarly, there are 12 picture cards in a deck of cards and

$$\Pr(B) = \frac{n_B}{n} = \frac{12}{52} = \frac{3}{13}.$$

Also, there are three picture cards that are diamonds in such a deck and

$$\Pr(A \cap B) = \frac{n_{A \cap B}}{n} = \frac{3}{52}.$$

Therefore,

$$\Pr(A \cup B) = \frac{13}{52} + \frac{12}{52} - \frac{3}{52} = \frac{22}{52} = \frac{11}{26}.$$

Thus, there is an $\frac{11}{26}$ chance that the drawn card is a picture card or a diamond. For the second part, we simply apply Theorem 1. That is,

$$\Pr(\overline{A \cup B}) = 1 - \Pr(A \cup B)$$

$$= 1 - \frac{11}{26} = \frac{15}{26}$$

Property 5.4.4 Additive Law

The **additive law** of probability can be easily extended to a finite number of events defined on the sample space. For example, let A, B, and C be any three events defined on the sample space S. The probability of the union of these events is given by

$$\Pr(A \cup B \cup C) = \Pr(A) + \Pr(B) + \Pr(C) - \Pr(A \cap B) - \Pr(A \cap C)$$

$$- \Pr(B \cap C) + \Pr(A \cap B \cap C)$$

$$= \frac{n_A}{n} + \frac{n_B}{n} + \frac{n_C}{n} - \frac{n_{A \cap B}}{n} - \frac{n_{A \cap C}}{n} - \frac{n_{B \cap C}}{n} + \frac{n_{A \cap B \cap C}}{n}.$$

The validity of the preceding expression of $\Pr(A \cup B \cup C)$ can be easily demonstrated by means of the accompanying Venn diagram (Figure 5.4) and the following reasoning. In calculating $\Pr(A)$ and $\Pr(B)$ we used the elements that are common to A and B in both cases; thus, we must subtract $\Pr(A \cap B)$ once. Similarly, when, we calculate $\Pr(A)$ and $\Pr(C)$ we must subtract the $\Pr(A \cap B)$ once. Again, for the case when we obtain $\Pr(B)$ and $\Pr(C)$, we must subtract the $\Pr(B \cap C)$ once. However, the probability of the common elements to all three events, $\Pr(A \cap B \cap C)$, has been subtracted more than once and should be added once for the balance.

If the events A, B, and C are mutually exclusive, then according to Axiom 3 we have (Figure 5.5)

$$\Pr(A \cup B \cup C) = \Pr(A) + \Pr(B) + \Pr(C)$$

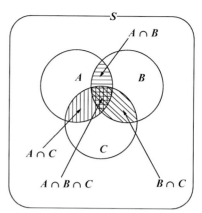

FIGURE 5.5 Venn diagram of event $(A \cup B \cup C)$.

EXERCISES

5.5.1. The results of certain experiments are delineated in the accompanying Venn diagram. Calculate the probability of the following events:
   a. the occurrence of event A,
   b. the occurrence of event C,
   c. the occurrence of event B,
   d. the occurrence of event A or B.

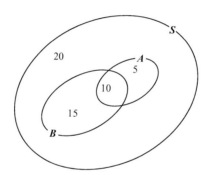

5.5.2. From the experiment: three fair coins are tossed, obtain the probability of the following events:
   a. Two heads appear or all the tosses are the same.
   b. Three heads appear or all the tosses are not the same.
5.5.3. From a single card is drawn from a standard deck of 52 playing cards, calculate the probability of the following events:
   a. Drawing a heart or an ace heart,
   b. Not obtaining a heart or an ace heart.
5.5.4. The probability that a student entering the snack bar at the student union will buy a soda is 75%. The probability that he or she will buy ice cream is 45%. The probability that a student will buy both ice cream and soda is 40%. What is the probability that a student will buy a soda, ice cream, or both?
5.5.5. Consider the experiment of rolling a fair die.
   a. What is the probability of observing an odd number or a number less than four?
   b. What is the probability of observing an even number or a number greater than four?
5.5.6. A box contains 10 operable, 4 defective, and 6 semi-operable electric shavers. Here, *semi-operable* means that the shaver can be classified between operable and defective. In appearance the shavers cannot be distinguished. A shaver is chosen from the box; obtain the probability that it is operable or semi-operable.
5.5.7. The experiment of tossing a coin and a die, determine the probability of the following events:
   a. Heads and an even number appear or a prime number appears.
   b. Tails and an odd number appear or a prime number appears.

5.5.8. In a class of 50 history majors, 10 elected a finite mathematics course, 15 elected a philosophy course, and five elected to take both mathematics and philosophy. A student is chosen from this class. What is the probability that he or she has elected the mathematics or philosophy course?

5.6 CONDITIONAL PROBABILITY

Suppose that from a deck of 52 cards we removed the three picture diamonds (jack, queen, and king), and from the remaining 49 cards selected a card at random. We want to know the probability of the chosen card being a picture card, with the additional condition that we know the card drawn is a spade. That is, we want to know the probability of a certain event under a specific condition that influences the answer to the posed question, problems of this type can be easily handled using the following definition:

5.6.1 Definition Conditional Probability

Let A and B be any two events of the sample space S. We define the **conditional probability** of event A, given that event B has occurred, denoted by $\Pr(A\,|\,B)$, by

$$\Pr(A\,|\,B) = \frac{\Pr(A \cap B)}{\Pr(B)}.$$

where $\Pr(B) > 0$.

That is, the conditional probability of event A, given that event B has occurred, is equal to the probability of the intersection of the events $A \cap B$ divided by the probability of B. Note that $\Pr(B)$ cannot be equal to zero because division of a number by zero is undefined and is thus not permissible. Also, is not meaningful to be asked to obtain the probability of an event under the condition of the impossible event. We can also write

$$\Pr(B\,|\,A) = \frac{\Pr(A \cap B)}{\Pr(A)}.$$

where $\Pr(A) > 0$.

We can easily compute conditional probabilities once the sample space S has been constructed.

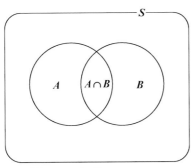

FIGURE 5.6 Venn diagram of the event $A \cap B$.

For example, in Figure 5.6, let the sample space consist of n simple events, n_A of which have the attribute associated with A, n_B of which have the attribute associated with B, and $n_{A \cap B}$ of which have the attributes associated with both A and B. Then, using the classical definition of probability, we have

$$\Pr(A\,|\,B) = \frac{\Pr(A \cap B)}{\Pr(B)} = \frac{\dfrac{n_{A \cap B}}{n}}{\dfrac{n_B}{n}} = \frac{n_{A \cap B}}{n_B}$$

and

$$\Pr(B\,|\,A) = \frac{\Pr(A \cap B)}{\Pr(A)} = \frac{\dfrac{n_{A \cap B}}{n}}{\dfrac{n_A}{n}} = \frac{n_{A \cap B}}{n_A}$$

We shall now answer the questions posed at the beginning of the section.

Example 5.6.1 Drawing a Single Card

Consider the experiment of drawing a single card from a deck of cards that is missing the three picture diamonds. We would like to find the probability of the following events:
1. The chosen card is a picture card, given that this card is a spade.
2. The drawn card is a spade, given that this card is a picture card.

Solution

Here, the sample space S consists of 49 simple events. Let A be the event that the chosen card is a picture card, and let event B denote that the chosen card is a spade; Event B consists of 13 sample points, since there are 13 spades in an ordinary deck of 52 cards. The probability of this event occurring is

$$\Pr(B) = \frac{n_B}{n} = \frac{13}{49}.$$

The event $A \cap B$ consists of three sample points, since there are only three picture cards that are spades. Hence,

$$\Pr(A \cap B) = \frac{n_{A \cap B}}{n} = \frac{3}{49}.$$

For the first part, applying the definition of conditional probability, we have

$$\Pr(A \mid B) = \frac{\Pr(A \cap B)}{\Pr(B)} = \frac{\frac{3}{49}}{\frac{13}{49}} = \frac{3}{13}.$$

Thus, if we draw a single card from such a deck and we are told that this card is a spade, there is a $\frac{3}{13}$ chance that the card selected is a picture card, Similarly, event A consists of nine sample points, since there are 12 picture cards in a 52-card deck and three are missing. Thus,

$$\Pr(A) = \frac{n_A}{n} = \frac{9}{49}.$$

For the second part, we have

$$\Pr(B \mid A) = \frac{\Pr(A \cap B)}{\Pr(A)} = \frac{\frac{3}{49}}{\frac{9}{49}} = \frac{3}{9} = \frac{1}{3}.$$

Note that $\Pr(A \mid B) \neq \Pr(B \mid A)$.

Example 5.6.2 Tossing a Fair Die ×2

Consider the experiment of rolling a pair of fair dice. We would like to know the probability of at least one of the dice showing a 3, given that we know a sum of seven has been observed.

Solution

The sample space of this experiment consists of 36 sample points. Let A be the event that the observed sum is 7 and let B denote the event that a 3 appears on at least one die. Here, event A consists of six sample points:

$$A = \{(1,6), (2,5), (3,4), (4,3), (5,2), (6,1)\}$$

and event B is made up of 11 simple events:

$$B = \{(3,1), (3,2), (3,3), (3,4), (3,5), (3,6)(6,3), (5,3), (4,3), (2,3), (1,3)\}.$$

Thus, the event $A \cap B$ consists of two sample events.

$$A \cap B = \{(3,4), (4,3)\}$$

Solution—cont'd

Applying the definition of conditional probability, we have

$$\mathbf{Pr}(B|A) = \frac{\mathbf{Pr}(A \cap B)}{\mathbf{Pr}(A)} = \frac{n_{A \cap B}}{n_A} = \frac{\frac{2}{36}}{\frac{6}{36}} = \frac{2}{6} = \frac{1}{3}.$$

Therefore, there is a $\frac{1}{3}$ chance of at least one of the dice showing a three if we are told the sum of seven has been observed.

The definition of conditional probability can be written as

$$\mathbf{Pr}(A \cap B) = \mathbf{Pr}(B)\mathbf{Pr}(A|B) = \mathbf{Pr}(A)\mathbf{Pr}(B|A)$$

by simply cross-multiplying. That is, if we are given the conditional probability of a given event along with either the $\mathbf{Pr}(A)$ or $\mathbf{Pr}(B)$, we can obtain the probability of the intersection of the two events, $A \cap B$.

Rule 5.6.1 Multiplicative

Let A and B be any two events on S. The probability of the intersection $A \cap B$ is

$$\mathbf{Pr}(A \cap B) = \mathbf{Pr}(A)\mathbf{Pr}(B|A) = \mathbf{Pr}(B)\mathbf{Pr}(A|B).$$

The **multiplicative law of probability** for a sequence of elements

$$\mathbf{Pr}(A_1 \cap A_2 \cap \ldots \cap A_n)$$
$$= \mathbf{Pr}(A_1)\mathbf{Pr}(A_2|A_1)\mathbf{Pr}(A_3|A_1 \cap A_2) \ldots \mathbf{Pr}(A_n|A_1 \cap A_2 \cap \ldots \cap A_{n-1})$$

Thus, for $n=3$; that is, three events, A_1, A_2, A_3, the multiplicative law is written

$$\mathbf{Pr}(A_1 \cap A_2 \cap A_3) = \mathbf{Pr}(A_1)\mathbf{Pr}(A_2|A_1)\mathbf{Pr}(A_3|A_1 \cap B_2)$$

Example 5.6.3 Effects of a Drug

A biochemist administers a particular drug to 40 white mice. From previous experience he knows that 15% or six of these mice will not be affected by the drug. Three of the 40 mice were selected at random by the biochemist to be used in a new experiment. We are interested in the probability that all three have been affected by the drug.

Solution

Let A_1, A_2, and A_3 be the events that the first, second, and third mice have been affected by the drug. Let n_A be the number of mice affected at the beginning of the experiment.
Here, the probability that the first mouse has been affected is

$$\mathbf{Pr}(A_1) = \frac{n_A}{n} = \frac{34}{40}.$$

The probability that the second mouse has been affected by the drug, given that the first has been affected, is

$$\mathbf{Pr}(A_2|A_1) = \frac{n_A - 1}{n - 1} = \frac{33}{39}.$$

That is, since we know that the first mouse selected has been affected by the drug there are only $40 - 1 = 39$ left to choose from; also, there were originally $40 - 6 = 34$ mice affected by the drug, and one has already been chosen, which leaves us with $34 - 1 = 33$ mice from which to choose the second mouse. Similarly, the probability that the third mouse has been affected by the drug, given that the first two have been affected, is given by

$$\mathbf{Pr}(A_3|A_1 \cap A_2) = \frac{n_A - 2}{n - 2} = \frac{32}{38}.$$

Continued

Solution—cont'd

Thus, applying the multiplicative law we obtain the probability of interest, $\Pr(A_1 \cap A_2 \cap A_3)$; that is,

$$\Pr(A_1 \cap A_2 \cap A_3) = \Pr(A_1)\Pr(A_2|A_1)\Pr(A_3|A_1 \cap A_2)$$

$$= \left(\frac{34}{40}\right)\left(\frac{33}{39}\right)\left(\frac{32}{38}\right)$$

$$= (0.85)(0.8462)(0.8421) = 0.606$$

Therefore, there is a 60.6% chance that all three mice selected have been affected by the drug.

5.6.2 Definition Independent Events

Let A and B be any two events of the sample space S. The event A is said to be **independent** of event B if

$$\Pr(A \cap B) = \Pr(A)\Pr(B)$$

That is, the probability of the intersection of the events $A \cap B$ is equal to the product of the probabilities of A and B.

When we say "two events are independent," we mean the occurrence or nonoccurrence of one event has no effect on the probability of the occurrence of the other. Likewise when we ask for the probability of some event A, given event B, we insinuate that we have information that B has occurred and now wish to know the probability that A will occur based on this knowledge. If A and B are independent, then it makes sense to assume that the probability of the occurrence of A is not affected by this knowledge.

Note that from the definition of conditional probability, if A and B are independent events, then we have

$$\Pr(A|B) = \frac{\Pr(A \cap B)}{\Pr(B)} = \frac{\Pr(A)\Pr(B)}{\Pr(B)} = \Pr(A),$$

That is, the probability of the occurrence of event A is not affected by the fact that event B has occurred.

Example 5.6.4 Testing a Specific Drug

Consider the experiment in which two individuals are given a specific drug. Let A and B represent these individuals. It has been previously observed that the probabilities that this drug will affect A and B are $\frac{1}{3}$ and $\frac{1}{4}$, respectively. What is the probability this drug will affect both A and B?

Solution

Here, we have $\Pr(A) = \frac{1}{3}$ and $\Pr(B) = \frac{1}{4}$. The probability that A is affected by the drug is not influenced by what happens to B. Thus, the behavior of A is independent of that of B, and the probability that the drug will affect both of them is

$$\Pr(A \cap B) = \Pr(A)\Pr(B) = \left(\frac{1}{3}\right)\left(\frac{1}{4}\right) = \frac{1}{12}.$$

We frequently want to find the conditional probability of certain event, *A*, given that a different event, *B*, has occurred at some prior time. For example, we might want to know the probability of our favorite football team winning on a specific Saturday, given that the team has won its games the preceding two Saturdays. Or, we might be interested in knowing the probability of a student passing the final examination in mathematics, given that he or she has passed all previous examinations. Answering questions of this type requires an application of the multiplicative law derived by Thomas Bayes, which we shall study in the next section.

EXERCISES

5.6.1. In the experiment: a single card is drawn form a standard deck of playing cards; calculate the probability of the following events:
  a. The selected card is an ace of diamonds, given that this card is a diamond.
  b. The drawn card is a heart when we have been told that this card is a picture card.

5.6.2. In the experiment: two fair dice are tossed and the sum appearing uppermost is observed; what is the probability that one of the dice is 2 if the sum is 6?

5.6.3. The probability that it will rain in Tampa, *T*, and Clearwater, *C*, Florida during any 24-h period was given by the Florida weather station to be 0.20 and 0.15, respectively. That is, $\mathbf{Pr}(T) = .20$ and $\mathbf{Pr}(C) = .15$. It was also given that the probability that it will rain simultaneously at both cities during a 24-h period is 0.08. Determine the probability of the following events:
  a. It will rain (or is raining) in Tampa, given that it is raining in Clearwater.
  b. It will rain in Clearwater, given that it is not raining in Tampa.
  c. It will not rain in Tampa, given that it is not raining in Clearwater.
  d. It will rain in both Tampa and Clearwater.

5.6.4. Consider the experiment of tossing a fair die. What is the probability of observing an odd number, given that the number that has appeared is greater than four?

5.6.5. The dean's advisory committee at a certain university consists of seven members of which two are students. A subcommittee of two members is to be selected from the advisory committee to study a particular problem. If the two individuals are selected at random, what is the probability that the students will be selected?

5.6.6. From Problem 5.6.1,
  a. What is the probability that the chosen card is an ace of diamonds?
  b. What is the probability that the drawn card is a picture heart?

5.6.7. Refer to Problem 5.6.3 and let A_1 and A_2 denote the events of the occurrence of rain in Tampa and Clearwater, respectively. Are A_1 and A_2 independent?

5.6.8. Two individuals, A_1 and A_2, are shooting at a target. The probabilities that A_1 and A_2 hit the target are 0.30 and 0.40, respectively. What is the probability that the target will be hit if A_1 and A_2 each shoot at the target?

5.6.9. The English majors at a certain university revealed the following information: 10% of them failed mathematics, 20% failed biology, and 5% failed both mathematics and biology. Determine the probability of the following events:
  a. An English major failed mathematics, given that he or she passed biology.
  b. The student passed biology, given that he or she passed mathematics.
  c. The student passed mathematics and it is known that he or she failed biology.

5.6.10. A secondary education class at a certain college consists of 24 students. Sixteen of them are females and eight are males. Three students are selected by the instructor to debate a certain issue. What is the probability that all three of the students are females? What is the meaning of the complement of this event?

5.6.11. A box contains two new and one used portable radios. The radios are numbered 1, 2, and 3, respectively. A second box contains only one new radio. Consider the following experiment: A radio is chosen from the first box and placed in the second box. Without being able to make a visual distinction between the two radios in the second box, a radio was chosen. What is the probability that the radio chosen from the second box is new? Furthermore, answer the same question if the radios were not numbered.

5.7 BAYES' RULE

Consider that in a department store we have two large boxes, each of which contains 50 portable electric hair dryers. We have been told that the first box contains 44 operable and six defective dryers and that the second box contains 46 operable and four defective dryers. We purchased a dryer from this store and we would like to know the probability that this dryer will be operable. The problem here is that we have not been told from which box the purchased dryer was selected. Of equal interest would be the following situation: A physician purchased a certain drug from two different companies, say, **A** and **B**. There is a slight variation in the effectiveness of the two preparations. The physician administered one of the drugs when treating a particular patient. The patient's condition has improved remarkably and we would like to know which of the two preparations the physician used. That is, assuming that the probability of the doctor choosing one batch or the other is the same, we wish to know the probability that the patient was helped by the drug from Company A. Questions of this type, which are quite common, would be answered using *Bayes' formula*.

Suppose, for instance, that we are concerned with the establishment of a certain committee within the student senate. The approval or disapproval of this committee is deferred until after the election of the president of the senate. Assume, that there are two candidates, say **X** and **Y**, running for the presidency and the probabilities that they will be elected are 0.35 and 0.65, respectively. Furthermore, if Candidate **X** is elected the probability that the committee will be established is 0.70 and if Candidate **Y** is elected the probability is only 0.30. The question of interest is: What is the probability that the committee will be established?

Let **A** represent the approval of the committee and let **B** be the event that Candidate **X** was elected. Here, we can write event **A** as the union of two mutually exclusive sets:

$$A = (A \cap B) \cup (A \cap \bar{B}),$$

as shown by the Venn diagram in Figure 5.7. Thus, applying Axiom 3 we have

$$\mathbf{Pr}(A) = [(A \cap B) \cup (A \cap \bar{B})]$$
$$= \mathbf{Pr}(A \cap B) + \mathbf{Pr}(A \cap \bar{B})$$

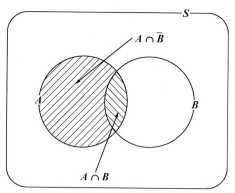

FIGURE 5.7 Venn diagram of event *A*.

Now, using the conditional definition of probability, we can write the preceding expression as

$$\mathbf{Pr}(A) = \mathbf{Pr}(B)\mathbf{Pr}(A \mid B) + \mathbf{Pr}(\bar{B})\mathbf{Pr}(A \mid \bar{B}).$$

We know that $\mathbf{Pr}(B)\mathbf{=0.35}$, which implies that $\mathbf{Pr}(\bar{B})\mathbf{=1-Pr}(B)\mathbf{=}$ $\mathbf{1-0.35=0.65}$, and that $\mathbf{Pr}(A\,|\,B)\mathbf{=0.70}$ and $\mathbf{Pr}(A\,|\,\bar{B})\mathbf{=0.30}$. Therefore, substituting these probabilities in our formula, we have

$$\mathbf{Pr}(A)\mathbf{=(0.35)(0.70)+(0.65)(0.30)}$$
$$\mathbf{=0.245+0.195=0.44}$$

Hence, the probability that the committee will be established under the given circumstances is 0.44.

A convenient way of describing the probability of the event A, $\mathbf{Pr}(A)$, is by the use of a *tree diagram*. The tree diagram for the preceding expression is shown in Figure 5.8.

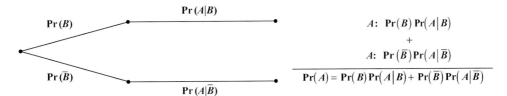

FIGURE 5.8   Tree diagram.

The branches of the tree show only those possible outcomes which lead to event A. The probability for each branch is obtained by multiplying the probabilities of the two stages and, finally, the $\mathbf{Pr}(A)$ is given by the sum of the probabilities of the two branches leading to A. This expression is sometimes called the **rule of elimination** and its general form is given below:

Rule 1 Elimination

Let $A_1, A_2, ..., A_n$ be mutually exclusive events of the sample space S, such that $A_1 \cup A_2 \cup ... \cup A_n = S$; and it is given that $\mathbf{Pr}(A_1) > 0$ for $i = 1, 2, ..., n$, Then for any event A we have

$$\mathbf{Pr}(A)\mathbf{=Pr}(A_1)\mathbf{Pr}(A\,|\,A_1)\mathbf{+Pr}(A_2)\mathbf{Pr}(A\,|\,A_2)\mathbf{+...+Pr}(A_n)\mathbf{Pr}(A\,|\,A_n)$$

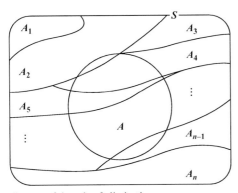

FIGURE 5.9   Venn diagram of the rule of elimination.

Note from Figure 5.9 that event A overlaps the mutually exclusive events A_i, $i = 1, 2, ..., n$ which constitute S and can be written as

$$A=(A\cap A_1)\cup(A\cap A_2)\cup...\cup(A\cap A_n),$$

where the events $A \cap A_i$ are mutually exclusive. Thus, applying Axiom 3 and then the definition of conditional probability, we have the rule of elimination. The rule of elimination may be visualized using a tree diagram, as shown in Figure 5.10.

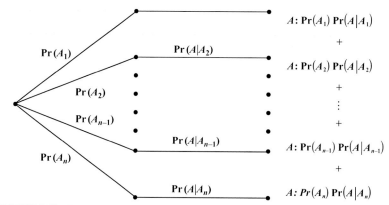

FIGURE 5.10 Tree diagram of the rule of elimination.

Example 5.7.1 Cutting Methods

A dress manufacturer uses three different cutting methods in his operation; let A_1, A_2, and A_3 represent these methods which produce respectively 40%, 35%, and 25% of the total number of dresses in the factory. The percentages of imperfectly sized dresses cut by these methods are 5%, 4%, and 3%, respectively. Suppose that a dress produced by this manufacturer was selected at random; what is the probability that the dress is sized imperfectly?

Solution

Let A be the event that the dress chosen is imperfect. We are given that $\mathbf{Pr}(A_1)=\mathbf{0.40}$, $\mathbf{Pr}(A_2)=\mathbf{0.35}$, and $\mathbf{Pr}(A_3)=\mathbf{0.25}$. Also, $\mathbf{Pr}(A|A_1)=\mathbf{0.05}$, $\mathbf{Pr}(A|A_2)=\mathbf{0.04}$, and $\mathbf{Pr}(A|A_3)=\mathbf{0.03}$. That is, the probability that the dress is imperfect, given that it was produced using method A_1, is 0.05; the probability that the dress is imperfect, given that it was made using method A_2, is 0.04; and so on. Thus, using the rule of elimination we have

$$\mathbf{Pr}(A)=\mathbf{Pr}(A_1)\mathbf{Pr}(A|A_1)+\mathbf{Pr}(A_2)\mathbf{Pr}(A|A_2)+\mathbf{Pr}(A_3)\mathbf{Pr}(A|A_3)$$

$$=(\mathbf{0.40})(\mathbf{0.05})+(\mathbf{0.35})(\mathbf{0.04})+(\mathbf{0.25})(\mathbf{0.03})=\mathbf{0.0415}$$

Hence, there is a 4.15% chance that the chosen dress is imperfect. This result is shown using the tree diagram in Figure 5.11.

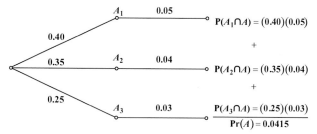

FIGURE 5.11 Tree diagram of Example 5.7.1.

To introduce another important law of probability; namely, **Bayes' formula**, suppose that in the preceding example we want to know the probability that the imperfect dress that was selected was produced by method A_2. That is, if the dress was selected at random and found to be imperfect, what is the probability that the dress was produced by method A_2? We assume here that each dress produced by the three methods had an equal chance of being selected. Thus, we must obtain the $\mathbf{Pr}(A_2|A)$,

Solution—cont'd

the probability that the chosen imperfect dress was produced using method A_2. Using the definition of conditional probability, we have

$$\Pr(A_2|A) = \frac{\Pr(A \cap A_2)}{\Pr(A)}.$$

The numerator, $\Pr(A \cap A_2)$, can be written

$$\Pr(A \cap A_2) = \Pr(A_2)\Pr(A|A_2).$$

We thus obtain the formula

$$\Pr(A_2|A) = \frac{\Pr(A_2)\Pr(A|A_2)}{\Pr(A)}.$$

In the example, we were given that $\Pr(A_2) = 0.35$ and $\Pr(A|A_2) = 0.04$. Also, we calculated the probability that the chosen dress was imperfect to be

$$\Pr(A) = 0.0415.$$

Substituting these probabilities in the formula for $\Pr(A_2|A)$, we have

$$\Pr(A_2|A) = \frac{(0.35)(0.04)}{0.0415} = 0.3374$$

Hence, the probability that the chosen imperfect dress was produced using method A_2 is 0.3374. Similarly, the probabilities that the imperfect dress was made by methods A_1 and A_3 are given by

$$\Pr(A_1|A) = \frac{\Pr(A_1)\Pr(A|A_1)}{\Pr(A)} = \frac{(0.40)(0.05)}{(0.0415)} = 0.4819$$

and

$$\Pr(A_3|A) = \frac{\Pr(A_3)\Pr(A|A_3)}{\Pr(A)} = \frac{(0.25)(0.03)}{(0.0415)} = 0.1807,$$

respectively. That is, the probabilities are 48.19% and 18.07% that the imperfect dress was made using methods A_1 and A_2, respectively.

These results can also be obtained from the tree diagram shown in Figure 5.10. The $\Pr(A_2|A)$ is simply the *ratio* of the probability of reaching event A along the branch passing through A_2 to the sum of the probabilities of reaching event A through all three branches of the tree. Similarly, we can obtain from the tree diagram the probabilities the imperfect dress selected was made using methods A_1 or A_3.

The formulas that we used to find the necessary probabilities of the posed problem are special cases of "Bayes" rule:

Bayes' rule: Let $A_1, A_2, \ldots, A_n$, be mutually exclusive events, such that $A_1 \cup A_2 \cup \ldots \cup A_n = S$, and let A be any other event of the sample space S. If $\Pr(A) > 0$, then for every $i = 1, 2, \ldots, n$ we have

$$\Pr(A_2|A) = \frac{\Pr(A_i)\Pr(A|A_i)}{\Pr(A_1)\Pr(A|A_1) + \Pr(A_2)\Pr(A|A_2) + \ldots + \Pr(A_n)\Pr(A|A_n)}$$

$$= \frac{\Pr(A_i)\Pr(A|A_i)}{\Pr(A)}$$

It is clear that $\Pr(A)$ is the formula we obtained for the rule of elimination and the numerator was obtained using the definition of conditional probability,

$$\Pr(A|A_i) = \frac{\Pr(A \cap A_i)}{\Pr(A_i)}, \quad i = 1, 2, \ldots, n$$

or

$$\Pr(A|A_i)\Pr(A_i) = \Pr(A \cap A_i).$$

We proceed to illustrate Bayes' rule returning to the problem at the beginning of this section.

Example 5.7.2 Operable Hair Dryers

Suppose that in a department store we have two large boxes, each of which contains 50 portable electric hair dryers. It is known that the first box contains 44 operable and six defective dryers and the second box contains 46 operable and four defective dryers. We are interested in obtaining the probability that a dryer purchased from this store will be operable. Also, if it is given that the dryer will be operable, we wish to find the probability that the dryer was chosen from Box 1. We shall assume that the probability of making a selection from each of the boxes is the same.

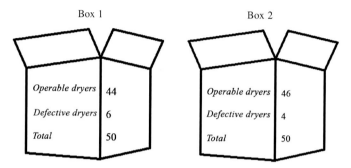

It is clear that we do not know from which box the purchased dryer was selected. If we had the information, for example, that the dryer was taken from Box 2, it would be easy to calculate the desired probability. That is, using the classical definition of probability, the desired probability would be simply $\frac{46}{50}$ in that case.

Solution

Let A_1 and A_2 be the events of selecting the first and second boxes, respectively; and let A be the event that an operable dryer was chosen. The first part of the problem can be solved using the elimination rule; that is, the event of selecting an operable hair dryer, A, can occur either with A_1 (Box 1) or A_2 (Box 2); that is

$$A = (A \cap A_1) \cup (A \cap A_2).$$

Since, the events $A \cap A_1$ and $A \cap A_2$ are mutually exclusive, applying Axiom 3, we have,

$$\Pr(A) = \Pr(A \cap A_1) + \Pr(A \cap A_2).$$

Using the definition of conditional probability, we can write

$$\Pr(A \cap A_1) = \Pr(A_1)\Pr(A \mid A_1)$$

and

$$\Pr(A \cap A_2) = \Pr(A_2)\Pr(A \mid A_2).$$

Thus,

$$\Pr(A) = \Pr(A_1)\Pr(A \mid A_1) + \Pr(A) = \Pr(A_2)\Pr(A \mid A_2)$$

the elimination rule we obtained previously. Since the probability of making a selection from each of the boxes is the same, we have

$$\Pr(A_1) = \Pr(A_2) = \frac{1}{2}.$$

Solution—cont'd

Also

$$\mathbf{Pr}(A\,|A_1)=\frac{n_0}{n}=\frac{44}{50}\ \text{and}\ \mathbf{Pr}(A\,|A_2)=\frac{n_0}{n}=\frac{46}{50}.$$

Hence, using the elimination rule, we obtain

$$\mathbf{Pr}(A)=\left(\frac{1}{2}\right)\left(\frac{44}{50}\right)+\left(\frac{1}{2}\right)\left(\frac{46}{50}\right)=\frac{22}{50}+\frac{23}{50}=\frac{45}{50}=\frac{9}{10}.$$

the probability that the purchased dryer is operable. For the second question, the probability that the operable dryer chosen was taken from the first box, $\mathbf{Pr}(A_1\,|A)$, is obtained using Bayes' rule. That is,

$$\mathbf{Pr}(A_1\,|A)=\frac{\mathbf{Pr}(A_1)\mathbf{Pr}(A\,|A_1)}{\mathbf{Pr}(A_1)\mathbf{Pr}(A\,|A_1)+\mathbf{Pr}(A_2)\mathbf{Pr}(A\,|A_2)}$$

$$=\frac{\mathbf{Pr}(A_1)\mathbf{Pr}(A\,|A_1)}{\mathbf{Pr}(A)}$$

$$=\frac{\left(\frac{1}{2}\right)\left(\frac{44}{50}\right)}{\frac{9}{10}}=\frac{\frac{22}{50}}{\frac{9}{10}}=\left(\frac{22}{50}\right)\left(\frac{9}{10}\right)=\frac{22}{45}.$$

Therefore, there is a $\frac{22}{45}=\mathbf{48.89\%}$ chance that the chosen operable hair dryer was selected from Box 1. Similarly, the probability that the chosen operable dryer was taken from Box 2 is given by

$$\mathbf{Pr}(A_2\,|A)=\frac{\mathbf{Pr}(A_2)\mathbf{Pr}(A\,|A_2)}{\mathbf{Pr}(A)}$$

$$=\frac{\left(\frac{1}{2}\right)\left(\frac{46}{50}\right)}{\frac{9}{10}}=\frac{\frac{23}{50}}{\frac{9}{10}}=\left(\frac{23}{50}\right)\left(\frac{10}{9}\right)=\frac{23}{45}.$$

Note that we are certain that the dryer was selected from one of the two boxes: That is

$$\mathbf{Pr}(A_1\,|A)+\mathbf{Pr}(A_2\,|A)=\frac{22}{45}+\frac{23}{45}=\mathbf{1},$$

the sure event.

The tree diagram of the preceding example is shown by Figure 5.12.

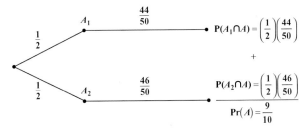

FIGURE 5.12   Tree diagram of Example 5.7.2.

The $\mathbf{Pr}(A_1\,|A)$ is simply the ratio of the probability of reaching event A along the branch passing through A_1 to the sum of the probabilities of reaching A through the branches A_1 and A_2. Similarly, we can obtain the $\mathbf{Pr}(A_2\,|A)$.

Example 5.7.3 Testing the Probability

In the testing center of a certain university, the probability that a student completed all the necessary courses prior to taking a standard examination is $\frac{2}{3}$. The probability that a student will pass the examination having taken all the required courses is $\frac{3}{4}$. The probability that he or she will pass the examination without having completed the required courses is only $\frac{1}{4}$. Suppose that a student successfully completed this standard examination. What is the probability that the student completed all the required courses prior to taking the examination?

Solution

Let A_1 and A_2 represent the events that a student completed and did not complete all the required courses prior to taking the examination, respectively; and let A be the event that a student passed the examination. Here we want the $\Pr(A_1|A)$; that is, using Bayes' Rule, we have

$$\Pr(A_1|A) = \frac{\Pr(A_1)\Pr(A|A_1)}{\Pr(A_1)\Pr(A|A_1) + \Pr(A_2)\Pr(A|A_2)}.$$

We are given that $\Pr(A_1) = \frac{2}{3}$, $\Pr(A_2) = 1 - \Pr(A_1) = 1 - \frac{2}{3} = \frac{1}{3}$, $\Pr(A|A_1) = \frac{3}{4}$, and $\Pr(A|A_2) = \frac{1}{4}$. Thus,

$$\Pr(A|A_2) = \frac{\left(\frac{2}{3}\right)\left(\frac{3}{4}\right)}{\left(\frac{2}{3}\right)\left(\frac{3}{4}\right) + \left(\frac{1}{3}\right)\left(\frac{1}{4}\right)} = \frac{\frac{1}{2}}{\frac{1}{2} + \frac{1}{12}} = \left(\frac{1}{2}\right)\left(\frac{12}{7}\right) = \frac{6}{7}.$$

Therefore, the probability is $\frac{6}{7} = 85.7\%$ that the student who passed the examination completed all the required courses. Similarly, the probability that the student who passed the examination did not complete all the required courses is

$$\Pr(A_2|A) = \frac{\Pr(A_2)\Pr(A|A_2)}{\Pr(A_1)\Pr(A|A_1) + \Pr(A_2)\Pr(A|A_2)}$$

$$= \frac{\left(\frac{1}{3}\right)\left(\frac{1}{4}\right)}{\left(\frac{2}{3}\right)\left(\frac{3}{4}\right) + \left(\frac{1}{3}\right)\left(\frac{1}{4}\right)}$$

$$= \frac{\frac{1}{12}}{\frac{1}{2} + \frac{1}{12}} = \left(\frac{1}{12}\right)\left(\frac{12}{7}\right) = \frac{1}{7}.$$

There is only a $\frac{1}{7} = 14.3\%$ chance that the student who passed the examination did not complete all the required courses.

EXERCISES

5.7.1. In Example 5.7.1, what is the probability that the selected dress is correctly sized?

5.7.2. A cola bottling plant employs three different methods to inspect their product prior to packaging for distribution. The first method is used to inspect 45% of the product, and the second and third methods 30% and 25%, respectively. It has been estimated, however, that the first method fails to detect defective bottles 6% of the time, while the second and third methods fail in the inspection only 3% and 2%, respectively. When a single bottle of this cola is bought by a customer,

   a. What is the probability that the bottle is defective?

b. What is the probability that the bottle is not defective?

c. Draw tree diagrams to illustrate (a) and (b).

5.7.3. Consider that we have in a truck leaving an electronics factory two large boxes, each of which contains 30 walkie-talkies. From previous experience it is known that the first box contains 26 walkie-talkies in excellent condition and four in poor condition. The second box contains 28 walkie-talkies in excellent condition and two in poor condition. A walkie-talkie was purchased. Assume that the probability of making a selection from each of the boxes is the same.

a. What is the probability that the purchased walkie-talkie is in excellent condition?

b. What is the probability that it is in poor condition?

c. Draw a tree diagram to illustrate the answers to (a) and (b).

5.7.4. Three different treatments are being employed by a large research hospital to cure a certain disease. We shall refer to these treatments as A_1, A_2 and A_3. Laboratory tests have so far revealed that treatment A_1 is effective ten out of 12 times; treatment A_2, 9 out of 12 times; treatment A_3, 8 out of 12 times. At the present stage of the study no one treatment is being given preference and each of the three is equally likely to be selected. A patient being treated by one of these methods asks the physician, "What is the probability that I will be cured"? Draw a tree diagram to illustrate the solution to the problem.

5.7.5. In Example 5.7.1, what is the probability that the imperfectly sized dress was cut by the first method of operation?

5.7.6. Example 5.7.2, what is the probability that the selected dryer is not operable?

a. What is the probability that the defective dryer was chosen from the first box?

b. Draw a tree diagram to illustrate the solution to the problem.

5.7.7. In Problem 5.7.2,

a. What is the probability that the non-defective bottle was inspected by the first method?

b. What is the probability that the defective bottle was inspected by the second method?

c. Draw a tree diagram to visualize the solution of the problem.

5.7.8. In Problem 5.7.3,

a. What is the probability that a purchased walkie-talkie in excellent condition was chosen from the second box?

b. What is the probability that it was chosen from the first box?

c. Draw a tree diagram illustrating (a) and (b) of the problem.

5.7.9. In Problem 5.7.4,

a. Given that the patient was cured, determine the probability that treatment A_1 was used.

b. Given that the patient was cured, determine the probability that treatment A_2 was used.

5.8 COMBINATORIAL TECHNIQUES

As we have seen throughout our discussion, the basis for obtaining probabilities of various events defined on the sample space depends primarily on the number of sample points or observations that constitute S. For example, if we know that in a given problem, the sample space was made up of n equally likely observations and n_A of them possess characteristic A or are favorable to event A, then the probability that event A will occur is given by

$$\Pr(A) = \frac{n_A}{n}.$$

Thus, it is important to know various counting techniques in order to be able to obtain the total number of observations that constitute the sample space.

In this section we shall introduce five basic rules for counting the sample points of S, Some of these counting techniques have been previously introduced; however, to enhance your understanding of the role they play in probability, we shall repeat certain ones. The usefulness of each rule will be illustrated by considering a number of examples.

Recall from Chapter 4 the main rules in counting: Multiplication Principle, Multiple Selection, Permutations, Combinations, and Partial Ordering.

Example 5.8.1 Rolling a Fair Die ×3

Consider the experiment of rolling a fair die three times. How many sample points are associated with this problem? What is the probability that the same number will appear in all three rolls?

Solution

Here, N is equivalent to the number of faces on the die and k the number of times we roll the die. Thus, according to **multiple selection rule**, the sample space S consists of

$$N^k = 6^3 = (6)(6)(6) = 216 \, sample \, points.$$

Hence, the sample space consists of 216 sample points. Let A be the event that the same number appears on each of the three rolls. There are six sample points on the sample space which are favorable to event A; that is,

$$A = \{(1, 1, 1), (2, 2, 2), (3, 3, 3), (4, 4, 4), (5, 5, 5), (6, 6, 6)\}.$$

Thus,

$$\Pr(A) = \frac{n_A}{n} = \frac{6}{216} = \frac{1}{36}.$$

Example 5.8.2 Defective Radios

A department store receives a shipment of 50 radios. It is known from previous experience that 10% of these radios will be defective. Ten radios were sold over a certain period of time.
1. What is the probability that all ten radios sold were operable?
2. What is the probability that half of the radios sold were operable and half defective?

Solution

Here, the ten radios sold will constitute our random sample. We assume that the radios were not tested prior to being taken by the customer. The sample space consists of

$$\binom{n}{k} = \binom{50}{10} = \frac{50!}{10!(50-40)!}$$

sample points.

For the first part of our problem, let A be the event that all ten radios sold are operable. There are $50-5=45$ operable radios in the shipment. Thus, the total number of sample points in S that are favorable to event A is given by $\binom{45}{10}$.

Therefore,

$$\Pr(A) = \frac{n_A}{n} = \frac{\binom{45}{10}}{\binom{50}{10}} = \frac{\frac{45!}{10!(45-10)!}}{\frac{50!}{10!(50-10)!}} = 0.311$$

Thus, there is only a 31.1% chance that all ten radios sold are operable.

For the second part, let B be the event that five of the radios sold are defective and five are operable. The number of ways of drawing five operable radios out of the 45 is given by $\binom{45}{5}$; the number of ways of drawing five defective radios out of the five defective is given by $\binom{5}{5}$. The total number of ways of achieving event B is given by applying the **multiplication principle**; that is

$$\binom{45}{5}\binom{5}{5}.$$

Solution—cont'd

Thus,

$$\Pr(B) = \frac{n_B}{n} = \frac{\binom{45}{5}\binom{5}{5}}{\binom{50}{10}}.$$

Note that the size of the sample space remains the same for both parts.

Example 5.8.3 Seven Random Cards

Suppose that seven cards are selected at random from an ordinary deck of 52 cards. We wish to obtain the probability that two of the cards are aces, one is a nine, one is a ten, one is a jack, one is a queen, and one is a king.

Solution

The sample space consists of $(n=52, k=7)$

$$\binom{n}{k} = \binom{52}{7}$$

sample points. Let A be the event that the seven cards drawn consist of two aces, one nine, one ten, one jack, one queen, and one king. There are four aces in an ordinary deck of cards and the number of ways of obtaining two aces is $\binom{4}{2}$. The number of ways of obtaining one nine, one ten, one jack, one queen, or one king is $\binom{4}{1}$. Thus, the total number of possible ways of achieving event A; that is, the total number of sample points in the sample space that are favorable to event A, is given by

$$\binom{4}{2}\binom{4}{1}\binom{4}{1}\binom{4}{1}\binom{4}{1}\binom{4}{1}.$$

Therefore

$$\Pr(A) = \frac{n_A}{n} = \frac{\binom{4}{2}\binom{4}{1}\binom{4}{1}\binom{4}{1}\binom{4}{1}\binom{4}{1}}{\binom{52}{7}} = 0.0000459$$

Hence, there is a 0.00459% chance that we will receive the desired combination of cards if seven cards are drawn.

Example 5.8.4 Defective Can Openers

A large box contains 30 electric can openers numbered 1 to 30. The can openers are drawn at random, one by one, and are divided among five customers. The manager of the store knows that the electric openers numbered 2, 7, 11, 17, and 21 are defective.
1. What is the probability that each customer will receive one defective can opener?
2. What is the probability that one customer will receive all five defective can openers?

Solution

The sample space according to Rule 5 for $k=30$ and $k_1=6, k_2=6, k_3=6, k_4=6, k_5=6$ consists of

$$\binom{k}{k_1, k_2, k_3, k_4, k_5} = \binom{30}{6, 6, 6, 6, 6} = \frac{30!}{6!\, 6!\, 6!\, 6!\, 6!}$$

sample points.

For the first part of the problem, let A be the event that each customer will receive one defective can opener. The total number of sample points favorable to event A can be reasoned as follows: There are five defective can openers which can be distributed to the five customers in

$$\binom{5}{1, 1, 1, 1, 1} = \frac{5!}{1!\, 1!\, 1!\, 1!\, 1!} = 120$$

ways, according to Rule 5, so that each customer will have one such can opener.

The remaining $30-5=25$ can openers can be divided among the five customers in

$$\binom{25}{5, 5, 5, 5, 5} = \frac{25!}{5!\, 5!\, 5!\, 5!\, 5!}$$

ways. Thus, according to Rule 1, there are

$$\binom{5}{1, 1, 1, 1, 1}\binom{25}{5, 5, 5, 5, 5}$$

sample points that are favorable to event A. Therefore

$$\Pr(A) = \frac{n_A}{n} = \frac{\binom{5}{1, 1, 1, 1, 1}\binom{25}{5, 5, 5, 5, 5}}{\binom{30}{6, 6, 6, 6, 6}}.$$

For the second part, let B be the event that all five defective can openers will be drawn by one customer. It is clear that the size of the sample space remains the same as in the first part. The total number of sample points that are favorable to event B can be reasoned as follows: The five defective can openers can be distributed to the five customers so that one of them will have all five defective can openers; that is, $\binom{5}{5}=1$. But, any one of the five customers can have all five defective openers; that is $\binom{5}{1}=5$.

The remaining $30-5=25$ can openers can be divided among the five customers in

$$\binom{25}{1, 6, 6, 6, 6} = \frac{25!}{1!\, 6!\, 6!\, 6!\, 6!}$$

ways. Thus, according to Rule 1 we have

$$\binom{5}{5}\binom{5}{1}\binom{25}{1, 6, 6, 6, 6}$$

sample points that are favorable to event B. Therefore, the probability that a single customer will receive all five defective can openers is

$$\Pr(B) = \frac{\binom{5}{5}\binom{5}{1}\binom{25}{1, 6, 6, 6, 6}}{\binom{30}{6, 6, 6, 6, 6}}$$

EXERCISES

5.8.1. Consider the experiment of casting four fair dice. How many sample points are associated with this problem?

5.8.2. Consider the experiment of tossing three fair coins. How many sample points are in the sample space?

5.8.3. In Problem 5.8.1, what is the probability that the same number will be observed on all four fair dice?

5.8.4. In Problem 5.8.2.

5.8.5. What is the probability that we observed three heads?

5.8.6. What is the probability of the complement of (a)?

5.8.7. Five basketball players, each a different height, are to be given a physical examination. If the players are called into the examining room one at a time, what is the probability that they will be called in order of decreasing height?

5.8.8. A debating class consists of ten students of varying ability. If the debating coach is to select six students at random, what is the chance that exactly four of the best six students would be among those selected?

5.8.9. Consider the experiment of drawing five cards from an ordinary deck of 52 cards. What is the probability that two of the cards are aces, two kings, and one a queen?

5.8.10. In Example 5.8.4, what is the probability that one customer will receive four defective can openers?

CRITICAL THINKING AND BASIC EXERCISE

5.1. A bag contains eight black balls and four white balls. A ball is drawn at random. What is the probability that it is white?

5.2. In a single throw of two dice, what is the probability of the following events:
   a. A total of nine is observed?
   b. A total other than nine is observed?

5.3. A philosophy class consists of 10 seniors, 20 juniors, and five sophomores. A student is chosen at random by the instructor to discuss a particular question. Find the probability of the following events:
   a. The chosen student is a senior.
   b. He or she is a sophomore.
   c. He or she is a graduate student.
   d. He or she is a junior.

5.4. Let A_1 and A_2 be events defined on the sample space S, If the probability of the occurrence of event A_1 is $\frac{1}{2}$; of event A_2, $\frac{1}{3}$; and of event A_1 and A_2, $\frac{1}{2}$; find the following probabilities:

a. $\Pr(A_1 \cup A_2)$	**f.** $\Pr(\overline{A_1} \cup A_2)$	**l.** $\Pr(A_1 \mid \overline{A_2})$
b. $\Pr(\overline{A_1})$	**g.** $\Pr(A_1 \cup \overline{A_2})$	**m.** $\Pr(\overline{A_1} \mid A_2)$
c. $\Pr(\overline{A_2})$	**h.** $\Pr(\overline{A_1} \cap A_2)$	**n.** $\Pr(\overline{A_1} \mid \overline{A_2})$
d. $\Pr(\overline{A_1} \cup \overline{A_2})$	**i.** $\Pr(A_1 \cap \overline{A_2})$	**o.** $\Pr(\overline{A_2} \mid \overline{A_1})$
e. $\Pr(\overline{A_1} \cap \overline{A_2})$	**j.** $\Pr(A_1 \mid A_2)$	
	k. $\Pr(A_2 \mid A_1)$	

5.5. A card is drawn at random from an ordinary deck of 52 cards. Obtain the probability of the following even:
   a. An ace or a king is observed.
   b. An ace or a diamond card is observed.

5.6. Twenty-five students are in a finite mathematics class. Fifteen of these students are males and ten are females. Eight of the males and four of the females have also registered for a philosophy course. A student is selected at random from the class. Obtain the probability of the following events:
   a. The chosen student is female or is registered for the philosophy course.
   b. The chosen student is female or is not taking the philosophy course.

5.7. An urn contains three red and two white balls. The balls are drawn one at a time without replacement. What is the probability that the first will be red, the second white, the third red, and so on alternately?

5.8. Suppose that from a deck of cards the three picture diamonds (jack, queen, and king) are removed, and from the remaining 49 cards we select one card at random. Determine the probability of the following events:
   a. The chosen card is a picture card, given that it is a spade.
   b. The chosen card is a spade, given that it is a picture card.

5.9. In Problem 5.8, what is the probability that the chosen card is a picture spade?

5.10. A sample poll of 100 voters revealed the following information concerning their opinions on three party candidates, A_1, A_2, and A_3, who are running for three different offices:

5.11. Refer to the information below:

10 in favor of both A_1 and A_2	35 in favor of A_1 or A_2 but not A_3
25 in favor of A_2 but not A_1 or A_3	65 in favor of A_2 or A_3 but not A_1
25 in favor of A_3 but not A_1 or A_2	0 in favor of A_1 and A_3 but not A_2, and 15 uncommitted

Determine the probability of the following events:

a. The voters favor all three candidates; A_1 and A_2 that is, $A_1 \cap A_2 \cap A_3$.

b. The voters favor candidates A_1 and A_3; that is, $A_1 \cap A_2$.

c. The voters favor candidates A_1 and A_3; that is, $A_1 \cap A_3$.

d. The voters favor candidates A_2 and A_3; that is, $A_2 \cap A_3$.

5.12. In Problem 5.11, are the events A_1, A_2, and A_3 independent?

5.13. A family owns two automobiles, an American and a foreign model. In any given year the probability that the family will be involved in an accident with the American model is 0.02 and with the foreign model 0.06. Determine the probability of the following events:

a. In a given year the family will be involved in automobile accidents with both models.

b. In a given year the family will not be involved in any automobile accidents.

c. In a given year the family will have an accident with one car but not both.

5.14. A lot in a certain warehouse contains 30 electrical generators of which six are known to be defective. On a particular day, three generators will be used. What is the probability that all three will be operable?

5.15. Consider a situation in which two guns are used to shoot at the same target. It has been observed that the probability that the first gun will hit the target is 0.6 and. the second 0.72. If one shot is fired from each gun, determine the probability that both shots will hit the target.

5.16. The defense lawyer in a criminal case intends to ask a group of five individuals to testify on behalf of the defendant. Three of the witnesses will speak the truth; the lawyer assumes that the other two will lie. If the lawyer selects the witnesses at random, what is the probability that they will alternate in true and untrue testimony?

5.17. Four cards are dealt at random from an ordinary deck of 52 cards, one at a time. What is the probability that all four cards are aces?

5.18. A lot contains 16 electric pencil sharpeners of which four are partially defective. Four of the sharpeners are drawn at random, one after the other. What is the probability that all four pencil sharpeners are in perfect condition?

5.19. On a given day a doctor delivered four babies of whom two were girls and two were boys. What is the probability that the first was a boy, the second a girl, the third a boy, and the fourth a girl?

5.20. A car rental agency purchases 40% of its tires from Company A_1, 35% from Company A_2, and 25% from Company A_3. It has been observed that 5% of the tires purchased from Company A_1 will be defective, 4% will be defective from Company A_2, and 3% from Company A_3. A tire is chosen at random at the garage of the rental agency to put on one of its cars. What is the probability that the chosen tire will be defective?

5.21. In Problem 5.20, the tire chosen at random was very carefully inspected and found to be defective. Obtain the probability of the following events:

a. The defective tire was purchased from Company A_1,

b. The defective tire was not purchased from Company A_2,

c. The defective tire was purchased from Company A_3.

5.22. The probability that a student will pass finite mathematics is 0.80, English composition 0.70, and both courses 0.60. What is the probability that the student will pass the mathematics or the English course?

5.23. Three automatic machines are used by an industrial complex to produce a certain sewing machine part. Machine A_1 produces 45% of the total, Machine A_2, 35% and Machine A_3, 20%. It is known from previous experience that 10% of the parts produced by Machine A_1 will not pass inspection, and for Machine A_2 and Machine A_3, 6% and 2%, respectively. If one part is chosen at random from the combined output, determine the probability that

a. the part will conform to the specifications,

b. the part was produced by Machine A_1.

5.24. Three men in turn each draw a card from an ordinary deck of 52 cards. If the cards are not replaced, what is the probability that the first man will draw the jack of diamonds, the second the queen of hearts, and the third the king of spades?

5.25. In a certain state 10% of the men and 6% of the women are alcoholics. Furthermore, men comprise 58% of the state's population. A person from the state is selected at random and found to be an alcoholic. What is the probability that the person is a woman?

5.26. Two urns contain 20 and 30 items, respectively. The first urn contains 15 good items and five defective items. The second urn contains 25 good items and five defective ones. An item is drawn at random from each urn. Obtain the probability of the following events:
   a. One of the items is defective and one is good.
   b. Both items are good.
   c. One item is good and one is defective and drawn from the second urn.

5.27. The probabilities that two weapon systems, A_1 and A_2, will hit a given target are 0.70 and 0.80, respectively. If each fires only one shot at the target,
   a. What is the probability that both weapon systems will hit the target?
   b. What is the probability that only one of them will hit the target?
   c. Given that only one weapon system hit the target, what is the probability that it was A_1?

5.28. Three weapon systems are shooting at the same target. Theoretically, each weapon is equally likely to hit the target; however, in actual practice it has been observed that the precision of the weapon systems is not the same; that is, the first weapon hits the target 8 out of 10 times, the second hits seven out of ten times, and the third hits only six out of 10 times. If each weapon system fires one shot at the target, determine the probability that the target will be hit.

5.29. In Problem 5.28, if we observe that the target was hit, what is the probability that the shot was fired by the second weapon system?

5.30. In Problem 5.28, given that the target was hit, which of the three weapon systems is more likely to have fired the shot?

5.31. Suppose we chose two items at random from a box containing ten items of which four were defective. Calculate the probability of the following events:
   a. Both items selected were defective.
   b. Both items selected were not defective.
   c. At least one item was defective.

5.32. Suppose that three cards are drawn at random from an ordinary deck of 52 cards. Find the probability of the following events:
   a. One card is a diamond and one a heart.
   b. Both cards are diamonds.

5.33. Suppose we chose three flashlight batteries at random from a box containing 20 batteries, five of which were dead. Determine the probability of the following events:
   a. All three of the batteries chosen were dead.
   b. All three were good.
   c. Only one was dead.

5.34. An urn contains ten balls of the same size that are numbered 1 to 10. Three balls are drawn at random without replacement. What is the probability that the sum of the numbers observed on the balls is even?

5.35. In Problem 5.34, determine the probability that the observed sum on the balls is even if the three balls are drawn one at a time with replacement.

5.36. A debating committee of five students is to be selected at random by the instructor of a political science class consisting of 14 girls and 16 boys. Determine the probability of the following events:
   a. The committee will consist of all boys.
   b. The committee will consist of three girls and two boys.
   c. The committee will consist of five girls.

5.37. A box in a wholesale jewelry store contains 20 watches numbered 1 to 20. The watches are drawn at random, one by one, and divided among four customers. Suppose that it is known that watches 4, 8, 12, and 16 are not operable. What is the probability that each customer will receive one defective watch?

5.38. In Problem 5.100, determine the probability that one customer will receive all four defective watches.

5.39. In Problem 5.100, determine the probability that two of the customers will receive two defective watches each and the others will receive no defective watches.

SUMMARY OF IMPORTANT CONCEPTS

We have discussed four definitions of probability; namely,

1. probability as a *measure of belief* (Definition 5.2.1),
2. the *classical definition* of probability (Definition 5.2.2),
3. the *relative-frequency definition* of probability (Definition 5.2.3), and
4. the *axiomatic definition* of probability (Definition 5.4.1).

Definitions:

5.2.1. **Personal probability** is a judgment about the probability of an event happening in a particular manner is often based on personal feelings or beliefs rather than on a mathematical analysis of the situation.

5.2.2. The **classical definition of the probability** of the event A is equal to the ratio of the number of ways in which event A may occur in a particular situation, denoted by n_A, and the total number of possible outcomes in a given situation, denoted by n. That is, $\Pr(A) = \dfrac{\text{number of elements attain attribute } A}{\text{total number of possible outcomes}} = \dfrac{n_A}{n}$

5.2.3. Let be the total number of identical trials in a given problem and n_B the number of occurrences of attribute or characteristic B. Here, *identical* means that each trial in a given situation is performed under identical conditions. The probability of B happening, $\Pr(B)$, or **relative frequency** is given by $\Pr(B) = \frac{n_B}{n}$, as n becomes very large

5.3.1. An **experiment** is the method by which an observation or a measurement is obtained.

5.3.2. A possible outcome of an experiment is called a **sample point**; the actual observations of an experiment constitute a **sample**.

5.3.3. A group of individual persons, objects, or items from which samples are taken for statistical measurement constitutes a **population**.

5.3.4. An **event** is a set of sample points.

5.3.5. An event that consists of only one sample point is called a **simple** or **elementary event**.

5.3.6. An event that consists of more than one sample point is called a **compound event**.

5.3.7. A set of all possible outcomes of an experiment is called a **sample space**. We shall denote the sample space by S.

5.3.8. Two events, A and B, of the sample space S are said to be **mutually exclusive** if $A \cap B = \varnothing$. That is, events A and B do not have any elements in common.

5.4.1. The **sample space** S is the set of all possible outcomes of our experiment. We have given a number of definitions associated with the sample space. Let $A_1, A_2, ..., A_n$ be events defined on the sample space S. The probabilities of the events, denoted by $\Pr(A_i)$, $i = 1, 2, ..., n$, are numbers that satisfy the following **axioms**:
1. $0 \leq \Pr(A_1)$
2. $\Pr(S) = 1$
3. $\Pr(A_1 \cup A_2 \cup ... \cup A_n) = \Pr(A_1) + P(A_2) + ... + \Pr(A_n)$, when $A_1, A_2, ..., A_n$ are mutually exclusive events, $\Pr(A_i)$.

5.6.1. Let A and B be any two events of the sample space S. We define the **conditional probability** of event A, given that event B has occurred, denoted by $\Pr(A|B)$, by $\Pr(A|B) = \frac{\Pr(A \cap B)}{\Pr(B)}$ where $\Pr(B) > 0$.

5.6.2. Let A and B be any two events of the sample space S. The event A is said to be **independent** of event B if $\Pr(A \cap B) = \Pr(A)\Pr(B)$. That is, the probability of the intersection of the events $A \cap B$ is equal to the product of the probabilities of A and B,

Properties:

5.5.1. If A is any event of the sample space S, then $\Pr(\bar{A}) = 1 - \Pr(A)$ where $\bar{A}$ denotes the **complement** of event A.

5.5.2. If $\varnothing$ is the impossible event (**empty set**) of the sample space S, then $\Pr(\varnothing) = 0$.

5.5.3. If A and B are any two events of the sample space S, then the probability of their union is given by $\Pr(A \cup B) = \Pr(A) + \Pr(B) - \Pr(A \cap B)$.

5.5.4. The **additive law** of probability can be easily extended to a finite number of events defined on the sample space. For example, let A, B, and C be any three events defined on the sample space S. The probability of the union of these events is giver by

$$\Pr(A \cup B \cup C) = \Pr(A) + \Pr(B) + \Pr(C) - \Pr(A \cap B) - \Pr(A \cap C)$$

$$-\Pr(B \cap C) + \Pr(A \cap B \cap C) = \frac{n_A}{n} + \frac{n_B}{n} + \frac{n_C}{n} - \frac{n_{A \cap B}}{n} - \frac{n_{A \cap C}}{n} - \frac{n_{B \cap C}}{n} + \frac{n_{A \cap B \cap C}}{n}.$$

Rules:

5.6.1. Let A and B be any two events on S. The probability of the intersection $A \cap B$ is $\mathbf{Pr}(A \cap B) = \mathbf{Pr}(A)\mathbf{Pr}(B|A) = \mathbf{Pr}(B)\mathbf{Pr}(A|B)$. The **multiplicative law of probability** for a sequence of elements $\mathbf{Pr}(A_1 \cap A_2 \cap ... \cap A_n) = \mathbf{Pr}(A_1)\mathbf{Pr}(A_2|A_1)\mathbf{Pr}(A_3|A_1 \cap A_2) ...\mathbf{Pr}(A_n|A_1 \cap A_2 \cap ...A_{n+1})$. Thus, for $n=3$; that is, three events, A_1, A_2, A_3, the multiplicative law is written $\mathbf{Pr}(A_1 \cap A_2 \cap A_3) = \mathbf{Pr}(A_1)\mathbf{Pr}(A_2|A_1)\mathbf{Pr}(A_3|A_1 \cap B_2)$

5.7.1. Let $A_1, A_2, ..., A_n$ an be mutually exclusive events of the sample space S, such that $A_1 \cup A_2 \cup ... \cup A_n = S$; and it is given that $\mathbf{Pr}(A_1) > 0$ for $i = 1, 2, ..., n$, Then for any event A we have $\mathbf{Pr}(A) = \mathbf{Pr}(A|A_1) + \mathbf{Pr}(A_2)\mathbf{Pr}(A|A_2) + \cdots + \mathbf{Pr}(A_n)\mathbf{Pr}(A|A_n)$ which reflects the **elimination** of A'.

5.7.2. **Bayes' rule**: Let A_1, A_2, and A_3 be mutually exclusive events, such that $A_1 \cup A_2 \cup A_3 = S$. Let A be any other event of S. If $\mathbf{Pr}(A_i) > 0$, then

$$\mathbf{Pr}(A_i|A) = \frac{\mathbf{Pr}(A_i)\mathbf{Pr}(A|A_i)}{\mathbf{Pr}(A_1)\mathbf{Pr}(A|A_1) + \mathbf{Pr}(A_2)\mathbf{Pr}(A|A_2) + \mathbf{Pr}(A_3)\mathbf{Pr}(A|A_3)}$$

REVIEW TEST

1. Which of the following cannot be the probability of an event?
   a. 0.0001
   b. $\frac{1}{3}$
   c. -0.25
   d. 1.00
   e. None of these
2. A box contains five red, two green and three blue marbles. If one ball is selected at random, find the probability that the marble will be:
   a. Green
   b. Red or blue
   c. Yellow
3. A box contains five red, two green and three blue marbles. If two ball is selected at random (without replacement), find the probability the marbles will be:
   a. Both red.
   b. The first is red and the second is green.
   c. One is red and the other is green.
4. A fair coin is tossed three times. Find the probability of getting three heads.
5. A fair die is tossed. What is the probability of getting a:
   a. Six.
   b. Seven.
   c. Three or less.
   d. Greater than two.
   e. An odd number less than 5.
6. Two fair dice are tossed. Find the probability that their dot sum will be:
   a. Six.
   b. Greater than eight.
   c. Three or less.
7. A single card is drawn from a standard deck of 52 cards. Find the probability that the card is:
   a. A heart or a king.
   b. A club or a jack.
   c. A face card.
8. Two cards are drawn from a standard deck of 52 cards without replacement. Find the probability that:
   a. Both are fives.
   b. The first is an ace and the second is a jack.

9. The probability of an event is $\dfrac{3}{7}$, what are the odds in favor of the event?

10. If the odds against an event are 7:1, find the probability that the event will occur.

11. If the odds in favor an event are 5:3, find the probability that the event will occur.

12. A box contains seven good fuses and three defective fuses. A single fuse is selected at random. What is the probability that it is not defective?

13. A box containing 20 pens contains two defective pens. A single pen is selected at random. What is the probability that it is not defective?

14. A shipment of 50 cars contains 13 used cars. Three cars have been purchased. What is the probability that:
    a. The probability that all three cars are used.
    b. The probability that all three cars are new.

15. The probability that a student passes French 101 is 0.59 and the probability that a student passes Humanities is 0.73. Assuming that the events are independent, what is the probability that:
    a. A student passes both courses?
    b. A student passes at least one course?

16. A survey was taken to determine which sports on the Wii are enjoyed by various age groups. The results are as follows:

Ages	Golf	Bowling	Archery	Ping-pong	Total
13-18	7	13	10	15	35
19-30	10	15	11	16	42
31-45	16	18	14	12	60
Total	33	46	35	43	137

   a. What is the probability that a person preferred Wii Golf?
   b. What is the probability that a person was in the 31-45 age group?
   c. What is the probability that a person surveyed was in the 13-18 age group given they preferred golf?

17. One fair die is rolled followed by the tossing of a fair coin. Find the probability of getting a head and an odd number on the die.

18. A container holds one dozen eggs of which two are bad. What is the probability of randomly selecting:
    a. A rotten egg?
    b. Two rotten eggs when selected without replacement?
    c. Two rotten eggs when selected with replacement?

19. A draw contains two pairs of black socks, three pairs of white socks and a pair of blue socks. The first sock selected is white without replacement. What is the probability that the second sock selected at random matches the first?

20. The A, K, Q, J and ten of hearts are shuffled and dealt out one at a time. What is the probability that the cards are dealt in order: 10, J, Q, K and A?

21. In a group of students, 37% are taking algebra, 43% are taking English and 21% are taking both. A student in this group is selected at random. Find the probability that the student is taking at least one of these two courses.

22. The probability that a student passes Math or English is 0.89, the probability that a student passes English is 0.63, and the probability of passing both is 0.57. What is the probability a student passes Math?

Practice problems for the CLAST exam:

23. A box contains three red, five white and two blue marbles. Two marbles are drawn from the box at random without replacement. What is the probability that neither is red?

   a. $\dfrac{7}{10} \times \dfrac{2}{10}$ b. $\dfrac{7}{10} \times \dfrac{7}{10}$ c. $\dfrac{7}{10} \times \dfrac{6}{9}$ d. $\dfrac{7}{10}$

24. The odds against Bob winning the election are 8:5. What is the probability that Bob will win the election?

   a. $\dfrac{8}{5}$ b. $\dfrac{5}{8}$ c. $\dfrac{5}{13}$ d. $\dfrac{8}{13}$

REFERENCES

Carnap, R., 1950. Logical Foundations of Probability. University of Chicago Press, Chicago.

Cramer, H., 1955a. The Elements of Probability Theory. Wiley, New York.

Drake, A.N., 1967. Fundamentals of Applied Probability Theory. McGraw-Hill, New York.

Feller, W., 1968a. 3rd ed. An Introduction to Probability Theory and Its Applications, vol. 1. Wiley, New York.

Parzen, E., 1960a. Modern Probability and Its Applications. Wiley, New York.

Tsokos, C.P., 1972b. Probability Distributions: An Introduction to Probability Theory with Applications. Wadsworth, Belmont, CA.

Tsokos, C.P., 1978d. Mainstreams of Finite Mathematics with Application. Charles E. Merrill Publishing Company, A Bell & Howell Company.

Tsokos, C.P., Wooten, R.D., 2011a. The Joy of Statistics Using Real-World Data. Kendall Hunt, Florida.

Wadsworth, G.P., Bryan, J.G., 1960. Introduction to Probability and Random Variables. McGraw-Hill, New York.

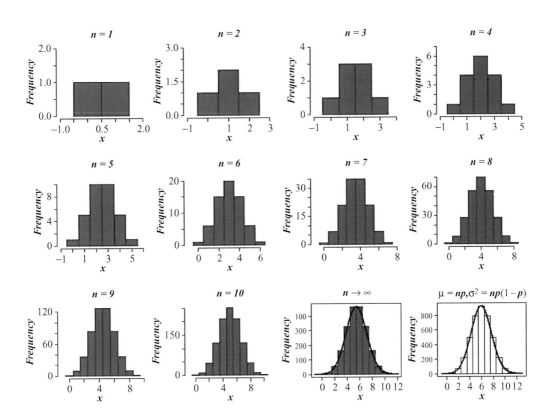

The discrete probability distribution approaches a continuous probability distribution as the sample size increases.

Chapter 6	Binomial Probability

*The merchant family, Bernoulli, of Basel, Switzerland, has produced scientists in each generation since the end of the seventeenth century. One of the distinguished members of this family was Jakob (also called James or Jacques), who pioneered work in differential and integral calculus and the field of probability theory. His work on the binomial distribution produced, among other achievements, Bernoulli''s law and Bernoulli trials. Jakob's initial studies were in theology, primarily because of his father's insistence, but he refused a church appointment and began lecturing on experimental physics. He became a mathematics professor at Basel University, and later served as rector there. His earliest papers concerned astronomy (especially the motions of comets); later papers explored infinite series and their summation (the Bernoulli numbers are named for him). Both Jakob and his brother, Johann (John, Jean) constantly exchanged correspondence with Leibniz (often with bitter rivalry), producing many of the fundamentals of calculus and the integration of ordinary differential equations. Jakob examined many special curves, including the catenaries (the curve of a hanging chain), the isochrones (the curve along which a body falls with uniform velocity), and the logarithmic spiral (that reproduces itself under various transformations). This last was such a delight to him that he willed that it be engraved on his tombstone with the inscription, **eadem mutata resurgo** ("Although changed, I shall arise the same").*

6.1 INTRODUCTION TO RANDOM VARIABLES

Discrete

 Separate or disconnected from others

 Discontinuous

 Using only algebra and arithmetic

Domain

 A field of action—in Math, the set of values assigned to the independent variable.

Range

 A space in which one may dwell—in Math, the set of values taken on by the dependent variable

Mathematics

 The science of numbers, there generalization and their abstractions.

Math

 The universal language of thought, usually an abbreviated form of the native language.

Formula

 A set form, a general fact—in Math, this rule is written in groups of symbols with associated meaning.

Observations generated by an experiment are classified as **quantitative** or **qualitative**. To observe measures in quantitative terms involves the assessment of numerical values. On the other hand, observations that are qualitative do not involve numerical measurements. For example, the measures of total rainfall on a given day at an airport, the consumption of coffee at the student union, and the number of visitors at Disney World during a certain period, among others, are quantitative observations. Weather descriptions, such as snowy, sunny, or rainy, or the description of a certain product as good, fair, or poor are qualitative observations.

We shall be concerned here primarily with quantitative measurements. In every study, statistics can be used to describe various discrete characteristics of the population of interest.

There are several real world problems involving a discrete response, such as how large is the average family in the United States? Such questions lead to us seeking various probabilities that a specific number will appear, among other interesting questions.

Our aim in this chapter is to introduce the basic concepts of discrete random variables and to illustrate the usefulness of the discipline by addressing some of the questions we mentioned above, among others.

In daily life, we are not interested, for the most part, in the possible outcomes of a given experiment, but only in a certain number associated with the experiment. For example, in the experiment of rolling two dice, our interest may only be in the total odd number of points showing (a quantitative measurement). Or in the experiment of tossing a pair of coins, our interest may be centered on the number of heads that will be obtained. To measure the particular outcomes of interest among all possible outcomes of an experiment we must incorporate the concept of a **discrete random variable**.

Before we proceed to define the meaning of and the role of a random variable in probability and applications, we shall briefly discuss the concept of a function, which plays a central role in real scientific investigations.

Definition 6.1.1 Function

A **function** is a relation between two variables, *x* and *y*, such that for each value of the independent variable, usually selected to be *x*, there is at most one value of the dependent variable, *y*. That is, "*y* is a function of *x*" when for each value of *x* in the domain, there is one value of *y* in the range. We express this relation by *y*=*f*(*x*).

Recall, we think faster than we speak and we speak faster than we write—therefore, we abbreviate everything. Now let us consider the notation used, similar to that used for the probability of an event, if we write,

"*y* is a function of *x*"

Then we can translate this

$$"\underline{y\ \text{is a}}\ \text{function}\ \underline{of}\ x"$$
$$y=\ \ f\ \ \ \ \ \ \ \ \ (\ x\ )$$

Hence, "*y* **is a function of** *x*" is written mathematically as *y*=*f*(*x*).

In mathematics and in many of the physical and social sciences we encounter basic **formulas**; formulas, like functions, are concise ways of expressing information symbolically. For example, if *r* is the radius of a circle, then the area of the circle is a function of *r*, *A*=*f*(*r*), and is given by $A=\pi r^2$. Thus, if we wish to determine the area of a circle, we must specify its radius. For each value of *r*, there is a corresponding value of the area; and the collection of these pairs of numbers is a function; that is, a function is a **rule** or **relationship** between values of *r* and *A*. Hence, we can say that the area of a circle is a function of its radius, or *A*=*f*(*r*).

A **random variable** is a function (rule) that assigns a number to each possible outcome of an experiment. In other words, it assigns a number to each point in the sample space. Some examples of **random variables** are the following:

1. *The number of heads in three flips of a coin. In 20 flips of a coin.*
2. *The height of a person chosen for a psychology experiment.*
3. *The number of traffic accidents in Miami during the month of January.*
4. *The number of defects in a foreign car.*
5. *The number of passes completed by a quarterback.*
   And many others.

Note that all of the preceding experiments yield a numerical measurement that may vary from sample point to sample point and hence is called a **random variable**; that is, we do not know which outcome will appear.

We have, then, the following definition:

Definition 6.1.2 Random Variable

A **random variable** (RV) is a numerically valued function defined over a sample space.

We shall use capital letters, say X or Y, to denote **random variables**, and the corresponding lowercase letters, x or y, to denote the possible values that they may assume. That is, if s is a point (an observation) in the sample space S and X is a **random variable**, then $X(s) = x$ is the numerical value of the random variable at s. Example 5.1 should further clarify the meaning.

S

$$X(s) = x$$

S

$\{H,H\}$ $\{H,T\}$

$\{T,H\}$ $\{T,T\}$

Example 6.1.1 Two Fair Coins

Consider the experiment of tossing two coins. There are four conceivable outcomes:

$$(\mathbf{H},\mathbf{H}),(\mathbf{H},\mathbf{T}),(\mathbf{T},\mathbf{H}), \text{ and } (\mathbf{T},\mathbf{T}).$$

Thus, the sample space consists of four points, or four possible outcomes of this experiment.

$$s_1 = (\mathbf{H},\mathbf{T}), \quad s_2 = (\mathbf{H},\mathbf{T}), \quad s_3 = (\mathbf{T},\mathbf{H}), \text{ and } s_4 = (\mathbf{T},\mathbf{T}).$$

Suppose that we are interested in the number of heads observed: Let X be a random variable that can assume the values x_1, x_2, x_3, and x_4, where

$$x_1 = \text{number of heads in } s_1,$$
$$x_2 = \text{number of heads in } s_2,$$
$$x_3 = \text{number of heads in } s_3, \text{ and}$$
$$x_4 = \text{number of heads in } s_4.$$

Solution

The random variable X can assume the values $x_1 = 2, x_2 = 1, x_3 = 1$, and $x_4 = 0$. That is, the numerical value of the random variable X at s denoted by $X(s)$ is as follows:

Observation in S	s_1	s_2	s_3	s_4
$X(s)$	2	1	1	0

Note: Here the sample space consists of all the possible values that the defined random variable may assume. Random variables are classified as either **discrete** or **continuous**.

Example 6.1.2 Tumor Size in Breast Cancer

Consider the observation of **tumor size** in breast cancer. There are a variety of possible outcomes, all of which range from the minimum of 0.5 cm to a maximum observed value of 4 cm.

Thus, the sample space consists of points in an open interval,

$$S = \{x | 0.5 \leq x \leq 4\}.$$

Suppose that we are interested in the number of patients with tumor size greater than **1**: Let X be a random variable that can assume the values between 0.5 and 4 cm.

Solution

The random variable X can assume the values between $x = 0.5$ and $x = 4$. Note: Here the sample space consists of all the possible values that the defined random variable may assume. This random variable is not discrete as illustrated by the fact that we cannot list all possible outcomes; thus, this random variable is continuous.

Hence, we proceed to define both **discrete random variables** and **continuous random variables**, which will be the subject of the next chapter.

Definition 6.1.3 Discrete RV

A **discrete random variable** is one that can assume only a finite, or countably infinite, number of distinct values. A discrete random variable can be easily identified by examining the nature of the values it may assume. In most practical problems, discrete random variables represent a count of sample points possessing a specific characteristic.

Some examples of a discrete random variable are:

1. The number of defective radios in a lot of 20.
2. The number of A's given in a history course.
3. The number of blondes at a certain California university.
4. The number of tournaments won by a tennis player.

Foreshadowing:

Quantitative variables can be discrete or continuous. For comparison, we proceed to formally define **continuous random variable** that will be the subject of study in this chapter.

Definition 6.1.4 Continuous RV

A **continuous random variable** is one that can assume any value in some interval or intervals of real numbers (and the probability that it assumes any specific value is zero).

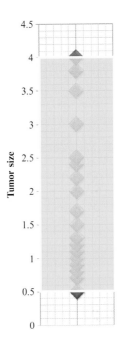

Countably infinite

 A set whose members can be arranged in an infinite sequence:

$$k_1, k_2, k_3, k_4, \ldots$$

in such a way that every member occurs in only one position; that is, the set is countable.

Continuum

 A coherent whole characterized as a collection

Here, the term continuous random variable implies that the variation takes place along a continuum. That is, it is a random variable, the set of possible values of which is an interval on the real axis.

As we mentioned, we will study continuous random variables in Chapter 7 in more detail. However, having discussed the concept of a **random variable**, our aim now is to learn how to calculate the probability that a given **random variable** will assume a specific value or set of values. This subject will be studied in the next section.

PROBLEMS

Basic Problems

Function: Is the relation a function?

6.1.1. $\{(-4, -3), (-8, -7), (6, -8), (-3, -9), (7, -5)\}$

6.1.2. $\{(0, 6), (3, 1), (-2, 5), (-6, 4), (6, 1), (9, 1)\}$

6.1.3. $\{(6, 9), (0, 9), (6, -8), (5, -8), (5, 5)\}$

6.1.4. $\{(100, 175), (18, 162), (117, 178), (121, -194), (10, 208), (128, 178)\}$

6.1.5. $\{(-6, -114), (-6, -76), (-48, -114)\}$

Discrete versus Continuous

6.1.6. List five examples of a discrete random variable.

6.1.7. Specify which of the following are discrete or continuous random variables:

    a. The amount of snow in the state of Maine for the month of January, 2010.

    b. The number of incoming calls through Verizon during a 12-h interval.

    c. The time required to finish taking an exam.

    d. The number of completed passes in a given football game.

6.1.8. Explain the difference between the following random variables:

    a. The *amount of rain* collected in a rain gage at a given site during a rain storm.

    b. The *total number of drops of rain* collected in a rain gage at a given site during a rain storm.

6.1.9. In the experiment of flipping four fair coins, we are interested in the number of heads. Construct a table to show the possible values that the random variable can assume.

6.1.10. In the experiment of tossing two fair dice, we are interested in the sum of the values showing uppermost. Construct a table to show the possible values that the random variable can assume.

6.1.11. Suppose we draw cards consecutively from a deck of cars until an ace is obtained. Let X be the number of cards drawn.

    a. Is X a random variable?

    b. If so, what type?

    c. What are the smallest and largest values X can attain?

Critical Thinking

6.1.12. Give two examples of a function.

6.1.13. List five examples of a random variable.

6.1.14. List five examples of a discrete random variable.

6.1.15. Specify which of the following are discrete or continuous random variables:

    a. The amount of snow in the state of Maine for the month of January, 1978.

    b. The number of incoming calls at the New York City telephone switchboard during a 12-h interval.

    c. The time required to finish recording a certain song.

6.1.16. Explain the difference between the following random variables:

    a. The amount of rain collected in a glass at a given site during a rainstorm.

    b. The total number of drops of rain collected in a glass at a given site during a rainstorm.

6.1.17. In the experiment of flipping three fair coins, let us say we are interested in the number of heads. Construct a table similar to that given in Example 4.1.1 to show the possible values that the random variable can assume.

6.1.18. Two dice are rolled and, since we are playing craps, we are interested in the total number of points showing. Construct a table showing the possible values that the random variable can assume in this problem.

6.1.19. Suppose we draw cards consecutively from a deck of cards until an ace is obtained. Let X be the number of cards drawn.
   a. Is X a random variable?
   b. If so, what type?
   c. What are the smallest and largest values X can attain?

6.1.20. Distinguish between a discrete random variable and a discrete probability distribution.

6.1.21. Distinguish between a continuous and discrete random variable.

6.2 DISCRETE RANDOM VARIABLES

As we mentioned in the previous section, a **discrete random variable** can easily be identified by examining the values it can assume. For example, an experiment consists of tossing three fair coins, and our interest is in the number of heads observed. Let X be the random variable that represents the number of heads, a discrete random variable that may take on a finite number of values (four). That is, $X = 0$, no head was observed; $X = 1$, one head was observed; and similarly, $X = 2$ and $X = 3$.

Having a **discrete random variable** that describes a certain problem, our next and most important objective is to learn how to obtain the probability that a **random variable** will assume one or another of the values. For example in the preceding problem we might be interested in obtaining the probability that two heads were observed; that is, $P(X = 2)$, or the probability that no head was observed, $P(X = 0)$. A function (rule) will be used to obtain these probabilities. Such a function we call a **probability distribution** for the discrete random variable. A probability distribution is a formula or table that provides the probability associated with each value that the discrete random variable may assume.

For example, let $x_1, x_2, \ldots, x_{10}$ be the values in a certain experiment that the discrete random variable X can assume. If $P(x)$ is the **probability distribution function** for the discrete random variable X, then we can obtain the following probabilities:

$$P(X = x_1) = P(x_1)$$
$$P(X = x_2) = P(x_2)$$
$$P(X = x_3) = P(x_4)$$
$$\vdots$$
$$P(X = x_{10}) = P(x_{10})$$

Note that if $x_1, x_2, \ldots, x_{10}$ are all the possible observations that constitute the **sample space**, then the sum of all the probabilities should add to one; that is,

$$P(x_1) + P(x_2) + \cdots + P(x_{10}) = 1,$$

the **sure event**. We formally define such a function (rule) $P(x)$ as follows:

$X = \{ x \mid x = 0, 1, 2, 3\}$

Tossing three fair coins.

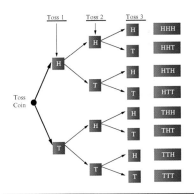

There are **ten sided dice** for this. The outcome is a discrete random variable.

The symbol ∀ (the upside down capital letter *A*, is read "for all"

P(x) is also referred to as the probability density function (pdf).

The symbol ∅ is the empty set or the impossible event.

$$P(\emptyset)=0$$
$$P(S)=1$$

Definition 6.2.1 Probability Distribution

Let *X* be a discrete random variable. A function (rule) *P(x)* is called the **probability distribution** of the random variable *X* if it satisfies the following two conditions:

1. $P(x) \geq 0$, for all *x*
2. $\sum_{\forall x} P(x) = 1$

This distribution is also referred to as the probability density function and is often denoted by $f(x)$, $f(x) = P(X=x)$.

The first condition of the definition simply says that when we evaluate *P(x)* with any of the values that the discrete random variable can assume, its probability will be greater than or equal to zero. It will equal zero if we substitute a value that the random variable cannot assume the impossible event, ∅. The second condition says that if we obtain the probability of every possible value that the random variable can assume, they must sum to one; that is, the sure event, *S*.

The following examples will further illustrate the meaning of the probability distribution of a **discrete random variable**.

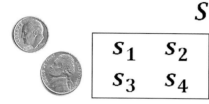

Example 6.2.1 Two Fair Coins

Consider the problem we discussed in Example 6.1.1. The sample points along with their respective probabilities are given in the following table:

Sample point	Random variable	Probability
$s_1 = (H, H)$	$x_1 = 2$	$P(s_1) = \dfrac{1}{4}$
$s_2 = (H, T)$	$x_2 = 1$	$P(s_2) = \dfrac{1}{4}$
$s_3 = (T, H)$	$x_3 = 1$	$P(s_3) = \dfrac{1}{4}$
$s_4 = (T, T)$	$x_4 = 0$	$P(s_4) = \dfrac{1}{4}$

Solution

You recall that $x_1 = 2$, $x_2 = 1$, $x_3 = 1$, and $x_4 = 0$ represent the number of possible heads in the experiment; that is, the values that the random variable *X* can take on. Thus,

$$P(X=0) = P(x_1) = \frac{1}{4},$$
$$P(X=1) = P(x_2) + P(x_3) = \frac{1}{4} + \frac{1}{4} = \frac{1}{2},$$

and

$$P(X=2) = P(x_4) = \frac{1}{4}.$$

For convenience, we can write this information in the form of a table:

x	0	1	2
P(x)	0.25	0.50	0.25

or

x	*P(x)*
0	0.25
1	0.50
2	0.25

Toss 1   Toss 2

Toss coin

H → H → HH
H → T → HT
T → H → TH
T → T → TT

Solution—cont'd

Note that the conditions of the definition of a probability distribution are satisfied.
That is,

1. For all values of x, $P(x) \geq 0$.
2. $\Sigma P(x) = 0.25 + 0.50 + 0.25 = 1.00$.

We can represent the discrete probability distribution of $P(x)$ in graphical form as shown in Figure 6.1. The spikes represent the probabilities that the random variable can assume their respective values.

Recall, there are two types of discrete data, qualitative (bar charts) and quantitative (histograms). This discrete random variable that is quantitative: $x = 0, 1, 2, \ldots$ and therefore, there are boundaries between these distinct observations which we use to construct a histogram shown in Figure 6.1.

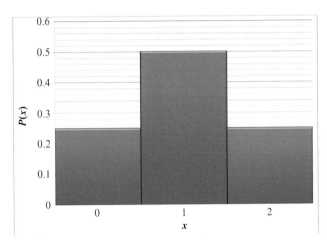

FIGURE 6.1 Histogram of probability distribution for the experiment two fair coins are tossed.

Thus, there is 25% chance that we will obtain no heads, 50% chance of obtaining one head, and 25% chance of obtaining two heads.

Example 6.2.2 Even Number on a Fair Die

Consider the experiment of rolling a single die. We are interested in observing an even number. Here the random variable X is discrete and can assume the values **2**, **4**, and **6**. We shall consider the outcomes **2**, **4**, and **6**, exclusively; if a **1**, **3**, or **5** occurs it will be as though the experiment did not take place. Thus, the probability distribution of the random variable X is given in the following table:

x	2	4	6
$P(x)$	1/3	1/3	1/3

or

x	1	2	3	4	5	6
$P(x)$	0	1/3	0	1/3	0	1/3

or

x	$P(x)$
2	1/3
4	1/3
6	1/3

If treated as a qualitative variable:
Bar Chart

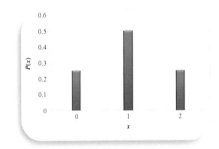

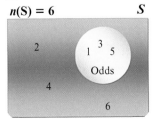

$n(S) = 6$ S

Continued

If treated as a qualitative variable:
Bar Chart

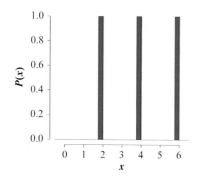

Histogram

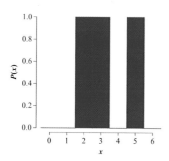

If treated as a qualitative variable:
Bar Chart

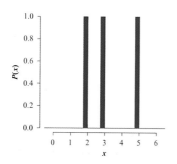

Example 6.2.2 Even Number on a Fair Die—cont'd

Note that $P(x) \geq 0$ and $\sum_{\forall x} P(x) = 1$. The graph of $P(x)$ is given in Figure 6.2.

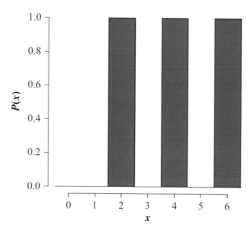

FIGURE 6.2   Histogram of discrete (quantitative) random variable.

Then, there is a $\frac{1}{3}$ probability of obtaining an even number in a single roll of a fair dice.

Example 6.5 Prime Number on a Fair Die

Consider the experiment of rolling a single die and assume that we are interested in observing a prime number, **2**, **3**, and **5**. Here, we let X be a random variable that represents the occurrence of a prime number; if a **1**, **4**, or **6** occurs it will be as though the experiment did not take place. Obtain the probability distribution of the random variable X.

Solution

The possible values that the random variable X can assume are **2**, **3**, and **5**, exclusively. Thus, the probability distribution of the discrete random variable X is

x	2	3	5
$P(x)$	1/3	1/3	1/3

We should mention that the probability distributions are also referred to in the literature as **discrete probability density functions**, **probability distribution functions**, and **probability mass functions**.

PROBLEMS

Basic Problems

6.2.1. Consider the experiment of rolling a single die. We are interested in observing an odd number. Indicate how we can introduce the concept of a random variable and state all the possible values it can assume.

6.2.2. Consider the experiment of tossing a fair coin three times and the number of heads observed. Construct a table to show the sample points in the sample space and all the possible values that the discrete random variable can assume.

6.2.3. Obtain the probability distribution of the discrete random variable in Problem 6.2.1.

6.2.4. Give a graphical representation of the probability distribution of Problem 6.2.1.

6.2.5. Calculate the probability distribution of the random variable in Problem 6.2.2.

6.2.6. Give a graphical representation of the probability distribution of Problem 6.2.2.

6.2.7. Suppose that a certain family has two children. Let X be the random variable that represents the number of girls in the family. Find the probability distribution of the random variable X.

6.2.8. Give a graphical representation of the probability distribution of the random variable X in Problem 6.2.7. Also show that the definition of the discrete probability distribution of the variable X is satisfied.

6.2.9. Consider an experiment of rolling a pair of dice. Let Y be a random variable that represents the number of spots observed. Find the probability distribution of the random variable Y.

6.2.10. Give a graphical representation of the probability distribution in Problem 6.2.9.

Experimental Probability Distributions: Give the outlined probability distribution in tabular form and answer the following questions

6.2.11. Last night, Alexis counted the number of instant messages she received. She received 4 from Luis, 3 from Alexis, 11 from Samuel, 5 from Christopher, and 30 from her best (and very chatty) friend Savannah. She turned on her computer. What is the probability that the first instant message she receives is from Savannah?

6.2.12. Morgan has taken 32 math quizzes this year. Of those, she scored above 90% on 4 of them and 80% or above on 15 of them. What is the probability that she will score from 80% to 90% on her next math quiz?

6.2.13. Austin counts the grasshoppers in the garden on Monday. He finds 35 big ones and 12 small ones. On Tuesday he counts them again. This time he counts a total of 99 grasshoppers. What is a reasonable prediction to make as to how many of the grasshoppers were large ones when he counted them on Tuesday?

6.2.14. The Ramirez family grew a large crop of sunflowers this year. As they were collecting the seeds and drying them, naturally they ate some to see how they tasted. In all they opened about 400 seeds. Some of them they could not eat (about 15) because there was no embryo inside. In two cases they could not eat the seed because there was a worm inside. Assuming these results are typical for the entire crop, as they eat the seeds throughout the year, what is the probability that any particular seed chosen at random will be edible? State your answer as a percent rounded to the nearest 10th of a percent.

6.2.15. Zachary and Aaron go to the mall parking lot on weekends to see if they can find any loose change. People tend to lose small amounts of money in the parking lot. Over the past year they have kept track of how much money they have found. They found 19 quarters, three 50-cent pieces, 23 dimes, 31 nickels, and 298 pennies. What is the probability that the next coin they find will be worth more than 10 cents? State your answer as a percent to the nearest percent.

Graphical Representations: For the given exercise, give a graphical representation of the probability distribution

6.2.16. Exercise 6.2.11

6.2.17. Exercise 6.2.12

6.2.18. Exercise 6.2.13

6.2.19. Exercise 6.2.14

6.2.20. Exercise 6.2.15

Theoretical Probability Distributions: Give the outlined probability distribution in tabular form and answer the given question

6.2.21. You are playing the "shell" game. In this game, there is an object (let's say a ball) hidden under one of five cups and you have to try and guess which cup it is under. Assuming the game is fair and there are five cups, what is the probability you will guess correctly on the first try?

6.2.22. If the odds of winning a particular card game are 1:4, approximately how many times can you expect to win if you play the game 900 times?

6.2.23. Anna is playing with a standard die. On her last roll she got a five. What is the probability she will get a five the next roll?

6.2.24. What is the probability of choosing a heart from a standard deck of 52 randomly arranged playing cards?

6.2.25. Suppose a family has four children. Let X be the random variable that represents the number of girls in the family. Find the probability distribution of the random variable X. State any assumptions you have made in solving this problem.

6.2.26. Rework Problem 6.2.25 assuming the odds of a girl are 1:3.

6.2.27. Consider an experiment of rolling a pair of fair dice. Let Y be a random variable that represents the number of dots observed uppermost. Find the probability distribution of the random variable Y.

Compound Events

6.2.28. Sarah shows her friend Christina a deck of cards. Assuming the cards in the deck are randomly distributed, what is the probability that Sarah draws an ace and does not replace it, and then draws another ace?

6.2.29. There are eight black marbles, nine red marbles and eight white marbles in a bag. The first marble selected at random from the bag is red. The red marble is not put back in the bag. What is the probability of pulling out a red marble on the next try?

6.2.30. Cameron does not know the answer to two questions on a multiple choice exam. The first question has four choices and the second question he does not know has seven choices. What is the probability that he will get both questions wrong?

6.2.31. What is the probability, that on two consecutive rolls of a die, first an odd number, then an even number will come up?

6.2.32. You have five pennies, seven nickels, and three dimes in a piggy bank. If you turn the bank upside down and shake it until a coin comes out of the slot, what is the probability that you will get out two pennies in a row?

6.2.33. Emily rolls a die twice. What is the probability she will get a three both times?

6.2.34. You have six pennies, seven nickels, and five dimes in a piggy bank. If you turn the bank upside down and shake it until a coin comes out of the slot, what is the probability that you will get out two pennies in a row?

Critical Thinking

6.2.35. Distinguish between a frequency distribution and a probability distribution.

6.3   EXPECTED VALUE AND VARIANCE

The **expected value** or **population mean**, sometimes called the mathematical **expectation**, of a **discrete random variable** is very important in characterizing its probability distribution. Originally, the concept of expected value was introduced with reference to games of chance where, if a player stood to win an amount k with a probability p, then his **expected value** or **expectation** was defined as the product $k \times p$. Let's look at a specific example.

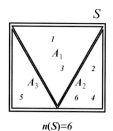

$n(S)=6$

Example 6.3.1 Games of Chance

A gambler tosses a fair die. If a 2, 4, or 6 appear, the gambler will be paid $2; if a 1 or a 3 turns up, he will lose $2; and if a 5 is showing, he will win $4. Naturally, the gambler wants to know his "**expected**" or "**average**" or "**mean**" winnings. Here, the payoffs are −2, 2, and 4 dollars and their respective probabilities are 1/3, 1/2, and 1/6. That is, let A_1 be the event that a 1 or a 3 is observed in a single throw of a die. Then

$$P(A_1) = \frac{n(A_1)}{n(S)} = \frac{2}{6} = \frac{1}{3}.$$

Example 6.3.1 Games of Chance—cont'd

Similarly, if we let A_2 and A_3 be the events that a 2, 4, or 6 is observed and a 5 occurs, respectively, then

$$P(A_2) = \frac{n(A_2)}{n(S)} = \frac{3}{6} = \frac{1}{2}$$

and

$$P(A_3) = \frac{n(A_3)}{n(S)} = \frac{1}{6}$$

where $n = n(S) = 6$ is the total number of sample points in the sample space. The **expected value** (or **average** or **mean**) is obtained by multiplying each payoff by its chance of occurrence and adding the results; thus,

$$\begin{aligned}\textbf{Expectation} &= (-2) \times P(-2) + (2) \times P(2) + (4) \times P(4) \\ &= (-2) \times P(A_1) + (2) \times P(A_2) + (4) \times P(A_3) \\ &= -2 \times \frac{1}{3} + 2 \times \frac{1}{2} + 4 \times \frac{1}{6} = 1.\end{aligned}$$

What this means is that if a player plays this game a large number of times, he can expect to win one dollar per toss of the die.

$$A_1 = \{1,3\}$$

$$A_2 = \{2,4,6\}$$

$$A_3 = \{5\}$$

When we speak of the expected value of a random variable X, we refer to an **ideal** or a **theoretical average**. However, if a given experiment is repeated many times, we would expect the average value of a random variable involved to be near its **expected value**. Here by average we simply mean the **arithmetic mean**, defined as follows:

Definition 6.3.1 Arithmetic Mean

The **arithmetic mean**, **average**, or **sample mean** of a set of n measurements $x_1, x_2, \ldots, x_n$ is equal to the sum of the measurements divided by n. We denote the average by $\bar{x}$, thus:

$$\bar{x} = \frac{\sum\limits_{i=1}^{n} x_i}{n} \quad \text{(Listed data)}$$

$$\bar{x} = \frac{\sum\limits_{i=1}^{k} (x_i f_i)}{n} \quad \text{(Frequency data)},$$

where f_i is the associated frequency for the set of distinct measures x_i, $i = 1, 2, \ldots, n$ and $\sum\limits_{i=1}^{k} f_i = n$.

The **Arithmetic Means** is also referred to as
⇒ **The Average**
⇒ **Sample Mean**
⇒ **Uniformly weighted mean**
⇒ **Standard mean**
⇒ **Population mean** (though unrealistic)

The **mean** or **average** is a descriptive measure of the **central tendency** or **central location** of the data or information. We should also mention that we refer to $\bar{x}$ as the **sample mean**. That is, it represents the arithmetic mean of the random sample of size n.

Data in frequency form is further manipulated and written in terms of the associated **relative frequencies** or **probabilities** as follows:

$$\begin{aligned}\bar{x} &= \frac{\sum\limits_{i=1}^{k} (x_i f_i)}{n} \\ &= \frac{x_1 f_1 + x_2 f_2 + \cdots + x_k f_k}{n} \\ &= \frac{x_1 f_1}{n} + \frac{x_2 f_2}{n} + \cdots + \frac{x_k f_k}{n} \\ &= x_1 p_1 + x_2 p_2 + \cdots + x_k p_k \\ &= \sum\limits_{i=1}^{k} (x_i p_i)\end{aligned}$$

where $p_i = P(x_i)$. In this form, the measure is called the **expected value**.

When we speak of the **expected value** of the random variable, in essence we mean a measure of its **central tendency**. We shall denote the **expected value** of a **discrete random variable** X by $E(X)$. Thus, we formally have the definition of **expected value** below.

Definition 6.3.2 Expected Value

Let X be a discrete random variable with probability distribution $P(x)$. Then the **expected value** of X is given by

$$E(X) = \sum_{i=1}^{k} (x_i p_i)$$

where $p_i = P(x_i)$ and $\forall_i$ is read "for all i."

The definition simply says that we sum the product of all possible values that the random variable X can take on, and their respective probabilities. Thus, in order to calculate the **expected value of a random variable**, we must know its **probability distribution**. For convenience we sometimes denote the expected value of a random variable by μ; that is, $E(X) = \mu$. We also refer to μ as the **population mean** that is being estimated by the **sample mean** $\bar{x}$.

Now we shall illustrate how the preceding definition is used to obtain expectations of random variables.

Example 6.3.2 Expected Value

Calculate the **expected value** of the random variable X, the probability distribution of which is given that

x	$P(x)$
0	0.25
1	0.50
2	0.25

Solution

Thus, we have

$$E(X) = \sum_{i=1}^{k} (x_i p_i) = 0 \times 0.25 + 1 \times 0.50 + 2 \times 0.25 = 1.$$

This can also be expressed in the alternative chart form below:

x	$P(x)$
0	0.25
1	0.50
2	0.25

It means that in tossing a fair coin twice, we **expect** to obtain one head. Note: This is also the most likely outcome.

Example 6.3.3 Expected Value

Obtain the expected value of the discrete random variable X, the probability distribution of which is given by Example 6.2.2.

Solution

The values that the random variable X can take on, along with its probability distribution, are

x	$P(x)$
2	1/3
4	1/3
6	1/3

Hence,

$$E(X) = \sum_{i=1}^{k}(x_i p_i) = 2 \times \frac{1}{3} + 4 \times \frac{1}{3} + 6 \times \frac{1}{3} = 4.$$

Alternatively,

x	$P(x)$	$xP(x)$
2	1/3	2/3
4	1/3	4/3
6	1/3	6/3
		$\sum_{i=1}^{k}(x_i p_i) = 4$

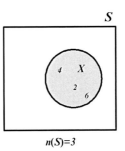

$n(S)=3$

Thus, we **expect** to observe a four on the average when observing the even numbers on a fair die. Note: Here, four is not the most frequent as they are all equally likely; however, four is the mean and the median.

Example 6.3.4 Games of Chance

Consider the experiment of a student rolling a fair die. If a prime number occurs, he pays his girlfriend that number of dollars; however, if a non-prime number is observed, she pays him that number of dollars. Find the expected value of the random variable that characterizes this game of chance.

Solution

Let X be the discrete random variable that represents our interest in the outcomes of the experiment. Here, X assumes the values $-2, -3, -5, 1, 4, 6$.

That is, the numbers $-2, -3,$ and -5 correspond to the fact that a prime number appears and the negative sign indicates that the student loses to his girlfriend that many dollars. Similarly, the non-prime numbers 1, 4, and 6 represent a possible outcome in the experiment and, at the same time, the amount the student receives from his girlfriend. Thus, the probability distribution of the discrete random variable, X, is given by

x	$P(x)$
1	1/6
4	1/6
6	1/6
−2	1/6
−3	1/6
−5	1/6

Prime Numbers:
2, 3, 5
Nonprime Numbers:
1, 4, 6

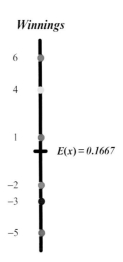

Winnings

Note that the conditions of a probability distribution are satisfied; that is, $P(x) \geq 0$ for all values of X and $\sum_{\forall x} P(X) = 1$. Therefore, the expected value of the game is

Continued

Solution—cont'd

$$E(X) = \sum_{\forall x} (x_i p_i) = 1 \times \frac{1}{6} + 4 \times \frac{1}{6} + 6 \times \frac{1}{6}$$
$$= -2 \times \frac{1}{6} - 3 \times \frac{1}{6} - 5 \times \frac{1}{6} = \frac{1}{6},$$

and the game is favorable to the student. That is, if this game is repeated many times, the girlfriend tends to lose and is expected to lose approximately 16-cent per game.

Example 6.3.5 Expected Value in Stocks

Weighing the odds when playing the stock market: investors make money when the price of stocks increase and lose money when the stock decreases in value. The probability that an investor will make a profit of $500 on a certain stock is 3/4 and the probability that he or she will take a loss of $250 is 1/4. Determine the expectation of the investor.

Solution

Here, the investor's expectation is obtained as follows:

$$E(X) = \sum_{\forall x} (x_i p_i) = 500 \times \frac{3}{4} - 250 \times \frac{1}{4} = \$312.50,$$

that is, the investor can expect a $312.50 profit as the amount expected is greater than the money at risk.

We shall now define the concept of the **variance** of a discrete **RV**, which is a measure of the **spread** or **dispersion** of the random variable around its expected value or mean. However, first we shall define the (true population) variance, σ^2, of a **discrete random variable**.

Definition 6.3.3 Population Variance

Let X be a discrete random variable with probability distribution $P(x)$. Then the **population variance** of X is the mean of the square error, in terms of expected values is given by

$$\sigma^2 = \text{Var}(X) = E\left[(X - \mu)^2\right]$$
$$= \sum_{\forall x} (x - \mu)^2 P(x) = V(X)$$

where $\forall x$ is read "for all x."

The **population variance**, σ^2, tells us how close to the population mean one can expect the value of a discrete random variable to be.

Alternatively, in terms of listed data,

$$\sigma^2 = \frac{\sum_{i=1}^{N} (x_i - \mu)}{N} = \frac{\sum_{i=1}^{N} \varepsilon_i^2}{N},$$

where $\varepsilon_i = x_i - \mu$ is the error term, collectively called the **residuals**.

Definition 6.3.4 Residuals

Let X be a discrete random variable with population mean μ, the residual errors, collectively called the **residuals**, are the differences between what is observed and what is expected: that is,

$$\varepsilon_i = x_i - \mu.$$

Note that the probability distribution relates only to the random variable X. Also the population variance σ^2 is a non-negative quantity, because no matter what x or μ may be, their difference is squared; that is, $(x-\mu)^2$ and $P(x)$ must always be greater than or equal to zero by definition. A large variance implies a large measure of variation between the values that the random variable assumes and its expected value (central tendency). Small variance implies that the values that the random variable can assume are very close to the mean; that is, there is a very small spread. Another concept commonly used as a measure of the dispersion of a random variable with respect to its population mean is the **population standard deviation**, defined as follows:

Definition 6.3.5 Population Standard Deviation

Let X be a discrete random variable with probability distribution $P(x)$. The square root of the variance of a discrete random variable is called the **standard deviation**, denoted by the lowercase Greek letter sigma

$$\sigma = \sqrt{V(x)} = \sqrt{\sigma^2}.$$

In actual practice, we are usually given a set of measurements, say x_1, $x_2, \ldots, x_n$, from which we calculate the sample mean $\bar{x}$ as an estimate of the theoretical expected value (population mean) μ. In addition, we want to obtain an estimate of the population variance σ^2 in order to measure the spread or variability of the measurements $x_1, x_2, \ldots, x_n$ about the sample mean $\bar{x}$. We shall define an estimate of the theoretical (population) variance σ^2 as follows:

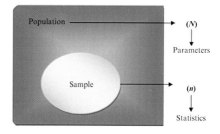

Definition 6.3.6 Sample Variance

The **sample variance** of a set of n measurements: $x_1, x_2, \ldots, x_n$, denoted by s^2, is the average of the square of the deviations of the measurements about their sample mean; that is,

$$s^2 = \frac{\sum_{i=1}^{n}(x_i - \bar{x})^2}{n-1}.$$

Thus, the positive square root of the sample variance is the sample standard deviation. Note that the variance is measured in terms of the square of the original measurements; that is, if the original measurements were in pounds, the variance would be expressed in square pounds. However, taking the square root of the variance, yielding the standard deviation, returns the variability measure of the data to the original units of measurement.

The **sample variance**, s^2 estimates the (true) **population variance**, σ^2.

The **standard sample deviation**, s, estimates the (true) **standard population deviation**, σ.

Sometimes in real world problems, we need to transform data by either adding or subtracting a value possibly to convert measurements into the same units. For example, weather stations in the United States record temperature in degrees Fahrenheit, whereas some other countries record temperature in degrees Celsius; hence to compare these data sets, we will convert them to the same scale. How does this affect the expected value and variance?

Property 6.3.1 Adding a Constant

Given a data set, $x_1, x_2, \ldots, x_n$, with an expected value of (population mean), $E(X)=\mu$, and variance, $V(X)=\sigma^2$. Let $y_1, y_2, \ldots, y_n$ be a manipulation of the original data such that $y=x+a$, where a is any constant, then $E(Y)=\mu+a$ and $V(Y)=\sigma^2$. Here, we expect a constant to be itself and not vary.

Example 6.3.6 Adding a Constant

In a quality control experiment, after weighing 50 widgets to test the variance in their weights, the technician realizes that the scale was not zeroed out and the weight of the small felt pad that is on the scale to set the widget on had been included in each measure. The average weight recorded was 15.2 oz with a variance of 0.15, and the small felt pad weighted 0.25 oz. Determine the mean weight of the widgets and the effect (if any) on the variance.

Solution

Let *x* represent the original recorded observations and *y* represent the desired weights of the widgets. Then we have $E(X)=15.2$ and $V(X)=0.15$, and we are looking for $E(Y)$ and $V(Y)$, where $y=x-0.25$. Hence, we have

$$E(Y)=E(X-0.25)=E(X)+E(-0.25)$$
$$=15.2-0.25=14.95$$

and

$$V(Y)=V(X-0.25)=V(X)+V(-0.25)$$
$$=15.2+0=0.15.$$

Hence, the mean weight of the widgets is 14.95 oz, where $14.95+0.25$ (the weight of the felt pad) $= 15.2$ oz (the mean weight observed). However, as the bias introduced by the felt pad is consistent and therefore does not vary. Thus, the variance in the weights of the widgets is the same variance we see in the original recordings, 0.15 oz.

Vary

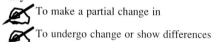 To make a partial change in

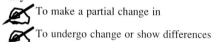 To undergo change or show differences

Property 6.3.2 Multiplying a Constant

Given a data set, $x_1, x_2, \ldots, x_n$, with an expected value of (population mean), $E(X)=\mu$, and variance, $V(X)=\sigma^2$. Let $y_1, y_2, \ldots, y_n$ be a manipulation of the original data such that $y=mx$, where *m* is any constant, then $E(Y)=m\mu$ and $V(Y)=m^2\sigma^2$. Recall, variance is a measure of error squared.

Example 6.3.7 Multiplying a Constant

Some online map searches can give directions to the nearest hundredth of a mile; however, this can vary, depending on which search engine you use and the restriction you place on the travel path (such as **shortest distance** or **shortest time**). Moreover, there is the option to have this information in either **miles** or **kilometers**. If the mean distance we are given is 1405.35 miles with a sample standard deviation of 55.86 miles, determine the mean distance and standard deviation in kilometers.

Solution

Let *X* represent the distances given in miles and *Y* represent the distances in kilometers. Then we have $E(X)=1405.35$ and $V(X)=55.86^2=3120.842$, and we are looking for $E(Y)$ and $V(Y)$, where $y=1.609344x$. Hence, we have

$$E(Y)=E(1.609334X)=1.609334E(X)$$
$$=2261.672$$

and

$$V(Y)=V(1.609334X)=1.609334^2V(X)$$
$$=2.589956\times3120.842=8082.843.$$

Solution—cont'd

and thus,

$$\sigma = \sqrt{V(Y)} = \sqrt{8082.843} = 89.9046.$$

Hence, the mean distance is approximately 2261.7 km and the variance in the travel distances is approximately 89.9 km.

Note: Here we have rounded to one significant digit; however, as these are all estimates, we can keep this accurate to four or five decimal places, but if we are going to do this, we could just simply select a smaller scale such as meters, as 0.001 of a kilometer is a meter.

Here, "$\equiv$" is read as **"equivalent to"**

1 mile $\equiv$ 1.609334 km
1 mile $\equiv$ 8 furlongs
1 furlong $\equiv$ 20 chains
1 chain $\equiv$ 22 yards
1 mile $\equiv$ 1760 yards
1 yard $\equiv$ 3 ft
1 ft $\equiv$ 12 inches
1 mile $\equiv$ 5280 ft
1 km $\equiv$ 1000 m
1 m $\equiv$ 100 cm
1 cm $\equiv$ 10 mm
1 dm $\equiv$ 10 cm

where dm is decimeter
1 m $\equiv$ 100 cm
among others

Example 6.3.8 Temperature Scales

There are three temperature scales in use today: **Fahrenheit**, **Celsius**, and **Kelvin**. The **Kelvin** temperature scale is the base of thermodynamic temperature measurements in the **International System (IS)**, and has an absolute zero. The **Fahrenheit** temperature scale is based on 32 for the freezing point of water and 212 for the boiling point of water. The **Celsius** temperature scale also uses the freezing point and boiling point of water, but uses 0 and 100, respectively. Hence, we can convert between degrees **Fahrenheit** and degrees **Celsius** using the linear equation:

$$F = \frac{9}{5}C + 32.$$

Given that the mean temperature in the United States is **11** °C with a standard deviation of 9 °C, determine the mean temperature and its standard deviation in degrees **Fahrenheit**.

Solution

Given $E(C) = 11$ and $V(C) = 9^2 = 81$, using the properties for both adding and multiplying data by a constant we have:

$$E(F) = E\left[\frac{9}{5}C + 32\right] = \frac{9}{5}E(C) + 32$$

$$= \frac{9}{5} \times 11 + 32 = 51.8$$

and

$$V(F) = V\left[\frac{9}{5}C + 32\right] = \left(\frac{9}{5}\right)^2 V(C) + 0$$

$$= \frac{81}{25} \times 81 + 262 = 44.$$

and thus,

$$\sigma = \sqrt{V(F)} = \sqrt{262.44} = 16.2.$$

Hence, the mean temperature in degrees Fahrenheit is approximately **51.8** °F (**11** °C) and the standard deviation is **16.2** °F (**9** °C).

PROBLEMS

Basic Problems

Find the expected value and standard deviation in the following raffles:

6.3.1.

	Value	Tickets
Grand prize	$4000.00	1
Second place prize	$500.00	20
Third place prize	$10.00	40
Fourth place prize	$5.00	200
Total number of tickets sold		10,000

6.3.2.

	Value	Tickets
Grand prize	$2000.00	1
Second place prize	$800.00	20
Third place prize	$40.00	60
Fourth place prize	$30.00	300
Total number of tickets sold		10,000

6.3.3.

	Value	Tickets
Grand prize	$3000.00	1
Second place prize	$500.00	10
Third place prize	$90.00	30
Fourth place prize	$40.00	200
Total number of tickets sold		10,000

6.3.4. A woman purchases a ticket for a certain raffle. The probability that she wins the first prize of $1000 is 0.0005 and the probability she wins the second prize is 0.005. What is the reasonable price for her to pay for the raffle ticket?

6.3.5. The probability that a businessman will make a profit of $400 on a given item during the Christmas holiday is 2/3 and the probability that he will take a loss of $50 is 1/3. Determine the businessman's expected profit.

6.3.6. The probability of the discrete random variable X is as follows:

x	−2	−1	0	1	2
$P(x)$	1/10	1/5	1/10	2/5	1/5

  a. Calculate the expected value of the random variable X.
  b. Obtain the variance of the random variable X.
  c. Determine the standard deviation of the discrete random variable X.

6.3.7. The probability of the discrete random variable X is as follows:

x	4	8	12	16	20
$P(x)$	1/8	1/6	3/8	1/4	1/12

  a. Determine the expected value of the random variable X.
  b. Obtain the variance of the random variable X.
  c. Calculate the standard deviation of the discrete random variable X.

6.3.8. A student rolls a fair die once. If an odd number is observed, he pays his roommate that number of dollars. However, if an even number is observed, his roommate pays him three dollars.

   a. Determine the probability distribution of the discrete random variable.

   b. Obtain the expected value of the random variable X.

   c. Interpret the meaning of the expectation of the discrete random variable.

   d. Calculate the variance of the random variable X.

   e. Find the standard deviation of the discrete random variable X.

Effects of adding and multiplying the data: Given $E(x)=\mu$ and $V(x)=\sigma^2$, let $y=f(x)$. Find $E(y)$ and $V(y)$.

  6.3.9. Given $E(x)=5$ and $V(x)=3$, let $y=x-2$.

 6.3.10. Given $E(x)=7$ and $V(x)=5$, let $y=2x$.

 6.3.11. Given $E(x)=2$ and $V(x)=2$, let $y=5x-3$.

 6.3.12. Given $E(x)=2$ and $V(x)=2$, let $y=\frac{x-2}{2}$.

 6.3.13. Given $E(x)=-4$ and $V(x)=5$, let $y=-4x+1$.

Critical Thinking

6.3.14. The probability that a businessman will make a profit of \$400 on a given item during the Christmas holidays is 2/3 and the probability that he might take a loss of \$50 is 1/3. Determine the businessman's expected profit.

6.3.15. A woman purchases a ticket for a certain raffle. The probability that she can win the first prize of \$1000 is 0.0005 and the probability that she can win the second prize of \$500 is 0.005. What is a reasonable price for her to pay for the ticket?

6.3.16. The probability distribution of the discrete random variable X is as follows:

x	−2	−1	0	1	2
$P(x)$	1/10	1/5	1/10	2/5	1/5

   a. Calculate the expected value of the random variable X and discuss its meaning.

   b. Obtain the variance of the random variable X and explain its usefulness.

   c. Determine the standard deviation of the discrete random variable X.

6.3.17. The probability distribution of the discrete random variable X is given below:

x	4	8	12	16	20
$P(x)$	1/8	1/6	3/8	1/4	1/12

   a. Determine the expected value of the random variable X and explain it usefulness.

   b. Obtain the variance of the discrete random variable X and discuss what information it conveys to you.

   c. Calculate the standard deviation of the random variable X. How can we improve it?

6.3.18. A student rolls a fair die once. If an even number is observed, he pays his roommate that number of dollars. However, if an odd number is observed, his friend pays him that number of dollars.

   a. Determine the probability distribution of the random variable that characterizes this situation and show that it satisfies the two basic properties.

   b. Obtain the expected value of the discrete random variable that characterizes the behavior of the random variable.

   c. Interpret the meaning of the expectation obtained in part (b).

   d. Calculate the variance of the random variable that characterizes the present situation.

   e. Find the standard deviation of the random variable.

The **Binomial probability distribution** is also known as the **Bernoulli probability distribution**.

Probabilities of Inequalities
The probability of...

 At most k success is $P(x \leq k)$

 Exactly k success is $P(x=k)$

 At least k success is $P(x \geq k)$

Note: in discrete probability distributions,
$$P(x<k)=P(x \leq k+1)$$

Pascal's Triangle is a well-known set of numbers, written in the shape of a triangle, which represent the primary coefficients in a binomial expansion

$$
\begin{array}{ccccccccc}
 & & & & 1 & & & & \\
 & & & 1 & & 1 & & & \\
 & & 1 & & 2 & & 1 & & \\
 & 1 & & 3 & & 3 & & 1 & \\
1 & & 4 & & 6 & & 4 & & 1 \\
 & & & & \vdots & & & &
\end{array}
$$

Where each consecutive row is formed by partial sums of the previous row

row 1 to row 2
0 1 0
0+1=**1** 1+0=**1**

row 2 to row 3
0 1 1 0
0+1=**1** 1+1=**2** 1+0=**1**
etc.

Hence, the fifth row is
1 5 10 10 5 1

6.4 THE BINOMIAL PROBABILITY DISTRIBUTION

We encounter many important real world problems that can be characterized by a **discrete random variable**, the probability distribution of which was developed by **Jakob Bernoulli**, 1654-1705.

We shall illustrate the importance and usefulness of Bernoulli/binomial probability distribution by stating several actual problems in **games of chance, sports, education, biology,** and **business**, as well as in other areas.

The discrete probability distribution that we use to answer these questions, among others, is the **binomial** or **Bernoulli probability distribution**. Before we formally discuss this probability distribution, we shall list the properties that constitute its applicability.

The applicability of the **binomial probability distribution** in any real world problem must follow the following properties:

Property 6.4.1 Binomial Experiment

In a **binomial** experiment, the follow must hold true:
a) It consists of n identical trials.
b) Each trial results in one of two outcomes; for example, success or failure, life or death, acceptance or rejection, etc.
c) The probability of a success in each trial is equal to p and remains the same for each of the n trials. The probability of a failure is $q=1-p$.
d) The trials are independent.

The discrete random variable that comes into play in such a situation is the number of successes that one observes in a specified number of trials. That is, we would like to obtain the probability that in n trials we will encounter exactly k $(k \leq n)$ **successes**, **at least k successes**, or **at most k successes**. The probability distribution that we use to answer these types of question concerning the behavior of this variable is the **binomial** or **Bernoulli probability distribution**—the subject of study in this section.

In order to enhance our understanding of the derivation of the **binomial probability distribution**, we shall first review the **binomial theorem** that you have studied in elementary algebra.

A **binomial** is an algebraic expression that consists of two terms: for example, $(a+b)$, $(3x-2y)$, and $\left(\frac{1}{2}x+\frac{2}{3}y\right)$, and so on. The **binomial theorem** gives the expansion of the binomial expression $(a+b)^n$, where n is any positive integer. For example, it is easy to verify that for $n = $ **2, 3, 4, and 5** we have

$$(a+b)^2 = a^2 + 2ab + b^2$$
$$(a+b)^3 = a^3 + 3a^2b + 3ab^2 + b^3$$
$$(a+b)^4 = a^4 + 4a^3b + 6a^2b^2 + 4ab^3 + b^4$$
$$(a+b)^5 = a^5 + 5a^4b + 10a^3b^2 + 10a^2b^3 + 5ab^4 + b^5$$

Note that in each case the expansion consists of $n+1$ terms and the exponent of a decreases by one while the exponent of b commences with zero and increases to n.

The coefficient of the first and last terms is one, while the coefficient of the second term corresponds to the exponent of a in the first term. The coefficient of the third term is the product of the coefficient of the second term and the exponent of a in the second term divided by two (the number of terms preceding it), and so on. In general, the next coefficient is the previous coefficient times the power of the first term divided by the second power plus one. This pattern suggests that we can write

$$(a+b)^n = \sum_{k=0}^{n} \binom{n}{k} a^{n-k} b^k$$

$$= \binom{n}{0} a^n b^0 + \binom{n}{1} a^{n-1} b^1 + \binom{n}{2} a^{n-2} b^2 +$$

$$\cdots + \binom{n}{n-1} a^1 b^{n-1} + \binom{n}{n} a^0 b^n,$$

where $\binom{n}{k}$ the number of ways to combine **k** out of **n** objects.

The proof of the theorem is by **mathematical induction**. The **binomial theorem** can also be expressed in terms of combinations is beyond the scope of this course. However, we will illustrate this point with several examples.

Example 6.4.1 Binomial Expansion

Expand $(a+b)^6$.

Solution

$$(a+b)^6 = \binom{n}{0} a^6 b^0 + \binom{6}{1} a^5 b^1 + \binom{6}{2} a^4 b^2 + \binom{6}{3} a^3 b^3 + \binom{6}{4} a^2 b^4$$

$$+ \binom{6}{5} a^1 b^5 + \binom{6}{6} a^0 b^6$$

$$= a^6 + 6a^5 b + 15a^4 b^2 + 20a^3 b^3 + 15a^2 b^4 + 6ab^5 + b^6.$$

If the sign in the binomial expression is negative, the expansion begins with a plus and then alternates its sign.

We shall now proceed to develop the **binomial probability distribution**. Consider an experiment that consists of a single trial $(n=1)$, the outcome of which is a **success, S**, or a **failure, F**, with probabilities **p** and $q=1-p$, respectively. Let the value $x=1$ be associated with sample point **S** (success) and $x=0$ with **F** (failure). Thus, the resulting probability distribution of the discrete random variable **X**, which may assume the values **0** or **1**, is as follows:

x	0	1
$P(x)$	q	p

That is, the probabilities of a success and of a failure are given by

$$P[X=1]=p$$

and

$$P[X=0]=q=1-p.$$

Note that the two conditions of properties of a probability distribution have been met:

$$\sum_{\forall x} P(x) = \sum_{x=0}^{1} P(x) = p + q = p + (1-p) = 1$$

and since $0 \leq p \leq 1$, we have $0 \leq q \leq 1$; that is,

$$P(x) \geq 0,$$

for $x=1$ or 0.

⇒ A **histogram** is representation of a frequency distribution using rectangles whose width represents the class width (class interval) and whose heights are proportional to relative frequencies. Hence, in order to make the transition from the discrete binomial probability distribution to the normal probability distribution and correction for continuity in Chapter 7, we will illustrate this discrete **RV x** using a histogram with boundaries 0.5 to the left of **x** and 0.5 to the right of **x**.

All a's
  aaaaaa
All but one a:
  aaaaab
  aaaaba
  aaabaa
  aabaaa
  abaaaa
  baaaaa
All but two a:
  aaaabb
  aaabab
  aabaab
  abaaab
  baaaab
  aaabba
  aababa
  abaaba
  baaaba
  aabbaa
  ababaa
  baabaa
  abbaaa
  babaaa
All but three a's:
  bbaaaa
  ⋮
Binomial Probability
  $P(Success)=p$
  $P(Failure)=q$
where $q=1-p$
Bar chart:
success/failure

Here, we have used an arbitrary probability; however, if we know the probability of success, we can compute these probabilities exactly.

For example, assume that there is a 50-50 chance of success. That is, in two independent trials, we expect one success and one failure. However, there are four equally likely possibilities:

Continued

Success, Success
$$P(SS)=0.5\times0.5=0.25$$
Success, Failure
$$P(SF)=0.5\times0.5=0.25$$
Failure, Success
$$P(FS)=0.5\times0.5=0.25$$
Failure, Failure
$$P(FF)=0.5\times0.5=0.25$$
The combination of one success and one failure: *SF* or *FS* is most probable occurrence:
$$P(SF\cup FS) =P(SF)+P(FS)$$
$$=0.25+0.25=0.5$$

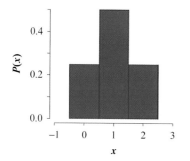

Solution—cont'd

Now consider an experiment that consists of two trials ($n=2$). The outcome of which is a **success**, S, or a **failure**, F, with probabilities p and $q=1-p$, respectively.

Here, the possible sample points or outcomes are **SS**, **SF**, **FS** and **FF**. That is, **SS** represents a success on the first and a success on the second trial, and so on. Thus, since the trials are independent, applying the multiplicative law of probability we have

$$P(\textbf{SS})=P(\textbf{S})P(\textbf{S})=p\times p=p^2$$
$$P(\textbf{SF})=P(\textbf{S})P(\textbf{F})=p\times q=pq$$
$$P(\textbf{FS})=P(\textbf{F})P(\textbf{S})=q\times p=qp$$
$$P(\textbf{FF})=P(\textbf{F})P(\textbf{F})=q\times q=q^2.$$

The random variable X, here representing the number of successes, can assume the values **0**, **1**, **2** with probabilities p^2, $2pq$, and q^2, respectively. Thus, the probability distribution is given by

x	0	1	2
$P(x)$	p^2	$2pq$	q^2

That is, the probability of one success in two trials is

$$P[X=1]=P(\textbf{SF}\cup\textbf{FS})=P(\textbf{SF})+P(\textbf{FS})$$
$$=pq+qp=pq+pq=2pq.$$

Note also that the properties of a probability distribution are satisfied:

$$\sum_{\forall x}P(x)=\sum_{x=0}^{2}P(x)=q^2+2pq+p^2$$
$$=(1-p)^2+2p(1-p)+p^2$$
$$=(1-2p+p^2)+2p-2p^2+p^2$$
$$=1,$$

and since $0\leq p\leq 1$, we have $0\leq q\leq 1$; that is,

$$P(x)\geq 0,$$

for $x=0$ or **1** or **2**.

We shall relate the preceding probability distribution to the binomial theorem. Expand the binomial expression

$$(p+q)^2=\binom{2}{0}p^0q^2+\binom{2}{1}p^1q^1+\binom{2}{2}p^2q^0$$
$$=q^2+2pq+p^2.$$

The coefficient of q^2 is $\binom{2}{0}$ represents the number of ways in which we can obtain **0** successes in **2** trials, and of course it is $\binom{2}{0}=\frac{2!}{0!(2-0)!}=1$ way; that is, **FF** failures in the first and second trials. The probability of this occurring is $q\times q=q^2$. Similarly, the coefficient of pq is $\binom{2}{1}$, represents the number of ways we can obtain **1** success in **2** trials, $\binom{2}{1}=\frac{2!}{1!(2-1)!}=2$ ways, **SF** and **FS**. The probability of this happening is $2pq$. Finally, the coefficient of p^2 is $\binom{2}{2}$ represents the number of ways in which we can obtain **2** successes in **2** trials, and of course it is $\binom{2}{2}=\frac{2!}{2!(2-2)!}=1$ way; that is, the probability of **SS**, success in the first and second trials, is p^2.

Similarly, if we consider an experiment with ($n=3$) trials satisfying the conditions of a probability distribution, we have the following sample points in the sample space along with their probabilities:

Solution—cont'd

$$P(\text{SSS}) = P(\text{S})P(\text{S})P(\text{S}) = p \times p \times p = p^3$$

$$P(\text{SSF}) = P(\text{S})P(\text{S})P(\text{F}) = p \times p \times q = p^2 q$$

$$P(\text{SFS}) = P(\text{S})P(\text{F})P(\text{S}) = p \times q \times p = p^2 q$$

$$P(\text{FSS}) = P(\text{F})P(\text{S})P(\text{S}) = q \times p \times p = p^2 q$$

$$P(\text{SFF}) = P(\text{S})P(\text{F})P(\text{F}) = p \times q \times q = pq^2$$

$$P(\text{FSF}) = P(\text{F})P(\text{S})P(\text{F}) = q \times p \times q = pq^2$$

$$P(\text{FFS}) = P(\text{F})P(\text{F})P(\text{S}) = q \times q \times p = pq^2$$

$$P(\text{FFF}) = P(\text{F})P(\text{F})P(\text{F}) = q \times q \times q = q^3$$

Thus, the discrete random variable X assumes the values **0**, **1**, **2**, and **3** and its probability distribution is

x	0	1	2	3
$P(x)$	q^3	$3pq^2$	$3p^2q$	p^3

It can easily be shown that the conditions of a probability distribution are satisfied: That is,

$$\sum_{\forall x} P(x) = \sum_{x=0}^{3} P(x)$$

$$= q^3 + 3p^2q + 3pq^2 + q^3$$

$$= (q+p)^3$$

$$= ((1-p)+p)^3 = 1^3 = 1,$$

and since $0 \le p \le 1$, we have $0 \le q \le 1$; that is,

$$P(x) \ge 0,$$

for $x = \mathbf{0}$ or **1** or **2** or **3**.

Again, we can relate this probability distribution to the binomial theorem. That is,

$$(p+q)^3 = \binom{3}{0}p^0 q^3 + \binom{3}{1}p^1 q^2 + \binom{3}{2}p^2 q^1 + \binom{3}{3}p^3 q^0$$

$$= q^3 + 3p^2q + 3pq^2 + p^3.$$

The coefficient of q^3 is $\binom{3}{0}$ represents the number of ways in which we can obtain **0** successes in **3** trials, and of course it is $\binom{3}{0} = \frac{3!}{0!(3-0)!} = \mathbf{1}$ way; that is, **FFF**. The probability of this happening is $q \times q \times q = q^3$.

Thus, we have established a pattern in these experiments. Along with their probability distributions, they can be expressed in terms of the binomial theorem as follows:

$$n=1, \quad P(x) = \binom{1}{x} p^x q^{1-x} \quad x = \mathbf{0}, \mathbf{1}$$

$$n=2, \quad P(x) = \binom{2}{x} p^x q^{2-x} \quad x = \mathbf{0}, \mathbf{1}, \mathbf{2}$$

$$n=3, \quad P(x) = \binom{3}{x} p^x q^{3-x} \quad x = \mathbf{0}, \mathbf{1}, \mathbf{2}, \mathbf{3}.$$

$$\vdots$$

The general case can therefore be written as

$$P(x) = \binom{n}{x} p^x q^{n-x}, \quad x = \mathbf{0}, \mathbf{1}, \mathbf{2}, \ldots, \mathbf{n},$$

where the values of x represent the possible number of successes in n trials, p and $q = (1-p)$ being the probability of a **success** or a **failure**, respectively, at each trial.

Histograms for the **binomial**(n,p) with Varying n, fixed p

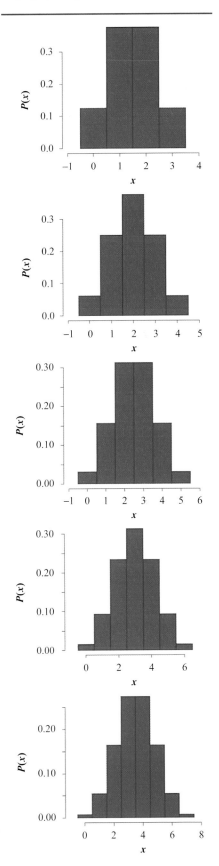

Note: $p=1$ is the sure event and $p=0$ is the impossible event, therefore, the binomial probability distribution restricts p to be greater than **0**, and less than **1**.

Logic

 The study of reasoning

 The science which deals with the criteria necessary for validity of inference

Histograms for the **binomial(n,p)** with Fixed **n**, varying **p**

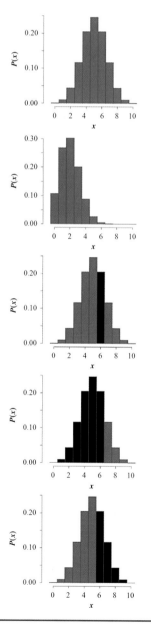

We shall now formally state the **binomial probability distribution**.

Definition 6.4.1 Binomial Probability Distribution

A discrete random variable X is said to have a **binomial** or **Bernoulli probability distribution** if its probability distribution is given by the following function:

$$P(x|n,p) = \binom{n}{x} p^x q^{n-x},$$

for $x=0,1,2,\ldots,n$, where $0<p<1$, $q=1-p$ and the vertical bar "|" is read "**such that**."

In the preceding discussion, we developed a pattern leading to the **binomial probability distribution**, also known as the **Bernoulli probability distribution**. We shall now give a summary of the assumptions that underlie the probability distribution and a shorter logic that leads to its formulation.

Consider an experiment consisting of **n** independent trials; the outcome of each trial is either a success or a failure with probabilities **p** and $q=(1-p)$, respectively. Let **0** denote a failure and **1** denote a success. If the random variable X denotes the number of successes among the **n** independent trials, the value that A: assumes will equal the number of **1**'s appearing. For example, in a sequence of **n** trials such as **0, 1, 1, 0, 1, 1, 1, 0, 0, 0** (here $n = 10$), we have a failure on the first trial, a success on the second and third trials, a failure on the fourth trial, and so on.

In such a sequence, the possible values that X can assume are $0,1,2,\ldots,n$. In order to obtain the probability that there are $X=r$ successes, we proceed as follows: The r **1**'s and $(n-r)$ **0**'s can occur in any one of

$$\binom{n}{r} = \frac{n!}{r!(n-r)!}$$

distinct orderings of the **n** trials. The probability of each such ordering is equal to the product of r **p**'s and $(n-r)$ **q**'s because all trials are independent.

Adding these probabilities, we obtain the **binomial probability distribution**: The probability of r successes is given by

$$P[X=r] = \binom{n}{r} p^r q^{n-r},$$

for $r=0,1,2,\ldots,n$ where $0<p<1$ and $q=1-p$.

It is easy to see that $P(x)$ is a probability distribution, it satisfies the basic two properties.

$$\sum_{\forall x} P(x) = \sum_{x=0}^{n} P(x) = q^n + np^{n-1}q + \cdots + p^n$$
$$= (q+p)^n$$
$$= ((1-p)+p)^n$$
$$= 1^n = 1,$$

where $0 \leq p \leq 1$. Moreover, we have,

$$P(x) \geq 0,$$

for $x=0, 1, 2, 3, \ldots, n$.

In actual situations in which the binomial probability distribution is applicable, the number of independent trials, **n**, and the probability of success **p** must be specified. However, we shall see in later chapters that the probability of "success," **p**, is estimated from a given set of data.

Common Binomial Calculations

There are usually three types of questions which are likely to be asked with respect to a random variable X that characterizes a worldwide phenomenon. These are as follows:

1. What is the probability that the discrete random variable X will equal a point r **exactly**? That is,

$$P[X=r]=P(r)$$

2. What is the probability that the discrete random variable X will assume values up to the point **a** (**at most r**)? That is,

$$P[X\leq r]=\sum_{\forall x\leq r} P(x)=\sum_{i=0}^{r}P(x).$$

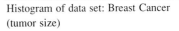

Histogram of data set: Breast Cancer (tumor size)

3. What is the probability that the discrete random variable X will assume values greater than or equal to the point **a** (**at least r**)? That is,

$$P[X\geq r]=\sum_{\forall x\geq r} P(x)=\sum_{i=0}^{r}P(x).$$

The following examples will illustrate the meaning and usefulness of the **binomial probability distribution** in all real world questions.

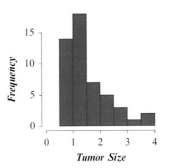

Example 6.4.2 Breast Cancer

Given that 28% of breast cancer patients have tumors of size greater than 1 cm, what is the probability that in a sample of 100 breast cancer patients, exactly 25 have a tumor size greater than 1 cm.

Note that in this problem that the four properties of the binomial probability distribution are satisfied.

Solution

Given that the probability of a tumor size greater than 1 cm is 28%, we have $p=0.28$, in a sample of 100 patients means $n=100$ and therefore, we have

$$P(25)=\binom{100}{25}0.28^{25}(1-0.28)^{75}=0.0731.$$

Therefore, given that 28% of breast cancer tumors have a size greater than 1 cm, in a sample of 100 patients; there is a 7.31% chance that exactly 25 patients have a tumor of size greater than 1 cm.

Example 6.4.3 Brain Cancer

Given that 40% of brain cancer patients are female, what is the probability in a sample of 10 brain cancer patients, that exactly five are female.

Histogram

Solution

Here you can reason that the four properties of the binomial probability distribution are satisfied. Thus, given that the probability of being female is 40%, we have $p=0.40$, in a sample of 10 patients, this means $n=10$ and hence

$$P(5)=\binom{10}{5}0.40^{5}(1-0.40)^{5}=0.2007.$$

Therefore, we can conclude that there is a 20.07% chance that exactly five of the brain cancer patients are female in a random sample of 10 patients.

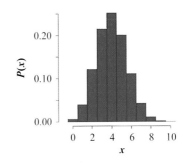

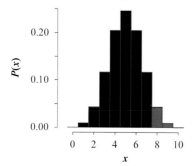

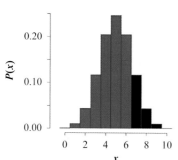

Complete exactly **twelve** passes
$X \sim \textbf{Binomial}(16, 0.62)$

$P[X=12] \approx \textbf{0.1224}$

Example 6.4.4 Ten Fair Coins

Consider the experiment of tossing a fair coin 10 times. We shall consider heads a success and tails a failure. What is the probability that:

(a) We will observe exactly seven successes?

(b) We will observe at most seven successes?

(c) We will observe at least seven successes?

Solution

Clearly, the probability of a head (a success) is one-half, $\left(p=\frac{1}{2}\right)$, $n=10$ and the assumptions that underlie the binomial probability distribution are satisfied.

The first part of the problem, the probability of exactly seven successes, is given by

$$P[X=7]=P(7)=\binom{10}{7}\left(\frac{1}{2}\right)^{7}\left(1-\frac{1}{2}\right)^{3}$$

$$=\binom{10}{7}\left(\frac{1}{2}\right)^{7}\left(\frac{1}{2}\right)^{3}$$

$$=\frac{10!}{7!(10-7)!}\left(\frac{1}{2}\right)^{10}$$

$$=\frac{15}{128}\approx 0.1172.$$

Thus, the probability that we will observe exactly 7 heads in 10 tosses of a fair coin is 0.1172.

Part (b), the probability of observing at most 7 heads in 10 trials, is given by

$$P[X \le 7]=\sum_{x=0}^{7}P(x)=\sum_{x=0}^{7}\binom{10}{x}\left(\frac{1}{2}\right)^{x}\left(1-\frac{1}{2}\right)^{n-x}$$

$$=\binom{10}{0}\left(\frac{1}{2}\right)^{0}\left(\frac{1}{2}\right)^{10}+\binom{10}{1}\left(\frac{1}{2}\right)^{1}\left(\frac{1}{2}\right)^{9}+\cdots+\binom{10}{7}\left(\frac{1}{2}\right)^{7}\left(\frac{1}{2}\right)^{3}$$

$$=\frac{121}{128}\approx 0.9453.$$

Hence, there is a 94.53% chance that in 10 flips of a fair coin we will observe at most seven successes (heads).

Part (c), the probability of observing at least 7 heads in 10 trials, is

$$P[X \ge 7]=\sum_{x=7}^{10}P(x)=\sum_{x=7}^{10}\binom{10}{x}\left(\frac{1}{2}\right)^{x}\left(1-\frac{1}{2}\right)^{10-x}$$

$$=\binom{10}{7}\left(\frac{1}{2}\right)^{7}\left(\frac{1}{2}\right)^{3}+\binom{10}{8}\left(\frac{1}{2}\right)^{8}\left(\frac{1}{2}\right)^{2}+\cdots+\binom{10}{10}\left(\frac{1}{2}\right)^{10}\left(\frac{1}{2}\right)^{0}$$

$$=\frac{176}{1024}\approx 0.1719.$$

Therefore, there is a 17.19% chance that we will observe at least seven successes in 10 flips of a fair coin.

Example 6.4.5 Completions in Football

A quarterback on a football team has a pass completion average of 0.62. If, in a given game, he attempts 16 passes, what is the probability that he will complete (a) exactly 12 passes and (b) more than half of his passes?

Solution

The assumptions that underlie the binomial probability distribution are satisfied; that is, the experiment consists of 16 identical trials and each trial (pass) results in a completion (success) or an incomplete pass (failure). The probability of the pass being completed on each try is equal to 0.62, and it remains the same for each of the 16 attempts. The probability of a failure is $1-p=1-0.62=0.38$, and the trials (pass attempts) are independent.

For Part (a), the probability that the quarterback will complete exactly 12 passes in 16 attempts is

$$P[X=12]=P(12)=\binom{16}{12}(0.62)^{12}(1-0.62)^{16-12}$$
$$=\frac{16!}{12!(16-12)!}(0.62)^{12}(0.38)^4$$
$$=0.1224$$

Thus, there is only a 12.24% chance that he will complete 12 passes in 16 tries.

For Part (b), the probability that he will complete more than half of his passes; that is, that the random variable X assumes values greater than eight, is

$$P[X>8]=P[X\geq9]=\sum_{x=9}^{16}\binom{16}{x}(0.62)^x(0.38)^{16-x}$$
$$=\binom{16}{9}(0.62)^9(0.38)^{16-9}+\binom{16}{10}(0.62)^{10}(0.38)^{16-10}+\cdots+\binom{16}{16}(0.62)^{12}(0.38)^{16-16}$$
$$=0.7701.$$

Therefore, the probability that the quarterback will complete more than half of his passes is 0.7701% or 77.10%.

Example 6.4.6 Expected Number of Children

A certain family has four children. Assume that the probability of any of the children being a boy is 1/2. What is the probability that there are two boys and two girls in the family?

Solution

Here, it can be reasoned that the assumptions that underlie the binomial probability distribution are satisfied.

For example the four children in the family will be treated as four independent trials of an experiment for which the probability of a success is 1/2. Thus, $n=4, p=1/2$, and $q=1-1/2=1/2$ the probability of two boys is

$$P[X=2]=P(2)=\binom{4}{2}\left(\frac{1}{2}\right)\left(1-\frac{1}{2}\right)^{4-2}$$
$$=\frac{4!}{2!(4-2)!}\left(\frac{1}{2}\right)^4$$
$$=0.375.$$

Therefore, there is a 37.5% chance that the family consists of two boys and two girls.

Example 6.4.7 Expected Number of Defects

A department store receives large lots of a certain product and applies the following inspection scheme for accepting the lots. Ten of the items are to be examined and the lot will be rejected if two or more are defective. If it is known from prior experience that each lot contains about 4% defective items, find the probability of (a) accepting the lot and (b) rejecting the lot.

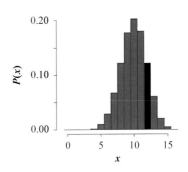

Wishful thinking for any **quarterback!**
$X \sim \text{Binomial}(16, 0.62)$

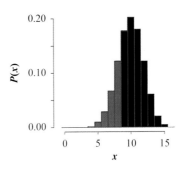

$P[X>8] \approx 0.7701$

Note, here to complete more than half the passes, 8 is not included, hence, in this discrete case,
$$P[X>8]=P[X\geq9]$$

$X \sim \text{Binomial}(4, 0.5)$

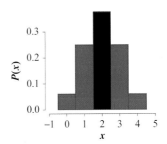

$P[X=2]=0.375$

$X \sim \textbf{Binomial}(10, 0.04)$

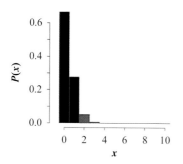

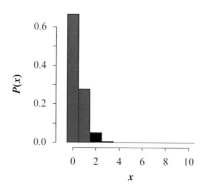

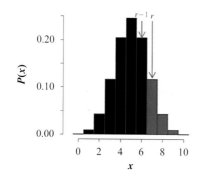

Solution

Here $n=10$ and $p=0.04$, the probability of observing a defective item on a single trial. Applying the binomial probability distribution, we obtain the desired probabilities.

For Part (a), the probability of accepting the lot is simply the probability that the random variable X assumes the values 0 or 1; a defective or non-defective-item. That is, if it assumes any value greater than one, the lot will be rejected. Thus,

$$P[X \leq 1] = \sum_{x=0}^{1} P(x) = \sum_{x=0}^{1} \binom{10}{x} (0.04)^x (1-0.04)^{10-x}$$

$$= P[X=0] + P[X=0]$$

$$= P(0) + P(1)$$

$$= \binom{10}{0} (0.04)^0 (0.96)^{10-0} + \binom{10}{1} (0.04)^1 (0.96)^{10-1}$$

$$= \frac{10!}{0!10!} (0.96)^{10} + \frac{10!}{1!9!} (0.04)(0.96)^9$$

$$= 0.6648 + 0.2770 = 0.9418.$$

Hence, there is a 94.18% chance of accepting the lot. That is, the probability that a lot will be accepted is 94.18%.

For Part (b), the probability of rejecting the lot is equal to one minus the probability of accepting the lot or the probability that we have observed two or more defective items in the lot. That is,

$$P[X \geq 2] = 1 - P[X < 2] = 1 - P[X \leq 1]$$

$$= 1 - 0.9418 = 0.0582.$$

That is, the probability that the lot will be rejected is 5.82%.

Note that in this case we are using the property that $P(A) = 1 - P(\overline{A})$; that is, to simplify the calculations for obtaining the desired probability, we use the fact that the $P[X \geq 2] = 1 - P[X \leq 1]$ that is equivalent to $1 - P[X < 2]$ since in a discrete probability distribution $P[X < r] = P[X \leq r-1]$. Thus, the probability of rejecting the lot is 0.0582.

It is important to note, that $P[X < r] = P[X \leq r-1]$ is not true in a continuous probability distribution. In fact, in a continuous probability distribution, $P[X=r] = 0$ and therefore, $P[X < r] = P[X \leq r]$. This will be further discussed in Section 5.6. The difference between "**less than**" and "**greater than**" as opposed to "**less than or equal to**" and "**greater than or equal to**" should also be noted.

Example 6.4.8 Sickle Cell Anemia

Sickle cell anemia, a hereditary disease, afflicts 1 out of 400 African Americans in this country. If a screening program for newborns is set up in a metropolitan hospital, what is the probability that exactly one case of sickle cell anemia will be detected among the first 25 African Americans infants born after the program's initiation?

Solution

Here, a newborn infant either does or does not suffer from the disease. The probability that an infant will have the disease is $p = \dfrac{1}{400}$. Assume that the 25 African American infants tested will represent that number of independent trials. Thus, the binomial probability distribution is applicable and

$$P[X=1] = P(1)$$

$$= \binom{25}{1}\left(\frac{1}{400}\right)^1 \left(1 - \frac{1}{400}\right)^{25-1}$$

$$\approx 0.0588.$$

Therefore, there is only a 5.88% chance that exactly one case of sickle cell anemia will be detected among the 25 babies.

We shall continue our study of the **binomial probability distribution** in the next section by examining two of its basic properties; namely, the **expected value** and **variance** of a random variable that is binomially distributed.

It should be noted that as p gets extremely small (that is, for rare events, the distribution becomes significantly skewed and extremely large samples are needed to focus in on the expected number of "successes." This special form of the binomial distribution is the **Poisson probability distribution**.

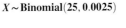

$X \sim \text{Binomial}(25, 0.0025)$

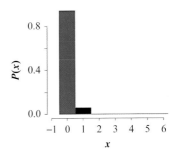

$P[X=1] \approx 0.0588$

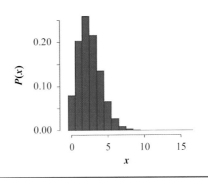

PROBLEMS

Basic Problems

Using a binomial probability distribution to find probabilities:

6.4.1. Given $X \sim \text{Binomial}(n=10, p=0.2)$, find $P(2)$.

6.4.2. Given $X \sim \text{Binomial}(n=100, p=0.2)$, find $P(20)$.

6.4.3. Given $X \sim \text{Binomial}(n=25, p=0.4)$, find $P(20)$.

At Most, At Least & Exactly!

6.4.4. State the assumptions that underlie the binomial probability distribution and give an example of a physical situation that satisfy these assumptions.

6.4.5. **Fair coins:** A fair coin is tossed 10 times. What is the probability we will observe:
  a. Exactly 6 heads?
  b. At most 6 heads?
  c. At least 6 heads?

6.4.6. **Bulls eye:** A man fires at a target six times; the probability of him hitting the bull's eye is 0.40 on each trial.
  a. What is the probability that the man will hit the target at least once?
  b. What is the probability that he will not hit the target at all?

6.4.7. **Batting average:** A baseball player's batting average is 0.310. If in a given game he bats four times, what is the probability that he will get
  a. No hits?
  b. At most two hits?
  c. At least two hits?

6.4.8. **Standard deck of cards**: A card is drawn and replaced four times from a standard deck of 52 cards. What is the probability that
   a. Four aces were drawn?
   b. Four diamonds were drawn?
   c. Four picture cards were drawn?
   d. Why is it necessary that the card was replaced?

6.4.9. **Birth**: Assuming that newborns are equally likely to be boys or girls, what is the probability that a family of six children will have at least two boys?

6.4.10. **Space travel**: Assuming that it is known that 99.8% of the launchings of satellites into orbit are successful. What is the probability that in the next five launchings there will be
   a. No mishaps?
   b. Exactly one mishap?
   c. At least one mishap?

6.4.11. **Passing statistics**: The probability that a history student will pass a statistics course is 0.80. What is the probability that out of 10 history majors enrolled in such a course
   a. At least five will pass?
   b. None will fail?

6.4.12. **Christmas-treeing**: When a student does not know the answers to a multiple choice test, they often randomly complete the test create strings of dark circles for each answer in the shape of Christmas tree lights. A student Christmas-trees a 10 question exam where each question had five options of which exactly one is correct.
   a. What is the probability that the student correctly answered exactly 7 questions?
   b. What is the probability that the student passed with 7 or more correct answers?
   c. What is the probability that the student answers at most 6 correctly?

6.4.13. **English alphabet**: The most frequently used letter in the English alphabet is E. The letter Z is the least frequently used letter. The letter E occurs 12.7% of the time whereas Z occurs only 0.07% of the time.
   a. What is the probability of at least two Zs occurring in a page containing 2500 characters?
   b. What is the probability of exactly two Es occurring in a sentence containing 25 characters?
   c. What is the probability of at least two Es occurring in a sentence containing 25 characters?
   d. What is the probability of exactly one Z occurring in a page containing 2500 characters?
   e. What is the probability of exactly two Zs occurring in a page containing 2500 characters?

Critical Thinking

6.1.14. State the assumptions that underlie the **binomial probability distribution** and give an example of a physical situation that satisfies these assumptions.

6.1.15. A fair coin is tossed 10 times. What is the probability that we will observe
   a. Exactly 6 heads?
   b. At most 6 heads?
   c. At least 6 heads?

6.4.16. A fair die is rolled six times. Consider a roll to be a success if a 1 or 6 is observed.
   a. What is the probability that we observe exactly two successes?
   b. What is the probability that we observe three failures? (A failure would be a 2, 3, 4, or 5 being observed.)

6.4.17. A man fires at a target six times; the probability of his hitting it is equal to 0.40 on each trial.
   a. What is the probability that the man will hit the target at least once?
   b. What is the probability that he will not hit the target at all?

6.4.18. A baseball player's batting average is 0.310. If in a given game he bats four times, what is the probability that he will get
   a. No hits?
   b. At most two hits?
   c. At least two hits?

6.4.19. A card is drawn and replaced four times from an ordinary deck of 52 cards and the sequence of colors is observed. What is the probability that
  a. Four red cards were drawn?
  b. Four black cards were drawn?
  c. Two red and two black cards were drawn?
  d. Why is it necessary to say that the card was replaced?

6.4.20. The basketball team of a certain university has a probability of 0.80 of winning whenever it plays a home game. The team will play 10 games at home. What is the probability that the team will win
  a. Exactly eight games?
  b. At most six games?

6.4.21. Assuming that newborns are three times more likely to be boys or girls, what is the probability that a family of six children will have at least two boys?

6.4.22. Assume that it is known that 95% of the launchings of satellites into orbit are successful. What is the probability that in the next four launchings there will be
  a. No mishap?
  b. Exactly one mishap?

6.4.23. The probability that a history major will pass a finite mathematics course is 0.90. What is the probability that out of 10 history majors enrolled in such a course
  a. At least five will pass?
  b. None will fail?

6.5 EXPECTED VALUE AND VARIANCE FOR A BINOMIAL RV

You will recall that in speaking of the **expected value** or **mean** of a random variable, we are referring to a measure of its **central tendency**. We shall now focus on obtaining the **expected value** of a discrete random variable that is being characterized by the **binomial probability distribution**.

For example, in a given problem to which the **binomial probability distribution** is applicable; that is, a situation in which we have: **identical and independent trials**, **each classified as a success or a failure with probability of success on each trial** μ, the **expected value** or **mean** of the random variable representing the number of successes observed can be interpreted as the number of successes we can "**expect**" in n trials.

Definition 6.5.1 Expected Value: Binomials

The **mean** or **expected value** of the discrete random variable X which is **binomially distributed** is given by

$$\mu = E(X) = np.$$

For example, if we inspect n components of a system in which each has a probability p of being operable, we could "**expect**" the total number of operable components to be close to np.

☑ Using summation notation, we can express this through the following manipulations:

PD = Probability Distribution

Expected
 To consider probable or certain
 To look forward to or await

Recall:

$$E(x) = \sum_{\forall x} x P(x)$$
$$x! = x(x-1)!$$
$$0 \times a = 0$$
$$p^x = p \times p^{x-1}$$
$$n! = n(n-1)!$$

$$E(x)=\sum_{x=0}^{n}xP(x)=\sum_{x=0}^{n}x\times\binom{n}{x}p^{x}(1-p)^{n-x}$$

$$=\sum_{x=0}^{n}x\times\frac{n!}{x!(n-x)!}p^{x}(1-p)^{n-x}$$

$$=\sum_{x=1}^{n}\frac{n!}{(x-1)!(n-x)!}p^{x}(1-p)^{n-x}$$

$$=\sum_{x=1}^{n}np\frac{(n-1)!}{(x-1)!(n-x)!}p^{x-1}(1-p)^{n-x}$$

$$=np\sum_{x=0}^{n}\frac{(n-1)!}{(x-1)!(n-x)!}p^{x-1}(1-p)^{n-x}$$

$$=np\sum_{y=0}^{n-1}\frac{(n-1)!}{y!(n-y-1)!}p^{y}(1-p)^{n-y-1}$$

$$=np\sum_{y=0}^{m}\frac{m!}{y!(m-y)!}p^{y}(1-p)^{m-y}$$

$$=np(1)=np$$

That is, the expected value in any **binomial distribution** is the probability of success (percent success) times the number of trials (sample size n).

Let:

$$y=x-1$$

then

$$x=1\Rightarrow y=0$$

and

$$x=n\Rightarrow y=n-1$$

Let:

$$m=n-1$$

Remember:

$\Rightarrow$ reads "implies"

We expect 60 families to have at least one girl. In reality we may not realize this expectation but we should be close to the number.

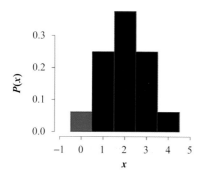

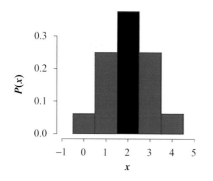

Example 6.5.1 Boys & Girls

Let us consider 64 families with four children each and assume all sex distributions to be equally probable. How many families would you expect to have (a) at least one girl and (b) exactly two girls?

Solution

For Part (a), the probability of at least one girl is obtained using the binomial probability distribution; that is,

$$P[X\geq 1]=1-P[X<1]$$

$$=1-P[X=0]$$

$$=1-\binom{4}{0}\left(\frac{1}{2}\right)^{0}\left(\frac{1}{2}\right)^{4}$$

$$=1-\frac{1}{16}=\frac{15}{16}$$

is the probability of each family having at least one girl. Thus, for $n=64$, $p=\frac{15}{16}$, and using the fact that the $E(X)=np$, we have the expected number of families having at least one girl:

$$E(X)=np=64\times\frac{15}{16}=60.$$

Therefore, we **expect** 60 families to have at least one girl.

Solution—cont'd

For Part (b), the probability of exactly two girls is obtained as follows:

$$P[X=2]=\binom{4}{2}\left(\frac{1}{2}\right)^2\left(\frac{1}{2}\right)^2$$

$$=\frac{6}{16}=\frac{3}{8}.$$

Thus, for $n=64$ and $p=3/8$, we have the expected number of families with two girls:

$$E(X)=np=64\times\frac{3}{8}=24.$$

Therefore, we **expect** to find that about 24 of the 64 families have two girls.

Example 6.5.2 Damaged Books

A bookstore receives a shipment of 50 books. Previous experience has indicated that about 10% of the books are damaged in transit. Find the expected number of damaged books.

Solution

Here, $n=50$ and $p=0.10$ and

$$E(X)=np=50\times0.1=5.$$

Thus, in a shipment of 50 books we can **expect** about five books to be damaged.

Example 6.5.3 Completions

Refer to Example 6.4.5. What is the number of passes that the quarterback is expected to complete?

Solution

The probability that he will complete a pass any time he attempts one is 0.62. If he throws the ball 16 times, we expect that he will complete about 10 passes. That is,

$$E(X)=np=16\times0.62=9.92.$$

Note that for this answer to be meaningful we must round it off to the nearest whole number—in this case 10.

As we mentioned previously, two other important characteristics of a random variable are its **variance** and **standard deviation**; that is, the measure of the "**spread**" or "**dispersion**" of the random variable around its expected value or central tendency. We proceed by stating the variance of a random discrete variable which is **binomially distributed**. The variance takes into account the sample size n and the probability of success p and the probability of failure $q=(1-p)$; recall, "and" means "multiply."

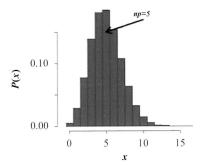

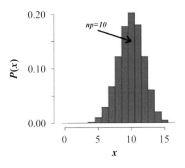

Remember everything in our society is subject to **variation.**

In real world problem we do not know p; thus, we estimate it.

$$N \ vs. \ n$$
$$\mu_{\hat{p}}=p\approx\hat{p}$$
$$\sigma_{\hat{p}}^2=npq \ \approx \ n\hat{p}\hat{q}$$
$$\sigma=\sqrt{npq} \ \approx \ \sqrt{n\hat{p}\hat{q}}$$
$$\hat{q}=\frac{x}{n}$$

x is the number of successes and n is the number of trials

$$\sigma^2 = 3.75 \approx 4$$
$$\Downarrow$$
$$\sigma = 1.94 \approx 2$$

Thus, we are more likely to have

$$\mu \mp 2$$
$$\Downarrow$$
$$np \mp 2$$
$$\Downarrow$$
$$60 \mp 2$$
$$\Downarrow$$
$$58 \, to \, 62$$

$$\sigma^2 = 15$$
$$\Downarrow$$
$$\sigma = 3.88 \approx 4$$

$$24 \mp 4$$
$$\Downarrow$$
$$20 \, to \, 28$$

it is more likely that the number of families that will have at least one girl as a member between

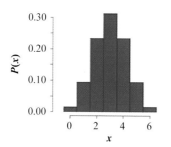

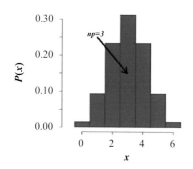

$$\mu = np = 3$$
$$\sigma^2 = 1.5$$
$$\Downarrow$$
$$\sigma = 1.22 \approx 1$$

Thus,

$$\mu \mp 1$$
$$\Downarrow$$
$$np \mp 1$$
$$\Downarrow$$
$$3 \mp 1$$
$$\Downarrow$$
$$2 \, to \, 4$$

Definition 6.5.2 Variance/Deviation in a Binomial

The **variance** of the discrete random variable X which is **binomially distributed** is given by

$$\sigma^2 = \text{Var}(X) = V(x) = npq,$$

and therefore, the **standard population deviation** is given by

$$\sigma = \sqrt{npq} = \sqrt{np(1-p)}.$$

The proof of this statement is similar to the proof that $E(X) = np$ and is recommended that the student think through the summation notation required in this proof. Interpreting this formulation, we have that the variances in the sample size in taking into account, not only, the probability of success, but the probability of failure.

Example 6.5.4 Family of Four

Find the variance and standard deviation of the random variable X characterized in Example 6.5.2; that is the number of girls in a family of four assuming that any given child could have been a boy or girl.

Solution

Here, for Part (a), $p = \frac{15}{16}$, $q = 1 - p = \frac{1}{16}$ and $n = 64$. Thus, the variance and standard deviation are given by

$$\sigma^2 = \text{Var}(X) = npq = 64 \times \frac{15}{16} \times \frac{1}{16} = 3.75$$

and $\sigma = \sqrt{3.75} \approx 1.94$.

This tells us that with such a large probability of that within a family of four there is at least one girl, we expect $np = 60$ out of 64 such families to have at least one girl, but this number will deviate by approximately 2; hence, it is more than likely that in a family of four, between 58 and 62 families have at least one girl.

For Part (b), $p = \frac{3}{8}$, $q = 1 - \frac{3}{8} = \frac{5}{8}$, and $n = 64$. Thus we have,

$$\sigma^2 = \text{Var}(X) = npq = 64 \times \frac{3}{8} \times \frac{5}{8} = 15$$

and $\sigma = \sqrt{15} \approx 3.88$.

This tells us that with such a large probability that within a family of four there are exactly two girls, we expect $np = 24$ out of 64 such families to have exactly two girls, and we expect this to deviate by approximately 4; hence, it is highly likely that out of 64 families with four children, between 20 and 28 families will have exactly two girls.

Thus, in both cases, since the **variance** or **standard deviation** is fairly small, we can expect most of the values that the random variable can assume to be located close to its **expected value**.

Example 6.5.5 Boys & Girls

Assume all sex distributions to be equally probable:
a. What is the expected number of girls in a family of six children?
b. What is the variance of the random variable that describes the present situation?
c. What is the probability that a family of six has the "expected" number of girls?

Solution

For Part (a), since the sexes are equally probable we have $p=q=0.5$. Thus, for $n=6$, we have

$$E(X)=np=6\times0.5=3.$$

For Part (b), the variance is given by

$$\mathbf{Var}(X)=npq=6\times0.5\times0.5=1.5$$

and

$$\sigma=\sqrt{15}\approx1.22.$$

Therefore, while in a family we expect to have three girls and three boys, this may deviate by 1.22; that is, we expect three boys and three girls, give or take approximately one.

For Part (c), the probability that the expected number of girls; that is, three, will be realized in the present family is as follows:

$$P[(X)=E(X)]=P[X=3]=\binom{6}{3}(0.5)^3(0.5)^{6-3}$$

$$=\frac{6!}{3!(6-3)!}(0.5)^6$$

$$=0.1875.$$

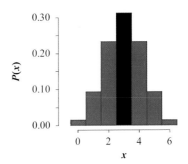

Thus, there is only an 18.75% chance that the expected number of girls will be realized in a family of six.

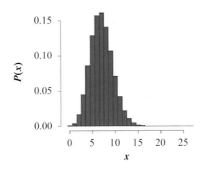

Example 6.5.6 Births in the U.S.

Children born to older women are more susceptible to premature births, with 14% of all women over the age of 40 giving birth at between 32 and 36 weeks of gestation. Given a group of 50 women, age 40 years or more,

a) What is the expected number of premature births?
b) What is the variance of the random variable that describes the present situation?
c) What is the probability that in this group of women, between 5 and 9 babies will be born prematurely?

$$\sigma^2=6.02$$
$$\Downarrow$$
$$\sigma=2.45\approx2.5$$
Thus,
$$\mu\mp2.5$$
$$\Downarrow$$
$$np\mp2.5$$
$$\Downarrow$$
$$7\mp2.5$$
$$\Downarrow$$
$$4.5\ to\ 9.5$$

Solution

For Part (a), since there is a 14%, $p=0.14$, chance that the baby is premature, with $n=50$, we have

$$E(X)=np=50\times0.14=7$$

For Part (b), the variance is given by

$$\mathbf{Var}(X)=npq=50\times0.14\times0.86=6.02$$

and $\sigma=\sqrt{6.02}\approx2.45$.

Therefore, in a group of 50 women over the age of 40 who give birth in the United States, we expect an average of 7 premature births with a 2.5 standard deviation. Therefore, on average, we expect between 4.5 and 9.5 premature births; however, on a discrete measure where the RV take on the integer values, we can round and estimate between 5 and 9 premature births in a group of 50 women over the age of 40.

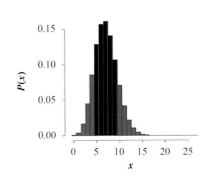

Continued

$$P[X=5] = \binom{50}{5}(0.14)^5(0.86)^{45} \approx 0.12857$$

$$P[X=6] = \binom{50}{6}(0.14)^6(0.86)^{44} \approx 0.15698$$

$$P[X=7] = \binom{50}{7}(0.14)^7(0.86)^{43} \approx 0.16063$$

$$P[X=8] = \binom{50}{8}(0.14)^8(0.86)^{42} \approx 0.14055$$

$$P[X=9] = \binom{50}{9}(0.14)^9(0.86)^{41} \approx 0.10678$$

Solution—cont'd

For Part (c), the probability that between 5 and 9 women, included, have premature babies is as follows:

$$P[5 \leq X \leq 9] = P(5) + P(6) + P(7) + P(8) + P(9)$$

$$= \binom{50}{5}(0.14)^5(0.86)^{50-5} +$$

$$\cdots + \binom{50}{9}(0.14)^9(0.86)^{50-9}$$

$$= 0.12857 + 0.15698 + 0.16063 + 0.14055 + 0.10678$$

$$= 0.69351.$$

Thus, there is an approximate 69% chance that in a group of 50 women who were over the age of 40 when they gave birth will have done so prematurely.

PROBLEMS

Basic Problems

Find the mean and variance in normal approximation of the binomial. Find μ and σ^2

6.5.1. $n=100, p=0.3$

6.5.2. $n=10, p=0.3$

6.5.3. $n=250, p=0.1$

6.5.4. $n=125, p=\dfrac{1}{5}$

6.5.5. $n=135, p=\dfrac{2}{7}$

6.5.6. **Damaged goods**: A drugstore received a shipment of 60 vials of a certain pharmaceutical. Previous experience indicates that such a shipment will result in about 2% of the vials being damaged. Find the expected number of undamaged vials in the shipment.

6.5.7. **Light bulbs**: The probability that a new light bulb will be defective is 0.005. If a lot of 5000 light bulbs was purchased from the same company, what is the expected number of defective light bulbs in the lot?

6.5.8. **Light bulbs**: Find the variance and standard deviation of the random variable described in the previous exercise. Interpret the meaning of the answer obtained.

6.5.9. **ACT**: Assume that 25% of all those who take a particular achievement examination obtain scores that indicate a need for remedial work. If the examination is given to 20 randomly chosen people, what is the expected number in need of remedial work?

6.5.10. **Lemons**: Suppose that 20% of the used cars of Lemons Automotive Company are defective. If 30 cars were sold in a given week, how many of these cars would you expect to be defective?

6.5.11. **Christmas-treeing**: When a student does not know the answers to a multiple choice test, they often randomly complete the test create strings of dark circles for each answer in the shape of Christmas-tree lights. A student Christmas-trees a 10-question exam where each question had five options of which exactly one is correct.

a. Find the mean and standard deviation for the number of correct answers.

b. Using the normal approximation and correction for continuity, what is the probability that the student correctly answered exactly seven questions?

c. Using the normal approximation and correction for continuity, what is the probability that the student passed with seven or more correct answers?

d. Using the normal approximation and correction for continuity, what is the probability that the student answers at most six correctly?

6.5.12. English alphabet: The most frequently used letter in the English alphabet is E. The letter Z is the least frequently used letter. The letter E occurs 12.7% of the time whereas Z occurs only 0.07% of the time.

   a. Find the mean and standard deviation for the number of times the letter E will be found in a written page of 2500 characters.

   b. In a passage with 2500 characters, is it unusual to find the letter E 180 times?

   c. How many Zs are expected in these 2500 characters?

   d. Using the normal approximation and correction for continuity, what is the probability of exactly one Z?

6.5.13. Large families: A family consists of 10 children. Assuming that the sexes are equally probable, find the expected number of boys in the family. What is the probability that the expected number of boys will not occur?

6.5.14. Modified coins: A certain coin has been modified in such a way that when it is tossed a head will turn up 66% of the time. Find the expected number of tails if the coin is tossed 100 times.

6.5.15. Standard deck of cards: From an ordinary deck of 52 cards a card is drawn at random, recorded, replaced and the deck is shuffled. This is done five times.

   a. What is the probability that all the cards are diamonds?

   b. What is the number of diamonds that one would expect?

   c. What is the probability of the expected number of diamonds?

6.5.16. M&M MARS: Proportion of **M&Ms** by variety according to M&M Mars Inc.

Variety	Red (%)	Orange (%)	Yellow (%)	Green (%)	Blue (%)	Brown (%)
Plain	13	20	14	16	24	13
Peanut	12	23	15	15	23	12
Dark	17	16	17	16	17	17
Almond	10	20	20	20	20	10
Crispy	17	16	17	16	17	17
Peanut butter	10	20	20	20	20	10

   a. M&M Mars Company claims that 14% of their **Plain M&Ms** are yellow. In a bag of 253 M&Ms, how many are expected to be yellow; that is, what is the mean number of yellow M&Ms? What is the standard deviation?

   b. M&M Mars Company claims that 20% of their **Almond M&Ms** are orange. Find the mean and standard deviation for the number of orange **M&Ms** in such a bag.

   c. What is your favorite type of **M&M** and what is your favorite color? Purchase a bag of **M&Ms** of your choice and count the number of **M&Ms** in your favorite color. Is this what you expected to find? Does the observed frequency fall within reason; that is, within two standard deviations of the mean?

   d. In a bag of **Peanut Butter M&Ms**, what is the probability that exactly 10 out of 100 **M&Ms** are red?

   e. What is the probability that at most 2 **M&Ms** are brown in a bag of **Crispy M&Ms**?

   f. In a bag of **Dark M&Ms**, what is the probability that eight or more **M&Ms** are not green?

Critical Thinking

6.5.17. A drugstore receives a shipment of 60 vials of a certain pharmaceutical. Previous experience indicates that such a shipment will result in about 2% of the vials being damaged. Find the expected number of undamaged vials in the shipment.

6.5.18. The probability that a new light bulb will be defective is 0.005. If a lot of 5000 light bulbs were purchased from the same company, what is the expected number of defective light bulbs in the lot?

6.5.19. A family consists of 10 children. Assuming that the sexes are equally probable, find the expected number of boys in the family. What is the probability that the expected number of boys will not occur?

6.5.20. A fair die is rolled 120 times. What is the expected number of aces?

6.5.21. A certain coin has been modified in such a way that when it is tossed a head will turn up 66% of the time. Find the expected number of tails, if the coin is tossed 100 times.

6.5.22. Find the variance and standard deviation of the random variable described in Problem 6.5.19. Interpret the meaning of the answers obtained.

6.5.23. Find and interpret the meaning of the standard deviation of the situation described in Problem 6.5.20.

6.5.24. From an ordinary deck of 52 cards a card is drawn at random, recorded, replaced and the deck shuffled. This is done five times.

    a. What is the probability that all cards are diamonds?

    b. What is the number of diamonds that one would expect?

6.5.25. Assume that 25% of all those who take a particular achievement examination obtain scores that indicate a need for remedial work. If the examination is given to 20 randomly chosen people, what is the expected number in need of remedial work?

6.5.26. Suppose that 20% of the used cars of Lemons Automobile Company are defective. If 30 cars were sold in a given week, how many of these cars would you expect to be defective?

CRITICAL THINKING AND BASIC EXERCISE

Basic Problems

6.1. Consider the experiment of rolling a single fair die. Let X be the random variable denoting the number that appears uppermost.

    a. Find the probability distribution of the random variable X.

    b. Give a graphical representation of $P(x)$.

    c. Show that the conditions of a probability distribution are satisfied.

6.2. A coin has been weighted in such a way that the probability of heads is 5/6 and the probability of tails is 1/6. The coin is tossed twice. Let X be the random variable representing the number of heads that occur.

    a. Find the probability distribution of the random variable X.

    b. Give a graphical representation of $P(x)$.

    c. Show that the conditions of a probability distribution are satisfied.

6.3. Find the expected value of the discrete random variable X, the probability distribution of which is

    a.

x	–2	–1	0	1	2
$P(x)$	0.1	0.2	0.4	0.2	0.1

    b.

x	2	4	8	16
$P(x)$	0.1	0.2	0.3	0.4

6.4. Find the expected value of the discrete random variable Y, the probability distribution of which is

    a.

y	1	2	3	6	12
$P(y)$	1/3	1/4	1/8	1/8	1/6

    b.

y	–4	–2	2	4
$P(y)$	1/6	1/3	1/6	1/3

6.5. The probability that a grocer will make a profit of $100 selling a certain type of fruit during the Thanksgiving holiday is 0.80, the probability that he will take a loss of $60 is 0.20. Determine the expectation of the grocer.

6.6. A card player is paid $4 if he draws a spade from an ordinary deck of 52 cards, he is paid $6 if he draws a picture card, and he loses $0.50 if any other card is drawn. What is the expected value of the gambling game?

Critical Thinking

6.7 Determine if the random variable is discrete or continuous:

    a. In a genetic study, to monitor the productivity of a newly developed pea pod, a researcher counts the number of peas per pod.

    b. In a genetic study, to monitor the rate of mutation, a researcher measures the time between spontaneous mutations in brown rats.

    c. In a genealogical study, to monitor the population growth rate, a census counting the number of children per female is performed every 10 years.

6.8. What is the expected value and variance in the following binomial distributions:

    a. $n=182, p=0.12$

    b. $n=73, p=0.28$

    c. $n=20, p=0.92$

6.9. Given the probability of success is 5%, what is the probability that you succeed twice in 15 trials.

6.10. What is the expected value and standard deviation in the following Poisson distribution:

    a. $\lambda=3.6$

    b. $\lambda=2.8$

    c. $\lambda=16.9$

    d. $\lambda=1.44$

6.11. Given that a litter size of the common house cat is 4.1, what is the probability that a cat has a litter of 8?

6.12. What is the expected value and standard deviation in the following geometric distribution:

    a. $n=182, p=0.12$

    b. $n=73, p=0.28$

    c. $n=20, p=0.92$

6.13. Given the probability of success is 5%, what is the expected number of trials before success is obtained and what is the expected standard deviation?

SUMMARY OF IMPORTANT CONCEPTS

Definitions

6.1.1. A **function** is a relation between two variables, x and y, such that for each value of the independent variable, usually selected to be x there is at most one value of the dependent variable, y. That is, "y is a function of x" when for each value of x in the domain, there is one value of y in the range. We express this relation by $y=f(x)$.

6.1.2. A **random variable** (RV) is a numerically valued function defined over a sample space.

6.1.3. A **discrete random variable** is one that can assume only a finite, or countably infinite, number of distinct values. A discrete random variable can be easily identified by examining the nature of the values it may assume. In most practical problems, discrete random variables represent a count of sample points possessing a specific characteristic.

6.1.4. A **continuous random variable** is one that can assume any value in some interval or intervals of real numbers (and the probability that it assumes any specific value is zero).

6.2.1. Let X be a discrete random variable. A function (rule) $P(x)$ is called the **probability distribution** of the random variable X if it satisfies the following two conditions:

1. $P(x) \geq 0$, for all x

2. $\sum_{\forall x} P(x) = 1$

This distribution is also referred to as the probability density function and is often denoted by $f(x), f(x)=P(X=x)$.

6.3.1. The **arithmetic mean**, **average**, or **sample mean** of a set of n measurements $x_1, x_2, \ldots, x_n$ is equal to the sum of the measurements divided by n. We denote the average by $\bar{x}$, thus:

$$\bar{x} = \frac{\sum_{i=1}^{n} x_i}{n} \quad \text{(Listed data)}$$

$$\bar{x} = \frac{\sum_{i=1}^{k} (x_i f_i)}{n} \quad \text{(Frequency data)},$$

where f_i is the associated frequency for the set of distinct measures x_i, $i=1,2,\ldots,n$ and

$$\sum_{i=1}^{k} f_i = n$$

6.3.2. Let X be a discrete random variable with probability distribution $P(x)$. Then the **expected value** of X is given by

$$E(X) = \sum_{i=1}^{k} (x_i p_i)$$

where $p_i = P(x_i)$ and $\forall x$ is read "for all x."

6.3.3. Let X be a discrete random variable with probability distribution $P(x)$. Then the population **variance** of X is the mean of the square error, in terms of expected values is given by

$$\sigma^2 = \mathrm{Var}(X) = E[(X-\mu)^2] = \sum_{\forall x} (x-\mu)^2 P(x) = V(X) = E[X^2] - (E[X])^2$$

where $\forall x$ is read "for all x."

6.3.4. Let X be a discrete random variable with population mean μ, the residual errors, collectively called the **residuals**, are the differences between what is observed and what is expected: that is, $\varepsilon_i = x_i - \mu$.

6.3.5. Let X be a discrete random variable with probability distribution $P(x)$. The square root of the variance of a discrete random variable is called the **standard deviation**, denoted by the lowercase Greek letter sigma $\sigma = \sqrt{V(x)} = \sqrt{\sigma^2}$.

6.3.6. The **sample variance** of a set of n measurements: $x_1, x_2, \ldots, x_n$, denoted by s^2, is the average of the square of the deviations of the measurements about their sample mean; that is, $s^2 = \dfrac{\displaystyle\sum_{i=1}^{n} (x-\bar{x})^2}{n-1}$.

6.3.7. A discrete random variable X is said to have a **binomial** or **Bernoulli probability distribution** if its probability distribution is given by the following function:

$$P(x|n,p) = \binom{n}{x} p^x q^{n-x},$$

for $x = 0, 1, 2, \ldots, n$, where $0 < p < 1$, $q = 1 - p$ and the vertical bar "|" is read "**such that**."

6.5.1. The **mean** or **expected value** of the discrete random variable X which is **binomially distributed** is given by $\mu = E(X) = np$.

6.5.2. The **variance** of the discrete random variable X which is **binomially distributed** is given by $\sigma^2 = Var(X) = V(x) = npq$, and therefore, the **standard population deviation** is give by $\sigma = \sqrt{npq} = \sqrt{np(1-p)}$.

Properties

6.3.1. Given a data set, $x_1, x_2, \ldots, x_n$, with an expected value of (population mean), $E(X) = \mu$, and variance, $V(X) = \sigma^2$. Let $y_1, y_2, \ldots, y_n$ be a manipulation of the original data such that $y = x + a$, where a is any constant, then $E(Y) = \mu + a$ and $V(Y) = \sigma^2$. Here, we expect a constant to be itself and not vary.

6.3.2. Given a data set, $x_1, x_2, \ldots, x_n$, with an expected value of (population mean), $E(X) = \mu$, and variance, $V(X) = \sigma^2$. Let $y_1, y_2, \ldots, y_n$ be a manipulation of the original data such that $y = mx$, where m is any constant, then $E(Y) = m\mu$ and $V(Y) = m^2 \sigma^2$. Recall, variance is a measure of error squared.

6.4.1. In a **binomial** experiment, the follow must hold true:

a. It consists of n identical trials.

b. Each trial results in one of two outcomes; for example, success or failure, life or death, acceptance or rejection, etc.

c. The probability of a success in each trial is equal to p and remains the same for each of the n trials. The probability of a failure is $q = 1 - p$.

d. The trials are independent.

Common: Binomial Calculations

There are usually three types of questions which are likely to be asked with respect to a random variable X that characterizes a worldwide phenomenon. These are as follows:

1. What is the probability that the discrete random variable X will equal a point r **exactly**? That is,

$$P[X=r]=P(r).$$

2. What is the probability that the discrete random variable X will assume values up to the point a (**at most** r)? That is,

$$P[X \leq r]=\sum_{\forall x \leq r} P(x)=\sum_{i=0}^{r} P(x).$$

3. What is the probability that the discrete random variable X will assume values greater than or equal to the point a (**at least** r)? That is,

$$P[X \geq r]=\sum_{\forall x \geq r} P(x)=\sum_{i=0}^{r} P(x).$$

Distribution	PDF	Means	Variance	Standard deviation
Binomial	$f(x\|n,p)=\binom{n}{x}p^x(1-p)^{n-x};$ $x=0,1,2,...,n$	$\mu=np$	$\sigma^2=np(1-p)$	$\sigma=\sqrt{np(1-p)}$

REVIEW TEST

1. Determine if the following relationships are functions:
   A. $\{(0,1),(1,2),(2,3)\}$
   B. $\{(0,1),(0,2),(0,3)\}$
   C. $\{(1,0),(2,0),(3,0)\}$
2. Specify which of the following are discrete or continuous measures:
   A. The amount of coffee used to brew a pot of coffee.
   B. The number of pistons in a car engine.
3. Which of the following are random variables:
   A. Length of an arbitrary race track.
   B. Length of a specific race track.
4. Three coins are tossed and the number of heads observed is recorded. List all possible outcomes in the experiment. Assuming the coins are fair, construct a table showing the theoretical probability distribution.
5. Suppose that a certain family has three children. Let X be the random variable that represents the number of girls in the family and p be the probability of having a girl. Find the probability distribution of the random variable X.
6. Give a graphical representation of the probability distribution outlined in Problem 5 above given $p=0.32$.
7. There are seven True/False questions on a quiz. Betty does not know the answers to any of them and "Christmas-trees" the exam, guessing each answer. What is the probability that she misses all seven questions?
8. The probability of the discrete random variable X is as follows:
   A. Calculate the expected value of the random variable X.
   B. Obtain the variance of the random variable X.
   C. Determine the standard deviation of the discrete random variable X.
9. Given $E(x)=\mu$ and $V(x)=\sigma^2$, let $y=-2x+5$. Find $E(y)$ and $V(y)$.
10. Given $X \sim Binomial(n=10,p=0.2)$, find $P(1)$.
11. Given $X \sim Binomial(n=10,p=0.2)$, find $P(X \leq 2)$.
12. Given $X \sim Binomial(n=10,p=0.2)$, what is the expected number of successes? What is the variance?
13. A fair die is rolled 30 times. Find the probability of getting exactly 10 fours.

REFERENCES

Feller, W., 1968b. 3rd ed. An Introduction to Probability Theory and Its Applications, vol. 1. Wiley, New York.

Mendenhall, W., 1971a. Introduction to Probability and Statistics, 3rd ed. Wadsworth, Belmont, CA.

Meyer, P.L., 1965a. Introduction to Probability and Statistical Applications. Addison-Wesley, Reading, MA.

Tsokos, C.P., 1972c. Probability Distributions: An Introduction to Probability Theory with Applications. Wadsworth, Belmont, CA.

Tsokos, C.P., 1978e. Mainstreams of Finite Mathematics with Application. Charles E. Merrill Publishing Company, A Bell & Howell Company.

Tsokos, C.P., Wooten, R.D., 2011b. The Joy of Statistics Using Real-World Data. Kendall Hunt, Florida.

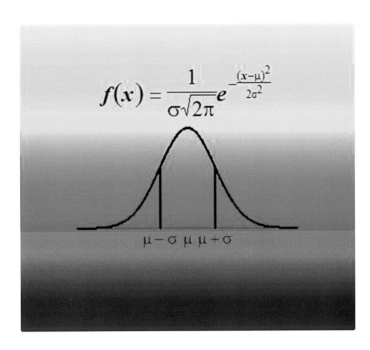

$$f(x) = \frac{1}{\sigma\sqrt{2\pi}} e^{-\frac{(x-\mu)^2}{2\sigma^2}}$$

$\mu - \sigma \quad \mu \quad \mu + \sigma$

The discrete probability distribution approaches a continuous probability distribution as the sample size increase.

Chapter 7 Normal Probability

German mathematician and physicist Carl Friedrich Gauss is sometimes called the "Prince of Mathematics." He was a child prodigy. At the age of 7, Gauss started elementary school and his potential was noticed almost immediately. His teachers were amazed when Gauss summed the integers from 1 to 100 instantly. At age 24, Gauss published one of the most brilliant achievements in mathematics, Disquisitiones Arithmeticae (1801). In it, Gauss systematized the study of number theory. Gauss applied many of his mathematical insights in the field of astronomy, and by using the method of least squares he successfully predicted the location of the asteroid Ceres in 1801. In 1820 Gauss made important inventions and discoveries in geodesy, the study of the shape and size of the earth. In statistics, he developed the idea of the normal distribution. In the 1832 he developed theories of non-Euclidean geometry and mathematical techniques for studying the physics of fluids. Although Gauss made many contributions to applied science, especially electricity and magnetism, pure mathematics was his first love. It was Gauss who first called mathematics "the Queen of the Sciences."

In Chapter 6, we introduced the **binomial** or **Bernoulli probability distribution**—the most useful distribution for **discrete random variables**. In this chapter, we shall focus on the **continuous random** variable and study the **normal** or **Gaussian Probability Distribution**—the most important distribution for continuous random variables and the most popular. A number of interesting applications will be described to illustrate the usefulness of this probability distribution.

7.1 INTRODUCING CONTINUOUS RANDOM VARIABLES

A **continuous random variable** can assume an uncountable number of values; that is, **continuous random variables** are associated with sample spaces representing a very large (infinitely large) number of sample points contained on a line interval. (You recall, on the other hand, that discrete random variables are restricted to taking on isolated values.) A continuous probability distribution is given by a continuous function

$$P(x) = f(x),$$

where $f(x)$ depends on the various parameters that characterizes the subject phenomenon.

We can formally define the **continuous random variable (RV)** as follows.

Definition 7.1.1 Continuous RV

A random variable which can assume an infinitely large number of values associated with the points on a line interval, and the probability of which is spread continuously over these points, is called a **continuous random variable**.

To obtain the probabilities that a random variable will assume values within a certain interval on the real axis, we must know the probability distribution of the continuous random variable. A rigorous and precise presentation of this topic is beyond the scope of this book; however, we shall give a descriptive presentation of continuous probability distributions and then concentrate our study on the famous and very useful **normal** or **Gaussian Probability Distribution**.

Let us begin with a discussion of a physical phenomenon. Consider putting an electrical system into operation at time, $t=0$, and considering the time that elapses until its failure (not predictable) as a random variable X that assumes values from the interval:

$$0 \leq X \leq T$$

Here, the random variable X may assume an infinitely large number of values Furthermore, let us assume that we have observed n such systems and have recorded the elapsed times until failure. Of course, we could interpret X as a discrete random variable and calculate the probabilities that it could assume certain isolated values. However, from a realistic point of view, it is more useful to deal with intervals rather than with isolated points. Let us divide the time interval $0 \leq X \leq T$ into r subintervals each of equal length:

$$t_0 \leq x < t_1, \quad t_1 \leq x < t_2, \quad t_{r-1} \leq x < t_r$$

such that,

$$0 < t_1 < t_2 < \cdots < T.$$

In a discrete sense, let x_r, represent the number of systems that have failed in the rth subinterval, such that

$$\sum_{i=1}^{r} x_i = n$$

is the total number of systems that have been observed. Thus, the number of systems that have failed in a given interval divided by the total number of systems that have been observed will be the probability that a system will fail between the two times. For example, if x_3 systems failed in the interval $t_2 \leq x < t_3$, then the probability that a system will fail in that interval is $\frac{x_3}{n}$. This ratio is also referred to as the **relative frequency**.

Thus, we can construct a graph by drawing rectangles, the bases of which coincide with the preceding **class intervals** and the areas of which are proportional to the class frequencies (see Figure 7.1). Such a graph is called a **probability histogram** and represents a *discretization* of a **continuous probability distribution**. That is, each area of the rectangle represents the probability, $\hat{p}_i = \frac{x_i}{n}$; where $i = 1, 2, \ldots, r$, the number of systems that failed between t_{i-1} and t_i, divided by the total number of systems that have been observed, n. It is clear that if we sum the areas (probabilities) of all the rectangles the sum will equal one; that is

$$\sum_{i=1}^{r} p_i = 1.$$

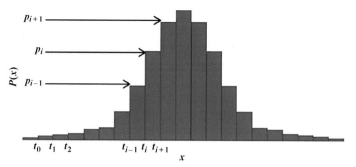

FIGURE 7.1 Histogram of probabilities between failure times.

Thus, we can estimate the probability that a system will fail in the interval t_0 to t_4 by adding the areas of the four rectangles over this interval. That is, if X is the random variable that represents the number of failures, then

$$P[t_0 \leq X \leq t_4] = P[t_0 \leq X \leq t_1] + P[t_1 < X \leq t_2] + P[t_2 < X \leq t_3] + P[t_3 < X \leq t_4]$$
$$= \frac{x_1}{n} + \frac{x_2}{n} + \frac{x_3}{n} + \frac{x_4}{n}$$
$$= p_1 + p_2 + p_3 + p_4.$$

Using this approach, we can obtain estimates of the probabilities that a continuous random variable is contained in an interval in terms of areas. These estimates can be improved by decreasing the widths of the intervals, which, in turn, increases the number of such intervals so that we can approximate the area under the rectangles by a continuous function, $f(x)$, as shown in Figure 7.2.

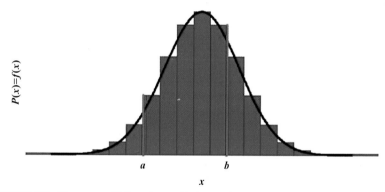

FIGURE 7.2 Histogram of failure times with the overlying normal approximation.

Thus, the probability that a continuous random variable X assumes a value between a and b,

$$P(a \leq x \leq b),$$

is simply the area under the curve $f(x)$ from point a to point b. This area (probability) is also shown by Figure 7.2. The continuous curve $f(x)$ is called the **probability distribution of the continuous random variable X.**

Note: here that the area under the curve, the function, $f(x)$ is equal to one, which corresponds to the sure event and if we add the areas of all the rectangles will be one (taking rounding errors into account).

In Section 7.2, we shall study one of the most famous probability distributions of a continuous random variable and illustrate its usefulness by applying it to a number of problems. This probability distribution is applicable as an exact or approximate characterization of practically any physical phenomenon.

PROBLEMS

Basic Problems

7.1.1 Determine in the following which variables are continuous or discrete random:
 (a) Number of characters in a document.
 (b) The amount of time it takes to make dinner.
 (c) The height of a palm tree.

7.1.2 Determine if the following are continuous or discrete random variables:
 (a) The length of a road trip.
 (b) The number of steps in a stair well.
 (c) The temperature of a nuclear reactor.

7.1.3 Determine if the following are continuous or discrete random variables:
 (a) The number of times one must try before finding success.
 (b) The age of a sea turtle.
 (c) The number of chess pieces remaining on a chess board.

Critical Thinking

7.1.4 Distinguish between discrete data and continuous data.

7.1.5 Distinguish between continuous and discretized data.

7.2 NORMAL PROBABILITY DISTRIBUTIONS

The **normal** or **Gaussian Probability Distribution** is most popular and important because of its unique mathematical properties which facilitate its application to practically any physical problem in the real world; if not for the data's distribution directly, then in terms of the **sampling distribution**, this will be the discussion in Section 7.3. It constitutes the basis for the development of many of the statistical methods that we will learn in the following chapters. The study of the mathematical properties of the **normal probability distribution** is beyond the scope of this book; however, we shall concentrate on its usefulness in characterizing the behavior of continuous random variables that frequently occur in daily experience.

The **normal probability distribution** was discovered by **Abraham De Moivre** in 1733 as a way of approximating the **binomial probability distribution** when the number of trials in a given experiment is very large. In 1774, **Laplace** studied the mathematical properties of the **normal probability distribution.** Through a historical error, the discovery of the **normal distribution** was attributed to **Gauss** who first referred to it in a paper in 1809. In the nineteenth century, many scientists noted that measurement errors in a given experiment followed a pattern (the normal curve of errors) that was closely approximated by this probability distribution. The **normal probability distribution** is formally defined as follows:

Definition 7.2.1 Normal Probability Distribution

A continuous random variable X is **normally distributed** or follows a **normal probability distribution** if its probability distribution is given by the following function:

$$f(x) = \frac{1}{\sigma\sqrt{2\pi}} e^{-\frac{(x-\mu)^2}{2\sigma^2}},$$

$$-\infty < x < \infty, -\infty < \mu < \infty, 0 < \sigma^2 < \infty.$$

The universally accepted notation $X \sim N(\mu, \sigma^2)$ is read as "the continuous random variable X is normally distributed with a population mean μ and population variance σ^2. Of course in real world problems we do not know the true population parameters, but we estimate them from the sample mean and sample variance. However, first, we must fully understand the normal probability distribution.

The graph of the normal probability distribution is a "**bell-shaped**" curve, as shown in Figure 7.3. The constants μ and σ^2 are the parameters; namely, "μ" is the population true mean (or expected value) of the subject phenomenon characterized by the continuous random variable, X, and "σ^2" is the population true variance characterized by the continuous random variable, X. Hence, "σ" the population standard deviation characterized by the continuous random variable X; and the points located at $\mu - \sigma$ and $\mu + \sigma$ are the points of inflection; that is, where the graph changes from cupping up to cupping down.

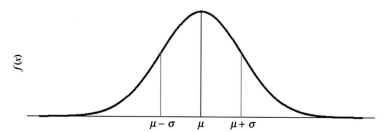

FIGURE 7.3   Normal probability with points of inflections $\mu - \sigma$ and $\mu + \sigma$.

The area under the bell-shaped curve is so disposed that it represents probability; that is, the total area under the curve is equal to one. The random variable X can assume values anywhere from minus infinity to plus infinity, but in practice we very seldom encounter problems in which random variables have such a wide range. The **normal curve graph of the normal probability distribution) is symmetric** with respect to the mean μ as the **central position**. That is, the area between μ and κ units to the left of μ is equal to the area between μ and κ units to the right of μ. This fact is illustrated in Figure 7.4.

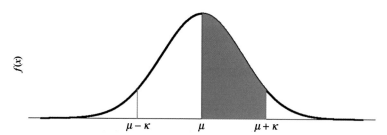

FIGURE 7.4   Normal probability from the center, μ to $\mu + \kappa$; that is, k above center.

$$P[\mu \leq X \leq \mu + \kappa]$$

There is not a unique **normal probability distribution**, since the mathematical formula of the graph depends on the two variables, the mean μ and the variance σ^2. Figure 7.5 is a graphical representation of the normal distribution for a fixed value of σ^2 with μ varying.

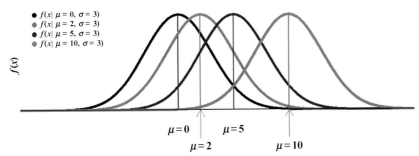

FIGURE 7.5   Normal probability distribution for fixed σ and varying μ.

You recall that the **variance** or **standard deviation** is a measure of the **spread** or "**dispersion**" of the random variable X around its expected value or central tendency, μ. Thus, σ^2 of the **normal distribution** determines the shape of the bell-shaped curve. Figure 7.6 is a graphical representation of the **normal distribution** for a fixed value of μ with varying σ^2. Thus, the expected value μ, locates the central tendency of the random variable, X, and the variance σ^2 determines the shape of the bell-shaped curve. That is, for small values of σ^2, the distribution is clustered close to the mean; as σ^2 increases, the distribution deviates away from the mean. Despite the fact that the shapes are different, the total area under each curve which represents probability is equal to one.

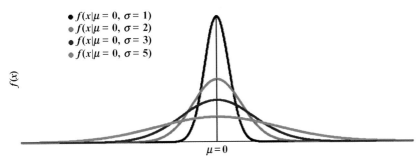

FIGURE 7.6   Normal probability distribution for fixed μ and varying σ.

PROBLEMS

Basic Problems

7.2.1 Illustrate the following normal curves indicating the points of inflection $\mu \mp \sigma$
    (a) $X \sim N\left(\mu = 5, \sigma^2 = 4\right)$
    (b) $X \sim N\left(\mu = 10, \sigma^2 = 4\right)$
    (c) $X \sim N\left(\mu = 15, \sigma^2 = 4\right)$

7.2.2 Illustrate the following normal curves indicating the points of inflection $\mu \mp \sigma$
    (a) $X \sim N\left(\mu = 10, \sigma^2 = 4\right)$
    (b) $X \sim N\left(\mu = 10, \sigma^2 = 9\right)$
    (c) $X \sim N\left(\mu = 10, \sigma^2 = 16\right)$

7.2.3 Illustrate the following normal curves indicating the points of inflection $\mu \mp \sigma$
    (a) $X \sim N\left(\mu = 25, \sigma^2 = 1\right)$
    (b) $X \sim N\left(\mu = 100, \sigma^2 = 4\right)$
    (c) $X \sim N\left(\mu = 200, \sigma^2 = 16\right)$

Critical Thinking

7.2.4 Distinguish between the bell-shaped normal probability distribution and the standard normal probability distribution.

7.2.5 Distinguish between a curve (probability distribution) and a symmetric probability distribution.

7.2.6 Discuss the use of the tilde "$\sim$" as it is used in statistics.

7.3   STANDARD NORMAL PROBABILITY DISTRIBUTIONS

As we see, there is a normal probability distribution for each pair of numerical values of μ and σ^2. Thus, we could generate numerous bell-shaped normal distributions, one for each possible data set. However, as it is difficult to calculate probabilities for each because of its mathematical complexity, we can overcome this difficulty by making the following substitution.

Let

$$z = \frac{x - \mu}{\sigma}$$

in the formula for the normal distribution to obtain

$$f(z) = \frac{1}{\sqrt{2\pi}} e^{-\frac{z^2}{2}}; \quad -\infty < z < \infty.$$

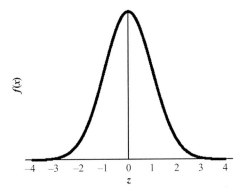

Definition 7.3.1 Standard Normal

The **standard normal probability distribution (standard normal curve)** has mean of zero, $E(Z) = 0$, and unit variance, $V(Z) = 1$ and will be denoted $Z \sim N(0, 1)$.

Hence, the transformation of the data moves the center of symmetry is to zero and changes the variance to one:

$$f(z) = \frac{1}{\sqrt{2\pi}} e^{-\frac{z^2}{2}},$$

$$-\infty < z < \infty.$$

The transformation into the standard normal distribution is shown by Figure 7.7.

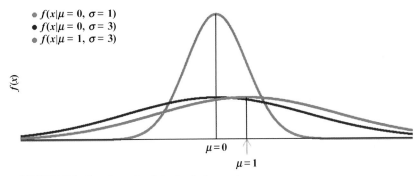

- $f(x | \mu = 0,\ \sigma = 1)$
- $f(x | \mu = 0,\ \sigma = 3)$
- $f(x | \mu = 1,\ \sigma = 3)$

FIGURE 7.7 Transformation into standard normal.

Therefore, when we have a continuous random variable X that is **normally distributed** with mean μ and variance σ, we can apply the substitution $z = \frac{x - \mu}{\sigma}$ and subsequently speak of the standard normal distribution, $z \sim N(0, 1)$.

Definition 7.3.2 Standard Score

The **standard score** is the transformation of the data given by:

$$z = \frac{x - \mu}{\sigma},$$

where $z \sim N(0, 1)$.

All of our real world problems that follow the Normal/Gaussian PDF are given in terms of $X \sim N(\mu, \sigma^2)$, but in order for us to solve them, we must reduce the data into the standard normal PDF, $Z \sim N(0, 1)$.

For example, suppose that we are interested in obtaining the probability that the random variable X assumes values between a and b; that is,

$$P(a \leq X \leq b)$$

where $X \sim N(\mu, \sigma^2)$.

This would be the same as the probability that the random variable Z assumes its values between $z_a = \frac{a - \mu}{\sigma}$ and $z_b = \frac{b - \mu}{\sigma}$; that is,

$$P(a \leq X \leq b) = P\left(\frac{a - \mu}{\sigma} \leq Z \leq \frac{b - \mu}{\sigma}\right).$$

Figure 7.8 is a graphical presentation of this relationship.

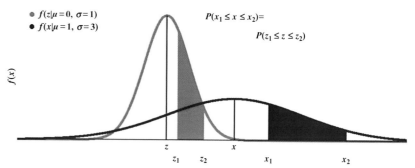

FIGURE 7.8 Standardizing the data.

Since we can always reduce the continuous **normally distributed random variable** X into the **standard normally distributed random variable** Z, the areas of the latter have been tabulated. Applying appropriate conversion formulas, the table can be used to find probabilities for any **normally distributed random variable.**

The table for probabilities to the center for the **standard normal probability distribution** is given in the Appendix. A sample, for illustrated purpose is given bellow.

The areas under the standard normal curve between the mean, $z = 0$ and $z = 2.09$, are recorded in Table 7.1.

Appendix Standard Normal

TABLE 7.1 Standard Normal Probability Distribution: To the Center $P(0 \leq Z \leq z)$

	0.00	0.01	0.02	0.03	0.04	0.05	0.06	0.07	0.08	0.09
0.0	0.0000	0.0040	0.0080	0.0120	0.0160	0.0199	0.0239	0.0279	0.0319	0.0359
0.1	0.0398	0.0438	0.0478	0.0517	0.0557	0.0596	0.0636	0.0675	0.0714	0.0753
0.2	0.0793	0.0832	0.0871	0.0910	0.0948	0.0987	0.1026	0.1064	0.1103	0.1141
0.3	0.1179	0.1217	0.1255	0.1293	0.1331	0.1368	0.1406	0.1443	0.1480	0.1517
0.4	0.1554	0.1591	0.1628	0.1664	0.1700	0.1736	0.1772	0.1808	0.1844	0.1879
0.5	0.1915	0.1950	0.1985	0.2019	0.2054	0.2088	0.2123	0.2157	0.2190	0.2224
0.6	0.2257	0.2291	0.2324	0.2357	0.2389	0.2422	0.2454	0.2486	0.2517	0.2549
0.7	0.2580	0.2611	0.2642	0.2673	0.2704	0.2734	0.2764	0.2794	0.2823	0.2852
0.8	0.2881	0.2910	0.2939	0.2967	0.2995	0.3023	0.3051	0.3078	0.3106	0.3133
0.9	0.3159	0.3186	0.3212	0.3238	0.3264	0.3289	0.3315	0.3340	0.3365	0.3389
1.0	0.3413	0.3438	0.3461	0.3485	0.3508	0.3531	0.3554	0.3577	0.3599	0.3621
1.1	0.3643	0.3665	0.3686	0.3708	0.3729	0.3749	0.3770	0.3790	0.3810	0.3830
1.2	0.3849	0.3869	0.3888	0.3907	0.3925	0.3944	0.3962	0.3980	0.3997	0.4015
1.3	0.4032	0.4049	0.4066	0.4082	0.4099	0.4115	0.4131	0.4147	0.4162	0.4177
1.4	0.4192	0.4207	0.4222	0.4236	0.4251	0.4265	0.4279	0.4292	0.4306	0.4319
1.5	0.4332	0.4345	0.4357	0.4370	0.4382	0.4394	0.4406	0.4418	0.4429	0.4441
1.6	0.4452	0.4463	0.4474	0.4484	0.4495	0.4505	0.4515	0.4525	0.4535	0.4545
1.7	0.4554	0.4564	0.4573	0.4582	0.4591	0.4599	0.4608	0.4616	0.4625	0.4633
1.8	0.4641	0.4649	0.4656	0.4664	0.4671	0.4678	0.4686	0.4693	0.4699	0.4706
1.9	0.4713	0.4719	0.4726	0.4732	0.4738	0.4744	0.4750	0.4756	0.4761	0.4767
2.0	0.4772	0.4778	0.4783	0.4788	0.4793	0.4798	0.4803	0.4808	0.4812	0.4817

Since the standard normal distribution is symmetrical about zero and the total area under the curve is equal to one, half of the area would lie to the right of zero and half to the left. Thus, if we know the areas under the curve from $z=0$ and $z=3.49$, we also know the areas under the curve from $z=0$ and $z=-3.49$. Note in Table 7.1 that z is in only correct to the second decimal place; the first decimal is recorded in the left-hand column and the second decimal place for z, corresponding to hundredths, is given across the top row. The corresponding probabilities are correct to the fourth decimal and in Table 7.1, are the probabilities to the center. However, it is also common practice to provide tables which give probabilities to the left, see Appendix.

The standard normal curves below illustrate how the probabilities are obtained by the various graphs.

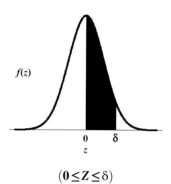

$$(0 \leq Z \leq \delta)$$

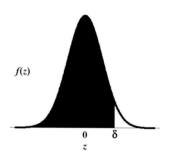

$$P(-\infty \leq Z \leq \delta)$$

Or

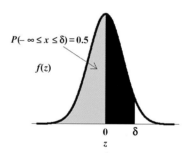

$$P(-\infty \leq Z \leq \delta) = 0.5 + P(0 \leq Z \leq \delta)$$

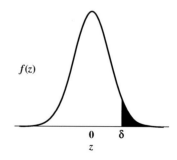

$$P(\delta \leq Z \leq \infty)$$

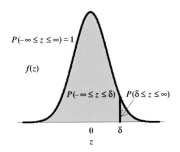

$$P(\delta \leq Z \leq \infty)=1-P(-\infty \leq Z \leq \delta)$$

Or

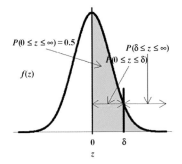

$$P(\delta \leq Z \leq \infty)=0.5-P(0 \leq Z \leq \delta)$$

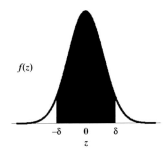

$$P(-\delta \leq Z \leq \delta)$$

Or

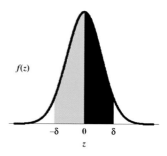

$$P(-\delta \leq Z \leq \delta) = 2 \times P(0 \leq Z \leq \delta)$$

This probability density function is well-defined, and hence as with the charts in the Appendix, we can create accurate probability charts depending on the number of decimals we take the standard z-score to.

Example 7.3.1 Standard Normal Probabilities

Using the chart, find the following probabilities:

(a) $P(0 \leq Z \leq 1.72)$
(b) $P(-1.81 \leq Z \leq 0)$
(c) $P(1.12 \leq Z \leq 2.1)$
(d) $P(-2.0 \leq Z \leq 0.5)$
(e) $P(Z \leq 2.57)$
(f) $P(Z \geq 1.97)$

Solution

For Part (a), using the chart which gives the probabilities to the center, we can just look up **1.7** in the left-most column and **0.02** in the first row (corresponding to $z = 1.7 + 0.02 = 1.72$). From this row and column, we can cross-hatch the table to find the area under the curve (a probability) of **0.4573**.

How to use the Appendix

Standard Normal Probability Distribution: To the Center $P(0 \leq Z \leq z)$

	0.00	0.01	**0.02**	0.03	0.04	0.05	0.06	0.07	0.08	0.09
0.0	0.0000	0.0040		0.0120	0.0160	0.0199	0.0239	0.0279	0.0319	0.0359
0.1	0.0398	0.0438		0.0517	0.0557	0.0596	0.0636	0.0675	0.0714	0.0753
0.2	0.0793	0.0832	↓	0.0910	0.0948	0.0987	0.1026	0.1064	0.1103	0.1141
⋮	⋮	⋮		⋮	⋮	⋮	⋮	⋮	⋮	⋮
1.6	0.4452	0.4463		0.4484	0.4495	0.4505	0.4515	0.4525	0.4535	0.4545
1.7	→		**0.4573**	0.4582	0.4591	0.4599	0.4608	0.4616	0.4625	0.4633
1.8	0.4641	0.4649	0.4656	0.4664	0.4671	0.4678	0.4686	0.4693	0.4699	0.4706

Solution—cont'd

Hence, we have

$$P(0 \leq Z \leq 1.72) = 0.4573,$$

which means that there is a 45.73% chance that the continuous random variable, z, will assume a value between 0 and 1.72.

For Part (b), by symmetry,

$$P(-1.81 \leq Z \leq 0) = P(0 \leq Z \leq 1.81)$$

hence using Appendix G1, we have

$$P(-1.81 \leq Z \leq 0) = P(0 \leq Z \leq 1.81) = 0.4649$$

This means that there is approximately 46.5% chance that the continuous random variable will assume a value between −1.81 (standard deviation from the mean) and 0 (standard deviations from the mean). Recall: z is the standard score, that is, the number of standard deviations from the mean.

For Part (c), using the chart to the center we have

$$P(1.12 \leq Z \leq 2.1) = P(0 \leq Z \leq 2.1) - P(0 \leq Z \leq 1.12) = 0.4821 - 0.3686$$
$$= 0.1135.$$

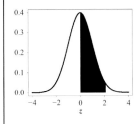 − 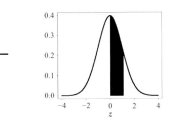 =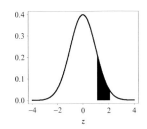

Hence, there is an 11.35% chance that the continuous random variable will assume a value between 1.12 and 2.10. For Part (d), we have

$$P(-2.0 \leq Z \leq 0.5) = P(0 \leq Z \leq 0.5) + P(-2.0 \leq Z \leq 0)$$
$$= P(0 \leq Z \leq 0.5) + P(0 \leq Z \leq 2.0)$$
$$= 0.1915 + 0.4772 = 0.6687$$

$$P(1.12 \leq Z \leq 2.10) = 11.35\%$$

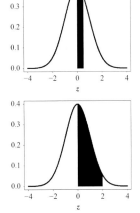

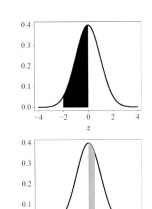

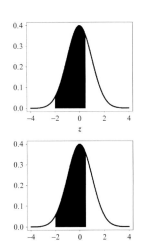

Hence, there is a 66.87% chance that the continuous random variable will assume a value between −2.00 and 0.50.

For Part (e), since the mean is the median we have,

$$P(Z \leq 2.57) = P(Z \leq 0) + P(0 \leq Z \leq 2.57) = 0.5 + 0.4949 = 0.9949$$

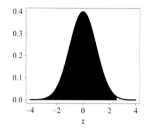 = 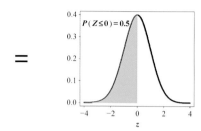 +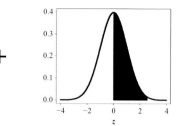

Hence, it is high likely that the data is less than 2.57 standard deviations above the mean; in fact, there is an approximately 99.5% that the continuous random variable will assume a value less than 2.57 standard deviation from the mean. Hence, there is only a 0.5% change that is will exceed this value.

For Part (f), $P(Z \geq 1.97)$, since the mean is the median, we have

$$P(Z \geq 1.97) = 0.5 - P(0 \leq Z \leq 1.97)$$
$$= 0.5 - 0.4765$$
$$= 0.0235$$

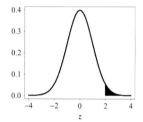 = 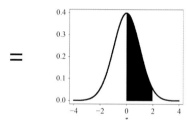 −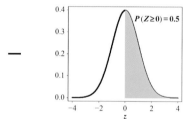

$$P(Z \geq 1.97) = 2.35\%$$

This means that there is a 2.35% chance that the continuous random variable, z, will assume a value between 0 and 1.97.

We should also mention the fact that we can find a value for the standard normal random variable **Z** such that the probability of a larger or a smaller value is equal to a certain area (probability); that is, we can use the chart in reverse (backward).

For example, find a value for **Z** such that the probability of a larger value is equal to 0.0517. The probability of a larger value is the area to the right of δ as shown by the graph below:

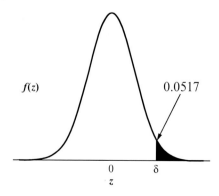

The area between zero (**0**) and δ must be **0.5000 − 0.0517** or **0.4483**.

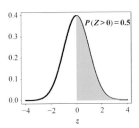 − 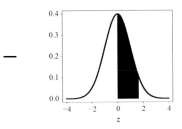 =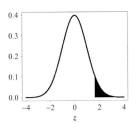

Thus, looking at Appendix G2, for the value of δ such that

$$P(0 \leq z \leq \delta) = 0.4483 \approx 0.4484$$

we find that $\delta = 1.63$.

Thus, there is a 5.17% chance that the random variable, z, will assume a value greater than 1.63.

How to use Appendix backward

Standard Normal Probability Distribution $P(0 \leq Z \leq z)$

STANDARD NORMAL PROBABILITY DISTRIBUTION $P(0 \leq Z \leq z)$									
	0.00	0.01	0.02	**0.03**	0.04	0.05	0.06	0.07	0.08	0.09
0.0	0.0000	0.0040	0.0080		0.0160	0.0199	0.0239	0.0279	0.0319	0.0359
0.1	0.0398	0.0438	0.0478		0.0557	0.0596	0.0636	0.0675	0.0714	0.0753
0.2	0.0793	0.0832	0.0871	↑	0.0948	0.0987	0.1026	0.1064	0.1103	0.1141
⋮	⋮	⋮	⋮		⋮	⋮	⋮	⋮	⋮	⋮
1.5	0.4332	0.4345	0.4357		0.4382	0.4394	0.4406	0.4418	0.4429	0.4441
1.6	←			**0.4484**	0.4495	0.4505	0.4515	0.4525	0.4535	0.4545
1.7	0.4554	0.4564	0.4573	0.4582	0.4591	0.4599	0.4608	0.4616	0.4625	0.4633

Before proceeding to illustrate the usefulness of the normal probability distribution, let us summarize how we use the standard normal areas for finding probabilities for the general normal variable. Let X be normal with mean μ and variance σ^2, $X \sim N(\mu, \sigma^2)$. Suppose that we are interested in finding the probability that a random selected value for X will be between a and b. That is,

$$P(a \leq X \leq b).$$

To obtain this probability, we must first calculate two numbers z_1 and z_2 using the following formulas:

$$z_1 = \frac{a - \mu}{\sigma} \quad \text{and} \quad z_2 = \frac{b - \mu}{\sigma}.$$

Then,

$$P(a \leq X \leq b) = P(z_1 \leq Z \leq z_2)$$

If we assume that $\mu = 74$, $\sigma^2 = 36$, $a = 69.8$, and $b = 78.8$, the probability that a value selected at random for X will be between 69.8 and 78.8 is obtained as follows: first we calculate

$$z_1 = \frac{a - \mu}{\sigma} = \frac{69.8 - 74}{6} = \frac{-4.2}{6} = -0.7$$

and

$$z_2 = \frac{b - \mu}{\sigma} = \frac{78.8 - 74}{6} = \frac{4.8}{6} = 0.8.$$

Then,

$$P(69.8 \leq X \leq 78.8) = P(-0.7 \leq Z \leq 0.8)$$
$$= P(-0.7 \leq Z \leq 0) + P(0 \leq Z \leq 0.8)$$
$$= P(0 \leq Z \leq 0.7) + P(0 \leq Z \leq 0.8)$$
$$= 0.2580 + 0.2580 = 0.5461$$

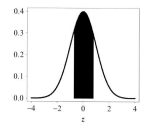 = 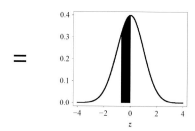 +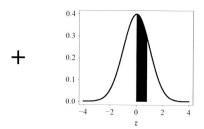

Hence, there is a 54.61% chance that the random variable, X, will assume a value between 69.8 and 78.8.

We shall illustrate some of the applications of the **normal distribution** with the following examples:

Example 7.3.2 Normal Probabilities

We may assume that scores, of a biology examination to be characterized by a continuous random variable, X, that is normally distributed with $\mu = 75$ and $\sigma^2 = 64$. What is the probability that

a) A student's score chosen at random will between 80 and 85?
b) A student's score chosen at random will be less than or equal to 85%?
c) A student's score chosen at random will above 90?

Solution

For Part (a), we need to obtain

$$P(80 \leq X \leq 85)$$

Here

$$z_1 = \frac{a - \mu}{\sigma} = \frac{80 - 75}{8} = \frac{5}{8} = 0.625 \approx 0.63$$

and

$$z_2 = \frac{b - \mu}{\sigma} = \frac{85 - 75}{8} = \frac{10}{8} = 1.25$$

Thus,

$$P(80 \leq X \leq 85) = P(0.63 \leq Z \leq 1.25)$$
$$= P(0 \leq Z \leq 1.25) - P(0 \leq Z \leq 0.63)$$
$$= 0.3944 - 0.2357 = 0.1587$$

Therefore, there is about a 15.87% chance that a student's score will fall between 80% and 85%.

For Part (b), we want to find

$$P(0 \leq X \leq 85).$$

Here,

$$z_1 = \frac{a - \mu}{\sigma} = \frac{0 - 75}{8} = \frac{75}{8} = -9.375$$

Solution—cont'd

and

$$z_2 = \frac{b-\mu}{\sigma} = \frac{85-75}{8} = \frac{10}{8} = 1.25.$$

Thus,

$$\begin{aligned} P(0 \le X \le 85) &= P(-9.375 \le Z \le 1.25) \\ &= P(-9.375 \le Z \le 0) + P(0 \le Z \le 1.25) \\ &\approx 0.5 + P(0 \le Z \le 1.25) \\ &= 0.5000 + 0.3944 = 0.8944 \end{aligned}$$

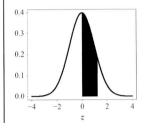 + 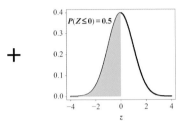 =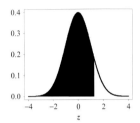

Note that $P(-9.735 \le Z \le 0)$ is not included in Table 7.1; however, practically 0.5 of the probability is contained in the interval from zero to 3.9. Thus, approximately 0.5 of the probability is contained in the interval from zero to 9.375. Hence, there is an 89.44% chance that the score of a student will be lower than or equal to 85%. Hence, there is approximately an 89.4% chance that a student's score will fall below 85%; that is, approximate 89.4% of the values this random variable takes is between the minimum of 0 and (above average), an 85.

For Part (c), we must calculate

$$P(X \ge 90) = 1 - P(X < 90)$$

First, we calculate

$$z = \frac{a-\mu}{\sigma} = \frac{90-75}{8} = \frac{15}{8} = 1.875.$$

Thus,

$$\begin{aligned} P(X \ge 90) &= 1 - P(X < 90) \\ &= 1 - P(Z < 1.875) \\ &= 1 - [P(Z < 0) + P(0 \le Z \le 1.875)] \\ &= 1 - [0.5 + P(0 \le Z \le 1.875)] \\ &= 0.5 - P(0 \le Z \le 1.875) \\ &= 0.5000 - 0.4699 = 0.0301. \end{aligned}$$

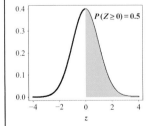 − 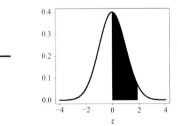 =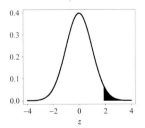

Therefore, there is only a 3.01% chance that the score will be a grade exceeding 90%.

Example 7.3.3 Minting Nickels

The thickness of the nickels minted in the Denver mint varies. We shall assume that the random variable, X, characterizing this phenomenon is normally distributed with mean

Continued

Example 7.3.3 Minting Nickels—cont'd

$\mu = \frac{1}{18}$ inch and $\sigma^2 = \frac{1}{49}$ inch. A nickel will activate a gumball vending machine and gives two gumballs if its thickness is between $\frac{1}{33}$ inch and $\frac{2}{33}$ inch. What is the probability that if a student randomly selects a nickel from his pocket and drops it into a machine he will get the gumball he paid for plus an additional gumball?

Solution

The machine will be activated if X, the thickness of the nickel, is between $\frac{1}{33}$ inch and $\frac{2}{33}$ inch. Thus, we need to obtain

$$P\left(\frac{1}{33} \le X \le \frac{2}{33}\right) = P(0.030 \le X \le 0.061).$$

We are given $\mu = \frac{1}{18} = 0.056$, $\sigma = \frac{1}{7} = 0.143$, and

$$z_1 = \frac{a - \mu}{\sigma} = \frac{0.030 - 0.056}{0.143} = \frac{-0.026}{0.143} = -0.182$$

and

$$z_2 = \frac{b - \mu}{\sigma} = \frac{0.061 - 0.056}{0.143} = \frac{0.005}{0.143} = 0.035$$

Hence,

$$
\begin{aligned}
P(0.030 \le X \le 0.061) &= P(-0.182 \le Z \le 0.035) \\
&= P(-0.182 \le Z \le 0) + P(0 \le Z \le 0.035) \\
&= 0.0714 + 0.0160 = 0.0874.
\end{aligned}
$$

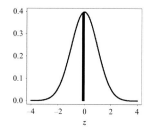 = 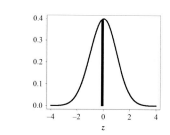 +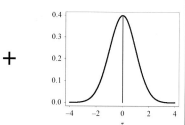

Therefore, there is only an 8.74% chance that the student will receive candy.

Example 7.3.4 Mean Weight of Man

Assume that the weights of male students, X, are normally distributed with mean weight of 160 pounds and standard deviation of 20. Out of 1000 men, how many students weigh
(a) between 150 and 170 pounds?
(b) more than or exactly 190 pounds?

Solution

Here, $\mu = 160$, $\sigma = 20$, and for Part (a)

$$z_1 = \frac{a - \mu}{\sigma} = \frac{150 - 160}{20} = \frac{-10}{20} = -0.5$$

and

$$z_2 = \frac{b - \mu}{\sigma} = \frac{170 - 160}{20} = \frac{10}{20} = 0.5.$$

Solution—cont'd

Hence,

$$\begin{aligned}
P(150 \leq X \leq 170) &= P(-0.5 \leq Z \leq 0.5)\\
&= P(-0.5 \leq Z \leq 0) + P(0 \leq Z \leq 0.5)\\
&= P(0 \leq Z \leq 0.5) + P(0 \leq Z \leq 0.5)\\
&= 2 \times P(0 \leq Z \leq 0.5)\\
&= 2 \times 0.1915 = 0.3830.
\end{aligned}$$

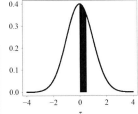 **+** 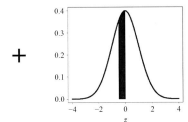 **=**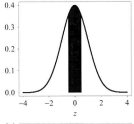

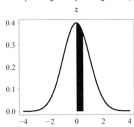 **+** 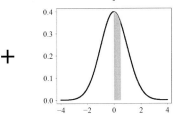 **=**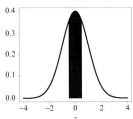

Therefore, 38.3% of the 1000 students, or 383 of them, weigh between 150 and 170 pounds.

For Part (b), we need to obtain the following probability:

$$P(X \geq 190) = 1 - P(X < 190)$$

First, we calculate

$$z = \frac{a - \mu}{\sigma} = \frac{190 - 160}{20} = \frac{30}{20} = 1.5$$

Thus,

$$\begin{aligned}
P(X \geq 190) &= 1 - P(X < 190)\\
&= 1 - P(Z < 1.5)\\
&= 1 - [P(Z < 0) + P(0 \leq Z \leq 1.5)]\\
&= 1 - [0.5 + P(0 \leq Z \leq 1.5)]\\
&= 0.5 - P(0 \leq Z \leq 1.5)]\\
&= 0.5000 - 0.4332 = 0.0668
\end{aligned}$$

Can be obtained from Appendix: chart to left.

Can be obtained from Appendix: chart to center

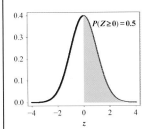 **−** 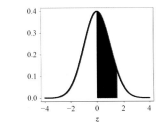 **=**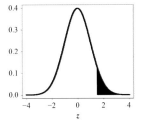

Hence, only 6.68% of the 1000 students, or about 67 of them, weigh more than 190 pounds.

Example 7.3.5 Seismology

In the "**time-term**" method of refraction seismology, depths from ground level to a refraction layer are estimated using the time of arrival of vibrations from a distant explosion. Suppose that, for a particular survey, the estimate of depth at each point is subject to a random error, which is assumed to be normally distributed with mean μ of 2 feet and variance of 4 feet. What is the probability that the estimated "depth" will be negative?

Solution

When we see that the estimated depth appears to fall above the ground, this will be in effect a negative depth. Therefore, we need to find

$$P(X<0).$$

We are given $X \sim N(\mu=2, \sigma^2=4)$, thus we have

$$z = \frac{a-\mu}{\sigma} = \frac{0-2}{2} = \frac{-2}{2} = -1$$

Thus,

$$
\begin{aligned}
P(X<0) &= P(Z<-1) \\
&= P(Z>1) \\
&= 1 - P(Z \leq 1) \\
&= 1 - [P(Z<0) + P(0 \leq Z \leq 1)] \\
&= 1 - [0.5 + 0.3413] \\
&= 0.1587.
\end{aligned}
$$

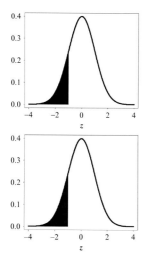

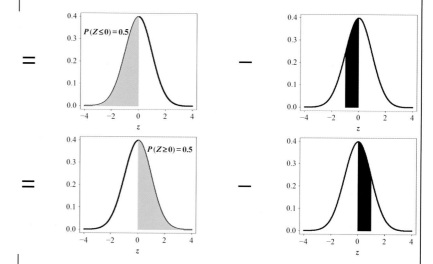

Thus, there is a 15.87% chance that the estimated "depth" will be negative. This information is important to seismologist for strategic planning purposes.

Example 7.3.6 Electric Lamps

A certain company annually uses a large number of electric lamps that burn continuously day and night. Assume that, under such conditions, the life of a lamp may be regarded as a continuous random variable, X, normally distributed about a mean of 70 days with a standard deviation of 20. On January 1st, the company put 10,000 new lamps into service. Approximately how many lamps would be expected to need replacement by February 1st of the same year?

Solution

Here, $\mu=70$, $\sigma=20$, and we need to find

$$P(0\le X\le 31),$$

that is, the probability that a lamp will need to be replaced during the period of January 1st to February 1st. Thus,

$$z_1=\frac{a-\mu}{\sigma}=\frac{0-70}{20}=\frac{-70}{20}=-3.5$$

And

$$z_2=\frac{b-\mu}{\sigma}=\frac{31-70}{20}=\frac{-39}{20}=-1.95.$$

Hence,

$$
\begin{aligned}
P(0\le X\le 31) &=P(-3.5\le Z\le -1.95)\\
&=P(-3.5\le Z\le 0)-P(-1.95\le Z\le 0)\\
&=P(0\le Z\le 3.5)-P(0\le Z\le 1.95)\\
&=0.4998-0.4744=0.0254.
\end{aligned}
$$

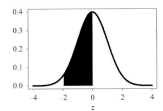 $-$ 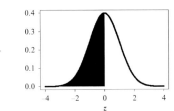 $=$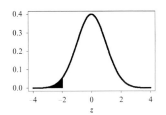

Therefore, the expected number of lamps that will need replacement is equal to 10,000 (0.0254) or 254 lamps.

Example 7.3.7 Corn Crops

A large agricultural complex is conducting a research project on the yields of corn using different types of fertilizers. A farm is partitioned into a large number of small experimental plots. Suppose that the yield of corn in pounds per plot is normally distributed around a mean of five pounds, $\mu=5$, with a standard deviation of 2 pounds, $\sigma=2$. Determine the probability that the yield of a plot will

(a) exceed 5 pounds

(b) be between 3.5 and 6.5 pounds.

Solution

Let X be a random variable denoting the yield of a plot. Here, for Part (a), we are interested in obtaining

$$P(X>5)=1-P(X\le 5),$$

Thus, for $\mu=5$, $\sigma=2$

$$z=\frac{a-\mu}{\sigma}=\frac{5-5}{2}=\frac{0}{2}=0,$$

and

$$
\begin{aligned}
P(X>5)&=P(Z>0)\\
&=0.5.
\end{aligned}
$$

Thus, there is a 50% chance that the corn yield of a plot chosen at random will exceed 5 pounds. For Part (b), we need to obtain

Continued

Solution—cont'd

$$P(3.5 \leq X \leq 6.5)$$

Hence,

$$z_1 = \frac{a-\mu}{\sigma} = \frac{3.5-5}{2} = \frac{-1.5}{2} = -0.75,$$

we must also consider

$$z_2 = \frac{b-\mu}{\sigma} = \frac{6.5-5}{2} = \frac{1.5}{2} = 0.75.$$

Thus,

$$\begin{aligned} P(3.5 \leq X \leq 6.5) &= P(-0.75 \leq Z \leq 0.75) \\ &= P(-0.75 \leq Z \leq 0) + P(0 \leq Z \leq 0.75) \\ &= 2 \times P(0 \leq Z \leq 0.75) \\ &= 2 \times 0.2734 = 0.5468 \end{aligned}$$

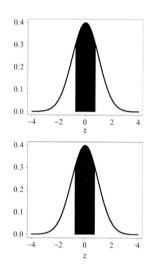

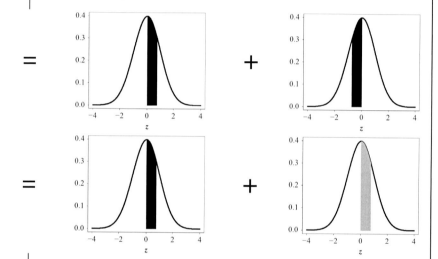

Therefore, there is a 54.68% chance that the yield of a plot chosen at random will be between 3.5 and 6.5 pounds of corn.

Example 7.3.8 Rainfall

In the United States, rainfall measurements are taken in three different regions: the northern region, central region and the southern region. Suppose that the amount of yearly rainfall is normally distributed with mean of 5 inches, $\mu = 5$, and standard deviation of 1 inch, $\sigma = 1$. Determine the probability that the rainfall will

(a) exceed 6.2 inches

(b) be between 3.5 and 6.5 pounds.

Solution

Let X be a random variable denoting the amount of rain. Here, for Part (a), we are interested in obtaining

$$P(X > 6.2) = 1 - P(X \leq 6.2),$$

Thus, for $\mu = 5$, $\sigma = 1$

$$z = \frac{a-\mu}{\sigma} = \frac{6.2-5}{2} = \frac{1.2}{2} = 0.60.$$

Solution—cont'd

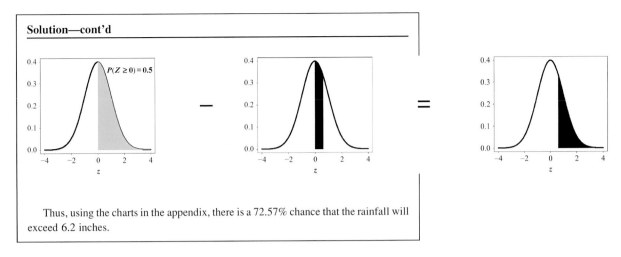

Thus, using the charts in the appendix, there is a 72.57% chance that the rainfall will exceed 6.2 inches.

PROBLEMS

Critical Thinking

7.3.1. Find the probability that a single value of the standard normal random variable **Z** will be between zero and 1.52.
7.3.2. What is the probability that the value the standard normal random variable **Z** assumes will be greater than 0.64?
7.3.3. Calculate the probability that the standard normal random variable **Z** will assume a value between 0.43 and 2.20.
7.3.4. Determine the probability that the standard normal random variable **Z** will assume a single value between -1.42 and 0.75.
7.3.5. Find a value for the standard normal variable **Z** such that the probability of a larger value is equal to .025.

Basic Problems

7.3.6. The random variable X is normally distributed with mean $\mu=50$ and $\sigma=10$. Find the following probabilities:
    (a) $P(X \leq 66)$
    (b) $P(X \leq 56)$
    (c) $P(43 \leq X \leq 63)$
    (d) $P(37 \leq X \leq 46)$
7.3.7. The random variable X is normally distributed with mean $\mu=0.06$ and standard deviation $\sigma=0.02$ Determine the following probabilities:
    (a) $P(X \geq 0.07)$
    (b) $P(0.07 \leq X \leq 0.079)$
7.3.8. If the random variable X is normally distributed with $\mu=50$ and $\sigma=10$, find k so that the $P(X \leq k)=0.1841$.
7.3.9. Assume that the height of 1000 students is normally distributed with mean height of 68 inches and standard deviation of 4 inches.
    (a) Determine the number of students with a height between 64 and 72 inches.
    (b) Find the number of students with a height exceeding the mean height.
7.3.10. Suppose that the test scores on a history examination are normally distributed with mean score of 74 and standard deviation of 10.
    (a) Find the probability that a student's test score chosen at random will be between 80 and 90.
    (b) Determine the probability that a student's test score will be lower than 60.
7.3.11. Suppose that the diameters of golf balls manufactured by a certain company are normally distributed with mean $\mu=1.94$ inches and standard deviation $\sigma=0.03$ inch. A golf ball will be considered defective if its diameter is less than 1.89 inches or greater than 1.99 inches. What is the percentage of defective balls manufactured by the company?
7.3.12. A noise voltage, X, which assumes values of x volts is normally distributed with mean $\mu=1$ volt and standard deviation $\sigma=0.9$ volt.
    (a) Determine the probability that the noise voltage is between -1 volts and 2 volts; that is, $P(-1 \leq X \leq 2)$.
    (b) Find the noise-voltage level such that the probability that this level is exceeded is 0.95.

7.4 NORMAL APPROXIMATION TO THE BINOMIAL

In Chapter 5, we studied the concept of a discrete random variable and its characterization by the **binomial or Bernoulli probability distribution**. That is, given n independent trials, the outcome of each trial is either a success or a failure, with probabilities p and $q = 1 - p$, respectively. Let X denote the number of successes that occur in the n trials. Thus, X may assume the values $0, 1, 2, \ldots, n$. You recall that the probability distribution which characterizes this phenomenon is the **binomial probability distribution** given by

$$P(x) = \binom{n}{x} p^x q^{n-x}, \quad x = 0, 1, 2, \ldots, n \quad 0 < p < 1$$

When the number of trials n in a given experiment is large, the calculation of various probabilities of the possible values X can assume becomes quite tedious. Since many practical problems involve large numbers of repeated trials, it is of interest to have a rapid technique for calculating the probabilities. Such a method is given by the **normal probability distribution**. That is, when n is sufficiently large, the normal curve can be used as an approximation to the binomial distribution for the calculation of probabilities for which the binomial is the correct probability distribution. It has been shown that the expected value or mean and variance of a discrete random variable that is binomially distributed are as follows.

Rule 7.4.1 Normal Approximation

To use **Normal Approximation**, given the binomial probability of success, p, and a given number of trials, n, then the expected number of successes is

$$\mu = np,$$

and the variance depends on the sample size, n, the probability of success, p, and the probability of failure, $(1-p)$; that is, $\sigma^2 = np(1-p)$ and $\sigma = \sqrt{np(1-p)}$.

Thus, in the normal probability distribution we need to substitute the mean and variance and use the approach of standardizing the normal curve as we have learned previously to obtain approximate probabilities of problems involving the binomial distribution. That is,

$$P(x) = \binom{n}{x} p^x (1-p)^{n-x} \approx \frac{1}{\sqrt{2\pi}\sqrt{np(1-p)}} e^{-\frac{1}{2}\left(\frac{x-np}{\sqrt{np(1-p)}}\right)^2},$$

for large n.

For example, suppose that we must find the probability that the number of successes in a binomial situation is between a and b.

The exact probability is given by

$$P(a \leq x \leq b) = \sum_{x=a}^{b} \binom{n}{x} p^x (1-p)^{n-x}.$$

To obtain an approximation to this probability using the normal curve, we proceed as follows: let

$$z_1 = \frac{a - \mu}{\sigma} = \frac{a - np}{\sqrt{np(1-p)}}$$

and

$$z_2 = \frac{b - \mu}{\sigma} = \frac{b - np}{\sqrt{np(1-p)}}$$

Hence,

$$P(a \leq X \leq b) \approx P(z_1 \leq Z \leq z_2)$$

$$= P\left(\frac{a - np}{\sqrt{np(1-p)}} \leq Z \leq \frac{b - np}{\sqrt{np(1-p)}} \right),$$

and for specific values for a, b, n, and p we can read the desired probabilities from the Appendix.

In the preceding discussion, we are approximating probabilities of a discrete distribution with a continuous distribution, and so we need to introduce a **correction factor for continuity**. In the discrete case, the probability is concentrated at the integers; but, when we approximate such probabilities with a continuous distribution, the corresponding probability is spread over a rectangle with base stretching from $-1/2$ to $+1/2$ on either side of the integer involved, as shown in figures to the left and below.

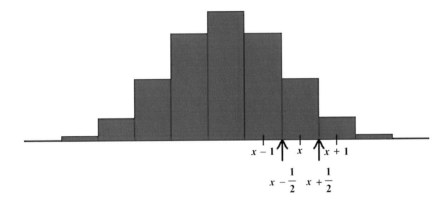

That is, if we were to approximate the probability at a point x, we would use the techniques studied in the previous section and find the area from $x - \frac{1}{2}$ to $x + \frac{1}{2}$. That is, we have

$$P(X = x) \approx P\left(\frac{x - \frac{1}{2} - np}{\sqrt{np(1-p)}} \leq Z \leq \frac{x + \frac{1}{2} - np}{\sqrt{np(1-p)}} \right)$$

as shown by

$$z_1 = \frac{x - \frac{1}{2} - np}{\sqrt{np(1-p)}} \quad \text{and} \quad z_2 = \frac{x + \frac{1}{2} - np}{\sqrt{np(1-p)}}.$$

Hence,

$$P(a \leq X \leq b) \approx P(z_1 \leq Z \leq z_2) = P\left(\frac{a-np}{\sqrt{np(1-p)}} \leq Z \leq \frac{b-np}{\sqrt{np(1-p)}}\right),$$

where $a = x - \frac{1}{2}$ and $b = x + \frac{1}{2}$.

Example 7.4.1 Normal Approximation

For example, in a problem in which binomial probability distribution is applicable with $p = \frac{1}{2}$, $n = 10$, the normal curve approximation with the correction factor is significantly closer to the exact probabilities, as shown in the following table:

x	Binomial Probabilities	Normal Approximation
	$n = 10, p = 0.5$	$\mu = 5, \sigma = 1.58$
0	0.0010	0.0020
1	0.0098	0.0112
2	0.0439	0.0435
3	0.1172	0.1145
4	0.2051	0.2045
5	0.2461	0.2482
6	0.2051	0.2045
7	0.1172	0.1145
8	0.0439	0.0435
9	0.0098	0.0112
10	0.0010	0.0020

This table is constructed as follows: we use the binomial probability distribution to obtain the exact probability of six successes in 10 trials; that is,

$$P(X=6) = \binom{10}{6}\left(\frac{1}{2}\right)^6\left(\frac{1}{2}\right)^4 = 0.2051.$$

Now we obtain the normal approximation of obtaining six successes in 10 trials; here, $\mu = np = 10 \times \frac{1}{2} = 5$ and $\sigma = \sqrt{npq} = \sqrt{10 \times \frac{1}{2} \times \frac{1}{2}} = 1.6$,

$$z_1 = \frac{x - \frac{1}{2} - \mu}{\sigma} = \frac{6 - \frac{1}{2} - np}{\sqrt{npq}} = \frac{6 - \frac{1}{2} - 5}{1.6} = \frac{0.5}{1.6} = 0.3125$$

and

$$z_2 = \frac{x + \frac{1}{2} - \mu}{\sigma} = \frac{6 + \frac{1}{2} - np}{\sqrt{npq}} = \frac{6 + \frac{1}{2} - 5}{1.6} = \frac{1.5}{1.6} = 0.9375.$$

Thus,

$$\begin{aligned}
P(a \leq X \leq b) &= P(z_1 \leq Z \leq z_2) \\
&= P(0.3125 \leq Z \leq 0.9375) \\
&= P(0 \leq Z \leq 0.9375) - P(0 \leq Z \leq 0.3125) \\
&= 0.3286 - 0.1241 = 0.2045
\end{aligned}$$

Example 7.4.1 Normal Approximation—cont'd

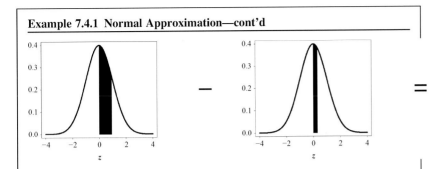

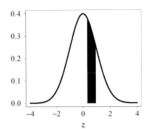

The difference between the exact probability, 0.2051 and the approximate probability using the normal probability distribution, 0.2045, is a very small difference of 0.0006.

Note that even for a small number of trials ($n = 10$) the normal curve gives a good approximation to the binomial probabilities. In general, when the value of p in the binomial is close to **1/2**, the approximations are better than when p is close to 0 or 1. This is because, for $p = 1/2$, the binomial distribution is symmetrical (see the preceding table), and, as we have seen, the curve of the normal distribution is symmetrical.

Note also, the assumptions are satisfied, that is, there are a count of at least five in each category; five expected successes and five expected failures. However, if this assumption had not been satisfied, the approximation would not have been as accurate.

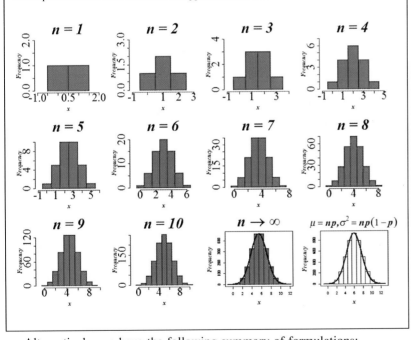

Alternatively, we have the following summary of formulations:

Property 7.4.1 Normal Approximation

When using **normal approximation** to the binomial, the standard score can be expressed in terms of the number of success or the relative frequency; and either with and without correction for continuity.

In terms of **expected number of successes**:

$$\mu = np, \ \sigma = \sqrt{np(1-p)}$$

and without correction for continuity

$$z = \frac{x - \mu}{\sigma}$$

and with correction for continuity

$$z = \frac{(x \pm 0.5) - \mu}{\sigma}$$

In terms of expected proportions:

$$\mu_p = p, \ \sigma_p = \sqrt{\frac{p(1-p)}{n}}.$$

and without correction for continuity

$$z = \frac{\hat{p} - p}{\sqrt{\frac{p(1-p)}{n}}}$$

and with correction for continuity

$$z = \frac{\left(\hat{p} \pm \dfrac{0.5}{n}\right) - p}{\sqrt{\dfrac{p(1-p)}{n}}}$$

PROBLEMS

Basic Problems

7.4.1. Let the random variable X be binomially distributed with $p = 0.6$ and $n = 10$. Evaluate the following probabilities:

7.4.2. Use the normal approximation both with and without correction for continuity to obtain the probabilities asked for in the previous exercise and compare your result with those of that problem.

7.4.3. During tests of reliability of a certain item, the probability of failure is 0.1. During a test of 100 items, determine the probability that the number of failures will be between 3 and 5.

7.4.4. A fair coin is tossed 15 times. Determine the probability that between 6 and 8 heads inclusive occurring
  (a) the binomial probability distribution.
  (b) the normal approximation without correction for continuity.
  (c) the normal approximation with correction for continuity.

7.4.5. An encyclopedia salesman has observed that, on the average, the probability of making a sale on the first contact is equal to 0.49. The salesman will contact 200 customers during a six-month period. What is the probability that the number of successful contacts will be between 100 and 110?

7.4.6. A fair die is tossed 200 times. Using normal approximation to the binomial, what is the probability that an ace (one) will appear between 34 and 36 times?

Critical Thinking

7.4.7. Suppose that a "loaded" coin is tossed a large number of times. If the probability of a head occurring at any given toss is 0.60, find the number of tosses required in order for the experiment to result in fewer than 20 tails, with probability of at least 0.95.

7.4.8. The probability of having a girl is 0.5, so, using the normal approximation, what is the probability that in a family of five will have exactly one boy? Compare this to the binomial probability distribution. Discuss.

7.4.9. If the probability of having a girl is 0.35, using normal approximation, what is the probability that in a family of five will have exactly one boy? Compare this to the binomial probability distribution. Discuss.

7.4.10. If the probability of having a girl is 0.35, using the normal approximation, what is the probability that in a family of five will have exactly three boys? Compare this to the binomial probability distribution. Discuss.

7.4.11. If the probability of finding success in a business venture is 0.05, using the normal approximation, what is the probability that in 200 business ventures the number of success
(a) will be between 10 and 15?
(b) will be exactly 10?
(c) will exceed 10?

CRITICAL THINKING AND BASIC EXERCISE

Basic Problems

7.1. Let the random variable X be normally distributed with mean 75 and standard deviation 10. Calculate the following probabilities:
(a) $P(X \leq 74)$
(b) $P(X > 68)$
(c) $P(72 \leq X \leq 76)$

7.2. Suppose that the weights of the football players in the NFL are normally distributed with a mean of 235 pounds and a standard deviation of 30 pounds. What is the probability that a football player chosen at random will weigh
(a) less than 240 pounds?
(b) between 225 and 245 pounds?

7.3. Assuming that the test grades of a biochemistry examination are normally distributed with mean 75% and standard deviation 10%. What is the probability that a test grade of a student selected at random will be
(a) lower than 80%?
(b) between 68% and 78%?
(c) greater than 96.7%, an A+?

7.4. A race car driver is experimenting with two different strategies to determine which will bring him the greatest success. The time it takes to cover the course is the main criterion by which he is judging the two strategies. Suppose that the time using Strategy A is normally distributed with a mean of 4 hours and a standard deviation of 0.8 hours; for Strategy B the distribution is also normal with a mean of 4.5 hours and a standard deviation of 0.4 hours. If our driver is to be a top contender to win the race, he must cover the course within 4.6 hours. Which strategy is the better choice?

7.5. The length of life of a plasma TV is known to be normally disturbed with a claimed mean of 100,000 hours to half-life or approximately 54 years at 3 hours per day with a standard deviation of 12 years. What is the probability that a plasma TV selected at random will operate for between 30 and 60 years.

7.6. Suppose the diameter of golf balls manufactured by a certain company is normally distributed with a mean of 1.90 inches and a standard deviation of 0.5 inches. A golf ball produced by this company will be approved for use in the US Open Golf Tournament if its diameter is within 0.08 inches from the mean. What is the probability that a golf ball of this kind chosen at random will not be acceptable for use the tournament?

7.7. Assuming the breaking strength of a woolen fabric is normally distributed with a mean of 100 pounds and a standard deviation of 8 pounds. A fabric of this type will successfully pass a government inspection if its breaking strength is greater than 90 pounds. What is the probability that a woolen fabric chosen at random will pass the inspection?

7.8. The probability that a freshman entering a certain university and eventually graduating is 0.66. Suppose that a freshmen class consists of 36,358 students enrolled in the University of South Florida in 2009. What is the probability that
(a) at least 2400 of the entering freshmen will graduate?
(b) the number of freshmen who will eventually graduate is between 1950 and 2250?

7.9. A basketball player succeeds on 82% of his foul shots. If during a basketball season he will shoot about 130 foul shots, determine the probability that he will be successful with at least 80 baskets.

7.10. Suppose that temperature fluctuation (measurements) during the winter months in Clearwater, Florida are normally distributed with a mean of 76 °F and standard deviation of 12 °F. What is the probability that on a selected windy day the temperature in Clearwater will be

(a) between 70 °F and 80 °F?

(b) higher than 80 °F?

7.11. About 10% of the pantyhose manufactured by company A do not pass inspection and are considered irregular or defective. What is the probability that in a random sample of 500 pantyhose produced by this process

(a) at most 50 will be defective?

(b) between 40 and 60 will not pass inspection?

7.12. Let X be binomially distributed with $n=100$, and $p=0.6$. Use the normal approximation to determine the following probabilities:

(a) $P(X \leq 70)$

(b) $P(58 \leq X \leq 62)$

7.13. Suppose that 5% of the pollution devices in the new automobiles are not effective after the first year of operation. If, following that year, we sampled 100 automobiles with such device, determine the probability that we would find fewer than 40 automobiles with ineffective pollution devices? (Assuming that owners did not bother to have these devices checked during the first year.)

Critical Thinking

7.14 Determine if the random variable is discrete or continuous:

(a) In an experiment, to monitor the productivity of a newly developed fuel, a researcher measures the liters produced by as single corn stock.

(b) In an experiment, the rate at which salts dissolve in water, the times are recorded from the moment the water is poured into the salt until no crystals can visually be seen.

(c) In an ecology study, to monitor trout in the St. Johns river, trout are counted and counted before being weighted, tagged and returned to the water.

7.15 Given that the data follows the normal probability distribution with mean of 100 and standard deviation of 15, determine the following probabilities:

(a) $P(x \leq 110)$ (d) $P(x \leq 80)$

(b) $P(x \geq 110)$ (e) $P(65 \leq x \leq 90)$

(c) $P(70 \leq x \leq 110)$ (f) $P(x \geq 70)$

7.16 Given that the data follows the normal probability distribution with mean of 100 and standard deviation of 15, find the value of a such that:

(a) $P(x \leq a)=0.90$ (c) $P(-a \leq x \leq a)=0.9$ (e) $P(-a \leq x \leq a)=0.95$

(b) $P(x \geq a)=0.90$ (d) $P(x \leq a)=0.95$

SUMMARY OF IMPORTANT CONCEPTS

Definitions:

7.1.1 A random variable which can assume an infinitely large number of values associated with the points on a line interval, and the probability of which is spread continuously over these points, is called a **continuous random variable**.

7.2.1 A continuous random variable X is **normally distributed** or follows a **normal probability distribution** if its probability distribution is given by the following function:

$$f(x) = \frac{1}{\sigma\sqrt{2\pi}} e^{-\frac{(x-\mu)^2}{2\sigma^2}},$$

$$-\infty < x < \infty, -\infty < \mu < \infty, 0 < \sigma^2 < \infty$$

7.3.1 The **standard normal probability distribution (standard normal curve)** has mean of zero, $E(z)=0$, and unit variance, $V(z)=1$ and will be denoted $z \sim N(0,1)$. Hence, the transformation of the data moves the center of symmetry is to zero and changes the variance to one:

$$f(z) = \frac{1}{\sqrt{2\pi}} e^{-\frac{z^2}{2}},$$

$$-\infty < z < \infty.$$

7.3.2 The **standard score** is the transformation of the data given by: $z=\frac{x-\mu}{\sigma}$, where $z \sim N(0, 1)$.

Rules:

7.4.1. To use **Normal Approximation**, given the binomial probability of success, p, and a given number of trials, n, then the expected number of successes is

$$\mu=np,$$

and the variance depends on the sample size, n, the probability of success, p, and the probability of failure, $(1-p)$; that is, $\sigma^2=np(1-p)$ and $\sigma=\sqrt{np(1-p)}$.

Properties:

7.4.1. When using **normal approximation** to the binomial, the standard score can be expressed in terms of the number of success or the relative frequency; and either with and without correction for continuity.

In terms of expected number of successes: $\mu=np$, $\sigma=\sqrt{np(1-p)}$; and without correction for continuity $z=\frac{x-\mu}{\sigma}$ and with correction for continuity $z=\frac{(x\pm0.5)-\mu}{\sigma}$.

In terms of expected proportions: $\mu_{\hat{p}}=p$, $\sigma_{\hat{p}}=\sqrt{\frac{p(1-p)}{n}}$ and without correction for continuity $z=\dfrac{\hat{p}-p}{\sqrt{\dfrac{p(1-p)}{n}}}$ and with correction for continuity $z=\dfrac{\left(\hat{p}\pm\dfrac{0.5}{n}\right)-p}{\sqrt{\dfrac{p(1-p)}{n}}}$.

REVIEW TEST

1. Which one of the following would not be expected to be normal?
    (a) The height of a rose bush
    (b) The yearly family income in the United States
    (c) The lifetime of a car battery
2. Which statement(s) is/are true regarding normally distributed data?
    (a) The mean equals the median equals the mode
    (b) The mean is positive.
    (c) The points of inflection are at $\mu\pm\sigma$
3. Given the distribution of IQ scores is normally distributed with a mean of 100 and a standard deviation of 15. What percentage of all scores lie within 1, 2, and 3 standard deviations from the mean?
4. On an exam, the scores are normally distributed with a mean of 145 and a standard deviation of 18. What percentage of the scores are
    (a) above 200
    (b) below 120
    (c) between 120 and 200
    (d) between 150 and 180
    (e) between 100 and 130
5. The wait time to be checked out at a local convenience store is normally distributed with a mean of 15 minutes and a standard deviation of 4 minutes. If there are 200 customers in a given day, answer the following questions:
    (a) What percent of customers waited between 10 and 25 minutes?
    (b) How many customers are expected to have waited between 10 and 25 minutes?
    (c) What percent of customers waited for more than 25 minutes?
    (d) How many customers are expected to have waited for more than 25 minutes?
6. A fair die is rolled 30 times. Find the probability of getting exactly 10 fours using normal approximation to the binomial.

REFERENCES

Cramer, H., 1955b. The Elements of Probability Theory and Some of Its Applications. Wiley, New York Almquist & Wiksell, Stockholm, 1954.

Goldberg, S., 1960. Probability: An Introduction. Prentice-Hall, Englewood Cliffs, NJ.

Monroe, M.E., 1951. Theory of Probability. McGraw-Hill, New York.

Parzen, E., 1960b. Modern Probability Theory and Its Applications. Wiley, New York.

Tsokos, C.P., 1972d. Probability Distributions: An Introduction to Probability Theory with Applications. Wadsworth, Belmont, CA.

Tsokos, C.P., 1978f. Mainstreams of Finite Mathematics with Application. Charles E. Merrill Publishing Company, A Bell & Howell Company.

Tsokos, C.P., Wooten, R.D., 2011c. The Joy of Statistics Using Real-World Data. Kendall Hunt, Florida.

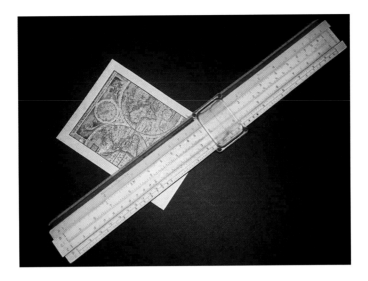

Numerical quantities focus on expected values, graphical summaries on unexpected values.

John Wilder Tukey

Chapter 8	**Basic Statistics**

John Wilder Tukey was a chemist turned topologist turned statistician and was one of the most influential statisticians of the past 50 years. He is credited with inventing the word software. He worked as a professor at Princeton University and was a senior researcher at AT&T's Bell Laboratories. He made significant contributions to the fields of exploratory data analysis and robust estimation. His works on the spectrum analysis of time series and other aspects of digital signal processing have been widely used in engineering and science. He coined the word bit, which refers to a unit of information processed by a computer. In collaboration with Cooley, in 1965, Tukey introduced the fast Fourier transform (FFT) algorithm that greatly simplified computation for Fourier series and integrals. Tukey authored or coauthored many books in statistics and wrote more than 500 technical papers. Among Tukey's most far-reaching contributions was his development of techniques for "robust analysis" and approach to statistics that guard against wrong answers in situations where a randomly chosen sample of data happens to poorly represent the population. Tukey also made significant contributions to the analysis of variance.

8.1 DATA ORGANIZATION

When data is first gathered, it is listed in some form; we refer to it as the **listed data**. **Listed data** is a sequence of information presented (listed) as it is observed; usually denoted: $x_1, x_2, \ldots, x_n$, where n is the sample size. Three basic forms of summarizing the frequency distribution of the given data include **Dot Plot**, **Stem-and-Leaf Plot**, and **Frequency Tables**.

We begin with a **Dot Plot**.

Definition 8.1 Dot Plot

A **dot plot** is one of the simplest ways to determine the distribution of numerical data— using a real number line, create a scale that includes the minimum and maximum data values, and then place a dot above each observed data point.

The simplest form of statistical plots best suited for smaller sample sizes; a useful aid in organizing numerical information.

Step-by-Step Dot Plot

Step 1. Construct a number line that contains both the minimum and maximum data and is partitioned into an appropriate scale.

Step 2. One at a time, as you run through the numbers, place a dot above the number line and above any previously drawn dot to create columns above each observed value as a count of frequency.

Example 8.1.1 Dot Plot of Listed Data

Given the **listed data**: 1, 4, 3, 5, 2, 3, 4, 3, 4, 3, draw the associated **dot plot**.

Solution

Step 1.   Draw a real number line which covers the full range of data, the **minimum** is 1 and the **maximum** is 5.

Step 2.   Then, one data value at a time, placing a dot over the observed value yields the following **dot plot**.

   Looking at the **dot plot** we can see that the most frequently observed value is 3 and it appears 4 times, the number 4 appears 3 times, and the remaining three observations: 1, 2, and 5 all occur once.

Dot plots are useful to help organize and describe **discrete quantitative** data sets with relative small **sample size** and little **variability**; however, for larger **quantitative** data sets, a **stem-and-leaf plot** is more appropriate.

Definition 8.1.2 Stem-and-Leaf Plot

A **stem-and-leaf plot** is completed in two stages. First the basic **stem** and **leaf** are constructed by breaking the observed data values into categories such as tens and hundreds (the **stem**) and then units digits or the remaining associated information (the **leaves**). This initial sort does give the overall **distribution**, but it is common practice to then sort the leaves. The **range** of a data set is the difference between the **minimum** and **maximum** value; or maximum minus minimum.

A four step procedure is given on how to structure a **stem-and-leaf plot**.

Step-by-Step Stem-and-Leaf Plot

Step 1.   Construct a column of stems—this is subjective to the range of numbers. For example, if numbers range from 100 to 150, then the hundreds and tens might be the stem and the units value the leaf.

Step 2.   Once the stems are in place, read the data values one at a time, placing each leaf on the appropriate stem.

Step 3.   Once the basic stem-and-leaf configuration is complete, sort each of the stems within their respective category.

Step 4.   ↺ If there is more than one data set being considered, we might extend this in both directions and create a back-to-back stem-and-leaf table. That is, repeat **Steps 1-3** for each set of comparable data.

Example 8.1.2 Stem-and-Leaf Plot

We have obtained the following data from an experiment: 193, 405, 330, 503, 244, 323, 487, 385, 454, and 391. Develop the **Stem-and-leaf plot**.

Solution

Since the data ranges over $(503 - 193) = 310$ values, we will assign the hundred's place value to be the **stem** and the tens and units values to be the **leaf**.

Step 1 Construct Stem	Step 2 Add Leafs		Step 3 Sort Leafs	
Stem	**Stem**	**Leaf**	**Stem**	**Leaf**
1	1	93	1	93
2	2	44	2	44
3	3	30, 23, 85, 91	3	23, 30, 85, 91
4	4	05, 87, 54	4	5, 54, 87
5	5	03	5	3

There are many ways to define a **stem-and-leaf plot**; however, there are a few underlying constructs. Thus, in steps 2 and 3 we see that most of the observations are in the third stem; that is, the most frequent category is one with numbers in the 300 s. There are no obvious gaps—thus we have low variability.

Given below are three rules that we follow constructing the **stem-and-leaf**.

Rule 8.1.1   Stem-and-Leaf Plots

A **stem-and-leaf plot** must follow the following rules:
1. The defined stems and leaves must be based on equal intervals.
2. All intermediate categories must be included even though there might be zero data values in a given category.
3. Including empty categories within the outlined range of data values appear as gaps and these gaps are indicative of variability in the data.

Example 8.1.3   Stem-and-Leaf Plot

A given experiment results in the following measurements: 5.63, 5.94, 5.73, 5.64, 5.69, 5.36, 5.88, 5.82, 5.59, and 5.67. Develop a **stem-and-leaf plot** of the data and interpret.

Solution

Since the data **ranges** over $(5.94 - 5.36) = 0.58$ values, which is less than one, we will be assigning the units and tenths place value to be the **stem** and the hundredths place value to be the **leaf**.

Step 1	Step 2		Step 3	
Stem	**Stem**	**Leaf**	**Stem**	**Leaf**
5.3	5.3	6	5.3	6
5.4	5.4		5.4	0
5.5	5.5	9	5.5	9
5.6	5.6	3, 4, 9, 7	5.6	3, 4, 7, 9
5.7	5.7	3	5.7	3
5.8	5.8	8, 2	5.8	2, 8
5.9	5.9	4	5.9	4

Solution—cont'd

In the above stem-and-leaf plot, we see that there is variability in the data as indicated by the gap, there are no observations starting with 5.4. However, this does appear most frequently in the 5.6's. The data ranges from 5.36 to 5.94, a maximum difference (range) of 0.58.

Definition 8.1.4 Frequency Table

A **frequency table** is simply a "t-chart" or two-column table which outlines the various possible outcomes and the associated frequencies observed in a sample.

Frequency tables are used for **discrete data** or **discretized data** (that is continuous data partitioned into distinct intervals which can be considered discretely).

The capital Greek letter sigma (Σ) will be used to represent the idea "to sum"; for example, as the sample size is the sum of the frequencies, it can be written $n = \sum f$. If there are k distinct outcomes with associated **frequencies** f_i for $i = 1, 2, \ldots, k$, then this notation can be extended to include more detail by including this count; that is, $n = \sum_{i=1}^{k} f_i = f_1 + f_2 + \cdots + f_k$. The lower script defines the counter, i, as well as the under script states the starting point (when simply counting, this is, 1) and the upper script defines the point of termination (in listed data this is simply n.)

Step-by-Step Frequency Table

Step 1. Construct a column of the different data values in ascending order.
Step 2. For each data value, count the number of times each value occurs.

Example 8.1.4 Frequency Table

The results of a given experiment are: 6, 4, 3, 5, 2, 3, 4, 3, 4, 3, 5, and 3 and the associated **Frequency Table** is given by:

Outcome	Frequency
2	1
3	5
4	4
5	2
6	1

Here we see that the most frequent data (the **mode**) is 3, the **minimum** is 2 and the **maximum** is 6.

From this **frequency table**, we can begin to visualize the **frequencies** or **frequency distribution** of the given data.

A common and useful method is to develop the frequency distribution of a given set of data.

Definition 8.1.4 Frequency Distribution

A **frequency distribution** is a tabulation of the values that one or more variables take in a sample which can be organized in a frequency table or illustrated graphically.

Frequency distribution is distinguished from probability distribution as the occurrence of each outcome is a count and not a percent. These counts can be manipulated by dividing by the total number of trials in the sample; that is, the sample size n, to obtain the **relative frequency distribution**. That is, the occurrence of each outcome in a percent or **proportion** of the total sample size; the **relative frequency**. Moreover, with a few additional calculations, both of these measures, the count that is **frequency**, f and the rate, $\frac{f}{n}$ that is **relative frequency** can be described cumulatively: and **cumulative relative frequency**.

Definition 8.1.5 Relative Frequency

A **relative frequency** is a measure between 0 and 1 that describes the frequency with which the values of the variable take relative to the sample size, n.

$$p_i = \frac{f_i}{n}$$

where

$$n = \sum_{i=1}^{k} f_i$$

and k is the distinct number of outcomes.

Definition 8.1.6 Probability Distribution

A **probability distribution** is a tabulation of the values that one or more variables take in a sample written in terms of **relative frequency**, **percentages**, or **proportions**. If these values follow a formula, then this formula will be denoted as $f(x)$.

Step-by-Step Frequency Table Extended

Step 1. Construct a column of the different data values in ascending order.

Step 2. For each data value, count the number of times each value occurs.

Step 3. ↳ Using the sum of the frequencies, we can verify that we have included all the data points as well as create a column for relative frequencies; that is, given that the **frequency** of a given outcome is f_i and the **sample size** (sum of frequencies) is n, then the **relative frequency** is $p_i = \frac{f_i}{n}$.

Step 4. ↳ Using the frequencies, we can also compute the cumulative frequencies; that is,

$$c_i = \sum_{j=1}^{i} f_j = f_1 + f_2 + \cdots + f_i$$

for $i = 1, 2, \ldots, k$, where k is the number of distinct outcomes.

Step 5. ↳ Using the cumulative frequencies, we can then compute the cumulative relative frequency

$$C_i = \frac{c_i}{n} = \sum_{j=1}^{i} p_j = p_1 + p_2 + \cdots + p_i$$

Note: $\sum_{i=1}^{k} p_i = 1$

Example 8.1.5 Frequency Table Extended

Using the data from the **frequency table** in Example 6.4, compute the **relative frequency**.

Solution

Since the total **sample size** is $n = \sum f = 1 + 5 + 4 + 2 + 1 = 13$, the **relative frequency** is the **frequency** of each of the **data** value divided by 13.

Step 1	Step 2		Step 3		
					Relative frequency
Outcome	Outcome	Frequency	Outcome	Frequency	frequency
x	x	f	x	f	p
2	2	1	2	1	$= 1/13 = 0.0769$
3	3	5	3	5	$= 5/13 = 0.3846$
4	4	4	4	4	$= 4/13 = 0.3077$
5	5	2	5	2	$= 2/13 = 0.1538$
6	6	1	6	1	$= 1/13 = 0.0769$
		13		13	

Above relative frequencies are given too many decimals, but this can be rounded or written as a percentage. For example, from the above table, we see relative to the data set, 38.4% of the information is in a single category, the mode 3.

The cumulative frequencies are created consecutively, adding additional counts as you progress, and you should always end with the total sample size n.

Step 4.

Outcome	Frequency	Cumulative frequency	
x	f	c_i	c_i
2	1	$C_1 = 1$	1
3	5	$C_2 = 1 + 5 = 6$	6
4	4	$C_3 = 6 + 4 = 10$	10
5	2	$C_4 = 10 + 2 = 12$	12
6	1	$C_5 = 12 + 1 = 13$	13
	13		

The cumulative relative frequencies are created in a similar way to the relative frequencies, except we are using the cumulative frequencies or summing the relative frequencies. Note: the last entry should be 1.000 or 100%.

Step 5.

Outcome	Cumulative frequency c_i	Cumulative relative frequency	C_i
2	1	$= 1/13 = 0.0769$	0.0769
3	6	$= 6/13 = 0.4615$	0.4615
4	10	$= 10/13 = 0.7692$	0.7692
5	12	$= 12/13 = 0.9231$	0.9231
6	13	$= 13/13 = 1.0000$	1.0000
	13		

Continued

Solution—cont'd

Otherwise, using the relative frequency, we have the following.

Outcome	Frequency	Relative frequency
2	1	0.0769
3	5	0.3846
4	4	0.3077
5	2	0.1538
6	1	0.0769

$C_1 = p_1$
$= 0.0769$

$C_2 = p_1 + p_2$
$= 0.0769 + 0.3846 = 0.4615$

$C_3 = p_1 + p_2 + p_3 = C_2 + p_3$
$= 0.4615 + 0.3077 = 0.7692$

$C_4 = p_1 + p_2 + p_3 + p_4 = C_3 + p_4 =$
$= 0.7692 + 0.1538 = 0.923$

$C_5 = p_1 + p_2 + p_3 + p_4 + p_5$
$= C_4 + p_5 =$
$= 0.923 + 0.0769 = 0.9999$

Note: *Here we have rounding errors as all the relative frequencies have been rounded to the fourth decimal place. Had we used the cumulative frequency divided by the total, we would not have this issue.*

Thus, we can conclude from the cumulative relative frequency that 46.15% of the outcomes in the experiment will be less than or equal to 3, whereas 92.3% of the outcomes are less than or equal to 5 and 100%, that is, all the data are less than or equal to 6.

EXERCISES

Critical Thinking

8.1.1. What is always the last entry in the cumulative frequency column? What is the last entry in the cumulative relative frequency column?

8.1.2. Distinguish between a frequency table and a stem-and-leaf plot.

8.1.3. Distinguish between a stem-and-leaf plot and a back-to-back stem-and-leaf plot.

8.1.4. Distinguish between a frequency table and a frequency distribution.

8.1.5. Distinguish between a frequency distribution and a probability distribution.

8.1.6. Distinguish between frequency and relative frequency.

8.1.7. Distinguish between frequency and cumulative frequency.

8.1.8. Distinguish between the relative frequency and the cumulative relative frequency.

Basic Problems

8.1.9. You are going to hold an ice cream party for the people in this survey.

Favorite ice cream flavor

Flavor	Number
Pistachio	⊞⊞ II
Coconut	⊞⊞ III
Chocolate	⊞⊞ ⊞⊞ II
Vanilla	⊞⊞ ⊞⊞
Banana nut	⊞⊞ I

a. If you can have three flavors of ice cream at your party, which flavors of ice cream would you bring?

b. What is the least popular ice cream flavor?

8.1.10. You are going to hold a bake sale, so you take the following survey.

Favorite cookie

Cookie	Number
Peanut butter	卌	
Chocolate chip	卌 |||
Animal crackers	卌
Ginger snaps	卌 卌	
Oatmeal	||||

a. How many people chose either animal crackers or chocolate chip?

b. What is the least popular cookie flavor?

c. How many people chose oatmeal as their favorite cookie?

8.1.11. You are interested in which animals are a favorite in your local zoo and take the following survey.

Favorite zoo animals

Animal	Number
Peafowl	10
Woodpecker	4
Kangaroo	7
Panda	9
Elephant	6
Cobra	5

a. How many people chose a mammal as their favorite animal?

b. List the zoo animals in order from the zoo animal with the fewest votes to the zoo animal with the most votes.

c. List the zoo animals in order from the zoo animal with the most votes to the zoo animal with the fewest votes.

8.1.12. You are going to hold an ice cream party for the people in this survey.

Favorite beverage

Beverage	Number
Lemonade	12
Root beer	10
Orange soda	5
Punch	9
Milk	2
Cola	4

a. How many people answered the survey?

b. List the beverages in order from the beverage with the fewest votes to the beverage with the most votes.

c. How many fewer people chose punch than chose root beer?

8.1.13. According to the dot plot above, what is the greatest number of games won?

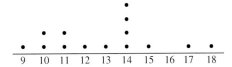

8.1.14. According to the dot plot above, what is the most frequent number of e-mails sent today?

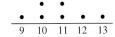

8.1.15. Given the data in the dot plot above, what is the most frequently occurring data value?

8.1.16. Construct a dot plot for the following data on number of games won by students: 1, 8, 2, 6, 1, 2, 3, and 1. How many students won more than three games?

8.1.17. Construct a dot plot for the following data on number of games won by students: 4, 2, 5, 7, 8, 6, 7, 5, 4, 5, and 2. What is the greatest number of games won?

8.1.18. Given the following stem-and-leaf for temperatures recorded in October.

Stem (tens)	Leaves (units)
4	8 5 2 7 4
5	1 0 3
6	0 2 5
7	4 2 0 1

   a. How many temperatures are recorded?
   b. What was the highest temperature recorded?
   c. What is the coldest (lowest) temperature recorded?

8.1.19. This stem-and-leaf plot shows Madison's scores from a video game.

Game scores

Stem	Leaves
19	2 6
20	2 8
21	2 6 5
22	4 6 3 1
23	2
24	0
25	
26	
27	0

   a. How many games scored at least 223?
   b. What is the range of the data set?
   c. What was the highest score?

8.1.20. Construct a back-to-back stem-and-leaf plot for the given data.

Group A	95	85	97	84	82	97	101	107	89
Group B	79	81	84	91	93	91	87	77	72

8.1.21. Given the data in listed form, create a frequency table:

4	1	0	4	9	3	9	5	3	0
3	6	4	1	4	6	8	7	6	4
3	8	3	0	6	9	2	6	4	9
8	7	5	6	6	1	8	5	2	1

What can you learn from the frequency table?

8.1.22. Draw the dot plot for the given listed data:

1	3	7	1	1	7	0	8	0	8
3	6	7	9	3	3	5	7	2	4
3	1	5	4	3	2	9	5	1	5
2	7	8	1	6	8	5	7	7	9

Can you obtain any useful information from the dot plot?

8.1.23. Put the given data into a stem-and-leaf form using the tens and hundreds as the stem and the units and tenths as the leaf.

104	128.2	93	88.5	92
113.2	98.5	89.9	90.5	96.5
104.7	70.8	124.5	86.6	77.2
102.6	101	101	93.4	99.7
94.7	87.6	106.9	111.5	92.9
102.8	129.4	99	88.6	117.2

What does the stem and leaf diagram tell you about the data? Can you recover the original data from the stem and leaf form?

8.1.24. Construct a frequency table for the listed data given below. Include the relative frequency, cumulative frequency, and cumulative relative frequency.

80	67	51	82	57
80	71	82	52	75
58	62	63	51	77
85	91	97	59	61
99	70	86	98	88
87	92	66	64	61

What information does the relative frequency with solution frequency and cumulative solutions frequency convey to you?

8.2 GRAPHICAL REPRESENTATIVE OF QUALITATIVE INFORMATION

We shall use the graphical and numerical techniques we learned in the previous section to introduce other useful graphical methods.

In the this section, the visual representations of the given information from a **frequency table** or **relative frequency table** are used for **qualitative data** and include such graphics as **bar charts, circle graphs**, and **contingency tables**.

One of the first ways to graphically illustrate **qualitative** data is using various forms of a **bar chart**. This can be done from **listed data** organized into a **frequency table**; the only difference is that this is not **quantitative** data, it is categorical.

Definition 8.2.1 Bar Chart

A **bar chart** is representation of a frequency distribution of a qualitative variable by way of rectangles whose heights are proportional to relative frequencies placed over each category.

When the categories are sorted by descending frequencies, the graph is a special form of the **bar chart** called a **Pareto Chart**.

Given below are three stages to obtaining a **bar chart** (**bar graph**).

Step-by-Step Bar Chart

Step 1. Organize the data values into a **frequency table** based on the outlined categories. Note: the data need to belong to one and only category. Determine **relative frequency** as needed.

Step 2. Place the **class marks** on a horizontal line; that is, mark out all the categories with tick marks as the **categories**. Then draw a perpendicular to the horizontal axis, draw a vertical axis and taking into consideration the maximum **frequency** or **relative frequency** and using a consistent scale mark out the **frequencies** or **relative frequency**.

Step 3. Finally, construct a bar (rectangular boxes) for each class, leaving **gaps** between the different categories as there are no boundaries between the various qualities characterized by the defined outcomes.

Example 8.2.1 Bar Chart

Consider a **survey** consisting of 129 people indicating their favorite fruit/snack summarized in the **frequency table** below. Graphically illustrate this information in a **bar chart** in terms of frequencies and discuss what this subject chart conveys.

Outcome	Frequency	Relative frequency (%)
A—Apples	32	24.8
B—Banana	51	39.5
C—Cantaloupes	26	20.2
D—Dole pineapple	11	8.5
F—Fruit & nut mix	9	7.0

Solution

Step 1. The frequency table is given.

Step 2. As the maximum frequency is 51, we have chosen a 10 point scale with a maximum count of 60 with the categories listed on the horizontal axis.

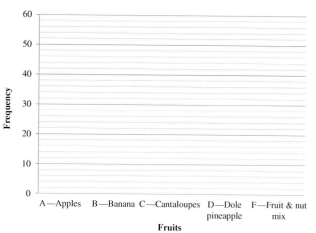

Solution—cont'd

Step 3. Using the information given above, the **bar chart** is given by:

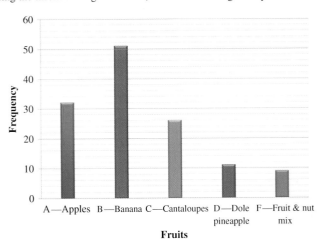

Thus, the **bar chart** illustrates that bananas represented by the largest bar is the most favored followed by the second largest bar that represents apples, etc. The least favored fruit is Fruit and Nut mix.

An alternative form of the **bar chart** is the **Pareto chart**.

Definition 8.2.2 Pareto Chart

A **Pareto chart** is a bar chart such that the categories have been sorted according to frequency (in descending order).

Given below is a step-by-step procedure to develop the **Pareto chart**.

Step-by-Step Pareto Chart

Step 1. Organize the data values into a **frequency table** based on the outlined categories. Note: the data needs to belong to one and only one category. Determine **relative frequency** as needed.

Step 2. Sort the categories by frequency, from highest to lowest count.

Step 3. Place the **class marks** on a horizontal line; that is, mark out all the categories with tick marks as the **class marks**. Then draw a line perpendicular to the horizontal axis, draw a vertical axis taking into consideration the maximum **frequency** or **relative frequency** and using a consistent scale mark out the **frequencies** or **relative frequency**.

Step 4. Finally, construct a bar (rectangular boxes) for each class leaving **gaps** between the different categories as there are no boundaries between the various qualities characterized by the defined outcomes.

Example 8.2.2 Pareto Chart

Consider the **survey** that consisted of 129 individuals in the previous example. Graphically illustrate this information in a **Pareto chart** in terms of frequencies and discuss what this subject chart conveys.

Outcome	Frequency	Relative frequency (%)
B—Banana	51	39.5
A—Apples	32	24.8
C—Cantaloupes	26	20.2
D—Dole pineapple	11	8.5
F—Fruit & nut mix	9	7.0

Solution

Step 1. The frequency table is given; therefore we must simply sort them by frequency (shown above).

Step 2. As the maximum frequency is still 51; hence we have the same scale as when we developed the **bar chart**.

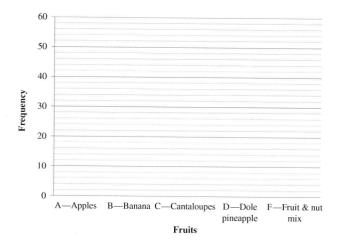

Step 3. Using the information given above, the **Pareto chart** is given by:

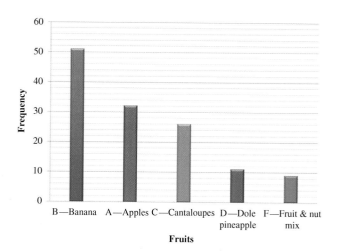

Thus, the **Pareto chart** more clearly illustrates that bananas represented by the largest bar is the most favored followed by the second largest bar that represents apples, etc. The least favored snack is Fruit and Nut mix.

The difference between a general bar chart and a **Pareto chart** is that the categories are re-organized in order of descending **frequencies**.

We encounter several real world problems that we are concerned with more than our qualitative variable. Data with more than one qualitative variable can be summarized in a **contingency table**.

Definition 8.2.3 Contingency Table

A **contingency table** is a frequency distribution for a two-way statistical classification; that is a matrix of information where one factor is represented by the row and the second factor is represented by the column and the count of individuals that belong to exactly one cell within the matrix that is one row and one column.

Below are three steps that we can follow in developing a **contingency table** for a given set of information.

Step-by-Step Contingency Table

Step 1. Organize the data values into a **table** which indicate the outcome for each individual as matched pairs.

Step 2. For each data value, count the number of times each of the possible matched pairs value occurs.

Step 3. Using these counts create a contingency table with the first outcome in the row and the second outcome in the column.

Example 8.2.4 Contingency Table

In a recent graduating class, we randomly selected 10 graduates and recorded their gender and type of degree they received. The information is given in the table bleow. Create a contingency table for the **degree** by **gender**.

Gender	Degree
F	BA
F	MA
F	BA
F	PhD
M	MA
M	MA
F	MA
M	BA
F	MA
M	MA

Solution

Step 1. As there are two **genders** and three possible **degrees**, there are $2 \times 3 = 6$ possible outcomes: namely, female with a BA, female with a MA, female with a PhD, male with a BA, male with a MA, and male with a PhD.

Gender	Degree
F	BA
F	MA
F	PhD
M	BA
M	MA
M	PhD

Continued

Solution—cont'd

Step 2. Then create a table which indicates the frequency of the possible outcomes.

Gender	Degree	Frequency
F	BA	2
F	MA	3
F	PhD	1
M	BA	1
M	MA	3
M	PhD	0

Step 3. Then construct the contingency table with one factor (**gender**) in the rows and the second factor (**degree**) in the columns:

	BA	MA	PhD	Row totals
F	2	3	1	6
M	1	3	0	4
Column totals	3	6	1	**Total 10**

Note: the total rows and the total columns is the same as shown above by the contingency table.

Thus, there are 6 females: 2 with a BA, 3 with an MA, and 1 with a PhD degree; there are 4 males: 1 with a BA, 3 with an MA, and 0 with a PhD degree, shown in the double bar chart below.

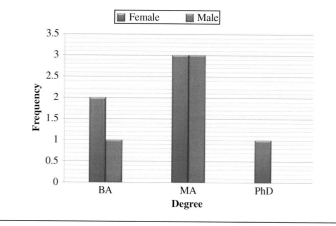

Example 8.2.5 Contingency Table

We conducted a survey that included the two qualitative variables: eye color and gender. The following contingency table below gives a summary of the information gathered.

	Brown	Blue	Green	Row totals
Male	9	3	7	19
Female	17	4	8	29
Column totals	26	7	15	**Total 48**

In this contingency table, we see that the majority of the females have brown eyes. Very few people have blue eyes, in fact, only 7 out of a total of 48. Brown eyes (with 26) appear to be more common than both blue eyes and green eyes combined (22), shown in the double bar chart below.

Example 8.2.5 Contingency Table—cont'd

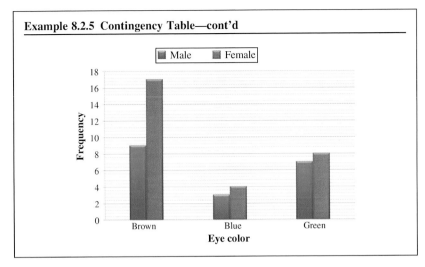

Another commonly used graphical display of **qualitative** data is using a **circle graph** (**pie chart**). This can be created from **listed data** organized into a **frequency table**, then the **relative frequency** is computed and then the **measure** of the angle in degrees is determined.

Definition 8.2.4 Circle Graph (Pie Chart)

A **circle graph** is a circle is divided into categories $C_1, C_2, C_3, ..., C_m$ where the size of each partition is proportional to the relative frequency (Figure 8.1).

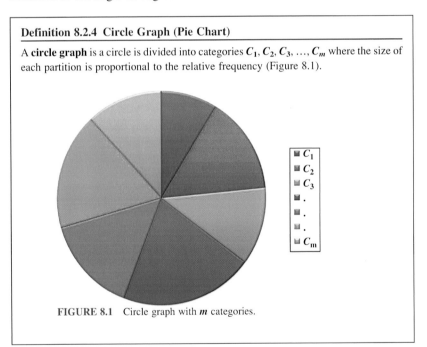

FIGURE 8.1   Circle graph with m categories.

We can easily structure a **pie chart** by following these steps.

Step-by-Step Contingency Table

Step 1. Construct a column of the different data values.

Step 2. For each data value, count the number of times each value occurs.

Step 3. Using the sum of the frequencies, we can verify that we have included all the data points as well as create a column for relative frequencies.

Step 4. Given that the **frequency** of a given outcome is f and the **sample size** (sum of frequencies) is n, the **relative frequency** is $p = \frac{f}{n}$. Apportion the circle using

Continued

Step-by-Step Contingency Table—cont'd

wedges to indicate the proportion, divide the circle by the same proportion. That is, given f_i is the **frequency** of the ith **outcome** and the **sample size**

$$n = \sum_{i=1}^{k} f_i,$$ then **relative frequency** of the ith **outcome** is $p_i = \frac{f_i}{n}$ and the

measure of the angle in degrees is $\alpha_i = p_i \times 360°$.

Example 8.2.6 Circle Graph

Develop a **pie chart** of the data given in Example 8.2.2. Extend the given frequency table and develop a **circle graph**.

Solution

Step 1. ⇓ Outcome	Step 2. ⇓ Count	Step 4. ⇓ Relative Frequency	Step 5 ⇓ Degree
A	32	32/129 = 0.248	0.248 × 360° = 89.3°
B	51	51/129 = 0.395	0.395 × 360° = 142.3°
C	26	26/129 = 0.202	0.202 × 360° = 72.6°
D	11	11/129 = 0.085	0.085 × 360° = 30.7°
F	9	9/129 = 0.070	0.070 × 360° = 25.1°

Step 3. ⇒ $n = \sum f$
          $= 129$

The five steps outlined in the chart above are used to construct the following circle graph.

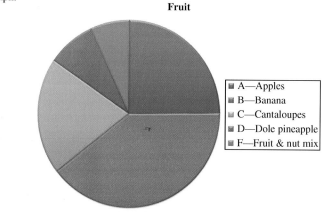

Fruit

- ■ A—Apples
- ■ B—Banana
- ■ C—Cantaloupes
- ■ D—Dole pineapple
- ■ F—Fruit & nut mix

Here we clearly see that bananas are the most favored of all fruits in the sample. Coming in second is apples with cantaloupes a close third; followed by Dole pineapple and lastly, fruit and nut mix.

EXERCISES

Critical Thinking

8.2.1. Distinguish between a bar chart and a Pareto chart.

8.2.2. Distinguish between a bar chart and a circle graph.

8.2.3. Distinguish between a frequency table and a contingency table.

8.2.4. Distinguish between a contingency table and a tree diagram.

Basic Problems

8.2.5. First construct a frequency table for the data and then construct a **bar chart** for the qualitative data. Construct a **Pareto chart**.

C	B	C	B	E	B	C	C	A	C
A	E	D	E	A	A	B	C	C	D
B	D	A	B	D	D	E	B	D	A
E	A	D	B	C	D	C	C	E	A
B	D	B	D	B	C	A	D	A	C
A	E	A	D	A	A	A	E	B	E

What useful information can be obtained by reading the **bar chart** and the **Pareto chart**?

8.2.6. First construct a frequency table for the given data obtained from tossing a die 25 times; then construct a bar chart for the discrete data that represents the number of dots uppermost on a die.

4	1	3	3	5
2	3	4	1	2
4	2	5	2	2
4	1	5	2	3
1	1	6	6	1
1	3	4	4	1

Does this die appear to be fair? What would be expected for a fair die?

8.2.7. Given the following bar chart for color of cars observed in a parking lot.

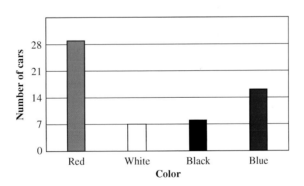

a. Which color has the least amount of cars in the parking lot?
b. How many more red cars are there than blue cars?
c. How many cars are there in the parking lot?
d. Construct the Pareto Chart.
e. What is the relative frequency of blue cars?

8.2.8. Given the following bar chart for points earned an a math test.

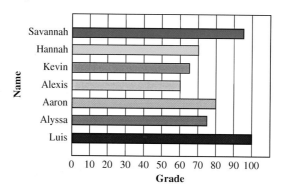

    a. Who had the highest grade?

    b. Which student(s) had a grade of a least 75 points?

8.2.9. Given the following bar chart for the number of minutes spent talking on the telephone by each child,

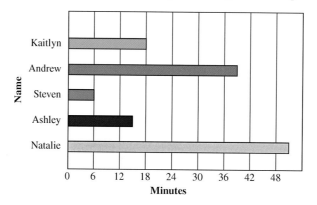

    a. How much longer did Andrew talk on the telephone than Kaitlin?

    b. Who was on the telephone for more than one-fourth of an hour?

8.2.10. Given the following pie chart for favorite season,

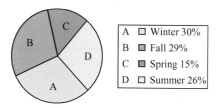

    a. If 300 individuals voted, how many people chose winter as their favorite season?

    b. What is the least popular season?

    c. Assuming 300 individuals voted, construct a bar chart using this information.

8.2.11. Given the following pie chart for how students get to school,

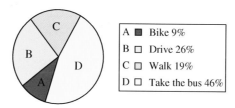

    a. If the school has 600 students, how many students walk to school?

    b. What fraction of the students drive to school?

8.2.12. Given the following pie chart for how many books each student read this year,

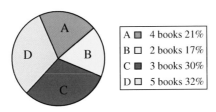

A ■	4 books 21%
B □	2 books 17%
C ■	3 books 30%
D □	5 books 32%

     a. If 379 students were surveyed, how many students read exactly 3 books?
     b. What fraction of the students read at least 3 books?

8.2.13. Given the following pie chart for what students drank for breakfast,

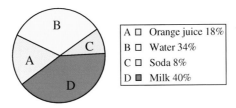

A □	Orange juice 18%
B □	Water 34%
C □	Soda 8%
D ■	Milk 40%

     a. What is the most popular drink?
     b. What fraction of the students drank milk for breakfast?

Contingency tables

8.2.14. Complete the following contingency table

	No effect	Positive effect	Negative effect	Total
Drug *A*	55		15	100
Drug *B*	40		31	100
Placebo		33		100
	160			300

8.2.15. Complete the following contingency table

	President	Vice president	Treasure	Total
Freshman		4		11
Junior	5		11	
Senior	13	7		24
		20	20	60

8.2.16. Construct a contingency table for the following survey data:

	Favorite colors				
Males	Red	Green	Blue	Blue	Green
	Purple	Orange	Blue	Green	
Females	Purple	Red	Yellow	Orange	Green
	Blue	Purple	Red		

Using the constructed contingency table, answer the following questions:
a) How many individuals where included in the survey?
b) How many males? How many females?
c) What is the most favored color?
d) What is the least favored color?
e) How many individuals favor red?

8.3   GRAPHICAL REPRESENTATION OF QUANTITATIVE INFORMATION

Visual representations information from a **frequency table** or **relative frequency table** for **quantitative data** include **histograms** and **frequency polygons** for **continuous** and **discretized data**.

We shall start discussing the concept of histograms, how we develop them and the visual interpretation it conveys to us about the phenomenon we are studying.

One of the first ways to graphically illustrate **quantitative** data is using various forms of a **histogram**. This can be done from **listed data** organized into a **frequency table** or with summaries information in the form of **grouped data**.

Definition 8.3.1   Histogram

A **histogram** is representation of a frequency distribution by way of rectangles whose width represents the class width (CW; class interval) and whose heights are proportional to relative frequencies.

Example 8.3.1   Histogram

Given the **listed data**: 1, 4, 3, 5, 2, 3, 4, 3, 4, 3 and the associated **frequency table**, illustrate the associated **histogram**:

Outcome	Frequency
2	1
3	5
4	4
5	2
6	1

Solution

As these are single digits, the digit is the **class mark** and the boundaries are half way between each set of digits as given in the table below.

Outcome	Frequency	Boundaries
2	1	(1.5, 2.5)
3	5	(2.5, 3.5)
4	4	(3.5, 4.5)
5	2	(4.5, 5.5)
6	1	(5.5, 6.5)
	13	

Hence, the **histogram** (similar to the **dot plot**) includes a real number line with a scale that includes the **minimum and maximum** data values, and then place a box above each observed class mark spanning the **CW** (the distance between **class boundaries**).

We proceed to put the information given in the above table into a sequence of rectangles as show on the next page. The rectangular base span one-half a unit to the left and one-half a unit to the right of the observed value, thus the base of each rectangle is one. The height is represented by the frequency of the observed outcome and hence, the area of the rectangle represents the proportion of times the observed outcome occurred. In terms of frequency, the height is number of times the class of outcomes occurs.

Solution—cont'd

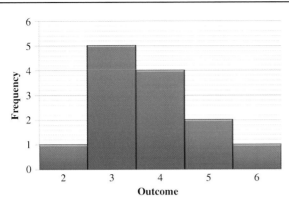

Here we see the tallest rectangle occurs for the data value 2; that is, the mode is 3. Moreover, we see that the distribution is not symmetric.

Recall: **Listed data** is a sequence of information presented as singletons; usually denoted: $x_1, x_2, \ldots, x_n$ where n is the sample size which is organized into a frequency table; however, when the sample size is large, we may not want to use each singleton, but rather use **grouped data**.

Grouped Data is data grouped into categories (**discretized**) by creating different classes based on the **range** of data values and a prior number of **classes**. These class intervals consist of a **lower class limit (LCL)** and an **upper class limit (UCL)**; that is, a class interval is an open interval on the real number line.

Definition 8.3.2 Class Limits

Given a **class interval** is an open interval on the real number line, each **class interval** consist of a lower value called the **LCL** and an upper value called the **UCL**.

By necessity, the class limits rounded to a given **tolerance**, that is, the decimal to which the data will be taken.

Definition 8.3.3 Class Width

CW is length of the **class interval** given by the **UCL** minus the **LCL**, plus one **tolerance**.

$$CW = (UCL - LCL) + T,$$

where T is the **tolerance**; that is, the unit to which data will be rounded before being placed in the various class intervals.

However, we are not given the **LCL** and **UCL**. Hence, to determine an approximate **CW**, use the **range** divided by a preset number of classes and round up. In general, given the number of classes, m, determine the CW; this CW must be such that the following rules are upheld.

Rule 8.3.1 Class Width

When determining the **CW**, follow these simple rules:
1. $m \times \text{class width} > \text{Range}$ where m is the number of intervals; that is, $CW \geq \dfrac{R}{m}$.
2. The first **LCL** $\leq$ **minimum**.
3. The last **UCL** $\geq$ **maximum**.
4. The **tolerance** is the difference between the **UCL** for one class and the **LCL** for the next class; the **tolerance** can be defined as the unit to which the **CW** is rounded.

Continued

Rule 8.3.1 Class Width—cont'd

---

5. There are no **gaps** between consecutive intervals; **gaps** may only exist if there is an interval with **frequency** zero.
6. The **CWs** are consistent in length.

Class	Lower class limit	Upper class limit
1	$LCL_1 \leq$ Minimum	$UCL_1 = LCL_1 + CW - T$
2	$LCL_1 = LCL_1 + CW$	$UCL_2 = LCL_2 + CW - T$
3	$LCL_3 = LCL_2 + CW$	$UCL_3 = LCL_3 + CW - T$
-	-	-
-	-	-
-	-	-
m	$LCL_m = LCL_{(m-1)} + CW$	$UCL_m = LCL_{(m-1)} + CW - \underline{T} >$ Maximum

---

Definition 8.3.4 Class Boundaries

Class boundaries are the end points of an open interval which contains the **class interval** such that the **lower class boundary (LCB)** is the **LCL** minus one-half the **tolerance** and the **upper class boundary (UCB)** is the **UCL** plus one-half the **tolerance**. Hence, from boundary to boundary is the full **CW**

$$LCB = LCL - 0.5T$$

and

$$UCB = UCL + 0.5T.$$

Class	Lower class boundary	Upper class boundary
1	$LCB_1 = LCL_1 - 0.5T$	$UCB_1 = UCL_1 + 0.5T$
2	$LCB_2 = LCL_2 - 0.5T$	$UCB_2 = UCL_2 + 0.5T$
3	$LCB_3 = LCL_3 - 0.5T$	$UCB_3 = UCL_3 + 0.5T$
-	-	-
-	-	-
-	-	-
m	$LCB_m = LCL_{(m-1)} - 0.5T$	$UCB_m = UCL_{(m-1)} + 0.5T$

Once these boundaries are established, we can determine the **frequency table** and then for graphical purposes, determine where to mark the horizontal axis of the histogram. These values will be either the **class boundaries** or the **class marks**, which are the midpoints of the **class intervals**.

Definition 8.3.5 Class Mark

Given a **class interval**, (**LCL, UCL**), the **class mark** is the value in the middle, in the center below the rectangle which represents the frequency of the various **class intervals**.

Class	Class mark
1	$CM_1 = (UCL_1 + LCL_1)/2$
2	$CM_2 = (UCL_2 + LCL_2)/2$
3	$CM_3 = (UCL_3 + LCL_3)/2$
-	-
-	-
-	-
m	$CM_m = (UCL_m + LCL_m)/2$

Now we have all the necessary information to graph the histogram.

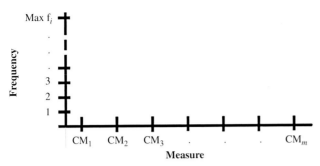

Step-by-Step Histogram

Step 1. Determine the **minimum (Min)** and **maximum (Max)** value of the given data and compute the **range**:

$$\text{Range} = \text{Max} - \text{Min}$$

Step 2. When constructing a histogram by hand, the easiest way to determine CW is:
Class width (CW) $=$ **Round(Range/m)** up to tolerance, where m is the desired number of classes. The degree to which you round is called the **tolerance**; hence, if you round to the integer then the **tolerance** is one whereas if you round to the first decimal place, the **tolerance** is 0.1.

Step 3. For convenience sake, let the first **LCL** be the minimum, the second **LCL** be the first **LCL** plus the **CW**, etc. The **UCLs** are the **LCLs** plus the **CW** minus the **tolerance**.

Step 4. Once the class intervals are defined, determine the midpoints of the intervals, these are the **class marks**. This information is for graphing the histogram and may be done after step 6.

Step 5. Next, the **class boundaries** are the **LCLs** minus half the **tolerance** and the **UCL** plus half the **tolerance**.

Step 6. Organize the data values into a **frequency table** based out the outlined **class intervals**. Determine **relative frequency**, **cumulative frequency**, and **cumulative relative frequency** as needed.

Step 7. Place the **class marks** on a real number line; to the left of all the **class marks**, perpendicular to the real line (horizontal axis), draw a vertical axis and taking into consideration the maximum **frequency** or **relative frequency** and using a consistent scale mark out the **frequencies**.

Step 8. Finally, using the **class boundaries** as the base and the **frequency** or **relative frequency** as the height, construct a bar (rectangular box) for each class.

Example 8.3.2 Histogram

A random sample of 50 pieces of 6 mm nylon rope are being used to measure breaking strength (**kN**). The resulting measurements are given in the table below. Develop a representative histogram with five classes and interpret it in context.

Data set

11.3	11	8.8	11.1	10.4
8.5	11.5	10.9	9.4	12.1
8.5	7.7	9.5	9.3	8.7
10	10.2	13.7	9.6	6.8
12.9	12.6	10.3	10.9	10.6
7.8	9.2	9.7	8.5	13
10.1	7.9	9.8	11.7	11.3
9.6	9.7	8.8	10.1	7.1
11.5	8.8	7.9	8.3	8.7
9	9.4	10.7	9.6	12

Solution

Step 1. With a **minimum** of 6.8 and a **maximum** of 13.7, the **range** is $(13.7 - 6.8) = 6.9$.

Step 2. With five classes, the **CW** must be greater than $(6.9/5) = 1.38$. Hence, with a **tolerance** of 0.1, let the **CW** be 1.4. Note: $5 \times 1.4 = 7 > 6.9$, the histogram will more than cover the range of information.

Lower limit	Upper limit
6.8	

Step 3. For convenience sake, let the first **LCL** be the minimum, 6.8.

Lower limit	Upper limit
6.8	
8.2	
9.6	
11	
$=11+1.4$	

Then using the **CW**, we can create the remaining lower limits: $6.8 + 1.4 = 8.2$, $8.2 + 1.4 = 9.6$, $9.6 + 1.4 = 11$ and $11 + 1.4 = 12.4$.

Lower limit	Upper limit
6.8	$=8.2-0.1$
8.2	
9.6	
11	
12.4	

The upper limits are off by 0.1, the tolerance and hence, we have the first upper limit as 8.1, which is the lower limit plus the **CW** minus the tolerance.

Lower limit	Upper limit
6.8	8.1
8.2	9.5
9.6	10.9
11	12.3
12.4	$=12.3+1.4$

From this point on, we can use the **CW**. Thus, the last entry for upper limits is 13.7. Note: the minimum is included in the first interval and the maximum of 13.7 is included in the last interval.

Step 4. The midpoints of these intervals is simply the average of each lower and upper limit, for example,

$$(12.4 + 13.7) = 13.05.$$

Lower limit	Upper limit	Class mark
6.8	8.1	7.45
8.2	9.5	8.85
9.6	10.9	10.25
11	12.3	11.65
12.4	13.7	$=(12.4+13.7)/2$

Solution—cont'd

Step 5. Next, the class boundaries are the limits plus or minus half the tolerance. For example, for the last interval, the **LCB** is $12.4 - 0.1/2 = 12.35$, etc.

Lower limit	Upper limit	Class mark	Lower boundary	Upper boundary
6.8	8.1	7.45	6.75	8.15
8.2	9.5	8.85	8.15	9.55
9.6	10.9	10.25	9.55	10.95
11	12.3	11.65	10.95	12.35
12.4	13.7	13.05	$=12.4-0.1/2$	13.75

Step 6. Organize the data values into a **frequency table** based on the outlined **class intervals**. This is a matter of simply counting the number of data points that fall within each interval.

Lower boundary	Upper boundary	Frequency
6.75	8.15	6
8.15	9.55	15
9.55	10.95	16
10.95	12.35	9
12.35	13.75	4

Determine **relative frequency**, **cumulative frequency**, and **cumulative relative frequency** as needed.

Step 7. Place the **class marks** on a real number line; and taking into consideration the maximum **frequency** or **relative frequency** and using a consistent scale mark out the **frequencies**.

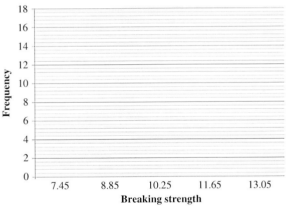

Step 8. Finally, using the **class boundaries** as the base and the **frequency** or **relative frequency** as the height, construct a bar (rectangular boxes) for each class.

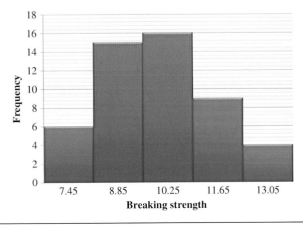

Continued

Solution—cont'd

A **CW** of 2 might also have sufficed—however, if you round too much, then there may be fewer classes than initially intended as well as the intervals should be centered over the full **range** of data values; therefore, the first **LCL** would not be the **minimum** observed data value.

Interpreting this graph we see the most frequently occurring interval is $(\mathbf{9.6}, \mathbf{10.9})$ with 16 data values falling within this class interval. However, with a count of 15, the interval $(\mathbf{8.2}, \mathbf{9.5})$ is also quite frequent. The least frequently occurring interval is $(\mathbf{12.4}, \mathbf{13.7})$, that is, a breaking strength greater than 12.4 is not as likely to occur as breaking strengths between 8.2 and 10.9. This data is somewhat symmetric, however, with only five classes, there does appear to be a skew to the right. The larger less likely values are pulling the mean greater than the median. Note: for this data set, the mean is 9.93 and the median is 9.7, which is consistent with what the data's graphical distribution implies.

Often, instead of constructing a histogram by hand calculations, modern day technology can be used for this purpose. For example, using the same data set illustrated in Example 2.10 with 8 classes; here the **CW** is taken to be approximately 1 which is greater than simply taking the range divided by the number of classes, rounded up; that is, $(6.9/8) = 0.8625$ which is approximately 0.9. Moreover, this computer-generated histogram did not clearly label the vertical axis as frequency; this will need to be edited. However, this is still a valid histogram.

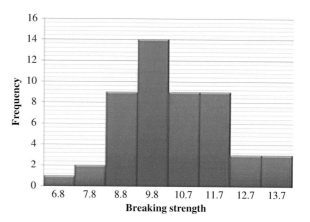

Finally, what is a good looking histogram—bell shaped, skewed to the left, skewed to the right, no gaps! These graphics where created in Excel, for more detail about using Excel to create graphics, see **Using Technology in Graphical Representations** at the end of the chapter.

EXERCISES

8.3.1. Given the minimum value observed is 12.7, the maximum value observed is 72.4, with a tolerance of 1 and five classes, what is the minimal **CW** required to construct a histogram. What are the class limits? What are the class boundaries?

8.3.2. Given the minimum value observed is 12.7, the maximum value observed is 72.4, with a tolerance of 0.1 and five classes, what is the minimal **CW** required to construct a histogram. What are the class limits? What are the class boundaries?

8.3.3. Given the minimum value observed is 12.7, the maximum value observed is 72.4, with a tolerance of 1 and six classes, what is the minimal **CW** required to construct a histogram. What are the class limits? What are the class boundaries?

8.3.4. First construct a frequency table for the data given below with five class intervals, include the class limits, the class boundaries, the class midpoints (marks) and then construct a histogram for the grouped data.

102.9	91.1	84.8	82.7	87.5
115.6	126.1	92.8	78.1	91.6
109.2	93.7	105.1	120.3	83.8
91.4	85.9	97.3	103.9	98.7
81.6	95.1	95.7	107.3	118
97.6	82.5	89.9	86.1	93.2
99.4	113.5	102.4	75.2	77.4
103.8	111.3	102.7	92	105
96.1	99.1	113.2	115	83.5
86.4	90.7	98.9	93.8	111
93.5	105.2	69.3	113.1	103.8
108.9	88.4	84.5	80.3	105
82	98.4	91	113	80.2
107.9	78.7	113	113.1	128.1
93	113	103	76.5	77.6
111	97.5	97.8	83.8	108
111.8	104.1	112.2	112.2	99.7
115	109.9	81.7	92	97.8

a) What useful information we can read from the histogram?

b) Can you recover the original data from the histogram?

8.3.5. First construct a frequency table for the given data using the singletons as the class marks, including the class boundaries and then construct a histogram for the listed data.

1	1	6	0	6	1	9	4	3	9
7	0	8	5	2	3	6	0	3	4
9	1	9	0	1	1	8	5	5	3
6	2	2	1	6	1	6	3	4	9
3	7	3	9	6	5	0	7	1	7
5	3	4	0	6	2	7	2	7	8

a) Discuss the useful information we can obtain from the histogram of the given data.

b) Can you recover the original data from the histogram?

c) What is the most frequent data observed?

8.3.6. First construct a frequency table for the given data using the singletons as the class marks, including the class boundaries and then construct a frequency polygon for the listed data.

6	9	5	3	0	2	0	6	7	5
8	2	8	2	5	0	1	9	7	5
2	8	9	2	4	6	9	3	1	8
9	1	9	9	2	8	9	9	0	0
0	1	2	3	7	9	4	7	9	6

Discuss the useful information we can obtain from the histogram of the given data.

8.3.7. First construct a frequency table for given data with five class intervals, include the class limits, the class boundaries, the class midpoints (marks) and then construct an ogive graph for the grouped data.

102.9	91.1	84.8	82.7	87.5
109.2	93.7	105.1	120.3	83.8
91.4	85.9	97.3	103.9	98.7
81.6	95.1	95.7	107.3	118
97.6	82.5	89.9	86.1	93.2
103.8	111.3	102.7	92	105
96.1	99.1	113.2	115	83.5
86.4	90.7	98.9	93.8	111
93.5	105.2	69.3	113.1	103.8
111	97.5	97.8	83.8	108
111.8	104.1	96.6	112.2	99.7

What useful information can we obtain from the ogive graph? Do you feel that five class intervals is the best number of intervals? If not, would you recommend more or less?

8.4 MEASURING CENTRAL TENDENCIES

Common central measures generically referred to as **central tendencies** or **averages** for **continuous measures** included the **mode**, the **median**, and **mean**. The **mode** of a measured **variable** in a **sample** is the value or **outcome** with the highest **frequency**.

Common Central Tendencies

Mode—the most frequent value.
Median—the middle of the ordered values.
Mean (Standard Arithmetic Mean)—the sum of the values divided by the number of values.
Expected Value—the mean evaluated using probabilities.
Trimmed Mean—the mean evaluated trimming off the extremes.
Weighted Mean—the mean evaluated using weights.

Definition 8.4.1 Mode (M)

The **mode** of data is the data value that occurs most frequently. We shall denote the mode by lowercase English letter m. As a measure of tendency in qualitative data, the **mode** is the most frequently occurring category.

Example 8.4.1 Mode

Given the **listed data**: 1, 4, 3, 5, 2, 3, 4, 3, 4, 3 and the associated **Frequency Table**, the **mode** of the **data** set is 3 with the highest **frequency** of occurrence, 4.

Outcome	Frequency	
1	1	
2	1	
3	4	← *mode*
4	3	
5	1	

The **mode** is the one **central tendency** that may fail to exist; however, it is the only measure of **central tendencies** that can be extended to **qualitative data**. To determine the **mode** of a **sample** data set, the data must first be sorted and associated **frequencies** must be considered. Note: the mode may fail to exist, such as with uniform data or with data where more than one data value occurs with maximum frequencies.

Step-by-Step Mode (m)

Step 1. Determine the associated frequencies of the observed outcomes.

Step 2. The mode is the most frequent data to model the center; that is, the most frequent data is the mode (if such values exist.)

Example 8.4.2 Tumor Size

In the **Breast Cancer** data set: 0.5, 0.5, 0.5, 0.7, 0.8, 0.8, 0.8, 0.9, 1, 1, 1, 1, 1, 1, 1.1, 1.1, 1.1, 1.2, 1.2, 1.2, 1.3, 1.3, 1.5, 1.5, 1.5, 1.5, 1.5, 1.5, 1.5, 1.5, 1.5, 1.5, 1.7, 2, 2, 2, 2, 2, 2, 2.2, 2.4, 2.5, 2.5, 2.5, 3, 3, 3, 3.5, 3.8, 4; what is the mode of the **tumor size**?

Solution

First, there are eighteen distinct observations ranging from 0.5 to 4 cm.

Step 1. Next, we construct the frequency table for these tumor sizes:

Tumor size	Frequency	
0.5	3	
0.7	1	
0.8	3	
0.9	1	
1	6	
1.1	3	
1.2	3	
1.3	2	
1.5	10	← *mode*
1.7	1	
2	6	
2.2	1	
2.4	1	
2.5	3	
3	3	
3.5	1	
3.8	1	
4	1	

Step 2. Reading the frequency table to the left, it is clear that 1.5 cm is the most frequently observed tumor size.

In **exploratory data analysis**, multiple **modes** can indicate **lurking variables** and **mixed probability distributions**; that is, the quantities may depend on a variable that has not been taken into account when the data was observed.

Continued

Solution—cont'd

As the mode may fail to exist, a second measure of central tendency is the **median**. The median defined to be a value such that 50% of the data is to the left of this value (that is, less than or equal to) and 50% of the data is to the right of this value (that is, greater than or equal to.)

Definition 8.4.2 Median (M)

The **median** of data set is a data value such that 50% of the information lies to the left and 50% of the information lies to the right. We shall denote the median by capital English letter M.

Note: This measure always exists and is not affected by outliers; that is, this measure is not sensitive to changes in the data or extremes.

In data sets, this number exists and is unique, but in the sample is only estimated and there this number need not be unique.

Example 8.4.3 Estimate of Median not Unique

Given a simple data set: 1, 2, 3, 4. Here, we would estimate the median to be 2.5; however, 50% of the data lies to the left and right of 2.4 as well. In fact, we can estimate the median to be any value between 2 and 3. Moreover, how accurate are the measures; if the data is more accurately, 1.2, 2.4, 2.8, and 3.9, then we would estimate the mean as 2.6. If the data is even more accurately, 1.24, 2.39, 2.84, and 3.87, then we would estimate the mean as 2.615. Hence, the more information that is given, the easier it is to determine the median.

Rule 8.4.1 Median

In general, to determine an estimate of the true **median**:
1. If there is an odd number of data values then take the middle of the order data.
2. If there is an even number of data values then trim all but the middle two values and take the midpoint, that is, the sum divided by two.

Given below is a two-step procedure for calculating the median.

Step-by-Step Median (M)

Step 1. Order the observed data in ascending order. Symbolically we shall distinguish the raw (original) data, x_i, from the ordered data, $x_{(i)}$ by using the grouping symbols (parenthesis) to indicate that this is position in the ordered data set as opposed to the observation order. Hence,

$$\mathbf{min} = x_{(1)} \leq x_{(2)} \leq \cdots \leq x_{(n)} = \mathbf{max}.$$

Step 2. Take the middle value or the sum of the two central values divided by two. That is, if there are an odd number of data, then a single data point will exist in the ordered data set; namely the data value in the $\left(\dfrac{n+1}{2}\right)$th position.

$$x_{(1)} \leq \cdots \leq x_{\left(\frac{n+1}{2}\right)} \leq \cdots \leq x_{(n)}$$
$$\uparrow$$
$$M$$

Step-by-Step Median (M)—cont'd

$$M = x_{\left(\frac{n+1}{2}\right)}.$$

For data sets with even sample size, this is the sum of the data in the $\left(\frac{n}{2}\right)$th position and the data in the $\left(\frac{n}{2}+1\right)$th position divided by two:

$$x_{(1)} \leq \cdots \leq x_{\left(\frac{n}{2}\right)} \leq x_{\left(\frac{n}{2}+1\right)} \leq \cdots \leq x_{(n)}$$

$$\uparrow$$
$$M$$

$$M = \frac{x_{\left(\frac{n}{2}\right)} + x_{\left(\frac{n}{2}+1\right)}}{2}$$

Step 3. Determine the associated frequencies of the observed outcomes

Step 4. The mode is the most frequent data to model the center; that is, the most frequent data is the mode (if such values exist.)

If fact, this value is such that at least 50% is greater than or equal to this value and at least 50% of the data is less than or equal to this value.

Example 8.4.4 Median with Odd Sample Size

In a given experiment we have observed the following data: 9, 8, 11, 7, 10, 9, 9, 7, and 19. Determine the median of the observed data set.

Solution

Step 1. First, this data must be put into ascending (descending) order:

$$7, 7, 8, 9, 9, 9, 10, 11, 19$$
$$\uparrow$$
$$M$$

Step 2. Since there are nine data points (odd sample size), the middle value is the fifth position which is nine (9). Thus, the median is the value 9. Moreover, nine (9) is also the mode and hence 2/3rds of the data are greater than or equal to this value as well as 2/3rds of the data are less than or equal to this value.

Example 8.4.5 Median with Even Sample Size

The results of a given experiment are:

$$12, 12, 5, 13, 11, 4, 10, \text{ and } 9.$$

Calculate the **median**.

Solution

Step 1. First this data must be put into ascending (descending) order:

$$4, 5, 9, 10, 11, 12, 12, 13$$
$$\uparrow$$
$$M = 10.5$$

Step 2. Since there are eight data points (even sample size), the middle value is between the fourth and fifth position which is between 10 and 11; while any value between 10 and 11 would suffice, standard practice is to take the midpoint 10.5 (that is, the sum of 10 and 11 divided by 2). Here, the median 10.5 is a value such that exactly 50% of the observed data is greater than this value and exactly 50% of the observed data is less than this value.

Example 8.4.6 Median Tumor Size

In the Breast Cancer data set provided below, what is the median of the **tumor size**?

Tumor size	Frequency
0.5	3
0.7	1
0.8	3
0.9	1
1	6
1.1	3
1.2	3
1.3	2
1.5	10
1.7	1
2	6
2.2	1
2.4	1
2.5	3
3	3
3.5	1
3.8	1
4	1

Solution

First, using the frequency table constructed, we can locate the middle data values by viewing the cumulative frequency.

Step 1.   In constructing the frequency table for these tumor sizes, we have the data in ascending order.

Step 2.

Tumor size	Frequency	Cumulative frequency
0.5	3	3
0.7	1	4
0.8	3	7
0.9	1	8
1	6	14
1.1	3	17
1.2	3	20
1.3	2	22
1.5	10	32
1.7	1	33
2	6	39
2.2	1	40
2.4	1	41
2.5	3	44
3	3	47
3.5	1	48
3.8	1	49
4	1	50

Solution—cont'd

Step 3. Now, using cumulative frequencies, we can locate the 25th and 26th data value, as there is an even number of observations. Here, the 22nd thru the 31st data value are 1.5; hence, the median is

$$M = \frac{1.5 + 1.5}{2} = 1.5.$$

Returning to averages in general, the **standard mean** gives each data value equal **weight**.

Definition 8.4.3 Standard Mean

The **standard mean** or **sample mean** of data is the sum of all data values divided by the total number of data values, denoted $\bar{x}$.

Using the capital Greek letter Sigma ($\sum$) to mean "the sum," we have

$$\bar{x} = \frac{1}{n} \sum_{i=1}^{n} x_i$$

where n is the sample size.

Note: This measure is the most sensitive.

While this measure is the most subject to change as it takes into account all the available information, for this reason, it is also the most important. Whereas the **mode** only considers the most frequent data value and it may fail to exist, the **median** uses one or two central data values; it is the (**sample**) **mean** that gives equal weight to each data value.

Given below is a two step procedure in calculating the **sample mean**, $\bar{x}$.

Step-by-Step Standard Mean

Step 1. Sum the values observed in a sample of size n.
Step 2. The mean is the sum divided by the total number of data points, n.

Using the capital Greek letter **Sigma** ($\sum$) to mean "the sum," these steps are commonly abbreviated by

$$\bar{x} = \frac{\sum x}{n}$$

which reads "the **sample mean** ($\bar{x}$) is the **sum** of the listed **data** ($\sum x$) **divided** by ($\div$) the **sample size** (n)"; hence, part of statistics is learning to understand the underlying language of **Math** upon which **Statistics** is built. This notation can be further elaborated on by including the counter; that is, the original data set is usually denoted by x_i for $i = 1, 2, 3, \ldots, n$ which extends the formulation to

$$\bar{x} = \frac{1}{n} \sum_{i=1}^{n} x_i$$

where the under-script is the starting point for the counter i and the upper-script is the ending point, namely the **sample size** n.

Notation

This formulation for the **arithmetic mean** can be manipulated and re-written

$$\bar{x} = \frac{\sum\limits_{i=1}^{n} x_i}{n} = \frac{x_1 + x_2 + \cdots + x_n}{n},$$

for those who do not like summation notation or alternatively,

$$\bar{x} = \frac{\sum\limits_{i=1}^{n} x_i}{n} = \frac{x_1 + x_2 + \cdots + x_n}{n} = \frac{1}{n} x_1 + \frac{1}{n} x_2 + \cdots + \frac{1}{n} x_n,$$

which illustrates that in the standard mean, each data points carries equal weight. This idea will be further discussed in weighted means.

Example 8.4.7 Standard Mean

Given four exams scores (listed): 98, 77, 72, and 94. Determine the sample mean of the given data.

Solution

Step 1. $\sum x = 98 + 77 + 72 + 94 = 341$

Step 2. $\bar{x} = \dfrac{341}{4} = 82.25$.

Hence, the samples mean (average test score) is 82.25.

Note: Throughout our study of statistics we shall use the sample mean, $\bar{x}$, to estimate the population mean. However, this is just one estimate of the population mean. In general, a statistical estimate of the population mean μ will be denoted $\hat{\mu}$ The use of the specific notation of the variable x coupled with the upper-script, namely the bar, is indicative not only that this is a statistical estimate (as it is English and not Greek) but also how the estimate was computed—that is, the sum of the data divided by the number of data.

We will not extend this notation to that of grouped data. In certain problems, we are given the data in the form of a frequency table. The sample mean is computed in the same way, however, as the organization of the information is no longer listed, but rather grouped, the notation changes.

Outcome x_i	Frequency f_i
x_1	f_1
x_2	f_2
$\vdots$	$\vdots$
x_k	f_k

Definition 8.4.4 Standard Mean (Frequency Data)

For data that has been grouped in terms of frequencies, the sample mean appears weighted as by the associated frequencies. Given the frequency table of k distinct outcomes: $x_1, x_2, \ldots, x_k$ with associated observed frequencies: $f_1, f_2, \ldots, f_k$ in a sample of size $n = \sum_{i=1}^{k} f_i$ is given in the frequency table above. Thus the formulation for the **sample mean** for **grouped data** becomes

$$\bar{x} = \frac{1}{n} \sum_{i=1}^{k} x_i f_i = \frac{x_1 f_1 + x_2 f_2 + \cdots + x_k f_k}{n},$$

where

$$n = \sum_{i=1}^{k} f_i.$$

Step-by-Step Standard Mean (Frequency Data)

Step 1. Organize the listed data into a frequency table.

Step 2. Compute the sum of the frequencies; this should sum to the total number of data values.

Step 3. Compute the products of the data value times the associated frequency. It might be helpful to create a new column in which to place this information.

Step 4. Sum the products.

Step 5. The mean is the sum of the products divided by the total number of data values.

Example 8.4.8 Mean using Frequency Table

In a rating poll, on a scale from 1 to 4, the question "How would you rate your senator?" was asked to a sample of 18 voters and the following information is obtained: 1, 2, 3, 1, 2, 4, 2, 4, 1, 1, 3, and 2. First place the information into a frequency table and then compute the sample mean.

Solution

Step 1. The frequency table is

x_i	1	2	3	4
f_i	4	3	3	2

Step 2. The total number of data points is given to be

$$n = \sum_{i=1}^{4} f_i = 4 + 3 + 3 + 2 = 12$$

Step 3. Compute the products:

x_i	1	2	3	4
f_i	4	3	3	2
$x_i f_i$	4	6	9	8

Step 4. Sum the products:

$$\sum_{i=1}^{4} x_i f_i = 4 + 6 + 9 + 8 = 27$$

Step 5. Calculate the mean:

$$\bar{x} = \frac{27}{18} = 2.25$$

That is, the sample mean of the given data is $\bar{x} = 2.25$. On a scale from 1 to 4, the rating of this senator is 2.25; if 2 is average, this senator rating is slightly above average.

Example 8.4.9 Mean Tumor Size in Breast Cancer

In the Breast Cancer data set (see Example 8.4.6), what is the sample mean of tumor size using frequented data?

Solution

Step 1. The first step, the frequency table is given in illustrated in the table, Step 3.

Step 2. The total number of data points is given as

$$n=\sum_{i=1}^{18} f_i = 50.$$

Step 3 ↘
Compute the products:

Tumor size	Frequency	$x_i f_i$
0.5	3	1.5
0.7	1	0.7
0.8	3	2.4
0.9	1	0.9
1	6	6
1.1	3	3.3
1.2	3	3.6
1.3	2	2.6
1.5	10	15
1.7	1	1.7
2	6	12
2.2	1	2.2
2.4	1	2.4
2.5	3	7.5
3	3	9
3.5	1	3.5
3.8	1	3.8
4	1	4
Sum	50	

Step 4 ↘
Compute the sum:

$x_i f_i$
1.5
0.7
2.4
0.9
6
3.3
3.6
2.6
15
1.7
12
2.2
2.4
7.5
9
3.5
3.8
4
82.1

Step 5 ↘
Sum divided by n.

Tumor size	Frequency	$x_i f_i$
0.5	3	1.5
0.7	1	0.7
0.8	3	2.4
0.9	1	0.9
1	6	6
1.1	3	3.3
1.2	3	3.6
1.3	2	2.6
1.5	10	15
1.7	1	1.7
2	6	12
2.2	1	2.2
2.4	1	2.4
2.5	3	7.5
3	3	9
3.5	1	3.5
3.8	1	3.8
4	1	4
Sum	50	82.1
Mean		1.642

Thus, the average (standard mean) is approximately 1.642 cm. This is skewed right by the value 4; that is, recall the median is 1.5 cm (as was the mode), however, the mean is greater than the median (to the right of) and hence, the distribution of tumor size in breast cancer is skewed.

Definition 8.4.5 Expected Value

For data that has been grouped in terms of proportions, the sample mean appears weighted as by the associated proportions.

Given the probability distribution of the k distinct outcomes: $x_1, x_2, \ldots, x_k$ namely: $p_1, p_2, \ldots, p_k$ in a sample of size n, where $\sum_{i=1}^{k} p_i = 1$. Thus the formulation for the **expected value** becomes

$$\bar{x}=\sum_{i=1}^{k} x_i p_i = x_1 p_1 + x_2 p_2 + \cdots + x_k p_k,$$

where $\sum_{i=1}^{k} p_i = 1$.

Notation

The formulation for grouped data can be further manipulated in terms of proportions; recall, $p_i = \dfrac{f_i}{n}$ hence

Notation—cont'd

$$\bar{x} = \frac{\sum\limits_{i=1}^{k}(x_i \cdot f_i)}{n}$$

$$= \frac{x_1 \cdot f_1 + x_2 \cdot f_2 + \cdots + x_k \cdot f_k}{n}.$$

$$= \frac{f_1}{n} \cdot x_1 + \frac{f_2}{n} \cdot x_2 + \cdots + \frac{f_k}{n} x_k$$

$$= p_1 \cdot x_1 + p_2 \cdot x_2 + \cdots + p_k \cdot x_k$$

However, in this form the mean is referred to as the **expected value** (E) of the variable (x); writing symbolically, this is

$$E(x) = \sum_{i=1}^{k} x_i \cdot p_i$$

or more simplistically,

$$E(x) = \sum xp,$$

where it is understood that this summation is over all observed values of the variable x.

Step-by-Step Expected Value

Step 1. Treating the relative frequencies as weights, organize the listed data into a "frequency" table as when computing a weighted mean.

Step 2. Compute the products of the data value times the associated relative frequency. It might be helpful to create a new column in which to place this information.

Step 3. Sum the products. The expected value is the sum of the products.

Example 8.4.10 Expected Age of Binge Drinkers

In the data set: Binge Drinking, what is the expected drinking age in the United States?

Solution

Step 1. The data is given in such a way that we have grouped data; that is, age is given as 12 or younger, 13 to 17, 18 to 25, and 26 or older. This is summary information, however, treating the relative frequencies as the weights, we can estimate the ages using the class marks, assuming the youngest possible age is 0 year (new born) and max 80 years old.

	12 or younger	13 to 17	18 to 25	26 or older
Total U.S.	22.82	10.1	42.02	21.2

Step 2. Therefore, we will use the ages: 6, 15, 21.5, and 53 to represent each group.

x_i	6	15	21.5	53
p_i	0.2282	0.1010	0.4202	0.2120
$x_i p_i$	1.3692	1.515	9.0343	11.236

Step 3. The sum is therefore 23.1514 years of age.

That is, under the assumption that the minimum age is 0 years old and the maximum is 80 years old, we can estimate the expected age of a binge drinker is about 23 years old. This seems to reason as 42.02% of all binge drinkers in the United States are between 18 and 25 years of age.

In the standard mean, all the information is included. Other measure of central tendency that does not use all the information is a **trimmed mean**. This measure is an extension of the **median** where a given percent of the central data is retained and the remaining outlining data values are trimmed.

Definition 8.4.6 $p\%$ Trimmed Mean

The $p\%$ **trimmed mean** is the sum of the remaining data values once $p\%$ of the information from the upper data values and the lower data values, divided by the remaining number of data values.

Let $x_{(i)}$ represent the ordered data, then

$$\underbrace{x_{(1)},\ldots,x_{(j)}}_{p\%},\underbrace{x_{(j+1)},\ldots,x_{(n-j)}}_{\Downarrow},\underbrace{x_{(n-j+1)},\ldots,x_{(n)}}_{p\%}$$

and

$$\hat{\mu}=\frac{x_{(j+1)}+x_{(j+2)}+\cdots+x_{(n-j)}}{n-2j},$$

where $j=p\times n$.

Here, the notation, $\hat{\mu}$ is an estimate of the population mean.

For example, a 10% **trimmed mean** is an average where within the ordered data, the center 80% of the data carry equal **weights** and the outer 10%'s (the lower 10% of the data values and the upper 10% of the data values) are trimmed, that is assigned zero weight. Note: only when j is an integer is this estimate easy to calculate; most software packages will compute this measure when j is not an integer apportion the information; however, this concept is higher than the material covered in this text.

Step-by-Step Trimmed Mean

Step 1. Order the observed data in ascending order as with the median.

Step 2. Trim the given percent of the information from each end. If this percent of the total is not exact, apportioning methods can be used or duplicate information.

Step 3. The trimmed mean is the sum of the remaining data divided by the number of remaining data values.

Example 8.4.11 25% Trimmed Mean

In a Statistics class, a quiz was given out of 15 possible points, a sample of 8 students is taken, and the following information is gathered: 12, 12, 5, 13, 11, 4, 10, and 9. Determine the 25% **trimmed mean**.

Solution

Step 1. As with the median, the data must be ordered:

$$4,5,9,10,11,12,12,13.$$

Step 2. As there are eight data values, 25% of the data is two and hence the lower two data values will be removed as well as the upper two data values:

$$\cancel{4,5},9,10,11,12,\cancel{12,13}.$$

Solution—cont'd

Step 3. This leaves four values which will be assigned equal weight; this is the same as summing the values and dividing by 4. Therefore we have,

$$\hat{\mu} = \frac{9 + 10 + 11 + 12}{4} = 10.5,$$

where $\hat{\mu}$ is a statistical estimate of the true population mean μ.

One estimate of the central tendency is 10.5. Note: this is also the value of the median; however, it is not the same as the standard arithmetic mean and the mode does not exist.

Example 8.4.12 25% Trimmed Mean

In a large class, a sample of four exams scores: 98, 77, 72, and 94 are taken; determine the 25% **trimmed mean**.

Note: since there are only four data values, the 25% trimmed mean is also the median.

Solution

Step 1. As with the median, the data must be ordered:

$$72, 77, 94, 98.$$

Step 2. As there are four values, 25% of the data is one and hence the minimum data value will be removed as well as the maximum data value:

$$\cancel{72}, 77, 94, \cancel{98}.$$

Step 3. This leaves two data values which will be assigned equal weight; this is the same as the median,

$$\hat{\mu} = \frac{77 + 94}{2} = 85.5,$$

where $\hat{\mu}$ is a statistical estimate of the true population mean μ. Moreover, as there where only four data values to start, the 25% trimmed mean is the median.

This idea of summing the middle values and taking an **average** can be extended to general means using various **weights**. **Weights** are numerical factors assigned to each observed value to make that outcomes effect on the **measure** representative of its importance or relevance of that observation as an estimate of the **population mean, μ**. The **median** is a **weighted mean** in that the **median** gives all the **weight** to a single value (when there is an odd number of data) or equally divides the **weights** fifty-fifty (when there is an even number of data); that is, the **median** is given by

$$M = x_{\left(\frac{n+1}{2}\right)} = 1.0 x_{\left(\frac{n+1}{2}\right)}$$

or

$$M = \frac{x_{\left(\frac{n}{2}\right)} + x_{\left(\frac{n}{2}+1\right)}}{2} = 0.5 x_{\left(\frac{n}{2}\right)} + 0.5 x_{\left(\frac{n}{2}+1\right)},$$

respectively.

Recall, bias is a consistent deviation of the statistics to one side of the parameter; hence, an **unbiased estimator** is a statistic that does not deviate from the parameter. That is, $E\left(\hat{\theta}\right)=\theta$. We shall now prove that weighted means are unbiased estimators of the population mean.

That is, given that the expected value of the data is the population mean, $E(x)=\mu$, prove that the statistic $\hat{\mu}=\dfrac{\sum x \cdot w}{\sum w}$ is an unbiased estimator of μ, where x is the variable of interest, the weights, w are fixed constants and hence $\sum w$ is also a constant.

Mathematical Reasoning

Let $W=\sum w$, then we have:

$$E(\hat{\mu})=E\left(\frac{\sum_{i=1}^{k}x_i \cdot w_i}{W}\right)=\frac{E\left(\sum_{i=1}^{k}x_i \cdot w_i\right)}{W}=\frac{\sum_{i=1}^{k}E(x_i \cdot w_i)}{W}$$

$$=\frac{\sum_{i=1}^{k}E(x_i)\cdot w_i}{W}=\frac{\sum_{i=1}^{k}\mu \cdot w_i}{W}=\frac{\mu \cdot w_1 + \mu \cdot w_2 + \cdots + \mu \cdot w_k}{W}$$

$$=\frac{\mu \cdot (w_1 + w_2 + \cdots + w_k)}{W}=\frac{\mu \cdot \sum_{i=1}^{k}w_i}{W}=\frac{\mu \cdot W}{W}=\mu.$$

In general, given a set of data values: $x_1, x_2, \ldots, x_k$, an unbiased estimator of the mean is any weighted mean

$$\hat{\mu}=\frac{\sum x \cdot w}{\sum w}$$

or

$$\hat{\mu}=\sum x \cdot q,$$

where $\sum q = 1$.

In certain situations it calls for us to weight a certain observation more than the others because of its importance. Thus, we introduce the weighted sample mean that will be used to estimate the true state of nature.

Definition 8.4.7 Weighted Mean

The weighted mean, $\hat{\mu}$, is an estimate of the population mean by which more relevant information is counted more frequently then less relevant information. Let w_i be the weight assigned to the data value x_i, then

$$\hat{\mu}=\frac{\sum_{i=1}^{n}x_i w_i}{W}=\frac{x_1 w_1 + x_2 w_2 + \cdots + x_n w_n}{W},$$

where $W=\sum_{i=1}^{n}w_i=w_1 + w_2 + \cdots + w_n$.

The five step procedure we use to calculate the **weighted mean**, $\hat{\mu}$, is given below.

Outcome x_i	Weights w_i
x_1	w_1
x_2	w_2
$\vdots$	$\vdots$
x_n	w_n

Step-by-Step Weighted Mean

Step 1. Determine the associated frequencies of the observed outcomes.

Step 2. The mode is the most frequent data to model the center; that is, the most frequent data is the mode (if such values exist).

Step 3. Treating the weights as frequencies, organize the listed data into a "frequency" table.

Step 4. Compute the sum of the weights:

$$W = \sum_{i=1}^{n} w_i.$$

Step 5. Compute the products of the data value times the associated weight. It might be helpful to create a new column in which to place this information.

$$x_1 w_1, x_2 w_2, \ldots, x_n w_n$$

Step 6. Sum the products

$$\sum_{i=1}^{n} x_i w_i$$

Step 7. The weighted mean is the sum of the products divided by the total sum of the weights.

$$\hat{\mu} = \frac{\sum_{i=1}^{n} x_i w_i}{W}$$

Example 8.4.13 Weighted Mean

Given four exam scores (listed): 98, 77, 72, and 94, determine a sample weighted mean using weights (linear weights):

$$1, 2, 3, \text{ and } 4.$$

SOLUTION

Step 1		Step 2			Step 3			Step 4			Step 5		
x	w	x	w		x	w	xw	x	w	xw	x	w	xw
98	1	98	1		98	1	98	98	1	98	98	1	98
77	2	77	2		77	2	154	77	2	154	77	2	154
72	3	72	3		72	3	216	72	3	216	72	3	216
94	4	94	4		94	4	376	94	4	376	94	4	376
			$\sum w = 10$			10		$\sum xw =$	844				844
												$\hat{\mu} =$	84.4

Solution

That is, weighting the later information more than the first creates an estimate of the true population mean of 84.4; comparing this to the trimmed mean and the median of 84.5. Recall that the sample mean was 85.25, whereas this weighted mean is 84.4, fairly close. Which of the two point estimates is better to estimate the true population mean, μ? This is a subject in inferential statistics.

In estimating the true mean, μ, of a given population of size N from a random sample of size n taken from the population with the sample mean or other weighted mean, we wish to determine how close is the statistic to the parameter; that is, how close is $\bar{x}$ or $\hat{\mu}$ to μ. This closeness we call the **margin of error**, and it is defined below.

Definition 8.4.8 Margin of Error

The **margin of error** is a statistical measure of amount of random sampling error, denoted by the capital Greek letter Epsilon

$$E = |\hat{\theta} - \theta|,$$

where this capital letter E reads "margin of error" and not "expected value"; however, these symbols look very much alike, for this reason, the margin of error may by denoted **ME**,

$$ME = |\hat{\theta} - \theta|.$$

Regardless of the notation used, the larger issue with the margin of error is that without the unknown parameter, it too must be estimated. However, this will be the subject of **confidence intervals**.

Example 8.4.14 Margin of Error

Let us assume that the true tumor size in breast cancer is $\mu = 1.5$ cm. The random sample, $n = 50$, reveals a sample mean of $\bar{x} = 1.642$ cm. Determine the margin of error, **ME**.

Solution

The margin of error here is $ME = |\bar{x} - \mu|$; hence, we have

$$ME = |1.642 - 1.5| = 0.142.$$

Thus, even with 50 pieces of information, we are off by 0.142 cm. However, is this significantly different?

EXERCISES

Basic Problems

Find the (a) mean, (b) median, and (c) mode of the following data sets.

8.4.1. 5.04, 1.10, 3.21, 0.57, 2.41, and 3.21.
8.4.2. 21, 21, 21, 51, 51, 51, 57, 57, and 81.
8.4.3. 1, 2, 3, 4, 5, 6, 7, and 9.
8.4.4. 23, 18, 23, 10, and 11.
8.4.5. 3, 24, 26, 26, and 21.
8.4.6. 21, 1, 21, 22, and 22.
8.4.7. 12, 5, 6, 5, and 12.
8.4.8. 16, 21, 22, 22, 10, and 29.
8.4.9. 96, 70, 60, 67, 55, 98, 62, 90, 90, and 90.
8.4.10. 37, −6, 70, 29, 42, 29, −14, 70, 37, and 81.
8.4.11.

8.4.12.

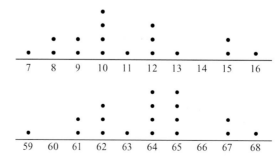

8.4.13.

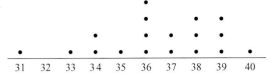

```
                                    •
                               •       •   •
                         •     •   •   •   •
         •     •   •   •   •   •   •   •   •   •
         31   32  33  34  35  36  37  38  39  40
```

8.4.14.

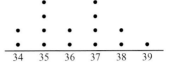

```
             •           •
             •           •
         •   •   •   •   •
         •   •   •   •   •   •
         34  35  36  37  38  39
```

8.4.15.

```
                         •
         •               •   •
         •               •   •
         •       •       •   •   •       •
         •       •   •   •   •   •   •   •   •
         1   2   3   4   5   6   7   8   9   10
```

8.4.16.

Stem	Leaves
5	3 1 4 8 9 4 2 7
6	8 5 0
7	9 2
8	7 5 1 7

8.4.18.

Stem	Leaves
5	3 1 4 8 9 4 2 7
6	8 5 0
7	9 2
8	7 5 1 7

8.4.19.

Stem	Leaves
4	8 8 3
5	7 8
6	7 3 6
7	0 5 7 6 9
8	6 0 1 2

8.4.17.

Stem	Leaves
12	7 2 7 3
13	6 2 9
14	
15	7 1
16	0
17	7 3 4

8.4.20.

Stem	Leaves
0	6
1	5 8 3

Find *x*.
8.4.21. *x*, 13, 14, 12, 6, and 21, given the mean is 13.
8.4.22. 21, *x*, 10, 9, 9, and 14, given the mean is 11.
8.4.23. 24, 16, 18, 24, 2, *x*, and 24, given the mean is 18.
8.4.24. 5, 13, 26, 13, 22, 21, and *x*, given the mean is 17.

8.4.25. x, 20, 20, 20, and 20, given the mean is 20.

8.4.26. 3, 10, x, 11, 8, 4, and 3, given the sample mean is 6 and the mode is 3.

8.4.27. 7, x, 24, 8, 5, 26, and 9, given the mode is 26 and the median is 9.

8.4.28. x, 12, 15, 8, 2, 12, and 15, given the mean is 11 and there are two modes: 12 and 15.

8.4.29. 11, 21, 17, x, 20, and 18, given the mode is 21 and the median is 19.

8.4.30. 47, 24, 190, 73, 13, 123, x, 139, 175, 83, 80, 97, and 38, given the mean is 94.

8.4.31. This stem-and-leaf shows the number of sit-ups each student did in class,

Stem	Leaves
0	7 7
1	9 8 4 7 8
2	1 6 2 5 6
3	5 9
4	6 6 4 6
5	6 1

   a. How many students are in the class?

   b. What is the median of the number of sit-up recorded? That is, the middle of the ordered data.

   c. What is the mode of the data; that is, the most frequently occurring data value.

   d. What is the mean of the data? Is the mean greater than or less than the sample mean; is the data skewed left, skewed right or symmetric?

   e. What is the range of the data; that is, the difference between the maximum and minimum data.

8.4.32. Given the following stem-and-leaf for Dana's scores from a video game,

Stem (hundreds and tens)	Leaves (units)
19	7 7
20	6 5
21	1
22	3 9
23	8 0 8
24	
25	2

   a. List the original data represented by the given stem-and-leaf plot.

   b. What is the mode of the data?

   c. What is the average of the scores recorded?

   d. What is the median of the scores recorded?

8.4.33. Given the following bar chart for points earned an a math test,

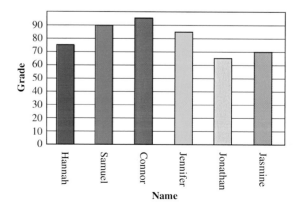

a. What is the average grade? Round your answer to the nearest tenth.

b. Who had the highest grade?

c. What is the median of the data?

Critical Thinking

8.4.34. Distinguish between the mean, median, and mode.

8.4.35. Distinguish between the population mean and the sample mean.

8.4.36. Distinguish between the standard mean and a weighted mean.

8.4.37. Distinguish between a weighted mean and a trimmed mean.

8.4.38. Distinguish between sample mean for listed data and the sample mean for grouped data in a frequency table.

8.4.39. Distinguish between the expected value and the margin of error.

8.4.40. Distinguish between the mean for grouped data in a frequency table and the expected value.

8.4.41. Find the mean, median, and mode of the given data.

   a) 5, 7, 6, 9, 11, 6, 18.

   b) 15, 17, 16, 19, 21, 16, 28.

   c) 15, 21, 18, 27, 33, 18, 54.

   What do all these estimates tell us about the central tendency?

8.4.42. First create a frequency table and the compute the **mean, median,** and **mode**.

   a) 1, 1, 1, 1, 2, 2, 2, 2, 2, 3, 3, 3, 3, 3, 4, 4.

   b) 12, 12, 12, 13, 15, 15, 15, 15, 17, 19.

   What is the usefulness of these estimates? Discuss.

8.4.43. Find the **weighted mean**, where the weights are: $w = 1, 2, 3$, and **4**, respectively.

   a) 79, 82, 67, 89

   b) 91, 78, 72, 92

   c) 90, 92, 94, 98

8.4.44. In problem 8.4.41-8.4.43 above, find the **weighted mean** and the **standard mean** and compare your results. Weights: $w = 1, 2, 2$, and **1**, respectively.

Compute the 25% trimmed mean for the following sets of data:

8.4.45. 52, 84, 67, 85, 58, 49, 65, and 75.

8.4.46. 2, 8, 4, 9, 5, 7, 6, 4, 5, 8, 4, 5, 7, 6, 5, and 9.

8.4.47. 82, 76, 85, 56, 74, 87, 87, 95, 72, 59, 72, and 95.

8.4.48. Given the following annual mutual fund returns:

−49.59	35.54	1.83	34.72
29.62	0.69	0.69	5.53
10.55	−42.33	−42.79	12.12
35.54	35.54	35.54	20.45
−44.21	−41.05	10.64	14.8

Compute the 10% trimmed mean for the following sets of data:

8.4.49. 68, 24, 52, 84, 67, 85, 58, 49, 65, and 75.

8.4.50. 8, 4, 9, 7, 6, 4, 5, 8, 4, 5, 7, 6, 5, and 9.

8.4.51. 85, 56, 74, 87, 87, 95, 72, 59, 72, and 95.

Compute the weighted mean for the following sets of data and associated weights:

8.4.52. **Data:** 5, 7, 9, 4; **Weights:** 1, 2, 3, 4.

8.4.53. **Data:** 5, 7, 9, 4; **Weights:** 1, 2, 2, 1.

8.4.54. **Data:** 5, 7, 9, 4; **Weights:** 1, 4, 9, 16.

8.4.55. **Data:** 4, 5, 7, 9; **Weights:** 1, 2, 3, 4.

8.4.56. **Data:** 4, 5, 7, 9; **Weights:** 1, 2, 2, 1.

8.4.57. **Data:** 4, 5, 7, 9; **Weights:** 1, 4, 9, 16.

8.4.58. Compare the weighted means in Exercises 3.52 through 3.54.

8.4.59. Compare the weighted means in Exercises 3.55 and 3.57.

Given the true mean is 6, determine the margin of error for the exercise indicated.

8.4.60. Exercise 8.4.52
8.4.61. Exercise 8.4.53
8.4.62. Exercise 8.4.54
8.4.63. Exercise 8.4.55
8.4.64. Exercise 8.4.56
8.4.65. Exercise 8.4.57

Mean of Means:

8.4.66. Compute the means of the various weighted means in Exercises 8.4.52–8.4.54: 6.2, 6.833, and 5.933. Compare this to the standard mean.

8.4.67. Compute the means of the various weighted means in Exercises 8.4.52–8.4.57: 6.2, 6.833, 5.933, 7.1, 6.167, and 7.7. Compare this to the standard mean. Discuss the effects of ordering the data before weighting the mean. What effect does this have on such weighted means?

8.5 MEASURING DEVIATION FROM THE CENTER

In estimating the true mean of a certain phenomenon of interest from a random sample, the quality of the estimate depends on how close or far the observations from a given experiment. The measure we used to measure these distances are called **deviations**. Thus, **deviations** are a measure of differences, for interval and ratio data, between the observed value and the population (true) mean.

Definition 8.5.1 Deviation

Deviation is a measure of differences between a set of data values and some fixed point such as the mean.

Common Deviations

Range—the difference between the minimum and maximum observed values.
Inner-quartile Range—the difference between the third and first quartile.
Mean Deviation—for comparison only.
Population Variance—true measure of variance.
Population Standard Deviation—true measure of deviation.
Sample Variance—the expected value of the differences observed between each observed value and the central tendency, squared. That is, this measure is the mean error squared.
Sample Standard Deviation—there are several measures of deviations, but the one used as the standard is the square root of the variance.

We have several measures of deviations; we begin with the most simplistic measure, the **range**.

Definition 8.5.2 Range (R)

The extent to which variation is possible; in a data set, this is defined to be the distance between the extremes;

$$Range = R = x_{(n)} - x_{(1)},$$

where $x_{(1)}$ is the minimum and $x_{(n)}$ is the maximum data value; that is, the maximum minus the minimum.

We can use two steps to determine the range.

Step-by-Step Range (R)

Step 1. Determine the maximum data value, $x_{(n)}$, and the minimum data value, $x_{(1)}$.

Step 2. Compute the range, $R = x_{(n)} - x_{(1)}$.

Special Case

In any **symmetric** data set, the maximum deviation in both directions from the population mean would be half the range.

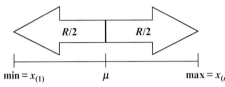

Hence, the most simplistic estimate of the population's **deviation**, σ is one-fourth of the **range;** that is, in a sample with at least two distinct observed values, an estimate of the population's deviation is given by

$$\hat{\sigma} = \frac{R}{4} = \frac{x_{(n)} - x_{(1)}}{4}$$

This estimate ensures that 100% of the data values fall within two deviations of the mean.

Example 8.5.1 Estimating Deviation Using R

Given the listed data: 7, 9, 10, 11, and 13, with a mean of 10; estimate the population's deviation.

Solution

$$\hat{\sigma} = \frac{13 - 7}{4} = 1.5$$

However, this **measure** is extremely sensitive to **outliers**; hence, for comparison a **mean deviation** is a statistically **valid measure** which becomes more **reliable** as the **sample size** increase.

Definition 8.5.3 Mean Deviation

The **mean deviation** is the averages of the absolute differences between a set of data values and the mean, that is,

$$\hat{\sigma} = \frac{\sum\limits_{i=1}^{n} d_i}{n},$$

where $d_i = |x_i - \bar{x}|$.

Step-by-Step Mean Deviation

Step 1. Given a sample of size $n : x_1, x_2, \ldots, x_n$, where the true population mean μ is known or specified, determine the deviation $d_i = |x_i - \mu|$.

Step 2. Sum the absolute differences:

$$\sum_{i=1}^{n} d_i$$

Step 3. Then take the mean deviation:

$$\hat{\sigma} = \frac{\sum\limits_{i=1}^{n} d_i}{n}$$

Example 8.5.2 Mean Deviation

Using the listed data from Example 8.5.1, calculate the **mean deviation**.

Solution

Step 1. First, determine the differences:

$$-3, \ -1, \ 0, \ 1 \text{ and } 3,$$

which illustrates why the absolute values must be taken—as otherwise, the sum of the deviation would be zero, indicating no deviation which is clearly not the case as there is deviation within this small sample and determining the absolution value of these differences, the deviations are:

$$3, \ 1, \ 0, \ 1 \text{ and } 3,$$

Step 2. $3 + 1 + 0 + 1 + 3 = 8$

Step 3. Then, the **mean deviation** is

$$\hat{\sigma} = \frac{3 + 1 + 0 + 1 + 3}{5} = \frac{8}{5} = 1.6$$

While unrealistic for comparison consider the mathematically defined definition of the population variance. Using the range, we estimated the deviation to be 1.5, using the mean deviation we estimated the deviation to be 1.6. Note, this point estimate is only slightly larger than the first estimate of the population's deviation; however, these first two estimates do not allow much leeway.

Hence, as an overestimate of the deviation ensures confidences in the measure as well as this measure is the maximum likelihood estimate of the **population's variance** (the mean deviation-squared).

Definition 8.5.4 Standard Population Variance

The **standard population variance** is the mean deviation-squared, given by

$$\sigma^2 = \frac{\sum_{i=1}^{N}(x_i - \mu)^2}{N}.$$

Note: Here it is assumed that the population mean, μ, is known; therefore, this measure assumes all information is known and hence, this is the true **population variance** denoted by the lowercase Greek letter sigma (squared). From which, the **population standard deviation** can be estimated by taking the square-root of the **variance**.

Definition 8.5.5 Standard Population Deviation

The **standard population deviation** is the square-root of the standard population variance; that is,

$$\sigma = \sqrt{\frac{\sum_{i=1}^{N}(x_i - \mu)^2}{N}}.$$

Step-by-Step Population Deviation

Step 1. Given a sample of size $n : x_1, x_2, \ldots, x_n$, where the true population mean μ is known or specified, determine the deviation or error $\varepsilon_i = |x_i - \mu|$

Step 2. Determine the error squared for each deviation found in step **1**: ε_i^2

Step 3. Sum the square differences: $\sum_{i=1}^{n} \varepsilon_i^2$

Step 4. Then take the mean square error:

$$\sigma^2 = \frac{\sum_{i=1}^{N} \varepsilon_i^2}{N}$$

Step 5. Finally, take the square root of the mean square error:

$$\sigma = \sqrt{\frac{\sum_{i=1}^{N} \varepsilon_i^2}{N}}$$

Example 8.5.3 Population Deviation

For the sake of comparison, calculate the **population variance** and determine the **population standard deviation** using the listed data in Example 8.5.1.

Solution

Step 1. Determine the differences: **–3, –1, 0, 1, 3**.

Step 2. Then square these deviations, **9, 1, 0, 1, 9**, which over emphasize larger deviations.

Step 3. Summing these values we get the sum of square errors, $\mathbf{SS = 9 + 1 + 0 + 1 + 9 = 18}$.

Step 4. Thus, the deviation-squared referred to as the **population variance** is
$$\sigma^2 = \frac{20}{5} = 4.$$

Step 5. Hence, an estimate of the population standard deviation is given by $\sigma = \sqrt{4} = 2$.

Note: this measure gives more leeway then either of the first two; however, this measure relies on several assumptions. The most unrealistic of which is that the true population mean is known. This measure is more likely to apply to a sample of size n and not the entire population. Hence, the variance or deviations can be estimated by:

1. The range over four—little to know leeway.
2. The mean deviation—some leeway, but is not the most likely estimate.
3. The standard population deviation—more leeway, but unrealistic due to required assumptions. Moreover, when we use the **sample mean** to estimate the true populations mean, this violation of the assumptions causes the statistic to be a **biased** estimate.

As we have mentioned in real world problems we do not know σ^2 or σ because it requires we known the actual value of N. Thus, from the random sample n from N we estimate σ^2 with the sample variance s^2 and σ with s.

Definition 8.5.6 Standard Sample Variance

The **standard sample variance** is the averages of the square differences between a set of data values and the mean accounting the degree of freedom, that is,

$$s^2 = \frac{\sum (x - \bar{x})^2}{n - 1}.$$

Definition 8.5.7 Standard Sample Deviation

The **standard sample deviation** is the square-root of the standard sample variance; that is,

$$s = \sqrt{\frac{\sum (x - \bar{x})^2}{n - 1}}.$$

Note that in defining s^2 and s we have divided by $(n-1)$ rather than n. The reason for this is to ensure that we have an unbiased estimate of σ. This fact will become more evident as we proceed to study inferential statistics. Given below is a six step procedure for calculating the sample variance, s^2, and sample standard deviation, s.

Step-by-Step Sample Deviation

Step 1. Given a sample of size $n : x_1, x_2, \ldots, x_n$, first determine an estimate of the population mean given by the standard sample mean: $\bar{x} = \dfrac{\sum_{i=1}^{n} x_i}{n}$.

Step 2. Then, estimate the differences (residual error):

$$\varepsilon_i = x_i - \bar{x}.$$

Step-by-Step Sample Deviation—cont'd

Step 3. Determine the estimated deviations-squared using the differences-squared:

$$\varepsilon_i^2 = (x_i - \bar{x})^2$$

Step 4. Calculate the sum of square error:

$$SS = \sum_{i=1}^{n} \varepsilon_i^2$$

Step 5. Then take the mean of these estimated deviation squared:

$$s^2 = \frac{\sum_{i=1}^{n} \varepsilon_i^2}{n-1} = \frac{\sum_{i=1}^{n} (x_i - \bar{x})^2}{n-1}$$

Step 6. Finally, the sample standard deviation is:

$$s = \sqrt{\frac{\sum_{i=1}^{n} (x_i - \bar{x})^2}{n-1}}.$$

Example 8.5.4 Sample Deviation

Using the listed data from Example 8.5.1, calculate the **sample variance** and determine the **sample standard deviation**.

Solution

Step 1. Determine the sample mean:

$$\bar{x} = \frac{7 + 9 + 10 + 11 + 13}{5} = 10.$$

Step 2. Determine the differences: **−3, −1, 0, 1, 3**. Determine the absolute value of these differences, the deviations are: **3, 1, 0, 1, 3**.

Step 3. Then square these deviations, **9, 1, 0, 1, 9**, which over emphasize larger deviations.

Step 4. Summing these values we get the sum of square errors,
$SS = 9 + 1 + 0 + 1 + 9 = 20$.

Step 5. Thus, the deviation-squared referred to as the **sample variance** is $s^2 = \frac{20}{4} = 5$.
Hence, an estimate of the population standard deviation is given by

$$s = \sqrt{5} \approx 2.24 > 2.$$

Note, this measure gives the most leeway in that $\sqrt{5} > 2$. Moreover, this measure is an unbiased estimate of the population variance; that is, we expect the statistic to estimate the parameter with bias. For this reason, the **standard sample deviation** is the most common **measures** of **deviation** used in **exploratory data analysis**.

We have four measures of deviations to compare that we use in determining the variability of a given real world problem. Here we have listed them in order of increased variability as illustrated in the previous, we have four calculations or estimations of the population deviation.

1. The **range over four**—little to no leeway was 1.5.
2. ✦ The **mean deviation**—some leeway was 1.6.

3. The **standard population deviation**—more leeway but unrealistic, is 2.
4. The **standard sample deviation**—most leeway, and an unbiased estimate is 2.24.

The following illustrate the three measure of variability in order of importance.

Important Deviations

1. *The **sample variance**, s^2, that measures the **variability** with respect to the **sample mean**.*
2. *The **sample standard deviation**, s, is simply the **square-root** of the **variance**.*
3. *The **range** of the data in a given random sample, **n**, is the **maximum** observed value (largest) minus the minimum observed value (smallest).*

Formulations

As with the **sample mean**, there are slight variations in the formulas depending on if the **data** is **listed**, grouped in terms of **frequency** or grouped in terms of **proportions** or **probabilities**.

The following diagram illustrates the key parameters within a population of size N and their corresponding estimated for the random sample of size n.

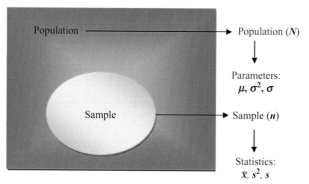

The table below summarizes the key formulas with respect to the population parameters with N and the corresponding estimates from the random sample, n taken from N.

Formulations—cont'd

Discrete Data	Listed Data	Grouped Data with Frequencies	Grouped Data with Proportions
Population mean	$\mu = \dfrac{\sum x}{N}$	$\mu = \dfrac{\sum (x \cdot f)}{n}, \; n = N = \sum f$	$\mu = \sum (x \cdot p), \; p = \dfrac{f}{n}, \; \text{and} \; \sum p = 1$
Sample mean	$\bar{x} = \dfrac{\sum x}{n}$	$\bar{x} = \dfrac{\sum (x \cdot f)}{n}, \; n = \sum f$	$\bar{x} = E(x) = \sum (x \cdot p), \; p = \dfrac{f}{n}, \; \text{and} \; \sum p = 1$
Population variance	$\sigma^2 = \dfrac{\sum (x-\mu)^2}{N}$	$\sigma^2 = \dfrac{\sum \left[(x-\mu)^2 \cdot f\right]}{n}, \; n = N = \sum f$	$\sigma^2 = \sum \left[(x-\mu)^2 \cdot p\right], \; p = \dfrac{f}{n}, \; \text{and} \; \sum p = 1$
Population standard deviation	$\sigma = \sqrt{\dfrac{\sum (x-\mu)^2}{N}}$	$\sigma = \sqrt{\dfrac{\sum \left[(x-\mu)^2 \cdot f\right]}{n}}, \; n = N = \sum f$	$\sigma = \sqrt{\sum \left[(x-\mu)^2 \cdot p\right]}, \; p = \dfrac{f}{n}, \; \text{and} \; \sum p = 1$
Sample variance	$s^2 = \dfrac{\sum (x-\bar{x})^2}{(n-1)}$	$s^2 = \dfrac{\sum \left[(x-\bar{x})^2 \cdot f\right]}{(n-1)}, \; n = \sum f$	$s^2 = \dfrac{n}{n-1} \sum \left[(x-\bar{x})^2 \cdot p\right], \; p = \dfrac{f}{n}, \; \text{and} \; \sum p = 1$
Sample standard deviation	$s = \sqrt{\dfrac{\sum (x-\bar{x})^2}{(n-1)}}$	$s = \sqrt{\dfrac{\sum \left[(x-\bar{x})^2 \cdot f\right]}{(n-1)}}, \; n = \sum f$	$s = \sqrt{\dfrac{n}{n-1} \sum \left[(x-\bar{x})^2 \cdot p\right]}, \; p = \dfrac{f}{n}, \; \text{and} \; \sum p = 1$

Alternative View

Another way to define variance is in terms of expected values; the variance is the difference between the expected value of the data-squared and the expected value of the data, squared. That is, the variance is a measure between the two operators, summing and squaring; that is, squared and then averaging versus averaging and then squaring:

$$V(x) = E(x^2) - [E(x)]^2$$

Thus, the expected value of the data-squared:

$$E(x^2) = \frac{x_1^2 + x_2^2 + \cdots + x_n^2}{n},$$

minus the expected value of the data, squared:

$$[E(x)]^2 = \left[\frac{x_1 + x_2 + \cdots + x_n}{n}\right]^2.$$

This variance is the difference between squaring then averaging and averaging followed by squaring.

One useful measure of the variability of a given set of data that characterizes a certain phenomenon is the **coefficient of variation (CV)**.

Definition 8.5.8 Coefficient of Variation

The **coefficient of variation** (**CV** or C_V) is a dimensionless measure, the ratio of the standard deviation to the mean; the coefficient of variation is an indication of the reliability of the measure, a measure of the dispersion of the data about the sample mean.

$$CV = \frac{s}{\bar{x}},$$

where $\bar{x}$ is the sample mean and s is the sample standard deviation.

If the sample mean is dominated by the standard deviation, then sample mean will not be reliable (good) estimate of the true mean. Thus, small values of *CV* will mean that the sample mean is a good estimate of the population mean.

Also, large values of the *CV* can lead to the inability to place any reliable assurance in your sample mean. A desirable *CV* value would be less than one, if greater than one it may indicate that our random sample n was not large enough to obtain a good estimate of μ.

Example 8.5.5 Lightning

When measuring the number of lightning strikes from a cloud to the ground, the **sample mean** was approximately 90,000 strikes a month and the **standard sample deviation** was approximately 100,000 strikes. This means within one standard deviation we expect between $-10,000$ and 190,000 lightning strikes. However, the $-10,000$ number of lightning strikes has no interpretation. Here, the **CV** is 1.11, which means that the deviation is larger than the value of the mean. Furthermore, this indicates that the data we are analyzing is "difficult data."

The *CV* is also useful in comparing the degree of variation between two data sets.

Example 8.5.6 CO$_2$ in the Atmosphere

When measuring the amount of carbon dioxide, CO_2 in the atmosphere, measurements are taken in 10 locations ranging from Alert, Alaska to the South Pole.

However, if you look at the time series graph, shown below, it is clear that there is greater variation in locations further north.

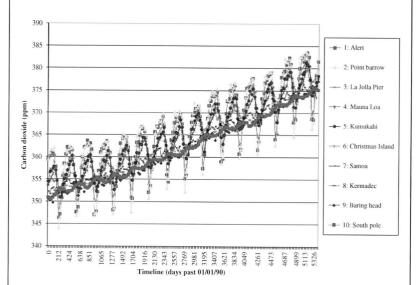

If we compare the *CVs* for **Alert, Alaska**, it is 0.0154, to that of the **South Pole**, 0.003774, there is four times the amount of variability; this is also seen in the range. The range of readings in **Alert, Alaska** is 17.52 and at the **South Pole**, 4.97. That is, carbon dioxide in **Alert, Alaska** has three-and-a-half times the range as those measured in the **South Pole**. This means that the data with smallest *CV* and range is the more reliable data for us to further analyze in determining the true mean.

EXERCISES

Basic Problems

Find **n**.

8.5.1. 11, 11, **n**, 13, 11, and 13, given the range is 2 and the median is 12.

8.5.2. 16, 9, 19, **n**, and 8, given the range is 16 and the mean is 10.

Find the range for the given set of data. Use the range to estimate the standard deviation using $\hat{\sigma} = \dfrac{R}{4}$.

8.5.3.

Stem	Leaves
10	6 1 5 1 1 4
11	3 1
12	0
13	3 6 7

8.5.4.

Stem	Leaves
7	5 7 9 6 1 4 9
8	7 1 4 2 3 4
9	7 6 0 6

8.5.5.

Stem	Leaves
40	4 5
41	6 2 0 7
42	2 4 6 0
43	9 8 3 0
44	5
45	9

Find the range, sample variance and sample standard deviations for the given data. Use the range to estimate the standard deviation using $\hat{\sigma} = \dfrac{R}{4}$ and compare to $\hat{\sigma} = s$.

8.5.6. 3, 15, 15, 15, 13, 26, and 18.

8.5.7. 12, 8, 16, 28, 18, 2, and 14.

8.5.8. 21, 28, 20, 24, 24, and 15.

8.5.9. 29, 15, 23, 25, 15, 22, and 18.

8.5.10. 20, 12, 24, 7, 28, and 11.

Find the range, variance, and standard deviations for the exercise indicated.

8.5.11. Exercise 8.5.6

8.5.12. Exercise 8.5.8

8.5.13. Exercise 8.5.10

8.5.14. Find the range and sample standard deviation for the following data sets:

a) 5, 7, 6, 9, 11, 6, 18

b) 15, 17, 16, 19, 21, 16, 28

c) 15, 21, 18, 27, 33, 18, 54

Compare the results in each case. What you conclude from each of the comparisons?

8.5.15. Find the range and sample variance and standard deviation for the following data sets:
   a) 1,1, 1, 1, 2, 2, 2, 2, 2, 3, 3, 3, 3, 3, 4, 4
   b) 12, 12, 12, 13, 15, 15, 15, 15, 17, 19
   Compare the results for (a) and (b) and what do you conclude from this comparison.

8.5.16. Find the range and sample variance and standard deviation for the following data sets:
   a) 79, 82, 67, 89
   b) 91, 78, 72, 92
   c) 90, 92, 94, 98
   Compare the results of $\hat{R}$ and s for each case and discuss their meaning.

Critical Thinking

8.5.17. Distinguish between the ranged and the standard deviation.

8.5.18. Distinguish between the range divided by four and the mean deviation.

8.5.19. Distinguish between the standard sample deviation and the population deviation.

8.5.20. Distinguish between the sample variance and the population variance.

8.5.21. Distinguish between sample size and degree of freedom.

8.5.22. Distinguish between the standard deviation and the CV.

8.5.23. **Degree of Freedom:** The mean of ten values is 75.0. Nine of the values are given to be 72, 75, 79, 81, 57, 98, 100, 76, and 65.

   Given Two Data Sets: find the sample mean and sample standard deviation for each of the two samples, then compare the two sets of results. Peter and Paul live in different cities. Each first Sunday of the month for 10 months, they recorded the temperature, in Fahrenheit, in their respective city. The temperatures that they recorded are given below.

   Peter's data: 50, 51, 50, 66, 60, 66, 71, 70, 72, 65, and 70.
   Paul's data: 72, 82, 76, 84, 77, 73, 70, 67, 67, 56, and 65.

8.5.24. What is the range of temperatures in Peter's data?

8.5.25. What is the mean, median, and mode for Peter's data? For Paul's data?

8.5.26. If you combined the data for both cities, what would be the mean, median, and mode for the combined data?

8.5.27. Peter added one more week to his data. The median of all of his data is now 62.8, rounded to the nearest tenth. What was the new temperature value that Peter added to the data?

8.5.28. Using Paul's data, what was the mean temperature in Celsius?

8.5.29. Paul calculated the median to be 71, however he forgot to include one number when calculating the median. Which number did Matthew forget?

8.5.30. What is the variance and standard deviation of Peter's data?

8.5.31. What is the variance and standard deviation of Paul's data?

8.5.32. What would be a good prediction for the temperature the next two months in Paul's city? Why?

8.5.33. If you could spend the next month in either city, whom would you want to visit? Why?

CRITICAL THINKING AND BASIC EXERCISE

Basic Problems

8.1. Compute the **mean, median,** and **mode** of the following data:

3	9	4	3	7	7	7	7	8	9
3	3	5	4	9	2	4	9	9	8
2	5	6	6	9	3	2	3	3	5
4	6	6	4	5	8	2	6	6	4
9	4	7	4	2	0	2	6	3	4
9	3	1	9	8	5	4	2	8	8

If you had to choose one of the three, which would you select to estimate μ?

8.2. Compute the linear weighted mean of the data: 70, 79, 83, and 97 with weight, weights: 1, 2, 3, and 4. Compare this to the standard sample mean and the 25% trimmed mean. In this example there is an obvious increase in the data values, how does this show in the three measures? Explain.

8.3. Compute the linear weighted mean of the data: 97, 79, 83, and 70, with weights: 1, 2, 3, and 4. Compare this to the standard sample mean and the 25% trimmed mean. In this example there is a relative decrease in the data values, how does this show up in the three measures, explain.

Critical Thinking

8.4. The standard sample mean, the 25% trimmed mean, the minimum, the maximum, the quartiles, and the percentiles are all types of what type of average? Which one do you believe is the best to estimate the true mean μ.

8.5. Compute the standard arithmetic mean for the data given in the frequency table below.

Outcome	Frequency
0	7
1	2
2	4
3	7
5	9
7	5

Explain the meaning and usefulness of the results.

8.6. Compute the standard sample variance and deviation for the data: 70, 79, 83, and 97.
Discuss their meaning and usefulness.

8.7. Compute the sample variance and standard deviation for the data: 40, 49, 53, and 80.
Explain their usefulness.

8.8. Compute the expected value for the data given in the relative frequency table below.

Outcome	Relative frequency
1.5	10%
2.2	34%
2.7	42%
3.4	11%
4.1	3%

Explain what the expected value means with respect to the given data.

8.9. Show that the standard sample mean is an unbiased estimator of the population mean; that is, given $E(x)=\mu$ and using summation notation prove that $E(\bar{x})=\mu$.

SUMMARY OF IMPORTANT CONCEPTS

Definitions:

8.1.1. A **dot plot** is one of the simplest ways to determine the distribution of numerical data—using a real number line, create a scale that includes the minimum and maximum data values, and then place a dot above each observed data point.

8.1.2. A **stem-and-leaf plot** is completed in two stages, first the basic **stem** and **leaf** are constructed by breaking the observed data values into categories such as tens and hundreds (the **stem**) and then units digits or the remaining associated information (the **leaves**). This initial sort does give the overall **distribution**, but it is common practice to then sort the leaves. The **range** of a data set is the difference between the **minimum** and **maximum** value; or maximum minus minimum.

8.1.3. A **frequency table** is simply a "t-chart" or two-column table which outlines the various possible outcomes and the associated frequencies observed in a sample.

8.1.4. A **frequency distribution** is a tabulation of the values that one or more variables take in a sample which can be organized in a frequency table or illustrated graphically.

8.1.5. A **relative frequency** is a measure between 0 and 1 that describes the frequency with which the values of the variable take relative to the sample size, n.

$$p_i = \frac{f_i}{n}$$

where

$$n = \sum_{i=1}^{k} f_i$$

and k is the distinct number of outcomes.

8.1.6. A **probability distribution** is a tabulation of the values that one or more variables take in a sample written in terms of **relative frequency**, **percentages** or **proportions**. If these values follow a formula, then this formula will be denoted as $f(x)$.

8.2.1. A **bar chart** is representation of a frequency distribution of a qualitative variable by way of rectangles whose heights are proportional to relative frequencies placed over each category.

8.2.2. A **Pareto chart** is a bar chart such that the categories have been sorted according to frequency (in descending order).

8.2.3. A **contingency table** is frequency distribution for a two-way statistical classification; that is a matrix of information where one factor is represented by the row and the second factor is represented by the column and the count of individuals that belong to exactly one cell within the matrix that is one row and one column.

8.2.4. A **circle graph** is comparable to the **divided bar chart** however, instead of a bar divided into categories where the size of each partition is proportional to the relative frequency; a circle is divided into categories $C_1, C_2, C_3, \ldots, C_m$, where the size of each partition is proportional to the relative frequency.

8.3.1. A **histogram** is representation of a frequency distribution by way of rectangles whose width represents the CW (class interval) and whose heights are proportional to relative frequencies.

8.3.2. Given a **class interval** is an open interval on the real number line, each **class interval** consist of a lower value called the **LCL** and an upper value called the **UCL**.

8.3.3. **CW** is length of the **class interval** given by the **UCL** minus the **LCL**, plus one **tolerance**; $CW = (UCL - LCL) + T$, where T is the **tolerance**; that is, the unit to which data will be rounded before being placed in the various class intervals.

8.3.4. **Class boundaries** are the end points of an open interval which contains the **class interval** such that the **LCB** is the **LCL** minus one-half the **tolerance** and the **UCB** is the **LCL** minus one-half the **tolerance**. Hence, from boundary to boundary is the full **CW LCB** $= LCL - 0.5T$ and **UCB** $= UCL + 0.5T$.

8.3.5. Given a **class interval**, (LCL, UCL), the **class mark** is the value in the middle, in the center below the rectangle which represents the frequency of the various **class intervals**.

8.4.1. The **mode** of data is the data value that occurs most frequently. We shall denote the mode by lowercase English letter m. As a measure of tendency in qualitative data, the **mode** is the most frequently occurring category.

8.4.2. The **median** of data set is a data value such that 50% of the information lies to the left and 50% of the information lies to the right. We shall denote the median by capital English letter M.

8.4.3. The **standard mean** or **sample mean** of data is the sum of all data values divided by the total number of data values, denoted $\bar{x}$. Using the capital Greek letter Sigma ($\sum$) to mean "the sum," we have $\bar{x} = \frac{1}{n} \sum_{i=1}^{n} x_i$, where n is the sample size.

8.4.4. For data that has been grouped in terms of frequencies, the sample mean appears weighted as by the associated frequencies. Given the frequency table of k distinct outcomes: $x_1, x_2, \ldots, x_k$ with associated observed frequencies: $f_1, f_2, \ldots, f_k$ in a sample of size $n = \sum_{i=1}^{k} f_i$ is given in the frequency table above. Thus the formulation for the **sample mean** for **grouped data** becomes $\bar{x} = \frac{1}{n} \sum_{i=1}^{k} x_i f_i = \frac{x_1 f_1 + x_2 f_2 + \cdots + x_k f_k}{n}$, where $n = \sum_{i=1}^{k} f_i$.

8.4.5. For data that has been grouped in terms of proportions, the sample mean appears weighted as by the associated proportions. Given the probability distribution of the k distinct outcomes: $x_1, x_2, ..., x_k$ namely: $p_1, p_2, ..., p_k$ in a sample of size n, where $\sum_{i=1}^{k} p_i = 1$. Thus the formulation for the **expected value** becomes

$$\bar{x} = \sum_{i=1}^{k} x_i p_i = x_1 p_1 + x_2 p_2 + \cdots + x_k p_k, \text{ where } \sum_{i=1}^{k} p_i = 1.$$

8.4.6. The p % **trimmed mean** is the sum of the remaining data values once p % of the information from the upper data values and the lower data values, divided by the remaining number of data values. Let $x_{(i)}$ represent the ordered data,

then $\underbrace{x_{(1)}, ..., x_{(j)}}_{p\%}, \underbrace{x_{(j+1)}, ..., x_{(n-j)}}_{\Downarrow}, \underbrace{x_{(n-j+1)}, ..., x_{(n)}}_{p\%}$ and $\hat{\mu} = \dfrac{\sum x \cdot w}{W}$, where $j = p \times n$.

8.4.7. The **weighted mean**, $\hat{\mu}$, is an estimate of the population mean by which more relevant information is counted more frequently then less relevant information. Let w_i be the weight assigned to the data value x_i, then

$$\hat{\mu} = \frac{\sum_{i=1}^{n} x_i w_i}{W} = \frac{x_1 w_1 + x_2 w_2 + \cdots + x_n w_n}{W}, \text{ where } W = \sum_{i=1}^{n} w_i = w_1 + w_2 + \cdots + w_n.$$

8.4.8. The **margin of error** is a statistical measure of amount of random sampling error, denoted by the capital Greek letter Epsilon $E = |\hat{\theta} - \theta|$, where this capital letter E reads "margin of error" and not "expected value"; however, these symbols look very much alike, for this reason, the margin of error may be denoted **ME**, $\mathbf{ME} = |\hat{\theta} - \theta|$.

8.5.1. **Deviation** is a measure of differences between a set of data values and some fixed point such as the mean.

8.5.2. The extent to which variation is possible; in a data set, this is defined to be the distance between the extremes; $\mathbf{Range} = \mathbf{R} = x_{(n)} - x_{(1)}$, where $x_{(1)}$ is the minimum and $x_{(n)}$ is the maximum data value; that is, the maximum minus the minimum.

8.5.3. The **mean deviation** is the averages of the absolute differences between a set of data values and the mean, that is,

$$\hat{\sigma} = \frac{\sum_{i=1}^{n} d_i}{n}, \text{ where } d_i = |x_i - \bar{x}|.$$

8.5.4. The **standard population variance** is the mean deviation-squared, given by $\sigma^2 = \dfrac{\sum_{i=1}^{N} (x_i - \mu)^2}{N}$.

8.5.5. The **standard population deviation** is the square-root of the standard population variance; that is,

$$\sigma = \sqrt{\frac{\sum_{i=1}^{N} (x_i - \mu)^2}{N}}.$$

8.5.6. The **standard sample variance** is the averages of the square differences between a set of data values and the mean accounting the degree of freedom, that is, $s^2 = \dfrac{\sum (x - \bar{x})^2}{n-1}$.

8.5.7. The **standard sample deviation** is the square-root of the standard sample variance; that is, $s = \sqrt{\dfrac{\sum (x - \bar{x})^2}{n-1}}$.

8.5.8. The CV or C_V is a dimensionless measure, the ratio of the standard deviation to the mean; the CV is an indication of the reliability of the measure, a measure of the dispersion of the data about the sample mean. $CV = \dfrac{s}{\bar{x}}$ where $\bar{x}$ is the sample mean and s is the sample standard deviation.

Rules:
8.1.1. A **stem-and-leaf plot** must follow the following rules:
    4. The defined stems and leaves must be based on equal intervals.
    5. All intermediate categories must be included even though there might be zero data values in a given category.
    6. Including empty categories within the outlined range of data values appear as gaps and these gaps are indicative of variability in the data.
8.3.1. When determine the **CW**, follow these simple rules:
    7. $m \times \mathbf{CW} \times \mathbf{Range}$ where m is the number of intervals; that is, $\mathbf{CW} \geq \dfrac{\mathbf{R}}{m}$.
    8. The first **LCL $\leq$ minimum**.
    9. The last **UCL $\geq$ maximum**.

10. The **tolerance** is the difference between the **UCL** for one class and the **LCL** for the next class; the **tolerance** can be defined as the unit to which the **CW** is rounded.

11. There are no **gaps** between consecutive intervals; **gaps** may only exist if there is an interval with **frequency** zero.

12. The **CWs** are consistent in length.

8.4.1. In general, to determine an estimate of the true **median**:

3. If there is an odd number of data values then take the middle of the order data.

4. If there is an even number of data values then trim all but the middle two values and take the midpoint, that is, the sum divided by two.

Common: Central Tendencies

❏ **Mode**—the most frequent value.

❏ **Median**—the middle of the ordered values.

❏ **Mean (Standard Arithmetic Mean)**—the sum of the values divided by the number of values.

❏ **Expected Value**—the mean evaluated using probabilities.

❏ **Trimmed Mean**—the mean evaluated trimming off the extremes.

❏ **Weighted Mean**—the mean evaluated using weights.

Common: Deviations

❏ **Range**—the difference between the minimum and maximum observed values.

❏ **Inner-quartile Range**—the difference between the third and first quartile.

❏ **Mean Deviation**—for comparison only.

❏ **Population Variance**—true measure of variance.

❏ **Population Standard Deviation**—true measure of deviation.

❏ **Sample Variance**—the expected value of the differences observed between each observed value and the central tendency, squared. That is, this measure is the mean error squared.

❏ **Sample Standard Deviation**—there are several measures of deviations, but the one used as the standard is the square root of the variance.

REVIEW TEST

1. A student spends $975 on rent, $75 on utilities, $200 on food and $100 on entertainment. In a circle graph showing this budget, how many degrees would there be in the section representing utilities?
For problems 2-6, consider the following set to tests scores:

$$70, 70, 71, 72, 74, 74, 75, 76, 76, 77, 78, 78, 78, 78, 79, 79, 79, 80, 80.$$

2. What is the mode?

3. What is the median?

4. What is the mean?

5. What is the range?

6. What is the standard deviation?

7. Which kind of average my fail to exist for a given data set?

8. Which kind of average is sensitive and would change for a given data set if one item in the set changed value?
Practice problems for the CLAST exam.

9. In a statistics class, half the students scored 78, most of the remaining students scored 92 except a few who scored 100. Which of the following statements is true about the distribution of scores?

a. The mode equals the median

b. The mean is less than the mode

c. The mean equals the median

d. The mean is greater than the median

10. Thirty students take a quiz and all score greater than a 5, but less than a 25. Which of the following is a reasonable estimate of the average score of the students?
   a. 25
   b. 30
   c. 4
   d. 18

11. A survey is done to determine how well people like Barbeque. Which of the following procedure would be most appropriate for selecting a statistically unbiased sample?
   a. Survey a barbeque restaurant
   b. Survey a burger joint
   c. Survey a college campus
   d. Survey a food court in a single mall

12. Forty individuals work for a company with salaries ranging from $400 per week to $1250 per week. Which of the following amounts is a reasonable estimate of the total weekly payroll?
   a. $1650
   b. $15,000
   c. $32,000
   d. $52,000

13. What is the mean of the data: 9, 4, 13, 8, 11, 2, 12, 9, and 4.
   a. 4
   b. 8
   c. 9
   d. 12

14. For each of the data sets, the given value of $\hat{\mu}$ represents the same type of measure of central tendency.

$$x: 5, 10, 15, 20, 25 \quad \hat{\mu} = 15$$
$$x: 5, 7, 10, 16, 29, 35 \quad \hat{\mu} = 13$$
$$x: 8, 9, 9, 9, 15 \quad \hat{\mu} = 9$$

What is the value of for the following data set?

$$x: 5, 5, 5, 7, 9, 11 \quad \hat{\mu} = \underline{\quad}$$

   a. 5
   b. 6
   c. 7
   d. 11

REFERENCES

Blackwell, D., 1969. Basic Statistics. McGraw-Hill, New York.

Brei man, L., 1973. Statistics: With a View Toward Applications. Houghton Mifflin, Boston.

Freund, J.E., 1964. Modern Elementary Statistics, 2nd ed. Prentice-Hall, Englewood Cliffs, NJ.

Li, Jerome C.R., 1968. Statistical Inference, vol. 1. Edwards Brothers, Ann Arbor, MI.

Lindgren, B.W., McElroth, G.W., 1966. Introduction to Probability and Statistics, 2nd ed. MacMillan, New York.

Mendenhall, W., 1971b. Introduction to Probability and Statistics, 3rd ed. Wadsworth, Belmont, CA.

Meyer, P.L., 1965b. Introduction to Probability and Statistical Applications. Addison-Wesley, Reading, MA.

Tsokos, C.P., 1978g. Mainstreams of Finite Mathematics with Application. Charles E. Merrill Publishing Company, A Bell & Howell Company.

Tsokos, C.P., Wooten, R.D., 2011d. The Joy of Statistics Using Real-World Data. Kendall Hunt, Florida.

There is no royal road to geometry.

EUCLID

Chapter 9 Geometry

Euclid was a Greek mathematician who is believed to have taught at Alexandria in Egypt during the time of Ptolemy. He was a contemporary of Archimedes (c. 287 BCE) and Plato (c. 347 BCE). A prominent mathematician, Euclid wrote the Elements, a treatise on geometry. Elements included compilations of works of earlier mathematicians resulting in Euclidian Geometry.

The Axiomatic methods outlined in depth by Euclid's Elements for proofs were threefold: clearly state the definitions of the underlying terms, give rules for the logical outcomes, and then judge the validity of the argument based on these assumptions or axioms.

Euclidian Geometry is the study of lines, angles, shapes, and areas in a plane and volume solids figures based on axioms and theorems.

FUNDAMENTAL AXIOMS

1. For any two points in a plane, there exists one and only one straight line between them.
2. A straight line can be indefinitely long.
3. A circle is all points equidistant from its center; the distance between the center and any point on the circle is the radius of the circle.
4. All right angles are equal (to 90°).
5. For any line **L** and point **p** not on the line, there exists one and only one line through **p** parallel (not intersecting) line **L**.

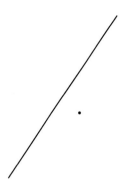

"geo" + "metric"
↓
Geometry

9.1 INTRODUCTION TO GEOMETRY

The concept of **geometry** originated from the word "**geo**" which means earth and "**metric**" meaning to measure. Thus, the main focus in **geometry** is to learn about measurement. For example, we need to measure **length**, **angles**, **area**, and **volume**, among other attributes. To accomplish this task, we shall begin by introducing several definitions that will be represented symbolically along with their interpretations. The table below summarizes the symbols that we will be using along with their description, meaning, and examples.

Symbol (abbr.)	Name	Meaning/ read as	Description	Example/description
l_i or L_i	ith line	"Line i"	A given line	l_1
$\parallel$	Parallel bars	"Is parallel to"	Used to relate lines	$l_1 \parallel l_2$
$\perp$	Perpendicular	"Is perpendicular to"	Used to relate lines	$l_1 \perp l_2$
$\angle$	Angle	"The angle"	Used to indicate an angle	$\angle A$
m	Measure of	"The measure of"	Used to indicate measure	$m\angle A$
π	Greek letter lowercase "pi"	Pronounced "pie"	$\pi = \dfrac{C}{d}$ or $\pi = \dfrac{A}{r^2}$	In the none repeating decimal that we named: approx. 3.14
$\circ$	Degree	"In degrees"	An angular measurement	$m\angle A = 90°$
rad	Radian	"In radians"	An angular measurement	$\pi\ rad \equiv 180°$
d		"Distance"	A linear measurement	Length l, width w, height h, base b; one dimension
A	Area	"Area"	A square measurement	Area, A; surface area, SA; two dimensions
V	Volume	"Volume"	A cubic measurement	Three dimensions

9.2   MEASUREMENTS

Measurement methods and skills have a variety of important uses in everyday life. Here we shall define four **types of measurements**: **distance**, one-dimensional object in one direction—lines or linear; **degree or radian**, turns made in an orientated two-dimensional object—measurement of the angle; **area**, surface contained in two-dimensional objects—measured in square units; and **volume**, three-dimensional objects—measured in cubic units.

Definition 9.2.1 Distance

On a straight line in one-dimensional space, the **distance** is the amount of space between two points.

Distance is used when measuring lines or one-sided figures (one dimension). Lines are measured linearly: in inches, feet, miles, or meters; abbreviated **in**, **ft**, **mi**, and **m**. Other names used to represent linear distance

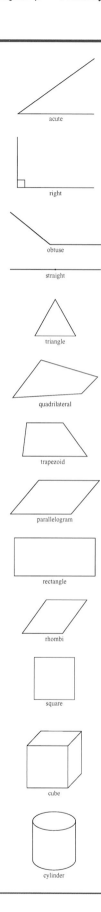

acute

right

obtuse

straight

triangle

quadrilateral

trapezoid

parallelogram

rectangle

rhombi

square

cube

cylinder

are: **length**, **width**, **height**, **base**, **depth**, and **perimeter**. Written mathematically, these words are abbreviated as follows: l, w, h, b, d, and P. Note: All of these lines are one-dimensional and are measured linearly.

Definition 9.2.2 Angle

In two-dimensional space, an **angle** is the amount of space between two intersecting lines or surfaces that meet.

Angles are two-sided figures measured in degrees (where $360°$ is a full rotation) or radians (where 2π is a full rotation). Angles measured in degrees are measured more like an idea rather than an object—it is between one-sided figures (one dimension) and three-sided figures (in two dimensions). Whereas angles measure in radians is equivalent to half the parameter (one-dimensional) of a unit circle (two-dimensional).

Definition 9.2.3 Area

In two-dimensional space, the **area** is the amount of space covered by a surface.

Moving into two dimensions, three-sided figures in a two-dimensional plane are measured in square units or **area**. In fact, this is true for all polygons (many sided objects) in two-dimensional space. Three and more-sided figures such as **triangles** (tri or three-sided figures), **quadrilaterals** (four-sided figures): **trapezoids**, **parallelograms**, **rectangles**, **rhombi**, **squares** are measured in terms of number of squares contained within each figure: inches squared, square feet, square miles, and meters squared; abbreviated in^2, ft^2, mi^2, and m^2. Here, area is abbreviated A and is two-dimensional. Each A will be formulated using the two perpendicular one-dimensional distances: length and width, base and height, etc. We will come back to this later and draw some pictures, but remember the key idea behind area is perpendiculars; which will be seen again in three-dimensional space and volume.

Definition 9.2.4 Volume

In three-dimensional space, the **volume** is the amount of space occupied by an object.

Volume is a measure in three-dimensional space. Objects such as **pyramids**, **rectangular solids**, and **cubes** are measured in terms of the number of cubes they contain: inches cubed, cubic feet, cubic miles, and meters cubed; abbreviated in^3, ft^3, mi^3, and m^3. In this text, volume is abbreviated V and will be formulated using the three individual one-dimensional distances; or in terms of a single one-dimensional distance and an orthogonal plane-area. (Orthogonal in multiple-dimensions is comparable to perpendicular in two dimensions.)

The accuracy of these measures is usually rounded to a specific decimal place using the conventional method. **Rounding conventionally** means if the digit to the left of the desired decimal is less than 5, then round down; otherwise, round up.

EXERCISES

Round the following to the indicated place value conventionally.

9.2.1. 1.42417 to the hundredths place
9.2.2. 2.71 to the units place
9.2.3. 3.1415 to the thousandths place
9.2.4. π Round to the second decimal place
9.2.5. e Round to the first decimal place
9.2.6. $\sqrt{2}$ Round to the tenths place
9.2.7. $\sqrt{5}$ Round to the third decimal place

Types of measurement

9.2.8. What unit of measurement would you use to measure a football field?
9.2.9. What unit of measurement would you use to measure a peanut?
9.2.10. What unit of measurement would you use to measure a pencil?
9.2.11. What unit of measurement would you use to measure a car?
9.2.12. What unit of measurement would you use to measure a penny?

Circle the correct answer   **Length   Area   Volume**

9.2.13. The distance from here to France is measured in:
miles   square miles   cubic miles
9.2.14. The area of a wall is measured in:
feet   square feet   cubic feet
9.2.15. The volume of a rectangular solid is measured in:
inches   square inches   cubic inches
9.2.16. The length of a line segment is measured in:
cm   square cm   cubic cm
9.2.17. The amount of water in a pool is measured in:
feet   square feet   cubic feet
9.2.18. The amount of space on a sheet of paper is measured in:
inches   square inches   cubic inches

9.3   PROPERTIES OF LINES

If you start with one line, there is not much of interest other than to describe it. It looks like a real number line but without the little hash marks: a one-dimensional movement: either forward or backward from any point.

Definition 9.3.1 Line

In one-dimensional space, a **line** is the long narrow mark or band. A **straight line** is a line that is true to its course and does not deviate or curve.

You can't criticize geometry. It is never wrong.
Paul Reand

One-Sided Figures: Lines

Placing a dot at an arbitrary point creates orientation; from this point there is forward and backwards. If each directional ray is considered individually, then this figure is a two-sided figure called a **straight angle**: two rays orientated as direct opposites. The type of angle always measures **180°**, and graphically illustrated below.

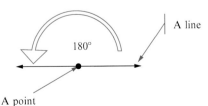

The type of angle always measures **180°**.

Two-Sided Figures: Angles

In general, an **angle** is formed by two rays from a given point in any two directions and is denoted by $\angle$; be careful, this symbol looks a lot like the inequality $<$, but their interpretations are very different.

Definition 9.3.2 Supplementary

Two adjacent angles (or **linear pairs**) are **supplementary** if they sum to 180°.

Therefore when you include the idea of angles we have that two angles—drawn adjacent—form a line, then these angles form **linear pairs**. Conceptually, such angles are call **supplementary** and quantitatively the sum of the measures of these two angles is the linear 180°. The point is now called the **vertex**. The diagram given below illustrates this concept.

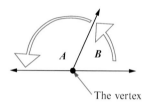

The vertex

Thus we can write it symbolically as
$$m \angle A + m \angle B = 180° \text{ or simply, } A + B = 180°.$$
The idea behind the second notation is that we know we are talking about "measuring angles," and we know we are measuring these angles in degrees; if you don't know this, the first notation is more descriptive.

In studying the behavior of straight lines, we focus on two main categories: **intersecting** and **non-intersecting**.

Definition 9.3.3 Intersecting Lines

Non-parallel lines, lines that cross paths at exactly one point are **intersecting lines**.

---

*Human knowledge in dark and uncertain; philosophy is dark, astrology is dark, and **geometry** is dark.*

John Jewel

Intersecting lines create four circular angles around the vertex; angles which form a full revolution or 360°. Idea: if you turn around once and you turn around again, then you are facing forward and the measure is twice that of turning around once: $2 \times 180°$.

The angles, which are adjacent, form linear pairs and the opposite angles (sometimes called the "**vertical angles**") are equal. This relationship is illustrated graphically below.

A B

C The vertex

Thus, we can write this relationship as

$$A + B = A + C = 180° \text{ and } B = C$$

or

$$m\angle A + m\angle B = m\angle A + m\angle C = 180° \text{ and } m\angle B = m\angle C.$$

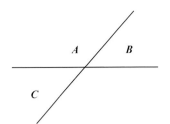

Definition 9.3.4 Perpendicular

Special cases of intersecting lines, lines are **perpendicular** if they form four right angles.

If these two lines intersect exactly at a **90°** then these two lines are said to be **perpendicular**; abbreviated ⊥.

In this specific type of intersection, these angles are equal; that is, if one of two angles that form a straight line is 90° and they are linear pairs then all four angles created are 90°. We illustrate it graphically by

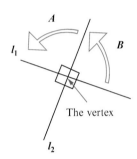

The vertex

Thus, we can express this relationship as

$$A + B = 180°, \ A = B = 90°, \text{ and } l_1 \perp l_2$$

or

$$m\angle A + m\angle B = 180°, \ m\angle A = m\angle B = 90°, \text{ and } l_1 \perp l_2$$

Therefore when you include the idea of **right** *angles*, 90°, when we have two angles—drawn adjacent—which form a right angle, then these angles are complementary. Conceptually, such angles are called **complementary** and quantitatively the sum of the measures of these two angles is the linear 90°.

Definition 9.3.5 Complementary

Two adjacent angles are **complementary** if they sum to 90°.

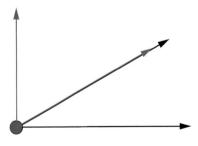

Non-intersecting lines are called **parallel** lines.

Definition 9.3.6 Parallel

Two lines are **parallel** if they never intersect.
They are called parallel lines and are abbreviated by two non-intersecting bars: ||.

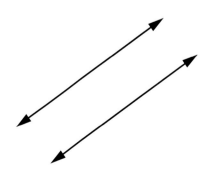

When we are given two parallel lines what happens when there is a third line which "cuts across" or "goes against"? Using the words "trans" as across and "verse" as against, we call this third line the **transverse** line.

Definition 9.3.7 Transverse

A **transverse** line is a line that cuts across other lines.

Illustrated below are two parallel lines and one transverse line which create eight angles labeled 1 through 8.

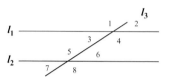

FIGURE 9.1   Two parallel lines and one transverse.

Written symbolically, we have $l_1 \| l_2$ and l_3 is the transverse.
From these three lines we can draw several relationships.

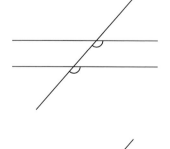

Property 9.3.1 Corresponding Angles

Corresponding angles are formed when a transversal line crosses parallel lines, and the corresponding incident angles are equal:

$$m \angle 1 = m \angle 5,$$
$$m \angle 2 = m \angle 6,$$
$$m \angle 3 = m \angle 7,$$
$$m \angle 4 = m \angle 8.$$

Property 9.3.2 Consecutive Angles

Consecutive angles are supplementary. This is clear with adjacent angles which form linear pairs:

$$m \angle 1 + m \angle 2 = 180°, \quad m \angle 1 + m \angle 3 = 180°, \text{etc.}$$

and incorporating property 9.3.1, consecutive angles between the two parallel lines are supplementary:

$$m \angle 3 + m \angle 5 = 180°, \text{etc.}$$

Property 9.3.3 Alternative Interior Angles

Alternate interior angles are formed when a transversal line crosses parallel lines and we consider the angles on opposite sides of the transverse line and interior to the parallels. Alternate interior angles are equal:

$$m\angle 3 = m\angle 6 \text{ and } m\angle 4 = m\angle 5.$$

Property 9.3.4 Alternative Exterior Angles

Alternate exterior angles are formed when a transversal line crosses parallel lines and we consider the angles on opposite sides of the transverse line and external to the parallels. Alternate exterior angles are equal:

$$m\angle 1 = m\angle 8 \text{ and } m\angle 2 = m\angle 7.$$

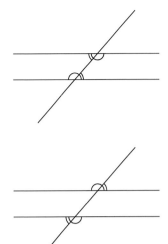

Note here, we can rename the numbers with letters.

Four or more lines do not bring any more interesting properties, but we sometimes use four lines to illustrate slightly more involved relationships.

Illustrated below are two parallel lines and two transverse lines which create 14 angles labeled *a* through *n* (Figure 9.2).

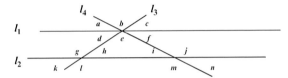

FIGURE 9.2   Two parallel lines and two transverse lines.

Written symbolically, we have $l_1 \| l_2$ and both l_3 and l_4 are transverses.

The above picture of four straight lines with 14 angles; given two of these angles we determine the remaining value using what might be classed the **trickle effect**: that is using the properties for a few lines and with a very limited amount of information, evaluate all measurements.

Example 9.3.1 Trickle Effect

Given that $n = 62°$ and $c = 48°$, what are the remaining values and why?

Lead One: given $n = 62° \rightarrow i = 62°$ (opposite angles are equal) $\rightarrow f = 62°$ (alternate interior angles are equal) $\rightarrow a = 62°$ (opposite angles are equal)

Another lead: given $n = 62°$ and $c = 48°$ and $a + b + c = 180° \rightarrow b = 70°$ (property of addition) $\rightarrow e = 70°$ (opposite angles are equal).

Moreover: given $c = 48° \rightarrow d = 48°$ (opposite angles are equal) $\rightarrow h = 48°$ (alternate interior angles are equal) $\rightarrow k = 48°$ (opposite angles are equal) $\rightarrow g = 132°$ (linear pairs) $\rightarrow l = 132°$ (opposite angles are equal).

Redraw this figure and label the measurements interpreted in the trickle effect.

EXERCISES

Answer TRUE/FALSE to the following: if false, give corrected statement(s).

9.3.1. The angle measuring 180° forms a straight line and is therefore called a straight angle.

9.3.2. Two angles which form a linear pair are supplementary and sum to 90°.

9.3.3. Two lines that intersect are perpendicular.

9.3.4. Two lines that intersect form equal opposite angles and consecutive angles form a linear pair.

9.3.5. Two parallel lines intersect exactly once.

9.3.6. Two angles are complementary if they form a right angle

9.3.7. The line that cuts across two parallel lines is called the transverse line.

One of the great challenges of modern cosmology is to discover what the geometry of the universe really is.

Margaret Gellar

9.4   PROPERTIES OF ANGLES

Measuring angles accurately is quite important in several applications that we experience in life. A single angle by itself is created by two lines (rays) leaving a vertex (point of origin) in two different directions. As the rays need not form a straight line, angles have a two-dimensional characterization of an object with two sides radiating for a vertex. The most commonly used directions are positive and negative on the real number line, north-south-east-west on maps, forward-backwards-up-down on Cartesians planes (positives and negatives again representing opposite directions), and clock-wise and counter-clock-wise on a polar plane. For example, consider a clock face: there are three radii, one length each for the seconds-minutes-hours (a completely different type of orientation—not in space, but in time). This polar plane has a vertex as the pole and consists of all rays with any given direction and magnitude—normally not drawn, just estimated. As in our example of a clock, we do have a few specific directions marked as well as their interpretations—furthermore, the initial direction from which everything is interpreted is up, marked with a 12. There is a lot of math involved in telling the time—this is why some people prefer digital.

For more accurate polar graphs you can purchase special paper, or a protractor (directions) and ruler (magnitude).

Note: scientists use the orientation of counter-clock-wise as the positive direction—moreover, the initial direction from which everything is interpreted is left—because the view from left to right looks like the horizon. Note: first, our orientation is the exact opposite orientation of what is assumed when telling time and second, our initial position is indicated as 0°—we consider looking up from the horizon as positive and looking down off the horizon as negative. We can illustrate this movement by the following diagram.

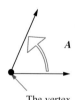

The vertex

or mathematically by $m \angle A > 0°$

In practice, we work with only four possible categories of angles excluding orientation. The categories are: **acute**, **right**, **obtuse**, and **straight**.

> **Common Types of Angles**
>
> **Acute**   like a cute kid is small, this type of angle measures such that: $0° < m \angle A < 90°$.
>
> **Right**   which stand "right" on its own and right in the middle of the line, this type of angle measures $m \angle A = 90°$.
>
> **Obtuse**   meaning large, think of the word obese—this type of angle measures such that: $90° < m \angle A < 180°$.
>
> **Straight**   this type of angle is the same as a line and measures $m \angle A = 180°$.

You recall that if there is more than one angle, then their angular sum may be supplementary or complimentary—or maybe not.

One interesting aspect is that when angles are considered in two or more dimensions. In two dimensions, all directional rays are perpendicular and in three dimensions, all directional rays are renamed **vectors** and each pair of **vectors** are perpendicular to each other. This perpendicular aspect in three or more dimensions is referred to as **orthogonal**, 90° between each intersecting lines.

> **Property 9.4.1 0° Angle**
>
> Two rays pointing in the same direction from the same initial point form a **0° angle**.
>
>

> **Property 9.4.2 180° Angle**
>
> Two rays pointing in opposite directions from the same initial point appearing as a straight line form a **180° angle**.
>
>

> **Property 9.4.3 90° Angle**
>
> A **right** angle is exactly **90° angle**.
>
>

> **Property 9.4.4 Complementary Angles**
>
> Two adjacent angles that when combined form 90°, then the angles are **complementary**.
>
>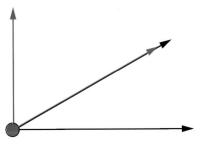

Property 9.4.5 Supplementary Angles

Two adjacent angles that when combined form 180°, then the angles are **supplementary**.

EXERCISES

Fill in the blank.

9.4.1. Acute measures between ____ and ____ degrees
9.4.2. Right measures exactly ____ degrees
9.4.3. Obtuse measures between ____ and ____ degrees
9.4.4. Straight measures exactly ____ degrees

Determine if the given angles are complementary, supplementary, or neither.

9.4.1. 45°, 50°
9.4.2. 45°, 45°
9.4.3. 45°, 150°
9.4.4. 145°, 35°
9.4.5. 25°, 65°

Determine the complementary angle to the given angle:

9.4.1. 70°
9.4.2. 45°
9.4.3. 15°
9.4.4. 5°
9.4.5. 25°

Determine the supplementary angle to the given angle:

9.4.1. 170°
9.4.2. 45°
9.4.3. 115°
9.4.4. 5°
9.4.5. 25°

Music is the arithmetic of sounds as optics is the geometry of light.

        Claude Debussy

"Tri" + "angle"
↓
Triangle

9.5   PROPERTIES OF TRIANGLES

Here we shall study the property of figures with three angles that constitute a triangle.

Definition 9.5.1 Triangle

In two-dimensional space, the **triangle** is a plane figure with three straight sides and three angles.

Three-Sided Figures: Triangles

Three angles, three sides—another word for three is "tri"—Tri-angles, abbreviated Δ. A graphical representation is given by

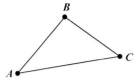

or we can write it mathematically as ΔABC.

Moreover, it can be proven that the sum of these three angles always equals $180°$. That is,

$$m \angle A + m \angle B + m \angle C = 180°$$

or

$$A + B + C = 180°$$

If the triangle has all **acute** angles, we call it an **acute triangle**; if it has a **right** angle, we call it a **right triangle**; and if it has an **obtuse** angle, we call it an **obtuse triangle**. Of all three types of triangles, the most important is the right triangle.

Pythagorean Theorem: A triangle is a right triangle if and only the sum of the square of the shorter sides equals the square of the larger. The shorter sides are called the legs and the larger (always opposite the right angle) is called the hypotenuse.

According to Pythagorean theorem, if a, b, and c are the lengths of the legs and the hypotenuse, respectively, then $a^2 + b^2 = c^2$.

A more interesting way to compare different triangles is in terms of how the angles relate; quantitatively there are exactly three possibilities: **no equal angles**, **exactly two equal angles**, or **three equal angles**. These three types of triangles are **scalene, isosceles**, and **equiangular**, respectively; even though we never use the word **equiangular**—we always say **equilateral**. "**Equi**" meaning equal and "**angular**" meaning in terms of angles, so why equilateral?

Consider $\angle A$ relative to the line segment between B and C. Note: we have to abbreviate this idea—a line segment between—too many words; we abbreviate this idea with a bar. For example: $\overline{BC}$ represents the line segment between B and C; illustrated below.

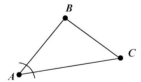

We can write it mathematically as $\angle A = \angle BAC$, where rays $\overrightarrow{AB}$ and $\overrightarrow{AC}$ are the sides of the angle radiating from the vertex A; indicated by the arc in the drawing above.

Now the question is how does this angle influence the side opposite? How does $\angle A$ relate to $\overline{BC}$. If $\angle A$ were increased, then $\overline{BC}$ would also increase and if $\angle A$ were decreased, then $\overline{BC}$ would also decrease. Hence, $\angle A$ relate to $\overline{BC}$ in ratio.

Arguably, the angle dictates the sides opposite relative to an internal ratio and hence equiangular would result in equilateral, "**lateral**" meaning pertaining to the side.

Properties of these three different types of triangles are as follows.

Common Types of Triangles

❏ **Scalene**
  (i). No equal angles
  (ii). No equal sides
  (iii). Comparisons in terms of angles are the same as comparisons in terms of sides.

  For example, given $m\angle A < m\angle B < m\angle C$ implies that $m\overline{BC} < m\overline{AC} < m\overline{AB}$, where m is still read "the measurement of."

❏ **Isosceles**
  (i). Two equal angles
  (ii). Two equal sides
  (iii). Comparisons in terms of angles are the same as comparisons in terms of sides.

  For example, given $m\angle A = m\angle B$ implies that $m\overline{BC} = m\overline{AC}$.

❏ **Equilateral**
  (i). Three equal angles
  (ii). Three equal sides
  (iii). Comparisons in terms of angles are the same as comparisons in terms of sides.

  For example, given $m\angle A = m\angle B = m\angle C$ implies that

  $$m\overline{BC} = m\overline{AC} = m\overline{AB}.$$

  And moreover, since $m\angle A + m\angle B + m\angle C = 180°$, we have

  $$m\angle A = m\angle B = m\angle C = 60°$$

Two triangles are only interesting if they look alike. "Look alike" is two words, so we use the word similar. Two or more triangles are similar if they have the same internal ratios.

Thus, we can proceed to state the formal definitions.

Definition 9.5.2 Similar

Triangles are **similar** if the share the same internal ratios. In terms of angles, this means that corresponding angles are equal, but in terms of sides, this simply means that the sides are proportional; like an enlargement of a triangle on a copier.

$\triangle ABC \sim \triangle DEF$ is read "triangle ABC is similar to triangle DEF."

The rules that govern **similar triangles** are given below.

Rule 9.5.1 Similar Triangles

Triangles are similar if they share these common ratios:

Rule 9.5.1 Similar Triangles—cont'd

Within triangles:

$$\frac{a}{b} = \frac{x}{y}$$

$$\frac{a}{c} = \frac{x}{z}$$

$$\frac{b}{c} = \frac{y}{z}$$

and between triangles:

$$\frac{a}{x} = \frac{b}{y}$$

$$\frac{a}{x} = \frac{c}{z}$$

$$\frac{b}{y} = \frac{c}{z}$$

If triangles are more than **similar**, and share all equal sides, then they are **congruent**. **Congruent** is a much stronger word than **similar**.

Definition 9.5.3 Congruent

Congruent means that corresponding angles are equal, as well as corresponding sides are equal; like a reproduction or regular copy of your triangle.

More than two triangles can be related in a similar fashion if they are all similar or congruent. Thus, the rules that drive congruent triangles are summarized below.

Rule 9.5.2 Congruent Triangles

Corresponding parts of **congruent triangles** are congruent.

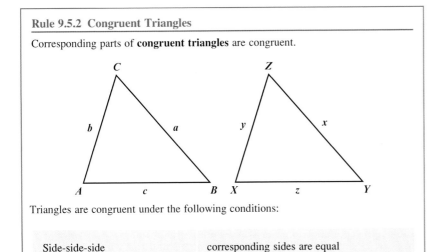

Triangles are congruent under the following conditions:

Side-side-side	corresponding sides are equal
	$a = x,\ b = y,\ c = z$
Angle-side-angle	$m \angle A = m \angle X,\ c = z,\ m \angle B = m \angle Y$
	$m \angle B = m \angle Y,\ a = x,\ m \angle C = m \angle Z$
	$m \angle C = m \angle Z,\ b = y,\ m \angle A = m \angle X$
	where $m \angle$ is read "the measure of the angle."
Side-angle-side	$a = x,\ \angle B = \angle Y,\ c = z$
	$b = y,\ \angle C = \angle Z,\ a = x$
	$c = z,\ \angle A = \angle X,\ b = y$

EXERCISES

For the following questions, define according to angle type: Acute, Right, and Obtuse

9.5.1. A triangle with interior angles measuring 60°, 30°, 90°.

9.5.2. A triangle with interior angles measuring 80°, 20°, 80°.

9.5.3. A triangle with interior angles measuring 45°, 20°, 115°.

9.5.4. A triangle with interior angles measuring 120°, 50°, 10°.

9.5.5. A triangle with interior angles measuring 45°, 45°, 90°.

For the following questions, define the above triangles in terms of: Scalene, Isosceles, and Equilateral.

9.5.6. Using the triangle in problem (9.5.2) above, if the shortest side measures 1 and the longest side measures 2, what is the middle side, the second leg?

9.5.7. Using the triangle in problem (9.5.5) above, if the shorter sides measures 1 what is the longest side—the hypotenuse.

For the following questions refer to the figure below $\Delta ABC \sim \Delta DEF$

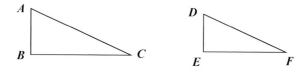

9.5.8. Given $AB = 6$, $BC = 10$, and $DE = 5$, find EF.

9.5.9. Given $AB = 6$, $BC = 10$, and $DE = 3$, find EF.

For the following questions, TRUE/FALSE.

9.5.10. $\dfrac{AB}{BC} = \dfrac{DE}{EF}$

9.5.11. $\dfrac{AB}{DE} = \dfrac{BC}{EF}$

9.5.12. $\dfrac{AB}{BC} = \dfrac{EF}{DE}$

9.5.13. $\dfrac{AB}{AC} = \dfrac{DE}{DF}$

9.5.14. $\dfrac{AB}{BC} = \dfrac{DE}{DF}$

For the following questions, fill in the inequality or equality, assuming

$$m \angle A > m \angle B > m \angle C$$

9.5.15. $m \angle F \underline{\qquad} m \angle E \underline{\qquad} m \angle D$

9.5.16. $\overline{AB} \underline{\qquad} \overline{AC}$

9.5.17. $\overline{DE} \underline{\qquad} \overline{DF}$

9.5.18. $\overline{AB} \underline{\qquad} \overline{BC}$

9.5.19. $\overline{FE} \underline{\qquad} \overline{DF}$

9.5.20. If $AB = DE$, then these triangles are more than just similar, they are $\underline{\qquad\qquad\qquad}$.

9.6 PROPERTIES OF QUADRILATERALS

We have learned about three-sided figures; now we shall study four-sided figures. The word "quad" meaning four and "lateral" again meaning pertaining to the sides; thus a four-sided figure is referred to as a **quadrilateral**.

Definition 9.6.1 Quadrilaterals

In two-dimensional space, the **quadrilateral** is a plane figure with four straight sides and four angles.

Four-Sided Figures

General quadrilaterals are not very useful, in that we don't divide land this way and we don't build functional structures this way. Only an artist would utilize such shapes and figures, and even then they are subject to different interpretations.

Quadrilaterals
↓
Four straight sides
&
Four angles

A typical quadrilateral is shown here:

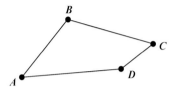

Unfortunately, we don't have an abbreviation for general quadrilateral; however, we should note that that the sum of the interior angles always equals $360°$. The idea is that if you draw a line from A to C, there are two triangles: hence $2 \times 180°$.

However, let's consider something a little more functional.

Definition 9.6.2 Trapezoid

In two-dimensional space, the **trapezoid** is a plane figure with four straight sides and four angles such that exactly one set of opposite sides are parallel.

Trapezoids are more interesting because two sides are parallel and their areas are easier to measure, but in terms of angles and sides, there are no other restrictions.

A trapezoid will look as shown below:

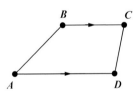

We can express this mathematically by $\overline{AD} \| \overline{BC}$.

Definition 9.6.3 Parallelogram

In two-dimensional space, the **parallelogram** is a plane figure with four straight sides and four angles such that both sets of opposite sides are parallel.

Extending this idea to both sets of opposite sides parallel, a typical parallelogram is as shown below.

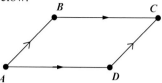

We can represent this using mathematical symbols as $\overline{AD} \| \overline{BC}$ and $\overline{AB} \| \overline{DC}$.

Given below are two properties of a parallelogram.

Property 9.6.1 Angles in Parallelograms

With respect to the angles in a parallelogram, opposite angles are equal:

$$\angle A = \angle C \quad \text{and} \quad \angle B = \angle D.$$

In the illustration below, we have used like markings to indicate which angles are equal.

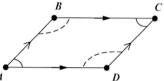

Property 9.6.2 Sides in Parallelograms

With respect to the **sides in a parallelogram**, opposite sides are parallel:

$$\overline{AD} \| \overline{BC} \quad \text{and} \quad \overline{AB} \| \overline{DC};$$

and opposite sides are equal:

$$\overline{AD} = \overline{BC} \quad \text{and} \quad \overline{AB} = \overline{DC}.$$

In the illustration below, we have used equal number of markings to indicate which sides are equal and directional arrows to demonstrate which lines move in that direction and are parallel.

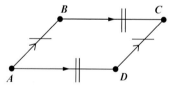

Opposite angles in a parallelogram are equal and **opposite sides** in a parallelogram are equal.

Extending the restrictions to include sides of equal length, the figure becomes a rhombus. A baseball diamond is shaped like a rhombus—the distance between bases is equal even if the field is not perfectly square.

Definition 9.6.4 Rhombus

In two-dimensional space, a **rhombus** is a type of parallelogram with four straight sides and four angles such that both sets of opposite sides are parallel and all sides are equal.

A rhombus can be graphically illustrated by

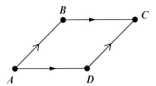

Property 9.6.3 Angles in a Rhombus

With respect to the **angles in a rhombus**, opposite angles are equal:

$$\angle A = \angle B \quad \text{and} \quad \angle C = \angle D.$$

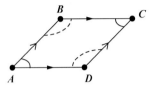

Property 9.6.4 Sides in a Rhombus

With respect to the **sides in a rhombus**, all sides are equal; that is,

$$\overline{AD} = \overline{BC} = \overline{AB} = \overline{DC};$$

and opposite sides are parallel:

$$\overline{AD} \| \overline{BC} \quad \text{and} \quad \overline{AB} \| \overline{DC}.$$

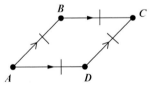

Opposite angles in a rhombus are equal and **all sides** in a rhombus are equal.

However, the diagonals of a rhombus are not equal; but they are perpendicular.

Property 9.6.5 Diagonals in a Rhombus

With respect to the **diagonals in a rhombus**, they are perpendicular to each other; that is,

$$\overline{AC} \perp \overline{BD}.$$

We now study the case when a figure consists of all right (equal) angles. Equiangular is all equal angles, and in the case of a four-sided figure, the four interior angles must all be right. Linguistically, saying "right-angles" is very close to "rect-angles" or **rectangle**.

Definition 9.6.6 Rectangle

In two-dimensional space, a **rectangle** is a type of parallelogram with four straight sides and four angles such that both sets of opposite sides are parallel and all angles are equal.

A schematic diagram of a rectangle is as shown below:

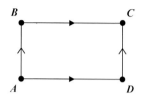

Property 9.6.6 Angles in a Rectangle

With respect to the **angles in a rectangle**, they are all equal to 90°; that is,

$$\angle A = \angle B = \angle C = \angle D = 90°.$$

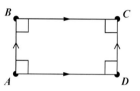

Property 9.6.7 Sides in a Rectangle

With respect to the **sides in a rectangle**, opposite sides are equal; that is,

$$\overline{AD} = \overline{BC} \quad \text{and} \quad \overline{AB} = \overline{DC};$$

and opposite sides are parallel:

$$\overline{AD} \| \overline{BC} \quad \text{and} \quad \overline{AB} \| \overline{DC}.$$

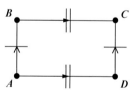

Because we have all angles are equal, each diagonal can be related to the sides by the Pythagorean Theorem, showing the diagonals are equal.

Property 9.6.8 Diagonals in a Rectangle

With respect to the **diagonals in a rectangle**, they are equal to each other; that is,

$$\overline{AC} = \overline{BD}.$$

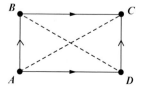

In summary, **all angles** in a rectangle are equal and **opposite sides** in a rectangle are equal and **diagonals** in a rectangle are equal.

A geometrical figure with both equal angles and equal sides is a **square**. A **square** is what we will base area—it is the perfect representation of perpendiculars.

Definition 9.6.6 Square

In two-dimensional space, a **square** is a type of parallelogram with four straight sides and four angles such that both sets of all sides are parallel and all sides are equal.

A graphical illustration of a square is given below:

Property 9.6.9 Angles in a Square

With respect to the **angles in a square**, they are all equal to 90°; that is,

$$\angle A = \angle B = \angle C = \angle D = 90°.$$

Property 9.6.10 Sides in a Square

With respect to the **sides in a square**, all sides are equal; that is,

$$\overline{AD} = \overline{BC} = \overline{AB} = \overline{DC};$$

and opposite sides are parallel:

$$\overline{AD}\|\overline{BC} \text{ and } \overline{AB}\|\overline{DC}.$$

Furthermore, not only are the diagonals equal, they are also perpendicular.

Property 9.6.11 Diagonals in a Square

With respect to the **diagonals in a square**, they are equal to each other and perpendicular; that is,

$$\overline{AC} = \overline{BD} \text{ and } \overline{AC} \perp \overline{BD}.$$

Definition 9.6.7 Kite

In two-dimensional space, a **kite** is a type of quadrilateral with four straight sides and four angles such that the two upper sides are equal and the two lower sides are equal.

A kite can be graphically illustrated by

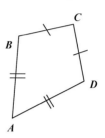

EXERCISES

TRUE/FALSE: if false, then give correct statement.

9.6.1. A general quadrilateral has four sides, with opposite sides parallel.

9.6.2. A trapezoid has four sides with exactly two parallel sides.

9.6.3. In a parallelogram, opposite angles and sides, respectively, are equal.

9.6.4. In a parallelogram, consecutive angles are complementary.

9.6.5. An equiangular quadrilateral is more commonly called a rectangle.

9.6.6. In a rectangle, the diagonals are perpendicular bisectors but are not equal.

9.6.7. In a rhombus, the diagonals are perpendicular bisectors and all sides are equal.

9.6.8. If the figure is a rhombus, then the figure is a parallelogram.

9.6.9. In a square, diagonals are not equal.

9.6.10. In a square, all angles and sides, respectively, are equal.

Name all four-sided figures with the following description:

9.6.11. Equal angles, but only opposite sides are equal

9.6.12. Opposite angles equal only and all equal sides

9.6.13. Opposite sides are parallel

9.6.14. All angles are equal

9.6.15. All sides are equal

9.6.16. Unequal sides

9.6.17. Unequal sides and one set of parallel sides

9.6.18. Equal diagonals

9.6.19. Unequal diagonals

9.6.20. Perpendicular diagonal

9.7 PROPERTIES OF POLYGONS

So far we have studied three and four-sided geometrical figures and now we wish to generalize this to figures with more than four sides. These figures are called **polygons**; this word is made up of the words "poly" meaning many and "gon" meaning sides.

Definition 9.7.1 Polygons

In two-dimensional space, the **polygon** is a plane figure with more than four straight sides with an equal number of angles.

Example 9.7.1 Pentagon

A **pentagon**—"pent" meaning five—is a five-sided figure. Note: not all sides are equal, nor are the interior angles.

Pentagon

Example 9.7.2 Hexagon

A **hexagon**—"hex" meaning six—is a six-sided figure.

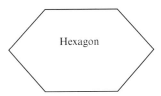

Hexagon

Others include the heptagon, octagon, nonagon, and decagon; seven, eight, nine, and ten-sided figures, respectively.

Definition 9.7.2 Regular Polygon

In two-dimensional space, the **regular polygon** is a polygon with all equal sides and all equal interior angles.

Example 9.7.3 Regular Pentagon

A **regular pentagon** is a five-sided figure with all equal sides.

Regular pentagon

Pentagon
Washington, D.C.

Example 9.7.4 Regular Hexagon

A **regular hexagon** is a six-sided figure with all equal sides.

Regular hexagon

Note: a polygon with n sides produces $(n-2)$ triangles, illustrated below. A five-sided figure produces three internal triangles.

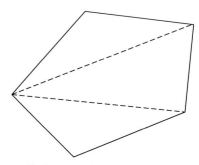

We have already studied sums of interior angles for three- and four-sided figures and the total measure of the interior angles. Recall, a triangle has total interior measure of 180° and any quadrilateral—being two triangles has a total interior measure of 360°.

Property 9.7.1 Sum of Interior Angles

In general, there will also be $(n-2)$ triangles when there are n sides in a figure; therefore since each triangle measures 180°, we can formulate the total interior measure for any polygon with n sides, S_n. That is, the **sum of interior angles** is given by

$$S_n = (n-2) \times 180°.$$

Definition 9.7.3 Perimeter

The distance around such straight edged figures is simply the sum of the individual distances and is called the **perimeter** and is abbreviated P.

Example 9.7.5 Perimeter

If there are three sides in a figure, then the perimeter is the sum of the three sides.

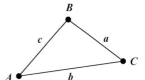

Then, by the definition the perimeter is $P = a + b + c$.

Example 9.7.5 Perimeter—cont'd

If the figure had four sides, then the perimeter is the sum of the four sides. That is,

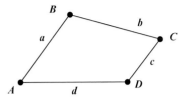

Then, by the definition the perimeter is $P = a + b + c + d$.

Thus finding perimeter of a figure depends on the number of straight edge sides. Notice, that we are writing straight-edged—if the figure is not straight edged, then it could be any smooth curve such as a circle. The distance around such a circle is called the **circumference**.

If finding the circumference of a circle, we need to first introduce the Greek letter π, used to represent the constant that is the ratio of a circles circumference to its diameter and is commonly approximated as 3.14159 and is sometimes referred to as "pi." It is an irrational number and it cannot be expressed exactly and consequently its decimal representation never ends. A formal definition of π is

$$\pi = \frac{\textit{Circumference of a circle}}{\textit{diameter of the circle}} = \frac{C}{d},$$

as shown by

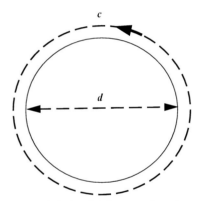

The circumference of a circle is slightly more than three times as long as its diameter.

Definition 9.7.4 Circumference

The distance around smooth curves is called the **circumference**; since these curve's perfection depends on it radiating the same in all directions from a centering point, circumference is often formulated in terms of the radius and is abbreviated C.

The **circumference** of a circle with fixed radius r is

$$C = 2\pi r,$$

where and π is approximately 3.1415.

Another way to formulate the circumference is in terms of the diameter; the diameter is the longest cord in a circle and measures twice the radius. That is,

$$C = \pi d \quad \text{and} \quad d = 2r$$

Note that everything is measured in distance or in one direction at a time.

EXERCISES

Distance: One-dimensional measurement

9.7.1. Find the perimeter of a square with sides measuring 6.
9.7.2. Find the perimeter of a square with sides measuring x.
9.7.3. Find the perimeter of a rectangle with length 2 and width 3.
9.7.4. Find the perimeter of a rectangle with length x and width y.
9.7.5. Find the perimeter of an equilateral triangle with side measuring 6.
9.7.6. Find the perimeter of an equilateral triangle with sides measuring s.
9.7.7. Find the perimeter of a quadrilateral with sides measuring 2, 4, 8, and 6.
9.7.8. Find the perimeter of a quadrilateral with sides measuring x_1, x_2, x_3, x_4.
9.7.9. Find the circumference of a circle with radius 7 ft.
9.7.10. Find the circumference of a circle with radius 12 cm.
9.7.11. Find the circumference of a circle with radius 15 mm.
9.7.12. Find the circumference of a circle with radius 40 in.
9.7.13. Find the circumference of a circle with diameter 7 ft.
9.7.14. Find the circumference of a circle with diameter 12 cm.
9.7.15. Find the circumference of a circle with diameter 15 mm.
9.7.16. Find the circumference of a circle with diameter 40 in.

9.8   AREAS AND SURFACE AREA

Recall that area measures the number of squares contained within each figure: inches square, square feet, square miles, and meters squared; abbreviated in^2, ft^2, mi^2, and m^2. Here, area is abbreviated as A and is two-dimensional. Moreover, each area will be formulated using the two perpendicular one-dimensional distances. We shall illustrate the learning about area with several pictures below.

Let's begin with learning how to find areas of commonly used geometric figures: **squares**, **rectangles**, **rhombus**, **parallelogram**, **trapezoid**, **quadrilateral**, and **triangles**. Then include circles.

Area

Rule 9.8.1 Area of Square

A square of any size is self-defining—if we stack s squares s times, then the area is $s \times s$. Therefore, if A is the **area of a square** with side length s then A is s^2; that is,

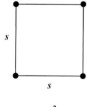

$$A = s^2.$$

Example 9.8.1 Area of Square

The area of a square that is **2 cm × 2 cm** is $A = 4$ cm^2. That is, there are four squares in the figure.

Square

$A = 2\ cm \times 2\ cm = 4\ cm^2$

Rule 9.8.2 Area of Rectangle

This idea can easily be extended to rectangles—if we stack (a base) b of squares h times to create a specific height, then the area of a **rectangle** is $b \times h$; therefore, if A is the area of a rectangle then it measures base times height. That is,

$$A = bh.$$

Example 9.8.2 Area of Rectangle

The area of a square that is **5 $in \times 2$ in** is $A = 10$ in^2. That is, there are 10 squares in the figure.

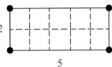

Rectangle

$$A = 2\ in \times 5\ in = 10\ in^2$$

Note: sometimes we use **length** and **width** of a rectangle and not base and height. Using the abbreviations for length and width, l and w, respectively, the formula for the area of a rectangle is simply $A = lw$.

This idea can further be extended to the rhombus—since the side is the average width of the figure, on average it is a rectangle.

Rule 9.8.3 Area of Rhombus

The average width is b and h is perpendicular to the upper and lower base; therefore, the area of a rhombus is $b \times h$. Therefore, if A is the **area of a rhombus** then it is measured base times height. That is,

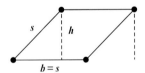

$$A = bh$$

Example 9.8.3 Area of Rhombus

The area of a rhombus with side **4** ft and height **5** ft is $A = 20$ ft^2.

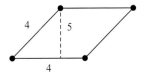

Continued

Example 9.8.3 Area of Rhombus—cont'd

Think of moving the triangle.

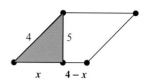

Resulting in a rectangle that is $4 \times 5 = 20$.

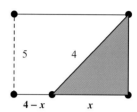

We can further visualize this in terms of averages—if we take a pair of scissors, then whatever we cut off, we will add it back in as shown by the diagram below.

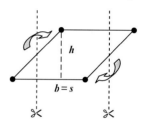

This idea can further be extended in the same way to the parallelogram. If A is the area of a parallelogram, then it is again measured base times height. That is,

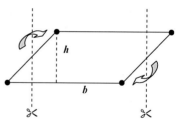

This approach of finding areas can be extended to the area of a trapezoid, then it is again measured average base times height.

Rule 9.8.4 Area of Trapezoid

In a trapezoid, the average base is the average of the lower base b and the upper base B; that is,

$$\text{Average base} = \frac{b+B}{2} = \frac{b+B}{2}$$

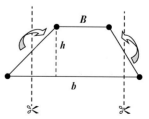

and the **area of a trapezoid** is obtained by

$$A = \frac{1}{2}(b+B)h.$$

Example 9.8.4 Area of Trapezoid

The area of a trapezoid with bases 10 in and 4 in, and height 5 in is

$$A = \frac{1}{2}(10+4) \times 5 = 35 \ in^2.$$

We proceed to study the area of a triangle.

Rule 9.8.5 Area of Triangle

The formula above directly leads to the formula for the area of any general triangle. From the diagram below, the average base is the average of the lower base **b** and the upper base 0; that is,

$$\text{Average base} = \frac{b+0}{2} = \frac{1}{2}$$

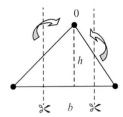

The **area of a triangle** is

$$A = \frac{1}{2} bh.$$

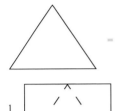

a triangle is half the
rectangle that contains it

Another way to visualize this is: we have congruent triangles and therefore, the triangle contained in the rectangle **b** by **h**, is exactly half the rectangle itself. That is, the area of a triangle is one-half the rectangle that contains it; one-half the base times the height.

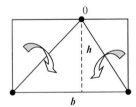

Either way, the formula is the same.

$$A = \frac{1}{2} bh.$$

Example 9.8.5 Area of Triangle

The area of a triangle with base 5 mm height 5 mm is

$$A = \frac{1}{2} \times 5 \times 5 = 12.5 \ mm^2$$

Another approach to obtaining the area of any triangle is by using Heron's Method.

Rule 9.8.6 Heron's Method

Heron's method for computing the area of any triangle with known sides a, b, and c can be computed as follows

$$A = \sqrt{s(s-a)(s-b)(s-c)}$$

where $s = \dfrac{a+b+c}{2}$.

If in a triangle, the height is not given, but the three sides, a, b, and c are known, we use the following approach to obtain the area. For a given triangle with sides a, b, and c, we calculate the perimeter P and let s be half the perimeter, $s = \frac{1}{2}P$; then the area is given by Heron's formula. That is, given a triangle

the area is given by

$$A = \sqrt{s(s-a)(s-b)(s-c)}, \quad s = \frac{1}{2}P \quad \text{and} \quad P = a+b+c$$

Therefore, if you have a general quadrilateral, you can then survey your figure and determine the various lengths, including the diagonal distance. Heron's formula can then be used twice and the areas then summed.

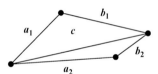

Example 9.8.6 Heron's Method

The area of a triangle with sides measuring 4 in, 7 in, and 5 in is determined by first calculating the perimeter, $P = 4+7+5 = 16$ inches and half the perimeter is $s = \dfrac{16}{2} = 8$; hence the area using Heron's method is

$$A = \sqrt{8 \times (8-4) \times (8-7) \times (8-5)} = 9.8 \text{ in}^2.$$

Rule 9.8.7 Area of a Circle

Area is measured using two perpendicular measurements. In general, the **area of a circle** is defined as its radius $\left(\dfrac{d}{2}\right)$ squared times π. That is,

$$A = \pi r^2.$$

If you find this hard to remember, consider it this way: a perfect circle, the general information given is the radius, but to compute area, we need a perpendicular and perpendicular to the radius is another radius. So we start with the radius square.

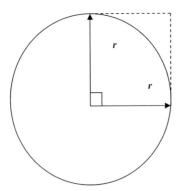

A square with side *r* has area *r*^{2}

However, this is obviously more than one-fourth the area of the entire circle; four squares would more than cover circle, illustrated below. So take the scissors and cut off the four corners, leaving about three squares to fill the circle. That is the area of a circle is about three squares. About three is 3.1415... and the square area is *r*^{2}.

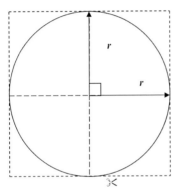

$$A = \pi r^2$$

Example 9.8.7 Area of a Circle

The area of a circle with radius of 6 mm is

$$A = \pi \times 6^2 = 36\pi \approx 104 \ mm$$

Next we study surface areas on three-dimensional objects. The areas are measured in the same fashion, but you have to use the scissors; for example, a rectangular solid has front and back, sides, a top, and bottom.

Rule 9.8.8 Surface Area

The **surface area** of flat-sided object in three-dimensional space is the sum of the areas present on each flat surface.

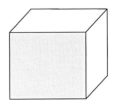

Three dimensions are very difficult to draw on paper because paper is only two-dimensional. So, we have to be a human camera—we have to imagine what a picture of a three-dimensional object might look like and draw it. Or, we could take two-dimensional paper and create a third-dimension by folding it. We will come back to this as well and draw visualized pictures of objects—that is, draw what a picture of the object might look like—and maybe build a few objects like origami. You will need tape with your scissors!

How to draw a rectangular solid: A rectangular solid looks like a shoebox and the picture of a shoebox is basically a rectangle with extended parallels to indicate the depth of the box and finally more parallels toward the back of the solid. Thus, we have parallels and perpendiculars to measure as shown below.

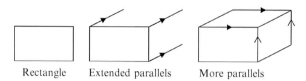

Rectangle   Extended parallels   More parallels

The surface of a rectangular solid consist of the front (rectangle drawn) and the back (which is implied), it could also be illustrated, but would have to be dotted since it is not visible from our point of view, but if you pick up the shoebox, it's there. Illustrated below are all six surfaces.

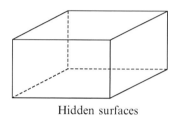

Hidden surfaces

Thus, we have the front and the back that have the same dimensions (height-length), two sides that have the same dimensions (height-width) and the top and the bottom, which again have the same dimensions (length-width); these areas on the surface additively are called the surface area: abbreviated by *SA*.

Furthermore, when we refer to these dimensions, we are talking about rectangles—all rectangles. They appear as parallelograms in the drawing, but remember this is a visual representation of a shoebox. If we take our scissors to it, the surface is as shown below. Create this illustration on a piece of paper and cut it out; then fold it and shape it to be the surface area of a rectangular solid.

This approach is referred to as Origami Design I:

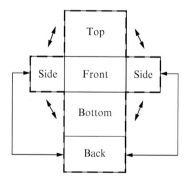

Thus, the surface area of a rectangular solid, SA, is given by

$$SA = 2lw + 2lh + 2wh.$$

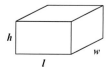

Example 9.8.8 Surface Area

The surface area of a rectangular solid that is **5 in × 2 in × 3 in** is $SA = (2 \times 5 \times 3 + 2 \times 5 \times 2 + 2 \times 3 \times 2)in^2 = 62\ in^2$. That is, there are 62 squares covering the surface of the figure.

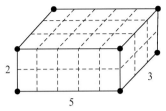

How would this change if we lost the lid of our shoebox? It is still surface area, but with only the bottom and not the top. The method should follow directly. If you would like to see it, fine but there should be no need to memorize it—knowing the area of the individual rectangles and how many we need is enough information. That is,

$$SA = lw + 2lh + 2wh$$

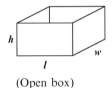

(Open box)

Definition 9.8.1 Right Circular Cylinder

A **right circular cylinder** is a cylinder with circular bases parallel to each other.

How to draw a right circular cylinder: We can visualize the geometry of a right circular cylinder by starting with two lines for the cylinder together with a circle drawn in perspective, which looks like an ellipse or oval as shown below.

Two parallel lines   Two ovals or ellipse

The surface of a right circular cylinder is a single rectangle wrapped around to form the cylinder and two circles, one on top and bottom. It could also have been illustrated with half the oval dotted since it is not visible from our point of view, but if it were a clear can you would see it there.

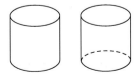

Rule 9.8.9 Surface of a Right Circular Cylinder

When we refer to the dimensions we have the base radius and the height. These circles appear as ellipses in the drawing; but remember this because it is our representation of a clear can. If we take our scissors to it, the surfaces are as shown below. Create this illustration on a piece of paper and cut it out. Then fold it and shape it to be the surface area of a right circular cylinder.

The **surface area, SA, of a right circular cylinder** is given by

$$SA = 2(\pi r^2) + (2\pi r)h.$$

This approach is referred to as Origami Design II:

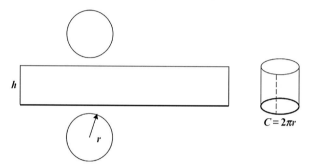

Example 9.8.9 Surface of a Right Circular Cylinder

The **surface area of a right circular cylinder** with radius **2** *in* and height **3** *in* is

$$SA = 2\pi \times 2^2 + 2\pi \times 2 \times 3 = 20\pi \approx 62.8 \ in^2.$$

The surface area of a sphere is also easy to calculate.

Rule 9.8.10 Surface of a Sphere

A sphere is a round ball. From the center, there is a fixed radius. Therefore, in all three perpendicular directions, the dimension is the same: *r*. And while this shape is easy to visualize, it is not easily constructed. It's like paint on a marble—smooth but not flat—so, no origami. But let's take a snapshot and at least write the formulated surface area.

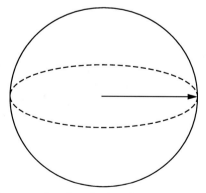

That is, the surface area of a sphere is given by

$$SA = 4\pi r^2.$$

Example 9.8.10 Surface of a Sphere

The surface area of a sphere with radius **2** *in* is

$$SA = 4\pi \times 2^2 = 8\pi \approx 25.12 \; in^2.$$

Common Measures of Area

Square	$A = s^3$
Rectangle	$A = lw$
Surface of a rectangular solid	$SA = 2lw + 2lh + 2wh$
Triangle	$A = \frac{1}{2}bh$
Heron's formula	$A = \sqrt{s(s-a)(s-b)(s-c)}$ $s = \frac{a+b+c}{2}$
Surface of a right circular cylinder	$SA = 2\pi r^2 + 2\pi rh$
Surface of a sphere	$SA = 4\pi r^2$

EXERCISES

Area: Two-dimensional measurement

9.8.1. Find the area of a square with sides measuring 6 in.

9.8.2. Find the area of a square with sides measuring x cm.

9.8.3. Find the area of a rectangle with length 2 ft and width 3 ft.

9.8.4. Find the area of a rectangle with length x cm and width y cm.

9.8.5. Find the area of a rectangle with length x cm and width y cm.

9.8.6. Find the area of a triangle with base x cm and height y cm.

9.8.7. Find the area of an equilateral with side measuring 6.2 mm.

9.8.8. Find the area of an equilateral with sides measuring **s** units.

9.8.9. Find the area of a quadrilateral with sides measuring 2 in, 4 in, 8 in, and 6 in.

9.8.10. Find the area of a quadrilateral with sides measuring x_1, x_2, x_3, x_4 measured in ft.

Surface area: Three-dimensional constructions

9.8.11. Find the surface area of a rectangular solid with length 4 units, width 13 units, and height 9 units.

9.8.12. Find the surface area of a rectangular solid with length 5 in, width 7 in, and height 8 in.

9.8.13. Find the surface area of a rectangular solid with length l ft, width w ft, and height h ft.

9.8.14. Find the surface area of a right circular cylinder with radius 2 mm and height 5 mm.

9.8.15. Find the surface area of a right circular cylinder with radius 4 units and height 7 units.

9.8.16. Find the surface area of a right circular cylinder with radius r cm and height h cm.

9.9 VOLUMES

In the final section of the basic aspects of geometry, we will learn how to find volumes of **cubes**, **rectangular solids**, **right circular cylinders**, **cones**, and **spheres**.

Recall, volume measures the number of cubes contained within each figurative space: inches cubed, cube feet, and meters cubed; abbreviated as in^3, ft^3, and m^3. Here, volume is abbreviated V and is three-dimensional. Moreover, each area will be formulated using the three perpendicular one-dimensional distances or a single one-dimensional object and an orthogonal two-dimensional object. Three-dimensional objects such as **pyramids**,

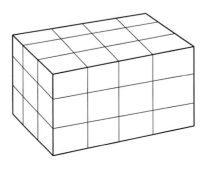

rectangular solids, and **cubes** are measured in terms of the number of cubes contained.

Here, volume *V* will be formulated using the three individual one-dimensional distances; or in terms of a single one-dimensional distance and an orthogonal plane-area. Note that orthogonal in multiple-dimensions is comparable to perpendicular in two dimensions.

Let's first extend a square to a cube—like a die. Then extend a rectangle to a rectangular solid and finally a circle to a sphere.

The basic cube will be the units by which we measure volume.

Rule 9.9.1 Volume of a Cube

A square of any size is self-defining—if we stack *s* squares *s* times, then the area is $s \times s$. When we stack these squares *s* deep, if *V* is the volume of the cube with side length *s* then *V* is s^3. That is,

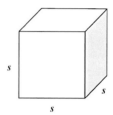

$$V = s^3.$$

Example 9.9.1 Volume of a Cube

The volume of a cube with side 5 *ft* is

$$V = 5^3 = 125 \, ft^3.$$

This idea can easily be extended to rectangles.

Rule 9.9.2 Volume of a Rectangle Solid

By stacking (a base layer) $l \times w$ of rectangles *h* times to create a specific height, then the **volume of a rectangular solid** is $l \times w \times h$. That is,

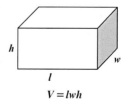

$$V = lwh$$

Example 9.9.2 Volume of a Solid

The volume of a rectangular solid that is **5 *in* × 2 *in* × 3 *in*** is

$$V = 5 \times 2 \times 3 = 30 \, in^3.$$

Example 9.9.2 Volume of a Solid—cont'd

Note: the volume of a cube and of the rectangular solid is the base area times the height. This idea can also be extended to shape with right side—this is what can be called the cookie cutter effect.

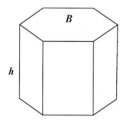

Rule 9.9.3 Volume of a Cookie Cutter Shape

Given the base area, **B**, of a figure with vertical height, **h**, a cookie cutter shape; the formula for the **volume of a cookie cutter shape** is the base area times the height; that is,

$$V = Bh$$

Example 9.9.3 Volume of a Solid

The volume of a cookie cut with a heart shape with base **16** cm^2 and a 1.5 cm depth is

$$V = 16 \times 1.5 = 24 \ cm^3.$$

Rule 9.9.4 Volume of a Right Circular Cylinder

Similar to the volume of a cookie cutter shape, the base area of a right circular cylinder is a circle and therefore we can formulate the **volume of a right circular cylinder** written as

$$V = \pi r^2 h.$$

Example 9.9.4 Volume of a Cylinder

The volume of a right circular cylinder with radius **2 *in*** and height **3 *in*** is

$$V = \pi \times 2^2 \times 3 = 12\pi \approx 37.58 \ in^3.$$

This formulated volume comes from the fact that a cone fits into a right circular cylinder—this is not a cookie cutter shape, but is part of the right circular cylinder.

Rule 9.9.5 Volume of a Cone

In fact, the **volume of a cone** is exactly one-third of the right circular cylinder; that is,

$$V = \frac{1}{3}\pi r^2 h.$$

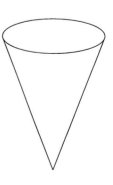

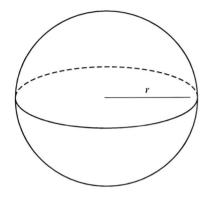

> **Example 9.9.5 Volume of a Cone**
>
> The volume of a right circular cone with radius **2 in** and height **3 in** is
>
> $$V = \frac{1}{3}\pi \times 2^2 \times 3 = 4\pi \approx 12.56 \ in^3.$$
>
> This volume comes from the fact that a sphere has only one measurement—the radius—this is not a cookie cutter shape, but is related to the radius cubed.

> **Rule 9.9.6 Volume of a Sphere**
>
> The **volume of a sphere** is about four cubes; that is,
>
> $$V = \frac{4}{3}\pi r^3.$$

> **Example 9.9.6 Volume of a Sphere**
>
> The volume of a sphere with radius **2 in** is
>
> $$v = \frac{4}{3}\pi \times 2^3 = \frac{64}{3}\pi \approx 66.99 \ in^3.$$

> **Common Measures of Volume**
>
> | | |
> |---|---|
> | Cube | $V = s^3$ |
> | Rectangular solid | $V = lwh$ |
> | Right circular cylinder | $V = \pi r^2 h$ |
> | Circular cone | $V = \frac{1}{3}\pi r^2 h$ |
> | Sphere | $V = \frac{4}{3}\pi r^3$ |

EXERCISES

Volume: Three-dimensional measurement

9.9.1. Find the volume of a cube with side edges measuring 6 mm.

9.9.2. Find the volume of a cube with side edges measuring *x* units.

9.9.3. Find the volume of a rectangular solid with length 2 in, width 3 in, and height 5 in.

9.9.4. Find the volume of a rectangular solid with length *x* units, width *y* units, and height *z* units.

9.9.5. Find the volume of an equilateral cookie with side measuring 6, see below.

Equilateral ⟶   ⟵ Cookie cutter

9.9.6. Find the volume of an equilateral cookie with sides measuring *s* cm, see above.

9.9.7. Find the volume of a right circular cylinder with radius 2 mm and height 5 mm.

Circular ⟶
Cylinder ⟶   ⟵ Cookie cutter
(Right) ⟶

9.9.8. Find the volume of a right circular cylinder with radius *r* units and height *h* units.

CRITICAL THINKING AND BASIC EXERCISE

9.1. **Definitions:** Define the following terms: **Area**, **Distance**, **Measurements**, **Pi**, and **Volume.**

Round the following to the indicated place value
9.2. 1.121314 to the hundredths place
9.3. 2.22222 to the units place
9.4. 7.13131313 to the thousandths place
9.5. 1.4197 to the second decimal place

Distance: One-dimensional measurement
9.6. Find the perimeter of a square with sides measuring 4 in.
9.7. Find the perimeter of a square with sides measuring y units.
9.8. Find the perimeter of a rectangle with length 5 cm and width 8 cm.
9.9. Find the perimeter of a rectangle with length l mm and width w mm.
9.10. Find the perimeter of an equilateral with side measuring 9 units.
9.11. Find the perimeter of an equilateral with sides measuring x ft.
9.12. Find the perimeter of a quadrilateral with sides measuring 1, 4, 9, and 12 measured in centimeters.
9.13. Find the perimeter of a quadrilateral with sides measuring a,b,c,d.

Area: Two-dimensional measurement
9.14. Find the area of a square with sides measuring 4 in.
9.15. Find the area of a square with sides measuring y units.
9.16. Find the area of a rectangle with length 5 mm and width 8 mm.
9.17. Find the area of a rectangle with length l units and width w units.
9.18. Find the area of an equilateral with side measuring 9 ft.
9.19. Find the area of an equilateral with sides measuring x units.
9.20. Find the area of a quadrilateral with sides measuring 1, 4, 9, and 12 measured in inches.
9.21. Find the area of a quadrilateral with sides measuring a,b,c,d measured in inches.

Surface area: Three-dimensional constructions
9.22. Find the surface area of a rectangular solid with length 3 in, width 11 in, and height 10 in.
9.23. Find the surface area of a rectangular solid with length 6, width 7, and height 9 units.
9.24. Find the surface area of a rectangular solid with length l, width w, and height h feet.
9.25. Find the surface area of a right circular cylinder with radius 3 mm and height 5 mm.
9.26. Find the surface area of a right circular cylinder with radius 4 in and height 10 in.
9.27. Find the surface area of a right circular cylinder with radius r and height h units.

Volume: three-dimensional measurement
9.28. Find the volume of a cube with side edges measuring 4 in.
9.29. Find the volume of a cube with side edges measuring s units.
9.30. Find the volume of a rectangular solid with length 12, width 13, and height 15 measured in inches.
9.31. Find the volume of a rectangular solid with length x, width y and height z measured in inches.
9.32. Find the volume of an equilateral cookie with side measuring 5 in and height of 4 in, see below.

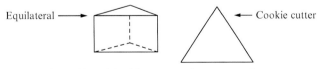

Equilateral ⟶   ⟵ Cookie cutter

9.33. Find the volume of an equilateral cookie with sides measuring x, see above.
9.34. Find the volume of a right circular cylinder with radius 3 and height 7.

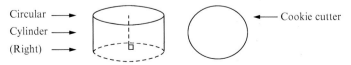

Circular ⟶
Cylinder ⟶   ⟵ Cookie cutter
(Right) ⟶

9.35. Find the volume of a right circular cylinder with radius r and height h.

Properties of Lines: One-Sided Figures

TRUE/FALSE: if false, give corrected statement(s).

9.36. The angle measuring 90° forms a straight line and is therefore called a straight angle.

9.37. Two angles which form a linear pair are supplementary and sum to 180°.

9.38. Two lines that are perpendicular must intersect.

9.39. Two lines that intersect form equal opposite angles and consecutive angles form a linear pair.

9.40. Two parallel lines never intersect

9.41. Two angles are complementary if form a straight angle (line)

9.42. The line that cuts across two parallel lines is called the transverse line.

Properties of Angles: Two-Sided Figures

Name the angle as Acute, Right, Obtuse, or Straight.

9.43. 90°

9.44. 70°

9.45. 135°

9.46. 180°

Properties of Triangles: Three-Sided Figures

For the following set of questions, define the triangle according to angle type: Acute, Right, and Obtuse

9.47. A triangle with interior angles measuring 45°, 45°, 90°.

9.48. A triangle with interior angles measuring 120°, 20°, 40°.

9.49. A triangle with interior angles measuring 30°, 120°, 30°.

9.50. A triangle with interior angles measuring 70°, 50°, 60°.

9.51. A triangle with interior angles measuring 60°, 60°, 96°.

For the following set of questions, define the above triangles in terms of: Scalene, Isosceles, and Equilateral.

9.52. A triangle with interior angles measuring 45°, 45°, 90°.

9.53. A triangle with interior angles measuring 120°, 20°, 40°.

9.54. A triangle with interior angles measuring 30°, 120°, 30°.

9.55. A triangle with interior angles measuring 70°, 50°, 60°.

9.56. A triangle with interior angles measuring 60°, 60°, 96°.

For the following set of questions refer to the figure below $\Delta ABC \sim \Delta DEF$

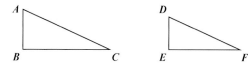

9.57. Given $AB = 4$, $BC = 5$, and $DE = 2$, find EF.

9.58. Given $AB = 6$, $BC = 7$, and $DE = 3$, find EF.

For questions 9.148 thru 9.152, TRUE/FALSE

9.59. $\dfrac{BC}{AB} = \dfrac{DE}{EF}$

9.60. $\dfrac{AC}{AB} = \dfrac{DF}{DE}$

9.61. $\dfrac{DE}{AB} = \dfrac{EF}{BC}$

9.62. $\dfrac{AB}{BC} = \dfrac{DF}{EF}$

9.63. $\dfrac{AB}{BC} = \dfrac{EF}{DE}$

For the following set of question, fill in the inequality or equality, assuming

$$m\angle A > m\angle B > m\angle C$$

9.64. $m\angle F$____$m\angle E$____$m\angle D$
9.65. $\overline{AB}$____$\overline{AC}$
9.66. $\overline{DE}$____$\overline{DF}$
9.67. $\overline{AB}$____$\overline{BC}$
9.68. $\overline{FE}$____$\overline{DF}$
9.69. If $AB = DE$, then these triangles are more than just similar, they are _____.

Properties of Quadrilaterals: Four-Sided Figures

TRUE/FALSE: if false, then give correct statement

9.70. A general quadrilateral has four sides.
9.71. A trapezoid has four sides with two sets of parallel sides.
9.72. In a parallelogram, opposite angles and sides, respectively, are equal.
9.73. In a parallelogram, consecutive angles are supplementary.
9.74. An equiangular quadrilateral is more commonly called a square.
9.75. In a rectangle, the diagonals are perpendicular bisectors and are equal.
9.76. In a rhombus, the diagonals are perpendicular bisectors and all sides are equal.
9.77. If the figure is a square, then the figure is a rhombus.
9.78. In a square, diagonals are equal.
9.79. In a square, all angles are right angles.
9.80. In a general polygon, the interior angles is $\dfrac{(n-2)\times 180°}{n}$, where n is the number of sides.

Properties of Polygons: More One-Dimensional Objects—Include Units

9.81. Find the sum of the interior angles of a regular polygon with eight sides; an octagon.
9.82. Find the number of sides in a regular polygon where each interior angle measures 120°.
9.83. Find the measure of each interior angle of a six-sided polygon; a hexagon.
9.84. Find the perimeter of a regular eight-sided polygon; an octagon with side measuring 10 m.
9.85. Find the perimeter of regular hexagon with side measuring 5 ft.
9.86. Find the perimeter of rectangle with length measuring 7 ft and height measuring 9 ft.
9.87. Find the perimeter of a square with side measuring 7 yards.
9.88. Find the circumference of a circle with radius measuring 3 in.
9.89. Find the circumference of a circle with diameter measuring 12 cm.
9.90. If the side of a square is increased by a multiple of 3, by what multiple does the perimeter increase?
9.91. If the side of a triangle is increased by a multiple of 3, by what multiple does the perimeter increase?
9.92. If the radius of a circle is increased by a multiple of 3, by what multiple does the circumference increase?
9.93. If the side of a general polygon is increased by a multiple of m, by what multiple does the perimeter increase?

Area/Surface Area: More Two-Dimensional Objects—Include Units

9.94. Find the area of a quadrilateral with sides measuring 4 m, 3 m, 5 m, and 8 m. Hint: Use Heron's Formula
9.95. Find the area of the right triangle with sides measuring 3 km, 4 km, and 5 km.
9.96. Find the area of a triangle with base measuring 17 in and height measuring 20 in.
9.97. Find the area of a rectangle with length measuring 8 ft and height measuring 5 ft.
9.98. Find the area of a circle with radius 5 in.
9.99. If the sides of a square are increased by a multiple of 3, by what multiple does the area increase?
9.100. If the sides of a rectangle are increased by a multiple of 6, by what multiple does the area increase?
9.101. If the radius of a circle is increased by a multiple of 10, by what multiple does the area increase?
9.102. If any polygon is magnified by a multiple of m, by what multiple does the area increase?

9.103. Find the surface area of the rectangular solid with length measuring 4 in, width measuring 10 in, and height measuring 5 in.

9.104. Find the surface area of the rectangular solid with length measuring 8 cm, width measuring 1 cm, and height measuring 6 cm.

9.105. Find the surface area of an open box with length measuring 8 cm, width measuring 1 cm, and height measuring 6 cm.

9.106. Find the surface area of a right circular cylinder with radius measuring 4 units and height measuring 10 units.

9.107. Find the surface area of a right circular cylinder with both the diameter and the height measuring 8 units.

9.108. Find the surface area of a sphere with radius measuring 3 yards.

Volume: More Three-Dimensional Objects—Include Units

9.109. Find the volume of a cube with edges measuring 5 cm.

9.110. Find the volume of a cube with edges measuring 3 in.

9.111. Find the volume of a rectangular solid with length measuring 8 cm, width measuring 11 cm, and height measuring 16 cm.

9.112. Find the volume of a rectangular solid with length measuring 28 cm, width measuring 15 cm, and height measuring 6 cm.

9.113. Find the volume of a right circular cylinder with diameter measuring 14 units and height measuring 7 units.

9.114. If the sides of a cube are increased by a multiple of 3, by what multiple does the volume increase?

9.115. If the sides of a rectangular solid are increased by a multiple of 3, by what multiple does the volume increase?

9.116. If the sides of a rectangular solid are increased by a multiple of m, by what multiple does the volume increase?

Project

9.117. Make your rectangular solid using the Origami Design I layout. You will need scissors, but don't tape it yet.

9.118. Find the surface area of a rectangular solid with length l, width w, and height h. Label each length, width, and height in your rectangular origami design and give the formula for the surface area of this rectangular solid.

9.119. Find the surface area of a rectangular solid with length l, width w, and height h. Label each length, width, and height in your rectangular origami design and give the formula for the surface area of this rectangular solid.

9.120. If removing the top alters this box, then how does this affect the formula; that is, what is the surface area of a box without a top.

SUMMARY OF IMPORTANT CONCEPT

Definitions

9.2.1. On a straight line in one-dimensional space, the **distance** is the amount of space between two points.

9.2.2. In two-dimensional space, an **angle** is the amount of space between two intersecting lines or surfaces that meet.

9.2.3. In two-dimensional space, the **area** is the amount of space covered by a surface.

9.2.4. In three-dimensional space, the **volume** is the amount of space occupied by an object.

9.3.1. In one-dimensional space, a **line** is the long narrow mark or band. A **straight line** is a line that is true to its course and does not deviate or curve.

9.3.2. Two adjacent angles (or **linear pairs**) are **supplementary** if they sum to 180°.

9.3.3. Non-parallel lines, lines that cross paths at exactly one point are **intersecting lines**.

9.3.4. Special cases of intersecting lines, lines are **perpendicular** if they form four right angles.

9.3.5. Two adjacent angles are **complementary** if they sum to 90°.

9.3.6. Two lines are **parallel** if they never intersect. They are called parallel lines and are abbreviated by two non-intersecting bars: ||.

9.3.7. A **transverse** line is a line that cuts across other lines.

9.5.1. In two-dimensional space, the **triangle** is a plane figure with three straight sides and three angles.

9.5.2. Triangles are **similar** if the share the same internal ratios. In terms of angles, this means that corresponding angles are equal, but in terms of sides, this simply means that the sides are proportional; like an enlargement of a triangle on a copier. $\Delta ABC \sim \Delta DEF$ is read "triangle **ABC** is similar to triangle **XYZ**."

9.5.3. Congruent means that corresponding angles are equal, as well as corresponding sides are equal; like a reproduction or regular copy of your triangle.

9.6.1. In two-dimensional space, the **quadrilateral** is a plane figure with four straight sides and four angles.

9.6.2. In two-dimensional space, the **trapezoid** is a plane figure with four straight sides and four angles such that exactly one set of opposite sides are parallel.

9.6.3. In two-dimensional space, the parallelogram is a plane figure with four straight sides and four angles such that both sets of opposite sides are parallel.

9.6.4. In two-dimensional space, a **rhombus** is a type of parallelogram with four straight sides and four angles such that both sets of opposite sides are parallel and all sides are equal.

9.6.5. In two-dimensional space, a rectangle is a type of parallelogram with four straight sides and four angles such that both sets of opposite sides are parallel and all angles are equal.

9.6.6. In two-dimensional space, a **square** is a type of parallelogram with four straight sides and four angles such that both sets of all sides are parallel and all sides are equal.

9.6.7. In two-dimensional space, a **kite** is a type of quadrilateral with four straight sides and four angles such that the two upper sides are equal and the two lower sides are equal.

9.7.1. In two-dimensional space, the **polygon** is a plane figure with more than four straight sides with an equal number of angles.

9.7.2. In two-dimensional space, the **regular polygon** is a plane figure is a polygon with all equal sides and all equal interior angles.

9.7.3. The distance around such straight edged figures is simply the sum of the individual distances and is called the **perimeter** and is abbreviated P.

9.7.4. The distance around smooth curves is called the **circumference**; since these curve's perfection depends on it radiating the same in all directions from a centering point, circumference is often formulated in terms of the radius and is abbreviated C. The **circumference** of a circle with fixed radius r is $C = 2\pi r$, where and π is approximately 3.1415.

9.8.1. A **right circular cylinder** is a cylinder with circular bases parallel to each other.

Properties

9.3.1. Corresponding angles are formed when a transversal line crosses parallel lines, and the corresponding incident angles are equal $m \angle 1 = m \angle 5$, $m \angle 2 = m \angle 6$, $m \angle 3 = m \angle 7$, and $m \angle 4 = m \angle 8$, Figure 9.1.

9.3.2. Consecutive angles are supplementary. This is clear with adjacent angles which form linear pairs: $m \angle 1 + m \angle 2 = 180°$, $m \angle 1 + m \angle 3 = 180°$, etc.; and incorporating property 9.3.1, consecutive angles between the two parallel lines are supplementary $m \angle 3 + m \angle 5 = 180°$, etc., Figure 9.1.

9.3.3. Alternate interior angles are formed when a transversal line crosses parallel lines and we consider the angles on opposite sides of the transverse line and interior to the parallels. Alternate interior angles are equal: $m \angle 3 = m \angle 6$ and $m \angle 4 = m \angle 5$, Figure 9.1.

9.3.4. Alternate exterior angles are formed when a transversal line crosses parallel lines and we consider the angles on opposite sides of the transverse line and external to the parallels. Alternate exterior angles are equal: $m \angle 1 = m \angle 8$ and $m \angle 2 = m \angle 7$, Figure 9.1.

9.4.1. Two rays pointing in the same direction from the same initial point form a **0° angle**.

9.4.2. Two rays pointing in opposite directions from the same initial point appearing as a straight line form a **180° angle**.

9.4.3. A right angle is exactly **90° angle**.

9.4.4. Two adjacent angles that when combined form 90°, then the angles are **complementary**.

9.4.5. Two adjacent angles that when combined form 180°, then the angles are **supplementary**.

9.6.1. With respect to the **angles in a parallelogram**, opposite angles are equal: $\angle A = \angle B$ and $\angle C = \angle D$.

9.6.2. With respect to the **sides in a parallelogram**, opposite sides are parallel: $\overline{AD} \| \overline{BC}$ and $\overline{AB} \| \overline{DC}$; and opposite sides are equal: $\overline{AD} = \overline{BC}$ and $\overline{AB} = \overline{DC}$.

9.6.3. With respect to the **angles in a rhombus**, opposite angles are equal: $\angle A = \angle B$ and $\angle C = \angle D$.

9.6.4. With respect to the **sides in a rhombus**, all sides are equal; that is, $\overline{AD} = \overline{BC} = \overline{AB} = \overline{DC}$; and opposite sides are parallel: $\overline{AD} \| \overline{BC}$ and $\overline{AB} \| \overline{DC}$.

9.6.5. With respect to the **diagonals in a rhombus**, they are perpendicular to each other; that is, $\overline{AC} \perp \overline{BD}$.

9.6.6. With respect to the **angles in a rectangle**, they are all equal to 90°; that is, $\angle A = \angle B = \angle C = \angle D = 90°$.

9.6.7. With respect to the **sides in a rectangle**, opposite sides are equal; that is, $\overline{AD} = \overline{BC}$ and $\overline{AB} = \overline{DC}$; and opposite sides are parallel: $\overline{AD}\|\overline{BC}$ and $\overline{AB}\|\overline{DC}$.

9.6.8. With respect to the **diagonals in a rectangle**, they are equal to each other; that is, $\overline{AC} = \overline{BD}$.

9.6.9. With respect to the **angles in a square**, they are all equal to 90°; that is, $\angle A = \angle B = \angle C = \angle D = 90°$.

9.6.10. With respect to the **sides in a square**, all sides are equal; that is, $\overline{AD} = \overline{BC} = \overline{AB} = \overline{DC}$; and opposite sides are parallel: $\overline{AD}\|\overline{BC}$ and $\overline{AB}\|\overline{DC}$.

9.6.11. With respect to the **diagonals in a square**, they are equal to each other and perpendicular; that is, $\overline{AC} = \overline{BD}$ and $\overline{AC}\perp\overline{BD}$.

9.7.1. In general, there will also be $(n-2)$ triangles when there are n sides in a figure, therefore since each triangle measures 180°, we can formulate the total interior measure for any polygon with n sides, S_n. That is, the **sum of interior angles** is given by $S_n = (n-2) \times 180°$.

Rules

9.5.1. Triangles are **similar** if they share these common ratios:

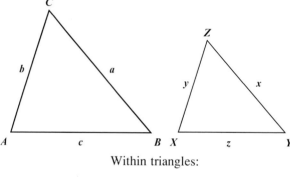

Within triangles:

$$\frac{a}{b} = \frac{x}{y} \qquad \frac{a}{c} = \frac{x}{z} \qquad \frac{b}{c} = \frac{y}{z}$$

and between triangles:

$$\frac{a}{x} = \frac{b}{y} \qquad \frac{a}{x} = \frac{c}{z} \qquad \frac{b}{y} = \frac{c}{z}$$

9.5.2. Corresponding parts of **congruent triangles** are congruent.

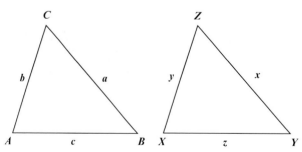

Triangles are congruent under the following conditions:

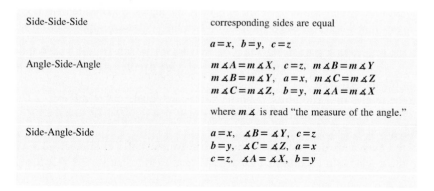

Side-Side-Side	corresponding sides are equal
	$a = x, \ b = y, \ c = z$
Angle-Side-Angle	$m\angle A = m\angle X, \ c = z, \ m\angle B = m\angle Y$
	$m\angle B = m\angle Y, \ a = x, \ m\angle C = m\angle Z$
	$m\angle C = m\angle Z, \ b = y, \ m\angle A = m\angle X$
	where $m\angle$ is read "the measure of the angle."
Side-Angle-Side	$a = x, \ \angle B = \angle Y, \ c = z$
	$b = y, \ \angle C = \angle Z, \ a = x$
	$c = z, \ \angle A = \angle X, \ b = y$

9.8.1. A square of any size is self-defining—if we stack s squares s times, then the area is $s \times s$. Therefore, if A is the **area of a square** with side length s then A is s^2; that is, $A = s^2$.

9.8.2. This idea can easily be extended to rectangles—if we stack (a base) b of squares h times to create a specific height, then the **area of a rectangle** is $b \times h$; therefore, if A is the area of a rectangle then it measures base times height. That is, $A = bh$.

9.8.3. The average width is b and h is perpendicular to the upper and lower base; therefore, the **area of a rhombus** is $b \times h$. Therefore, if A is the area of a rhombus then it is measured base times height. That is, $A = bh$.

9.8.4. In a trapezoid, the average base is the average of the lower base b and the upper base B; that is, average base $= \dfrac{b+B}{2}$; and the **area of a trapezoid** is obtained by $A = \frac{1}{2}(b+B)h$.

9.8.5. The formula above directly leads to the formula for the area of any general triangle. From the diagram below, the average base is the average of the lower base b and the upper base 0; that is, average base $= \dfrac{b+0}{2} = \dfrac{1}{2}$. The **area of a triangle** is $A = \frac{1}{2}bh$.

9.8.6. **Heron's method** for computing the area of any triangle with known sides a, b, and c can be computed as follows $A = \sqrt{s(s-a)(s-b)(s-c)}$, where $s = \dfrac{a+b+c}{2}$.

9.8.7. Area is measured using two perpendicular measurements. In general, the **area of a circle** is defined as its radius $\left(\dfrac{d}{2}\right)$ squared times π. That is, $A = \pi r^2$.

9.8.8. The **surface area** of flat-sided object in three-dimensional space is the sum of the areas present on each flat surface.

9.8.9. The **surface area, SA, of a right circular cylinder** is given by $SA = 2(\pi r^2) + (2\pi r)h$.

9.8.10. The surface area of a sphere is given by $SA = 4\pi r^2$

9.9.1. A square of any size is self-defining—if we stack s squares s times, then the area is $s \times s$. When we stack these squares s deep, if V is the **volume of the cube** with side length s then V is s^3. That is, $V = s^3$.

9.9.2. By stacking (a base layer) $l \times w$ of rectangles h times to create a specific height, then the **volume of a rectangular solid** is $l \times w \times h$. That is, $V = lwh$.

9.9.3. Given the base area, B, of a figure with vertical height, h, a cookie cutter shape; the formula the **volume of a cookie cutter shape** is the base area times the height; that is, $V = Bh$.

9.9.4. Similar to the volume of a cookie cutter shape, the base area of a right circular cylinder is a circle and therefore we can formula the **volume of a right circular cylinder** written as $V = \pi r^2 h$.

9.9.5. In fact, the **volume of a cone** is exactly one-third of the right circular cylinder; that is, $V = \dfrac{1}{3}\pi r^2 h$.

9.9.6. The **volume of a sphere** is about four cubes; that is, $V = \dfrac{4}{3}\pi r^3$.

Common: Types of Angles

Acute a cute kid is small, so this type of angle measures such that: $0° < m\angle A < 90°$.
Right which stand right on its own and right in the middle of the line, this type of angle measures $m\angle A = 90°$.
Obtuse meaning large, think of the word obese—this type of angle measures such that: $90° < m\angle A < 180°$.
Straight this type of angle is the same as a line and measures $m\angle A = 180°$.

Common: Types of Triangles

Scalene
  i. No equal angles
  ii. No equal sides
  iii. Comparisons in terms of angles are the same as comparisons in terms of sides

Isosceles
  i. Two equal angles
  ii. Two equal sides
  iii. Comparisons in terms of angles are the same as comparisons in terms of sides

Equilateral
  i. Three equal angles
  ii. Three equal sides
  iii. Comparisons in terms of angles are the same as comparisons in terms of sides

Common: Measures of Distance

Perimeter of triangle	$P = a + b + c$
Perimeter of rectangle	$P = 2b + 2h$
Circumference	$C = 2\pi r$

Common: Measures of Area

Square	$A = s^3$
Rectangle	$A = lw$
Surface of a rectangular solid	$SA = 2lw + 2lh + 2wh$
Trapezoid	$A = \dfrac{1}{2}(B + b)h$
Triangle	$A = \dfrac{1}{2}bh$
Heron's formula	$A = \sqrt{s(s-a)(s-b)(s-c)} \quad s = \dfrac{a+b+c}{2}$
Surface of a right circular cylinder	$SA = 2\pi r^2 + 2\pi rh$
Surface of a sphere	$SA = 4\pi r^2$

Common: Measures of Volume

Cube	$V = s^3$
Rectangular solid	$V = lwh$
Right circular cylinder	$V = \pi r^2 h$
Circular cone	$V = \dfrac{1}{3}\pi r^2 h$
Sphere	$V = \dfrac{4}{3}\pi r^3$

REVIEW TEST

1. Round the measurement 52.49 in to the nearest inch.
2. Round the measurement 3847 pounds to the nearest ten pounds.
3. Round the measure of the line segment AB to the nearest 1/4 in. $\overline{\text{A} \qquad\qquad\qquad \text{B}}$
4. What is the perimeter of the right triangle with legs length 6 and 8 cm.
5. A spherical tank of radius 2 yards can be filled for $100. How much would it cost to fill a spherical tank of radius 6 yards?

6. The lines shown in the diagram below lie in the same plane. TRUE/FALSE

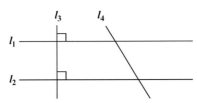

   i. Lines l_1 and l_2 are horizontal and parallel
   ii. Lines l_1 and l_3 are horizontal and parallel
   iii. Lines l_1 and l_3 are perpendicular
   iv. Lines l_3 and l_4 are intersecting lines

7. Determine the angles in the given diagram.

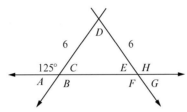

8. Determine the angles in the given diagram. Lines l_1 and l_2 are parallel.

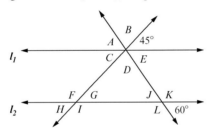

9. Can the sides of a right triangle be 6, 8, and 11?

10. Find the measure of angle A.

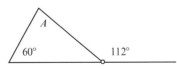

11. Name the geometric figure(s) that possesses all of the following characteristics:
   i. quadrilateral
   ii. opposite sides are parallel
   iii. diagonals are not equal

12. Name the geometric figure(s) that possesses all of the following characteristics:
   i. quadrilateral
   ii. opposite sides are parallel
   iii. all angles equal

REFERENCE

Miller, D., 1942a. Popular Mathematics: The Understanding and Enjoyment of Mathematics. WM. H. Wise & CO., INC, New York.

$$X + 7 = 10$$

$$X \quad = 10 - 7$$

$$X \quad = 3$$

Knowledge is a subset of both that which is true and believed.

PLATO

Chapter 10 Arithmetic and Algebra

Diophantus (200 AD)

Diophantus, 3rd Century AD, an Alexandrian Greek mathematician, is called the father of algebra. He has written a series of thirteen (13) books called "Arithmetica" that deals in solving algebraic equations. He studied at the University of Alexandria where he introduced the syncopated style of algebraic writing, he would write complex algebraic problems in symbolic form with one equation and several unknowns.

Diophantine equations are equations where the unknowns are restricted to integer values. Diophantus asked, given a value of n, how many distinct integer solutions exist?

*In the equation below, for **n = 4**, there are three distinct solutions: (5, 20), (6, 12) and (8, 8).*

$$\frac{1}{x} + \frac{1}{y} = \frac{1}{n}$$

The Diophantine equation below was the subject of Fermat's Last Theorem.

$$x^n + y^n = z^n$$

*When **n = 2**, the above equation is equivalent to the Pythagorean theorem which holds many solutions; however, for other values of n, are there infinitely many solutions, a finite solution, or none at all.*

Epitaph:

"Here lies Diophantus.
God gave him his boyhood one-sixth of his life;
One twelfth more as youth while whiskers grew rife;
And then yet one-seventh 'ere marriage begun.
In five years there came a bouncing new son;
Alas, the dear child of master and sage,
After attaining half the measure of his father's life, chill fate took him.
After consoling his fate by the science of numbers for four years, he ended his life."

10.1 INTRODUCTION TO THE REAL NUMBER SYSTEM

The real number system is a subset of the complex number system shown below:

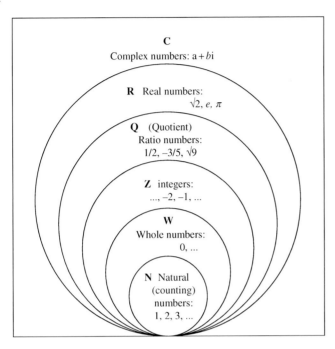

A real number is a value that represents a quantity along a continuum; that is, $\mathbb{R}$, the set of real numbers, consist of all values or points on an infinitely long number line.

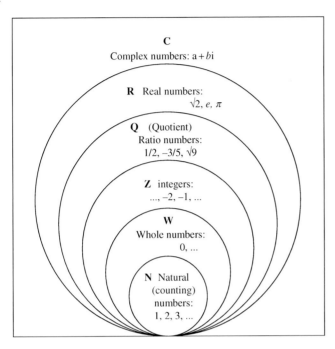

Real number line

This number line can be broken down to the various subsets: Rational, $\mathbb{Q}$, and Irrational, $\overline{\mathbb{Q}}$. The Rational numbers are divided into Integers, $\mathbb{Z}$, and Fractional or non-Integers, $\overline{\mathbb{Z}}$. Integers are divided into Natural numbers, $\mathbb{N}$, and negative Integers including zero, $\overline{\mathbb{N}}$. However, the creation of this number system was created in reverse.

Natural numbers are numbers used for counting. In ancient times, shepherds would send their sheep out to pasture and as the sheep exited the pens into the fields, the shepherd would tie a knot in a piece of rope. Then, when the sheep were gathered at the end of the day, the knots would be removed one by one and if there were any knots remaining at the end of the day, there were sheep lost in the thicket that needed to be found. Therefore, it is natural to state counting with one.

Definition 10.1.1 Natural Numbers

Natural numbers are numbers used for counting and ordering objects.

$$\mathbb{N} = \{1, 2, 3, 4, \ldots\}$$

Number

 One of a series of things distinguished by quantity or marked with numerals

 To count over one by one

Archival Note

A stone carving from Karnack dating about 1500 BC and on display at the Louvre in Paris, is one of the first known uses of a numeral system.

Archival Note

The use of a zero digit in place-value notation dates back to the Babylonians in 700 BC.

0
Zero
Nothing
Nota
Empty

Returning to our shepherds, if there were no knots left in the rope, then it was safe to assume that there were no lost sheep. This is the idea of **zero**. Both a number and a numerical digit, **zero** fills several central roles in arithmetic and algebra; in arithmetic, **zero** is used as a placeholder and in algebra it is the additive identity.

Definition 10.1.2 Zero

The word **zero** has had many derivations, but comes from the Arabic word *sifr* meaning "nothing" or "empty."

0

The natural numbers in conjunctions with zero constitute the set of whole numbers.

Definition 10.1.3 Whole Numbers

Whole numbers are numbers used for counting objects in a given set including the empty set, i.e., zero.

$$\mathbb{W} = \{0, 1, 2, 3, 4, \dots\}$$

Some say that **whole numbers** may refer to the **natural numbers**, the **natural numbers** including **zero** or possibly the **natural numbers**, **zero** and the **negatives** of all natural numbers. However, all distinguish whole numbers form fractions or fractional numbers such as 5.2.

The idea of positive and negatives numbers includes direction and magnitude. Magnitude is the numerical value or distance from the zero and direction is either to the left or right of zero. With positive values showing to the right of zero and negative numbers appearing to the left of **zero**; these whole numbers to the left and right of zero, including **zero**, are referred to as the **integers**.

Archival Note

*The letter $\mathbb{Z}$ for Integers stands for **Zahlen** which is German for "numbers."*

Definition 10.1.4 Integers

The word **integer** comes from Latin meaning 'untouched'. **Integers** are positive or negative whole numbers or natural numbers including zero.

$$\mathbb{Z} = \{\dots, -3, -2, -1, 0, 1, 2, 3, 4, \dots\}$$

The set of integers, while having many uses, is incomplete since fractions and fractional (non-whole) numbers are not included; the set of integers does not cover the continuum.

Real Number Line

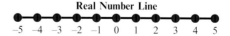

To fill in part of the number line, one set of numbers that can be added to the integers are all the fractional values; that is, all **rational** numbers.

Definition 10.1.5 Rational Numbers

Rational numbers are the numbers which can be expressed as the quotient or fraction with integers (with the non-zero denominator) or as the quotient of an integer and a natural number.

$$\mathbb{Q} = \left\{ \frac{m}{n} \middle| m \in \mathbb{Z} \text{ and } n \in \mathbb{N} \right\}$$

However, not all real numbers can be written as a ratio; for example, $\sqrt{2}$.

Proof that $\sqrt{2}$ is irrational:

Assume to the contrary that $\sqrt{2}$ is **rational** and there exist *m* and *n* such that the fractional expression $\frac{m}{n}$ is in reduced form. Hence, we have

$$\sqrt{2} = \frac{m}{n},$$

which implies (by squaring both sides and multiply both sides by n^2) that $m^2 = 2n^2$. Which means that m^2 is even, and therefore, it can be shown that *m* is also even. Let $m = 2k$, then we also have (using substitution and simplification) that $2n^2 = 4k^2$. Therefore, if $\sqrt{2}$ is rational, then both *m* and *n* are even and will reduce; a contradiction to the fact that $\frac{m}{n}$ is in reduced form. Hence, $\sqrt{2}$ is not rational or is **irrational**.

Definition 10.1.6 Irrational Numbers

Irrational numbers are the numbers which cannot be expressed as a ratio or rational number. The irrational numbers include transcendental numbers that cannot be represented as terminating or repeating decimals.

All these values in union form the set of **Real numbers**.

Definition 10.1.7 Reals

The set of **real numbers** are those rational and irrational numbers which constitute the continuum from $-\infty$ to ∞. That is, the union of the sets $\mathbb{R} = \mathbb{Q} \cup \overline{\mathbb{Q}}$

Real Number Line

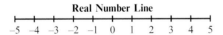

$$-5 \quad -4 \quad -3 \quad -2 \quad -1 \quad 0 \quad 1 \quad 2 \quad 3 \quad 4 \quad 5$$

Example 10.1.1 Natural Numbers

Determine which set each number belongs to the various sets and complete the following chart:

Real	Rational	Irrational	Integer	Whole	Natural
5					
$\sqrt{9}$					
$\sqrt{7}$					
π					
e					
$0.\overline{1234}$					
$\frac{2}{7}$					

$\mathbb{N} = \{1,2,3,\dots\}$
$\mathbb{W} = \{0,1,2,3,\dots\}$
$\mathbb{Z} = \{\dots-3,-2,-1,0,1,2,3\}$
$\mathbb{Q} = \{m/n | m \dots \text{SEE DEFINITION}$
Irrationals $= \mathbb{Q}'$

Solution

5   The first number **5** is one of the counting numbers, therefore, since **5** is a natural number, it is also a whole number, and an integer and a rational number as $5 = \frac{5}{1} = \frac{10}{2} = ...$

$\sqrt{9}$ The second number $\sqrt{9}$ is actually equal to **3**, which is one of the counting numbers, therefore, since **3** is a natural number, $\sqrt{9}$ is also a whole number, an integer and a rational number as $\sqrt{9} = 3 = \frac{3}{1} = \frac{6}{2} = ...$

$\sqrt{7}$ The third number $\sqrt{7} \approx 2.64575...$, a non-terminating and non-repeating decimal; therefore, $\sqrt{7}$ is irrational.

π The fourth number $\pi \approx 3.1415...$, a non-terminating and non-repeating decimal; therefore, π is irrational.

e The fifth number $e \approx 2.71...$, Euler's number, is a non-terminating and non-repeating decimal; therefore, e is irrational.

x The sixth number $x = 0.\overline{1234}$, is a repeating decimal and is therefore rational.

Proof that $0.\overline{1234}$ is rational:

Consider the value $x = 0.\overline{1234}$ and (since there are four repeating decimals), $10^4 x$:

$$10,000x = 1234.\overline{1234}$$

$$-x = -0.\overline{1234}$$

$$\overline{9999x = 1234}$$

$$x = \frac{1234}{9999}.$$

The last number $\frac{2}{7}$, is an already written as a ratio and is therefore rational.

	Real	Rational	Irrational	Integer	Whole	Natural
5		✓		✓	✓	✓
$\sqrt{9}$		✓		✓	✓	✓
$\sqrt{7}$			✓			
π			✓			
e			✓			
$0.\overline{1234}$		✓				
$\frac{2}{7}$		✓				

Now that we have our real number system, there are several basic operations that can be performed on this system. For example, addition, subtraction, multiplication, division, etc.

Definition 10.1.8 Operators

An **operator** is a function representing a mathematical operation. This can range from **conjunction** and **disjunction** in logic, to **intersection** and **union** in sets, to **addition**, **subtraction**, **multiplication**, **division**, and **exponentiation** in algebra.

Addition, **subtraction**, **multiplication**, **division**, and **exponentiation** in algebra are all binary operations; that is, they take two values form the real number system and manipulate these numbers to create another real number.

Definition 10.1.9 Addition

Addition is the mathematical operator that represents combining collections or sets of objects together into a larger collection or set; signified by the plus sign (+).

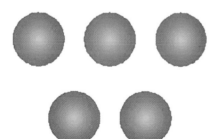

$3 + 2 = 5$

However, everything that can be done algebraically should be able to be undone. In mathematics, every operation we perform, can be inverted and the action undone. For addition, this operator is subtraction.

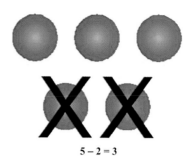

5 − 2 = 3

Definition 10.1.10 Subtraction

Subtraction is the inverse operation of addition; signified by the minus sign (−).

In general, the use of a dash or a negative sign indicates the inverse operation. We will see various forms of this sign in division (the inverse operation of multiplication) and with roots (the inverse operation of power or exponentiation).

Now we return to addition, in that addition leads directly to multiplication. For example, instead of writing $2 + 2 + 2 + 2$, that is, **2** added to itself four times, becomes **4×2**.

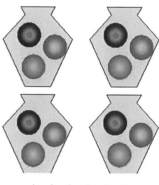

Definition 10.1.11 Multiplication

Multiplication is the mathematical operation of scaling one number by another; signified by the cross symbol or multiplication sign (×).

$$4 \times 3 = 3 + 3 + 3 + 3$$
$$= 4 + 4 + 4 = 12$$

This idea is readily seen when we are scaling by whole numbers as multiplication can be thought of as copies of the original set; however, we will extend this idea to fractions in the next section.

Definition 10.1.12 Division

Division is the inverse operation of addition; signified by the division sign (÷) or (as the division sign is not readily available on the keyboard), the backslash (/) or more appropriately, a division line as used in fraction; for example, $1 \div 2 = \frac{1}{2}$ is used.

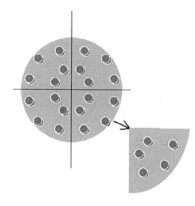

Note: −, ÷, the backslash (/) and the division bar are all forms of the dash.

Returning to multiplication, as addition leads to multiplication, multiplication leads to exponentiation. For example, instead of writing **2×2×2×2**, that is, **2** multiplied by itself four times, becomes 2^4. Note: whereas addition and multiplication have unique symbols, exponentiation is signified by its location, a superscript to the right of the base value.

$$20 \div 4 = \frac{20}{4} = 5$$

Definition 10.1.13 Exponentiation

Exponentiation is the mathematical operation involving two numbers, the base value, b, and the exponent, n. This heading covers both **powers** (n is a positive integer) and **roots** (n is one over a positive integer). Written b^n, when n is a positive integer, we have a **power** of the base,

$$b^n = \underbrace{b \times b \times \cdots \times b}_{n \ times}.$$

Hierarchy of Basic Operations (Including Powers).

	Addition	→	Multiplication	→	Powers
Notation	+		×, ·, () or positional		Strictly positional
	⇓		⇓		⇓
Inverse operation	Subtraction −		Division ——, ÷		Roots $/$, $\sqrt[m]{\ }$

Note: the symbol $\sqrt{\ }$ is a radical sign; it is NOT the square root sign. In general, if a number is placed on the radical sign such as **m**, as in our chart above this sign is read the **m**th root. Only if no number is written is this radical sign assumed to mean square root; the opposite of squaring.

Addition, Subtraction, Multiplication and Division can also be written in stack notation:

$$\begin{array}{r} 6 \\ +2 \\ \hline \end{array}$$

$$\begin{array}{r} 6 \\ \times 2 \\ \hline \end{array}$$

$$\begin{array}{r} 6 \\ -2 \\ \hline \end{array}$$

$$\begin{array}{r} 6 \\ /\,2 \end{array}$$ (BUT USE DIVISION SIGN)

Example 10.1.2 Basic Operators

Simplify the following operations:

a) $6+2$ c) 6×2 e) 6^2

b) $6-2$ d) $6\div2$

Solution

a) $6+2=8$

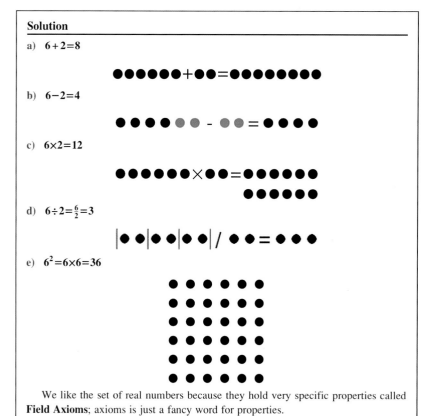

b) $6-2=4$

c) $6\times2=12$

d) $6\div2=\frac{6}{2}=3$

e) $6^2=6\times6=36$

We like the set of real numbers because they hold very specific properties called **Field Axioms**; axioms is just a fancy word for properties.

Rule 10.1.1 Sign Effects in +

When adding real numbers the **sign effect** on positive and negative values is as follows:
Like signs: Add and keep like sign.

$$(+4)+(+5)=+(4+5)=+9$$
$$(-4)+(-5)=-(4+5)=-9$$

Opposite signs: Subtract and keep sign of larger absolute value

$$(+4)+(-5)=-(5-4)=-1$$
$$(-4)+(+5)=+(5-4)=+1$$

Note: when you have opposite signs, you always subtract the smaller value from the larger value and follow the rule for the sign of the resulting value.

Rule 10.1.2 Sign Effects in ×

When multiplying positive and negatives real numbers the **sign effect** on positive and negative values is as follows:

Interpreting to the right as forward and to the left as negative, multiplying or dividing by a negative is like turning round **180°**.

$$(-1)\times(+1)=-1$$

Continuing with this geometric description, multiplying by a negative twice is that same as doing a 360 turn, facing the same (positive direction) as you started.

$$(-1)\times(-1)=+1$$

Hence, an even number of negatives multiply to be positive and an odd number of negatives multiply to be negative.

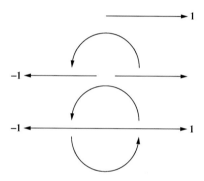

Using the two rules above, one can consider direction (positive or negative) to be separate from magnitude (value or distance from zero).

Field Axioms of Addition and Multiplication

1. Closure
2. Associative
3. Communicative
4. Identity
5. Inverse
6. Distributive

1. Closure

The set of real numbers are closed. That is if you take any two real numbers and add or multiply them, the resulting number is still a real number. Note: not all set or collections of numbers have this property.

Property 10.1.1 Closure

A set has **closure** if a and b are real numbers, then $a+b$ and $a\times b$ are also real numbers. If we let the symbol $\in$ stand for "is an element of" then this idea can be abbreviated further mathematically as follows:

if $a,b \in \mathbb{R}$, then $a+b \in \mathbb{R}$ and $a\times b \in \mathbb{R}$.

$a, b \in \mathbb{R}$

$a + b \in \mathbb{R}$

$ab \in \mathbb{R}$

Example 10.1.3 Closure

The real numbers are closed under addition; for example, **2** is an element of the real numbers and **3** is an element of the real numbers, in addition, $2 + 3 = 5$ is also an element of the real numbers. The real numbers are closed under multiplication; for example, **2** is an element of the real numbers and **3** is an element of the real numbers, in addition, $2 \times 3 = 6$ is also an element of the real numbers.

2. Association

In English, to associate means to connect with or keep company with, to relate to: under addition and multiplication, real numbers can also choose whom they keep company with or relate to; they can also re-associate at any time.

Property 10.1.2 Associative

In the additive expression: $(a + b) + c$, the number a is associating with the number b, but in the expression: $a + (b + c)$, the number b as been re-associated with the number c. And these associations are equivalent; similarly with multiplication, the expression $(a \times b) \times c$ is equivalent to $a \times (b \times c)$; the **associative** properties are written mathematically:

$$(a + b) + c = a + (b + c)$$

and

$$(a \times b) \times c = a \times (b \times c)$$

Note: In the multiplication above the operation of multiplication could have been written: $(ab)c = a(bc)$, where the multiplication is understood because of the positional notation used: a written next to b, as well as the by the use of parenthesis.

Example 10.1.4 Associative

The operators, addition and multiplication, are associative; $(2 + 5) + 4 = 2 + (5 + 4)$, they both simplify down to **11**. This ability to re-associate is also true for multiplication; $(2 \times 5) \times 4 = 2 \times (4 \times 5)$, the multiply out to be **10**, 10×4 or 2×20.

3. Communicative

In English, commute means to exchange one with another or to move, to go back and forth as in the word commuter: under addition and multiplication, real numbers can also commute or move.

Property 10.1.3 Commutative

In the additive expression: $a + b$, the number a is in the first position with the number b in the second position, but in the expression: $b + a$, the number a as been commuted with the number b. And these expressions are equivalent; similarly with multiplication $a \times b$ is equivalent to $b \times a$; the **commutative** properties are written mathematically:

$$a + b = b + a$$

and

$$a \times b = b \times a.$$

Example 10.1.5 Commutative

Under addition, $2+3=3+2=5$ and under multiplication, $2\times3=3\times2=6$.

$a, b \in \mathbb{R}$
then $a+b = b+a$
$ab = ba$

4. Identity

In English, identity means to make the same, and is related to the word identify meaning to recognize or point out: under addition and multiplication, real numbers can also be identified. Under addition, the identity is zero or **0**; and under multiplication, this is unity or **1**.

Property 10.1.4 Identity

In the additive expression: $a + 0$, when nothing is added to the number a, the number is not changed; it is the same number a and therefore **0** is the **additive identity**; similarly with multiplication, $a\times1$ is just one number a, which is equivalent to a and therefore **1** is the **multiplicative identity**; the **identity** properties are written mathematically:

$$a+0=0+a=a$$

and

$$a\times1=1\times a=a.$$

Example 10.1.6 Identity

Under addition, $2+0=0+2=2$ and under multiplication, $2\times1=1\times2=2$.

5. Inverse

In English, inverse means opposite in order or relation: under addition and multiplication, real numbers can be inverted or taken back to their identity. Under addition, the identity is **0**; hence the inverse is the negative of the value. Under multiplication, the identity is **1** hence the inverse is the reciprocal of the value.

Property 10.1.5 Inverse

In the additive expression: $a + (-a)$, $(-a)$ is the opposite (**additive inverse**) of a, which means if you add both a and $(-a)$, you have in essence have added nothing. Equivalently since subtraction is the opposite of addition, in the expression $a-a$ if it is understood that you added a and then subtracted a, you have in essence have added nothing. Hence, both of these expressions are equivalent to adding **0**.

Similarly with multiplication, in the expression $a\times\frac{1}{a}$, $\frac{1}{a}$ is the reciprocal (**multiplicative inverse**) of a, which means multiplying by both is in essence just identifying a given expression. Equivalently, since using power notation $a=a^1$ and division is the opposite of multiplication, the reciprocal can be denoted by $\frac{1}{a}=a^{-1}$; that is in the expression $a\times a^{-1}$, if it is understood that you multiplied and divided by a, you have in essence multiplied by **1**.

Therefore, the **inverse** properties are written mathematically:

$$a + (-a)=a-a=0$$

and

$$a\times\frac{1}{a}=a\times a^{-1}=1.$$

$$a + {}^-b = a - b$$

Example 10.1.7 Identity

Under addition, $2 + (-2) = 2 - 2 = 0$ and under multiplication,

$$2 \times \frac{1}{2} = 1.$$

$$-(-a) = a$$

Example 10.1.8 Identity

Under addition, $-5 + [-(-5)] = -5 + 5 = 0$ and under multiplication,

$$\frac{1}{5} \times \frac{1}{1/5} = \frac{1}{5} \times 5 = 1.$$

6. Distributive

In English, to distribute means to issue or dispense equally: in terms of addition and multiplication, since addition leads to multiplication, multiplication can be distributed equally over addition.

Property 10.1.6 Distribution

In the expression $a \times (b + c)$ the parentheses indicate that the multiplication by the number a is affecting both the number b in addition to number c; that is, this expression is equivalent to $a \times b$ in addition to $a \times c$. The number a needs to be **distributed** to both the number b in addition to the number c.

Similarly, this relation of distribution can also be seen between multiplication and powers: since multiplication leads to powers, powers can be distributed equally over multiplication. In the expression $(a \times b)^n$ the parentheses indicate that the power on n is affecting both the number a and the number b. We will learn in counting that "and" means "multiply" and therefore this expression is equivalent to a^n and b^n. The power b needs to be **distributed** to both the number a and the number b.

That is, the **distributive** properties are written mathematically:

$$a \times (b + c) = a \times b + a \times c$$

and

$$(a \times b)^n = a^n \times b^n.$$

Example 10.1.9 Distribute

Multiplication over addition,

$$5 \times (7 + 2) = 5 \times 7 + 5 \times 2$$

and powers over multiplication,

$$(3 \times 11)^2 = 3^2 \times 11^2.$$

We have discussed entirely the operations of addition and multiplication and their properties. We have also briefly introduced subtraction and division.

Rule 10.1.3 Subtractions vs Addition

Subtraction is the addition of the negative; written mathematically:

$$a - b = a + (^-b)$$

Example 10.1.10 − vs +

Subtraction as addition:

$$5 - 3 = 5 + (^-3) = 2.$$

Rule 10.1.4 Division vs Multiplication

Division is the multiplication of the reciprocal; written mathematically:

$$a \div b = a \times \frac{1}{b}.$$

Example 10.1.11 ÷ vs ×

Division as multiplication:

$$16 \div 4 = 16 \times \frac{1}{4} = 4.$$

Therefore, the properties of addition and multiplication can be extended to that of subtraction and division. Just be careful of what is being undone and realize that double negatives (just like in English) are a positive.

Field Axioms for Subtraction and Division

1. **Closure**: if $a, b \in \mathbf{R}$, then $a - b, a \div b \in \mathbf{R}$.
2. **Associative**
   Subtraction

$$(a - b) - c = [a + (-b)] + (-c) = a + [(-b) + (-c)] = a - (b + c)$$

   The number a is subtracting b and then subtracting c, that is the number a is subtracting b in addition to c.
   Division

$$(a \div b) \div c = a \div (b \times c)$$

   The number a is being divided by b and then divided by c, that is the number a is being divided by b and c. Recall, "and" means "multiply."
3. **Commutative**
   Subtraction

$$a - b = a + (-b) = (-b) + a = -b + a$$

   Be careful when playing with subtraction, if it helps, always write as addition of the negative or opposite.
   Division

$$a \div b = a \times \frac{1}{b} = \frac{1}{b} \times a.$$

   Be careful when playing with division, if it helps, remember that division is multiplication by the reciprocal.

Zero exist between the positive and negatives and can be consider both + and −.

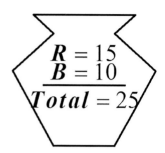

4. **Identity**

Subtraction

Since $0 = +0 = -0$, we have $a + 0 = a - 0 = a$

Division

Since $\frac{1}{1} = 1$, we have $a \div 1 = a \times 1 = a$.

5. **Inverse**

Subtraction

Since subtraction is adding the opposite, we have

$$a - a = a + (-a) = 0.$$

Division

Since dividing is multiplying by the reciprocal, we have

$$a \div a = a \times \frac{1}{a} = 1.$$

6. **Distributive**

Subtraction

Multiplication distributes over subtraction the same as over addition; that is,

$$a \times (b - c) = a \times b - a \times c.$$

Division

Powers distribute over division the same as over multiplication; that is,

$$(a \div b)^n = a^n \div b^n.$$

We have already introduced powers and the distributive property of powers over multiplication and division; we have also briefly discussed the inverse operation of powers, namely roots: collectively these operations are called exponentials. Recall also, that we discuss powers and roots of an arbitrary number a, this number is call the base of the exponential and for this reason is often abbreviated as b instead of a.

Recall: powers describe the number of times n a number a is multiplied by itself. For example: $4^3 = 4 \cdot 4 \cdot 4$, the power of 3 describes the number of times that the number 4 is multiplied by itself. And roots, the inverse operation of powers asks the reverse question: what number multiplied by itself m times (the root) is the number a; for example: $64^{1/3} = 4$, since 4 multiplied by itself 3 times is 64. Note: the 1 on the top of the bar or dash is simply a placeholder, just like 1 is used as a placeholder when taking the reciprocal of an integer.

And just as there are rational numbers that are not simple reciprocals of integers, we can have exponentials that are not just powers or roots, but a combination of both. For example, $64^{2/3}$ means the third (or cube) root as well as the power of two (or square.) In general, just as in multiplication by a fraction, it does not matter which is preformed first.

For example, written mathematically: in multiplication by a fraction, we see

$$12 \times \frac{5}{4} = \frac{12}{1} \times \frac{5}{4} = \frac{60}{4} = 15$$

or

$$12 \times \frac{5}{4} = \frac{12}{4} \times 5 = 3 \times 5 = 15.$$

And with an exponential, we have

$$8^{2/3} = \left(8^2\right)^{1/3} = 64^{1/3} = 4$$

or

$$8^{2/3} = \left(8^{\frac{1}{3}} \right)^2 = 2^2 = 4.$$

In some case one order is easier to perform than the other: for example in the example $64^{2/3}$, **64** squared is a large number and the cube root of large numbers are not easy to see. However, the cube root of **64** is **4** and **4** squared is **16**; this is much easier to compute without a calculator. The can be expressed mathematically as

$$64^{2/3} = \left(64^{\frac{1}{3}} \right)^2 = 4^2 = 16$$

Property 10.1.7 Exponentials

The **exponential** properties reduce multiplication between like bases to addition in the exponent:

$$a^x \times a^y = a^{x+y};$$

reduces division between like base to subtraction:

$$\frac{a^x}{a^y} = a^{x-y};$$

reduces powers of a base to a power to multiplication:

$$(a^x)^y = a^{x \times y}.$$

Moreover, $a^1 = a$ and $a^0 = 1$.

In the table below, find a comparison of properties between exponentials/multiplication and multiplication/addition.

Property of exponentials with like bases	Comments regarding exponential properties	Comparative property of multiplication
$a^n \cdot a^m = a^{n+m}$	Multiplication of (exponentials) taken down a level to addition	$n \cdot a + m \cdot a = (n+m) \cdot a$
$a^{\frac{1}{q}} \cdot a^{\frac{1}{m}} = a^{\frac{1}{q} + \frac{1}{m}}$		$\frac{1}{q} \cdot a + \frac{1}{m} \cdot a = \left(\frac{1}{q} + \frac{1}{m} \right) \cdot a$
$a^{\frac{\ell}{q}} \cdot a^{\frac{n}{m}} = a^{\frac{\ell}{q} + \frac{n}{m}}$		$\frac{\ell}{q} \cdot a + \frac{n}{m} \cdot a = \left(\frac{\ell}{q} + \frac{n}{m} \right) \cdot a$
$\frac{a^n}{a^m} = a^{n-m}$	Division (of exponentials) taken down a level to subtraction	$n \cdot a - m \cdot a = (n-m) \cdot a$
$\frac{a^{\frac{1}{q}}}{a^{\frac{1}{m}}} = a^{\frac{1}{q} - \frac{1}{m}}$		$\frac{1}{q} \cdot a - \frac{1}{m} \cdot a = \left(\frac{1}{q} - \frac{1}{m} \right) \cdot a$
$\frac{a^{\frac{\ell}{q}}}{a^{\frac{n}{m}}} = a^{\frac{\ell}{q} - \frac{n}{m}}$		$\frac{\ell}{q} \cdot a - \frac{n}{m} \cdot a = \left(\frac{\ell}{q} - \frac{n}{m} \right) \cdot a$
$(a^n)^m = a^{n \cdot m}$	Exponentials (of exponentials) taken down a level to multiplication	
$\left(a^{\frac{1}{q}} \right)^{\frac{1}{m}} = a^{\frac{1}{q \cdot m}}$		
$\sqrt[m]{\sqrt[q]{a}} = \sqrt[mq]{a}$		
$\left(a^{\frac{\ell}{q}} \right)^{\frac{n}{m}} = a^{\frac{\ell \cdot n}{q \cdot m}}$		

In the table below, find a comparison of properties between exponentials/multiplication and multiplication/addition.

Property of exponentials with unlike bases	Comments regarding exponential properties	Comparative property of multiplication/exponentials
$(a \cdot b)^n = a^n \cdot b^n$ $\sqrt[m]{a \cdot b} = \sqrt[m]{a} \cdot \sqrt[m]{b}$ $(a \cdot b)^{\frac{n}{m}} = a^{\frac{n}{m}} \cdot b^{\frac{n}{m}}$ $\sqrt[m]{a^n \cdot b^n} = \sqrt[m]{a^n} \cdot \sqrt[m]{b^n}$	Distributive property of powers over multiplication	$n \cdot (a+b) = n \cdot a + n \cdot b$
$\left(\frac{a}{b}\right)^n = \frac{a^n}{b^n}$	Distributive property of powers over division	$n \cdot (a-b) = n \cdot a - n \cdot b$

Property of Exponentials Defined Properties	Comments Regarding Exponential Properties	Comparative
$a^1 = a$	Exponential Identity	Multiplication of one a is a
$a^0 = 1$	Extension of Multiplicative Identity	Multiplication of zero a's is like multiplication by 1

Property 10.1.8 Order of Operation

The **order of operation** is the order in which the operations are performed in an expression. This is done in order of complexity from the most complex operations down to addition and subtraction: **parenthesis**, **exponents**, **multiplication**, **division**, **addition**, and **subtraction**.

Example 10.1.12 Order of Operations

Compute $2 \times (5-3) + 4^2 \times (4 \div 2)$.

Solution

$$2 \times (5-3) + 4^2 \times (4 \div 2) = 2 \times 2 + 4^2 \times 2$$
$$= 2 \times 2 + 16 \times 2 = 4 + 32 = 36.$$

EXERCISES

10.1.1. Determine which set each number belongs to the various sets and complete the following chart:

Real	Rational	Irrational	Integer	Whole	Natural
3					
$\sqrt{7}$					
$\sqrt{25}$					
$\frac{\pi}{2}$					
e^2					
$0.\overline{12}$					
$\dfrac{2}{\sqrt{7}}$					

10.1.2. Solve each of the following problems. Show your work.

    a. Roy planted 45 acres of wheat, 52 acres of corn, and 37 acres of peas. How many acres did he plant altogether?

    b. In the first five games of a football season, Chris gained 251 yards, 145 yards, 121 yards, 163 yards, and 164 yards. What is his total yardage?

10.1.3. Addition:

    a. 7 + 8 =

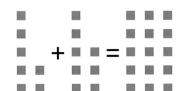

    b. 1 + 6 =

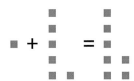

    c. 2 + 7 =

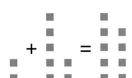

    d. 9 + 4 =

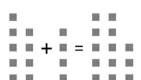

10.1.4. Addition:

    a.
$$\begin{array}{r} \square\ 6 \\ +\ 7 \\ \hline \square\ \square \end{array}$$

    d.
$$\begin{array}{r} 2\ 6 \\ +\ 7 \\ \hline \square\ \square \end{array}$$

    b.
$$\begin{array}{r} 1\ 5 \\ +\ 8 \\ \hline \square\ \square \end{array}$$

    e.
$$\begin{array}{r} 1\ 8 \\ +\ 7 \\ \hline \square\ \square \end{array}$$

    c.
$$\begin{array}{r} 1\ 3 \\ +\ 9 \\ \hline \square\ \square \end{array}$$

10.1.5. Addition:

a.
```
  5 6
+   7
─────
□ □
```

d.
```
□ 4 6
+ 5 9
───────
□ □ □
```

b.
```
  7 5
+   8
─────
□ □
```

e.
```
  9 8
+   7
─────
□ □
```

c.
```
□ 5 9
+ 7 4
───────
□ □ □
```

10.1.6. Addition:

a.
```
  1 5 9
+   7 4
───────
□ □ □
```

d.
```
  7 0 9
+   9 6
───────
□ □ □
```

b.
```
  6 4 6
+   5 9
───────
□ □ □
```

e.
```
  6 5 9
+   7 4
───────
□ □ □
```

c.
```
  3 4 7
+   7 7
───────
□ □ □
```

10.1.7. Solve each of the following problems. Show your work.
  a. A family started a trip with a full tank of gas and an odometer reading of 45,021. At the end of the trip, the odometer read 46,321. How many miles was the trip?
  b. A box of printer ink contains 12 cartridges. There are 7 cartridges left. How many cartridges have been used?
  c. A board $3\frac{3}{4}$ feet long was cut from a 12-foot board. How many feet were left?
  d. How much change do you receive from a $20 bill if the purchase totals $13.14?

10.1.8. Subtraction:

a.
```
□ 5 9
−   3 4
───────
□ □ □
```

d.
```
□ 8 9
−   9 6
───────
□ □ □
```

b.
```
□ 4 6
−   3 9
───────
□ □ □
```

e.
```
□ 5 9
−   7 4
───────
□ □ □
```

c.
```
□ 7 7
−   7 7
───────
□ □ □
```

10.1.9. Subtraction:

a.
```
  1 5 9
-   7 4
  ☐ ☐ ☐
```

d.
```
  7 0 9
-   9 6
  ☐ ☐ ☐
```

b.
```
  6 4 6
-   5 9
  ☐ ☐ ☐
```

e.
```
  6 5 9
-   7 4
  ☐ ☐ ☐
```

c.
```
  3 4 7
-   7 7
  ☐ ☐ ☐
```

10.1.10. Solve each of the following problems. Show your work.

a. A mess sergeant wants to serve $\frac{1}{4}$— pound hamburgers to 30 soldiers. How many pounds of hamburger are needed?

b. A copy of a figure 3 inches wide and 4 inches long is enlarged to 15 inches wide. How long is the enlargement?

c. Todd's car cost $9780. If his insurance company insures the car for 90% of its cost, how much will he receive for the car if it is totaled?

d. A cab company has 12 cars that need tunes-ups. Each car requires 8 spark plugs. How many spark plugs are required to complete the task at hand?

10.1.11. Multiplication:

a.
```
  ☐ 6
× 7
  ☐ ☐
```

d.
```
  2 6
× 7
  ☐ ☐
```

b.
```
  1 5
× 8
  ☐ ☐
```

e.
```
  1 8
× 7
  ☐ ☐
```

c.
```
  1 3
× 9
  ☐ ☐
```

10.1.12. Multiplication:

a.
```
  5 6
× 7
  ☐ ☐
```

d.
```
  ☐ 4 6
× 5 9
  ☐ ☐ ☐
```

b.
```
  7 5
× 8
  ☐ ☐
```

e.
```
  9 8
× 7
  ☐ ☐
```

c.
```
  ☐ 5 9
× 7 4
  ☐ ☐ ☐
```

10.1.13. Solve each of the following problems. Show your work.

    a. Deb wants to save $360 this year for holiday presents. How much should she save each month?

    b. One inch is approximately 2.54 cm. How many inches are in 6 cm?

    c. If pens cost $2.50 for 10, how much will 130 pens cost?

10.1.14. Division:

    a. Divide 25 dots into groups of 5.

$$\bullet$$

    Determine the number of groups: _____ $25 \div 5 =$

    b. Divide 24 raindrops into groups of 4.

$$\blacklozenge$$

    Determine the number of groups: _____ $24 \div 4 =$

    c. Divide 17 dots into groups of 3.

$$\maltese\maltese\maltese\maltese\maltese\maltese\maltese\maltese\maltese\maltese\maltese\maltese\maltese\maltese\maltese\maltese\maltese$$

    Determine the number of groups: _____ and the number of snowflakes left over _____ $17 \div 3 =$ _____ **R** _____

10.1.15. Division: Fill in the missing numbers

    a. $2 + 2 + 2 + 2 + 2 + 2 =$ ____ ____ groups of the number ____ ____ $\div 6 = 2$

    b. $8 + 8 =$ _____ ____ groups of the number ____ _____ $\div 2 = 8$

    c. $7 + 7 + 7 + 7 + 7 =$ _____ ____ groups of the number ____ _____ $\div 5 = 7$

10.1.16. Division: Solve

    a. Samuel saw 12 bicycles parked at school in three equal groups locked to bike racks. How many bicycles are in each group?

    b. There are 48 apples in a bucket and you pull them out into 6 equal groups. How many apples are there in each group?

    c. Kaitlin stacked 35 library books into 7 equal stacks; how many books are in each stack?

10.1.17. Division: Divisibility by 2—a number is divisible by 2 if the number ends in an even number: 2, 4, 6, 8, or 0; that is, even. Determine if the given numbers are even.

    a. 52   **c. 24**   **e. 125**

    b. 81   **d. 17**   **f. 44**

10.1.18. Division: Divisibility by 3—a number is divisible by 3 if the digits' sum is divisible by 3. Determine if the numbers in the Problem 10.1.17 are multiples of three.

10.1.19. Division: Divisibility by 4—a number is divisible by 4 if the last two digits are divisible by 4. Determine if the numbers in the exercise Problem 10.1.17 are multiples of four.

10.1.20. Division: Divisibility by 5—a number is divisible by 5 if the number ends in a 5 or 0. Determine if the numbers in the exercise Problem 10.1.17 are multiples of 5.

10.1.21. Division: Divisibility by 9—a number is divisible by 9 if the digits sum is divisible by 9. Determine if the numbers in the exercise Problem 10.1.17 are multiples of three.

10.1.22. Simplify

    a. $-10 - 2$

    b. $-15 - (-2)$

    c. $(-2)(17)$

    d. $-\dfrac{144}{-6}$

10.1.23. Powers: Simplify
    a. 9^0 d. $(-9)^3$ f. $9^{3/2}$
    b. 9^2 e. $\left(\frac{1}{9}\right)^2$
    c. 9^{-1}

10.1.24. Powers: Simplify
    a. 4^0 d. $(-4)^3$ f. $4^{3/2}$
    b. 4^2 e. $\left(\frac{1}{4}\right)^2$
    c. 4^{-1}

10.1.25. Powers: Simplify
    a. 5^0 d. $(-6)^2$ f. $64^{3/2}$
    b. 7^2 e. $\left(\frac{3}{7}\right)^2$
    c. 2^{-1}

10.1.26. Powers: Simplify
    a. a^0 d. $(-x)^3$ f. $x^{6/2}$
    b. $a^n b^n$ e. $\left(\frac{3}{b}\right)^2$
    c. $a^2 a^5$

10.1.27. Roots: Write each perfect square using the power 2.
    a. **169**   c. **25**   e. **121**
    b. **361**   d. **529**   f. **400**

10.1.28. Roots: Find the square root of each value given in exercise 11.22.

10.1.29. Roots: Find the square root.
    a. $\sqrt{144}$ c. $\sqrt{49}$ f. $\sqrt{841}$
    b. $\sqrt{36}$ d. $\sqrt{900}$
                         e. $\sqrt{324}$

10.1.30. Roots: Find the value of **n**.
    a. $\sqrt{n}=1$ c. $\sqrt{n}=9$ f. $n^2=256$
    b. $\sqrt{n}=11$ d. $n^2=49$
                         e. $n^2=225$

10.1.31. Complete each of the equation: Simplify.
    a. $4+4+4+4=$ d. $a \times a^3 =$
    b. $e+e+e+e+e=$ e. $9g+7g+6g=$
    c. $3 \times 3^2 =$ f. $9f \times (4f \times 6f)=$

10.1.32. Complete each of the equation: Simplify.
    a. $3 \times 3^2 \times 3^3 =$
    b. $a \times a^3 \times a^5 =$
    c. $7j+(7j+6j)=$
    d. $5ab \times (4a \times 6b)=$

10.1.33. Determine which operation is being performed:
    a. $9 \times (4 \times 5)=(9 \times 4) \times 5$
    b. $6 \times 12=12 \times 6$
    c. $10+2=2+10$
    d. $2 \times 10+5 \times 10=(2+5) \times 10$
    e. $7+(8+3)=(7+8)+3$
    f. $23 \times 4=11 \times 4+12 \times 4$

10.1.34. Determine which operation is being performed:
    a. $10 \times 11-5 \times 11=(10-5) \times 11$
    b. $7 \times 3=3 \times 7$
    c. $8 \times 11+3 \times 11=(8+3) \times 11$
    d. $12+8=8+12$
    e. $6 \times 9-3 \times 9=(6-3) \times 9$
    f. $(12+9)+2=12+(9+2)$

10.2   BASIC ARITHMETIC

Now that we have introduced the basic idea of a base number and its properties, we will introduce several uses for various bases; the most common base being base 10. In general, we have ten fingers, we have ten toes, and we count in base 10; base 10 is the **common base**. However we use a variety of bases every day: computers use base two (sequences of zeros and ones or charges and non-charges); minutes and seconds are base 60 (60 seconds is a minute and 60 minutes is an hour, but then the base number changes when changing hours to days, days to weeks, weeks to months and months to years). This progression concludes with base 10 again, when going from years to decades (10 years), to centuries (100 years) to millenniums (1000 years.)

Note in the above discussion of various bases, it is assumed that you are familiar with the base ten notation. We abbreviated sixty as 60, assuming you understand that the zero is a place holder describing the number of singletons and the position of the 6 indicated groups of ten; that is 60 is the abbreviation for 6 groups of ten. Similarly, 10 (ten) is the abbreviation for 1 group of ten, 100 (a hundred) is the abbreviation for 10 groups of ten (or 10^2) and 1000 (a thousand) is the abbreviation for 100 groups of ten (or equivalently 10 groups of a hundred or 10^3)

Conversely, if we wanted to represent a fraction in terms of base ten, we will need to indicate that we do not have multiples of ten, but pieces of a singleton. In base ten, these pieces are described with "th" as a suffix. That is one out of ten pieces is a tenth, one out of a hundred pieces is a hundredth, and one out of a thousand pieces is a thousandth. But how do we abbreviate this idea?

Definition 10.2.1 Place Value

Positional notation or **place-value notation** is a method of representing numbers. A form of coding, this is part of the number system that uses symbols or position for the different orders of magnitude. In the decimal system, we have the "ones place," "tens place," "hundreds place," etc. (all powers of base 10).

There are many different coding systems for numbers as illustrated below. Modern day mathematics uses the Arabic numeral system with base 10 in conjunction with a decimal system. That is, the coding system has ten symbols known as digits: namely, 0, 1, 2, 3, 4, 5, 6, 7, 8, and 9. The decimal in this system will indicate the power of the base by its relative position to the decimal point.

Counting

English	Zero	One	Two	Three	Four	Five	Six	Seven	Eight	Nine
Arabic	0	1	2	3	4	5	6	7	8	9
Mayan	⊂⊃	•	••	•••	••••	▬	▬•	▬••	▬•••	▬••••
Roman		I	II	III	IV	V	VI	VII	VIII	IX
Chinese rods		丨	丨丨	丨丨丨	丨丨丨丨	丨丨丨丨丨	T	⊤	⊤	⊤
Chinese		一	二	三	四	五	六	七	八	九
Hindi	०	१	२	३	४	५	६	७	८	९
Babylonian	𒑊	𒐕	𒐖	𒐗	𒐘	𒐙	𒐚	𒐛	𒐜	𒐝
Egyptian		I	II	III	IIII	˥	⌐	⌐	=	⌐

In base ten, this distinction between the powers of ten and the fractions of ten is a decimal point. It looks like a period, similar to the dot used to indicate multiplication; however, it is written on the lower line and the dot used to indicate multiplication is written above the line toward the middle of the line or median of the number. Remember, this is an abbreviated language; if you write a symbol in the wrong position, you may inadvertently write something incorrectly.

Example 10.2.1 5468.125 Base 10

The value 5468.125 can be written as

$$5000 + 400 + 60 + 8 + 0.1 + 0.02 + 0.05$$
$$= 5 \times 10^3 + 4 \times 10^2 + 6 \times 10^1 + 8 \times 10^0$$
$$+ 1 \times 10^{-1} + 2 \times 10^{-2} + 5 \times 10^{-3}$$

10^3	10^2	10^1	10^0	.	10^{-1} $= \dfrac{1}{10}$	10^{-2} $= \dfrac{1}{100}$	10^{-3} $= \dfrac{1}{1000}$
$=1000$	$=100$	$=10$	$=1$		$=0.1$	$=0.01$	$=0.001$
5	4	6	8	.	1	2	5

An older computer base system is "**octal**," or base eight. As only eight symbols are required to write a number in base 8, we will use the first eight digits: 0, 1, 2, 3, 4, 5, 6, 7 and 8.

Example 10.2.2 5258 Base 10

Convert the value **5258** (*base 10*) into *base 8*.

Solution

To remove the powers of 8, divide the number by 8 until you have a remainder less than 8.

$$5258 \div 8 = 657R2$$
$$657 \div 8 = 82R1$$
$$82 \div 8 = 10R2$$
$$10 \div 8 = 1R2$$
$$5258\,(base\,10) = 1 \times 8^4 + 2 \times 8^3 + 2 \times 8^2 + 1 \times 8^1 + 2 \times 8^0 = 12,212\,(base\,8)$$

Example 10.2.3 1258.12 Base 8

Determine the value **1258.12** (*base 8*) as *base 10*.

Solution

8^3	8^2	8^1	8^0	8^{-1}	8^{-2}	8^{-3}
$=512$	$=64$	$=8$	$=1$	$=\dfrac{1}{8}$	$=\dfrac{1}{64}$	$=\dfrac{1}{512}$
1	2	5	8	1	2	0

$$1\times512 + 2\times64 + 5\times8 + 8\times1 + 1\times\frac{1}{8} + 2\times\frac{1}{64}$$

$$=512 + 128 + 40 + 8 + \frac{1}{8} + \frac{1}{32} = 688 + 0.125 + 0.03125$$

$$=688.1575\,(base\,10)$$

Another common base used in computer program (computer engineering and computer graphics) is hexadecimal or base sixteen. As sixteen symbols are required to write a number in base 16 and we need a single solitary "digit" that stands for "ten," "eleven," "twelve," "thirteen," "fourteen," and "fifteen"; hence, we will use the first ten digits: 0, 1, 2, 3, 4, 5, 6, 7, 8, and 9 as well as the letters: A (10), B (11), C (12), D (13), E (14) and F (15).

Example 10.2.4 6283 Base 16

Convert the value **7258** (*base* **10**) into *base* **16**.

Solution

To remove the powers of 16, divide the number by 16 until you have a remainder less than 16.

$$7258 \div 16 = 392R11$$

$$392 \div 16 = 24R8$$

$$24 \div 16 = 1R8$$

$$5258\,(base\,10) = 1\times16^3 + 8\times16^2 + 8\times16^1 + B\times16^0 = 188B\,(base\,16)$$

Example 10.2.5 A258 Base 16

Determine the value $A258$ (*base* **16**) as *base* **10**.

Solution

16^3	16^2	16^1	16^0	16^{-1}
$=4096$	$=256$	$=16$	$=1$	$=\dfrac{1}{16}$
A	2	5	8	0

$$10\times4096 + 2\times256 + 5\times16 + 8\times1$$
$$=40,960 + 512 + 80 + 8 = 41,560\,(base\,10)$$

As a convention, we will use base 10 and decimal notation. However, this number system has its drawbacks. For example, extremely large numbers or very small numbers such as 0.00000000000016; such numbers can be more easily be written in **scientific notation**.

Definition 10.2.2 Scientific Notation

A number is said to be in **scientific notation** if there is one non-zero number before the decimal, times a power of the base 10.

Scientific notation is an alternative way to write extremely large or extremely small numbers and is commonly used by mathematicians, engineers, and scientists (among others) and is an optional setting on most calculators.

Example 10.2.6 Scientific Notation

Write the given values in scientific notation:

a.   **200**   d.   **7,250,000**
b.   **5000**   e.   **0.000000000015**
c.   **−54,000**

Solution

a. $200 = 2.0 \times 10^2$
b. $5,000 = 5.0 \times 10^3$
c. $-54,000 = -5.4 \times 10^4$
d. $7,250,000 = 7.25 \times 10^6$
e. $0.000000000015 = 1.5 \times 10^{-11}$

An alternative to decimals are **fractions**.

Fractions can be represented several ways:

$$a/b$$
$$a \div b$$
or
$$b\sqrt{a}$$

Definition 10.2.3 Fraction

A **fraction** is a part of a whole or ratio of two numbers, the numerator (number of parts) divided by the denominator (how many parts in a whole).

➤ A **proper fraction** is one such that the numerator is less than the denominator.

➤ An **improper fraction** is one such that the numerator is greater than or equal to the denominator.

➤ Proper fractions added to whole numbers are called mixed numbers or **mixed fractions**.

A fraction in proper form represents a rational number between zero and one; $x \in \mathbb{Q} \cap (0, 1)$. To put this proper fraction into **reduced form**, divide the numerator and denominator by the **greatest common factor**. This can be done in stages if the larger number is not initially recognized or used, continue dividing until the numerator and denominator are **relatively prime**.

Definition 10.2.4 Greatest Common Factor

The **greatest common factor** (**GCF**) or **greatest common divisor** (**GCD**) is the largest possible positive integer that divides evenly into the numerator and the denominator.

Example 10.2.7 GFC Between 8 and 12

As $8 = 4 \times 2$ and $12 = 4 \times 3$, therefore the **GCF** is **4**.

To determine the **GCF**, we can use the prime factorization of the number and consider the common factors.

Definition 10.2.5 Prime/Composite

A **prime number** is a natural number greater than 1 that has 1 as its only positive divisor.

$$\{2, 3, 5, 7, 11, 13, 17, 23, \ldots\}$$

The opposite of a prime number is a **composite number**:

$$\{4 = 2 \times 2, 6 = 2 \times 3, 8 = 2 \times 2 \times 2, \ldots\}$$

Hence, **prime factorization** is the decomposition of a composite number into smaller primes.

All primes less than 1000:

2, 3, 5, 7, 11, 13, 17, 19, 23, 29, 31, 37, 41, 43, 47, 53, 59, 61, 67, 71, 73, 79, 83, 89, 97, 101, 103, 107, 109, 113, 127, 131, 137, 139, 149, 151, 157, 163, 167, 173, 179, 181, 191, 193, 197, 199, 211, 223, 227, 229, 233, 239, 241, 251, 257, 263, 269, 271, 277, 281, 283, 293, 307, 311, 313, 317, 331, 337, 347, 349, 353, 359, 367, 373, 379, 383, 389, 397, 401, 409, 419, 421, 431, 433, 439, 443, 449, 457, 461, 463, 467, 479, 487, 491, 499, 503, 509, 521, 523, 541, 547, 557, 563, 569, 571, 577, 587, 593, 599, 601, 607, 613, 617, 619, 631, 641, 643, 647, 653, 659, 661, 673, 677, 683, 691, 701, 709, 719, 727, 733, 739, 743, 751, 757, 761, 769, 773, 787, 797, 809, 811, 821, 823, 827, 829, 839, 853, 857, 859, 863, 877, 881, 883, 887, 907, 911, 919, 929, 937, 941, 947, 953, 967, 971, 977, 983, 991, 997

Example 10.2.8 GCF Using PF

As $8 = 2 \times 2 \times 2$ and $12 = 2 \times 2 \times 3$, therefore the prime factorizations have two 2's in common and therefore the GCF is $2 \times 2 = 4$.

Example 10.2.9 Reduce Fractions

Write the numerator and detonator in terms of its prime factorization and reduce the fraction.

a. $\dfrac{15}{48}$ b. $\dfrac{15}{42}$ c. $\dfrac{15}{25}$ d. $\dfrac{5}{14}$ e. $\dfrac{16}{20}$

Solution

a. $\dfrac{15}{48} = \dfrac{3 \times 5}{2 \times 2 \times 2 \times 2 \times 3} = \dfrac{5}{2^4} = \dfrac{5}{16}$

b. $\dfrac{15}{42} = \dfrac{3 \times 5}{2 \times 3 \times 7} = \dfrac{5}{2 \times 7} = \dfrac{5}{14}$

c. $\dfrac{15}{25} = \dfrac{3 \times 5}{5 \times 5} = \dfrac{3}{5}$

d. $\dfrac{5}{14} = \dfrac{5}{2 \times 7} = \dfrac{5}{14}$ is in reduced form.

e. $\dfrac{16}{20} = \dfrac{2 \times 2 \times 2 \times 2}{2 \times 2 \times 5} = \dfrac{2 \times 2}{5} = \dfrac{4}{5}$

Solution—cont'd

This process can be done in reverse to create fractions in higher terms; instead of dividing numerator and denominators by a positive integer, we can simply multiply by any number. This is useful when finding a common denominator.

Definition 10.2.6 Common Multiples

A **common denominator** (**CD**) is a common multiple of the denominator; the **least common denominator** is the **least common multiple** (**LCM**) between the given numbers.

Example 10.2.10 LCM Between 8 and 12

The multiples of **8** are **8, 16, 24, 32,** ... and the multiples of **12** are **24, 48, 72,** ..., therefore the **LCM** is **24**.

Example 10.2.11 Summing Fractions

Two-thirds of a cheese pizza and half a pepperoni pizza remain. How much pizza is left?

Solution

As a third of a cheese pizza is $\frac{1}{3}$ and half a pepperoni pizza is $\frac{1}{2}$, the total amount of pizza remaining is

$$\frac{2}{3}+\frac{1}{2}=\frac{4}{6}+\frac{3}{6}=\frac{7}{6}.$$

That is, there is more than one full pizza remaining; one full pizza (6/6) and one-sixth of a full pizza (1/6) assuming the pizza was cut into six equally-sized pieces.

The above answer is in improper form. To change an improper fraction to a whole number or mixed number, divide the denominator into the numerator and write the whole number with the remainder over the original denominator and reduce if necessary.

Example 10.2.12 Mixed Form

Write the following numbers in mixed form.

a. $\frac{12}{7}$ b. $\frac{15}{9}$ c. $\frac{27}{9}$

Solution

a. $\frac{12}{7}=1\frac{5}{7}$

$$\begin{array}{r} 1\ R5 \\ 7\overline{)12} \\ 7 \\ \hline 5 \end{array}$$

Continued

Solution—cont'd

b. $\dfrac{15}{9}=1\dfrac{6}{9}=1\dfrac{2}{3}$

$$\begin{array}{r} 1 \text{ R6} \\ 9\overline{)15} \\ -9 \\ \hline 6 \end{array}$$

c. $\dfrac{27}{9}=3\dfrac{0}{9}=3$

$$\begin{array}{r} 3 \text{ R0} \\ 9\overline{)27} \\ -27 \\ \hline 0 \end{array}$$

Rule 10.2.1 Operators in Fractions

Operators in Fractions: When adding fractions, the fractions must have a common denominator.

$$\frac{a}{b}+\frac{c}{d}=\frac{ad}{bd}+\frac{bc}{bd}=\frac{ad+bc}{bd}$$

However, when multiply we multiply top to top and bottom to bottom; that is,

$$\frac{a}{b}\times\frac{c}{d}=\frac{ac}{bd}.$$

When divided by a fraction, we multiply by the reciprocal,

$$\frac{a}{b}\div\frac{c}{d}=\frac{a}{b}\times\frac{d}{c}=\frac{ad}{bc}.$$

Example 10.2.13 × & ÷ Fractions

Simplify.

a. $\dfrac{3}{4}\times\dfrac{1}{2}$

b. $3\dfrac{2}{5}\times1\dfrac{7}{8}$

c. $\dfrac{2}{7}\div\dfrac{5}{6}$

d. $6\dfrac{1}{2}\div\dfrac{5}{8}$

Solution

a. $\dfrac{3}{4}\times\dfrac{1}{2}=\dfrac{3\times1}{4\times2}=\dfrac{3}{8}$

b. $3\dfrac{2}{5}\times1\dfrac{7}{8}=\dfrac{17}{5}\times\dfrac{15}{8}=\dfrac{17}{1}\times\dfrac{3}{8}=\dfrac{51}{8}=6\dfrac{3}{8}$

c. $\dfrac{2}{7}\div\dfrac{5}{6}=\dfrac{2}{7}\times\dfrac{6}{5}=\dfrac{12}{35}$

d. $6\dfrac{1}{2}\div\dfrac{5}{8}=\dfrac{13}{2}\div\dfrac{5}{8}=\dfrac{13}{2}\times\dfrac{8}{5}=\dfrac{13}{1}\times\dfrac{4}{5}=\dfrac{52}{5}=10\dfrac{2}{5}$

Note: all fractions can be written as a decimal. This can be done readily when the denominator is 10; 100; 1000, etc. The number of decimal places is given by the power of ten in the denominator. For example,

"four-tenths" is equivalent to $\dfrac{4}{10}=0.4$

"five-hundredths" is equivalent to $\dfrac{5}{100}=0.05$

Solution—cont'd

"42-thousandths" is equivalent to $\dfrac{42}{1000}=0.042$.

Otherwise, fractions can be divided out to find their decimal equivalent. For example,

$$\frac{3}{4}=\frac{75}{100}=0.75$$

or

$$\frac{1}{3}=0.333333=0.\overline{3}.$$

The second fraction shown above is a repeating decimal as indicated by the bar over the three in the second expression.

Rule 10.2.2 Operators in Decimals

Operators in Decimals:

➤ When **adding** or **subtracting decimals**, first line up the decimal points underneath each other; add or subtract decimals in the same way as with whole numbers; then bring down the decimal point straight down into the answer.

➤ When **multiplying**, perform the operation without the consideration of the decimal, as if the numbers are whole and then count over the total number of decimals in each number combined in the final result.

➤ When **dividing** decimals, move the decimal place to the left the same number of places in the numerator as the denominator; or use scientific notation.

Example 10.2.14 × & ÷ Decimals

Simplify.
a. 0.75×0.5 c. $0.45 \div 0.15$
b. 5.15×1.2 d. $7.5 \div 0.25$

Solution

a. 0.75×0.5

$$
\begin{array}{rr}
 & 7\ \ 5 \\
\times & 5 \\
\hline
 & 2\ \ 5 \\
3\ \ & 5 \\
\hline
3\ \ 7\ \ & 5 \\
\end{array}
$$

Hence, since there are two decimals in the first number and one is the second; there are a total of three in the final answer, **0.375**.

b. 5.15×1.2

$$
\begin{array}{rrrr}
 & \overset{1}{1} & \\
5 & 1 & 5 \\
\times & 1 & 2 \\
\hline
1 & 0 & 3 & 0 \\
5 & 1 & 5 & \\
\hline
6 & 1 & 8 & 0 \\
\end{array}
$$

Continued

Solution—cont'd

Hence, since there are two decimals in the first number and one is the second; there are a total of three in the final answer, **6.18**.

c. **0.45 ÷ 0.15**

$$\frac{0.45}{0.15} = \frac{45}{15} = 3$$

d. **7.5 ÷ 0.25**

$$\frac{7.5}{0.25} = \frac{750}{25} = 30$$

Definition 10.2.6 Ratio/Proportion

A **ratio** is a comparison of two numbers (preferably whole numbers) by division. Ratios can be expressed in three distinct ways:

$$\frac{a}{b} \text{ or } a \text{ to } b \text{ or } a : b,$$

where a and b are in reduced form.

A **proportion** is a statement that two ratios are equal.

Example 10.2.15 Using Ratios

The scale on a road map is 1 inch for every 80 miles. If the distance between two towns measures 1.75 inches; how many miles apart are the towns?

Solution

$$\frac{1 \text{ inch}}{80 \text{ miles}} = \frac{0.25 \text{ inches}}{20 \text{ miles}} = \frac{1.75 \text{ inches}}{140 \text{ miles}}$$

The towns are 140 miles apart.

As rational numbers are real numbers, they can be represented in base 10 in decimal form or fraction; however, when measuring ratios between 0 and 1, we might give the number as a percentage instead.

Definition 10.2.8 Percent

Percent means "hundredth" and is denoted by the percent sign %.

To change a number written in decimal form into a percentage, move the decimal point two places to the right and include the percent sign to the right of the new number. To change a number written in percentage form into a decimal, move the decimal point two places to the left and remove the percent sign.

Example 10.2.16 Percentages

Change the given decimal to a percent or percent to decimal:

a. **0.67**   d. **1.2%**

b. **0.002**   e. **5%**

c. **5**   f. **$5\frac{1}{2}\%$**

Solution

a. $0.67 = 67\%$

b. $0.002 = 0.2\%$

c. $5 = 500\%$

d. $1.2\% = 0.012$

e. $5\% = 0.05$

f. $5\frac{1}{2}\% = 5.5\% = 0.055$

Example 10.2.17 Fraction to Percent

Change the given fraction to a percent or percent to fraction:

a. $\dfrac{3}{4}$

b. $\dfrac{2}{3}$

c. $\dfrac{5}{2}$

d. 1.5%

e. 3%

f. $2\frac{1}{2}\%$

Solution

a. $\dfrac{3}{4} = \dfrac{3}{4} \times \dfrac{100}{1}\% = 75\%$ or $\dfrac{3}{4} = 0.75 = 75\%$

b. $\dfrac{2}{3} = \dfrac{2}{3} \times \dfrac{100}{1}\% = 66.\overline{6}\% \approx 66.7\%$ or $\dfrac{2}{3} = 0.66\overline{6} \approx 66.7\%$

c. $\dfrac{5}{2} = \dfrac{5}{2} \times \dfrac{100}{1}\% = 250\%$ or $\dfrac{5}{2} = 2.5 = 250\%$

d. $1.5\% = \dfrac{1.5}{100} = 0.015$

e. $3\% = \dfrac{3}{100} = 0.03$

f. $2\frac{1}{2}\% = 2.5\% = \dfrac{2.5}{100} = 0.025$

Example 10.2.18 Using Percentages

The Thompson family owns a restaurant with a market value of $225,500. Property in the area is assessed at 75% of the market value and the tax rate is $3 per $100 assessed value. How much is the tax on the property?

Solution

$$\textbf{Assessed value} = \textbf{75\% of Market value}$$

$$\textbf{Assessed value} = 0.75 \times \$225,500 = \$169,125$$

$$\$169,125 \times \dfrac{\$3}{\$100} = \$5073.75$$

That is, the property tax in this area for a restaurant worth $225,500 is $5073.75.

EXERCISES

10.2.1. Write the value of the underlined digit in words.

    a. 30.<u>5</u>   **c.** <u>1</u>81.2   **e.** <u>3</u>2.9

    b. <u>1</u>.62   **d.** 9.3<u>1</u>   **f.** 8215.1<u>2</u>5

10.2.2. Write each number in standard form.

    a. 8 hundred + 4 tens + 2 ones   **d.** four-tenths

    b. Eighty   **e.** 500 + 50 + 0.1

    c. Three hundred five

10.2.3. Write each decimal in words.

    a. 0.3   **d.** 0.89

    b. 57.80   **e.** 13.75

    c. 4.4

10.2.4. Convert the given number to base 10

    a. 1053 (base 6)   **c.** 324 (base 5)

    b. 27,510 (base 8)   **d.** $AB5$ (base 16)

10.2.5. Convert the given number to base 8

    a. 1053 (base 10)   **d.** $AB5$ (base 16)

    b. 27,510 (base 10)

    c. 324 (base 5)

10.2.6. Write the given number to scientific notation

    a. 1015,000   **d.** 125.45

    b. 0.05   **e.** 7451.25

    c. 19.521

10.2.7. Write the given number in standard decimal form:

    a. 1.24×10^4 **d.** 1.2×10^5

    b. 3.54×10^{-2} **e.** 3.42×10^{-7}

    c. 7.4×10^{-1}

10.2.8. Give the GCF for the following sets of numbers.

    a. 24, 15   **d.** 12, 18, 36

    b. 24, 60   **e.** 100, 400, 750

    c. 15, 60

10.2.9. Give the reduced form of the fraction.

    a. $\dfrac{12}{48}$ **d.** $\dfrac{5}{13}$

    b. $\dfrac{15}{90}$ **e.** $\dfrac{62}{93}$

    c. $\dfrac{18}{21}$

10.2.10. Sum and simplify.

    a. $\dfrac{1}{2} + \dfrac{4}{7}$ **d.** $\dfrac{2}{9} + \dfrac{4}{12}$

    b. $\dfrac{7}{9} + \dfrac{4}{6}$ **e.** $-\dfrac{1}{3} + \dfrac{4}{9}$

    c. $\dfrac{12}{17} + \dfrac{9}{17}$

10.2.11. Add $\dfrac{3}{8}, \dfrac{5}{12}$, and $\dfrac{7}{16}$.

10.2.12. Add $15\dfrac{1}{2}$ and $3\dfrac{5}{9}$.

10.2.13. Subtract and simplify.

    a. $\dfrac{1}{2} - \dfrac{4}{7}$ d. $\dfrac{2}{9} - \dfrac{4}{12}$

    b. $\dfrac{7}{9} - \dfrac{4}{6}$ e. $-\dfrac{1}{3} - \dfrac{4}{9}$

    c. $\dfrac{12}{17} - \dfrac{9}{17}$

10.2.14. Subtract $2\dfrac{5}{12}$ from **5**.

10.2.15. Subtract $2\dfrac{4}{9}$ from $6\dfrac{7}{9}$.

10.2.16. Solve the following problems. Show your work.

    a. An electrician needs $20\dfrac{3}{4}$ feet of wiring for one job and $32\dfrac{3}{8}$ feet for another job. How many feet of wiring are needed for both jobs?

    b. Michael worked $8\dfrac{3}{4}$ hours on Monday, $7\dfrac{3}{8}$ hours on Tuesday, $8\dfrac{1}{4}$ hours on Wednesday, **9** hours on Thursday and $7\dfrac{1}{2}$ hours on Friday. How many hours did he work during the week?

    c. Mrs. Miller is sewing 4 costumes for a school play. Each costume requires $1\dfrac{3}{4}$ yards of fabric. How much fabric did Mrs. Miller use in all?

10.2.17. Multiply or divide.

    a. $\dfrac{2}{3} \times \dfrac{5}{6}$ c. $\dfrac{4}{9} \times \dfrac{3}{4}$ e. $7 \div \dfrac{7}{9}$

    b. $3\dfrac{1}{2} \div 1\dfrac{3}{4}$ d. $2\dfrac{1}{8} \div 1\dfrac{1}{2}$ f. $1\dfrac{1}{5} \times 2\dfrac{3}{4}$

10.2.18. Solve the following problems. Show your work.

    a. A recipe for meat loaf serving 8 people calls for 2 pounds of hamburger. How much hamburger is needed to serve a family of 4?

    b. A cookie recipe calls for 2 1/2 cup of sugar and makes 120 mini cookies. How much brown sugar is needed for 360 mini cookies?

    c. Given wooden floorboards come in widths of 1/2 foot (6 inches), how many floorboards will be need to cover a hallway that is 66 inches wide?

10.2.19. Add or Subtract.

    a. $0.9 + 0.7$ d. $1.4 - 0.7$

    b. $0.5 + 0.9 + 0.7$ e. $6.54 - 2.345$

    c. $1.78 + 0.154 + 5 + 1.05$ f. $3.74 - 1.154$

10.2.20. Multiply or Divide.

    a. 0.9×0.7 d. $1.4 \div 0.7$

    b. $0.5 \times 0.9 \times 0.7$ e. 6.54×2.345

    c. $1.78 \div 5$ f. $3.74 \div 1.154$

10.2.21. Solve the following problems.

    a. Danielle purchased 6 shirts that were on sale for $12.49 each. What was the total purchase price?

    b. Ben earned $380.25 for working 39 hours last week. What was his rate per hour?

    c. Boxes of macaroni and cheese are on sale 5 for $1. How much do 12 boxes cost?

    d. Paint is on sale for $8.79, regular $10.99 per gallon; how much money is saved on 4 gallons of paint?

10.2.22. Solve the following problems:

    a. If a flash drive cost $9.75 per gigabit, how much do 4 gigabits cost?

    b. Nicholas can type 65 words per minute. How long will it take him to type a 1000 word report?

    c. How much paint is needed for 3200 square feet if one gallon covers 800 square feet?

10.2.23. Complete the chart. Reduce fraction to lowest terms.

	Fraction	Decimal	Percent
a.	$\dfrac{1}{5}$		
b.	$\dfrac{1}{3}$		
c.		0.25	
d.		2.75	
e.			9%
f.			0.5%
g.			2.25%
h.	$1\dfrac{3}{8}$		
i.		0.375	
j.			250%

10.2.24. Solve.
  a. 9 is what percent of 45?
  b. 36 is what percent of 72?
  c. What is 4% of 400?
  d. What is 12% of 108?
  e. 12 is what percent of 72?
  f. 25% of what number is 80?

10.2.25. Solve the following problems.
  a. Of 120,000 eggs in a chicken coop, 97% are not expected to hatch but rather be eaten. How many of the eggs are expect to be eaten?
  b. There are 140 new jobs available in Tampa this month. Thirty-two percent of these jobs are blue collar jobs. How many blue collar jobs are available in Tampa this month?
  c. How much money do you save on a lawn chair originally priced at $22.50 if you purchase this item on sale at 30% off?
  d. A flat screen TV was $2545 and is now selling for $1875. What is the percent saving?
  e. An item is marked as 33% off, and is now selling for $5.28. What was the original price?

10.2.26. The Carter family owns an orange groove with a market value of $275,000. Land in the area is assessed to be 45% of market value. The tax rate is $1.65 per $100 assessed value. How much is the tax on the land?

10.3 PATTERN RECOGNITION

Pattern recognition is the first step in solving many various problems. As with machine learning, pattern recognition assigns labels to input values, such as classification, in order to better understand the nature and order of things. Pattern recognition is based on logic. In logic, there are two methods of reasoning: deductive and inductive.

Deductive reasoning starts with the theory, followed by the construction of a hypothesis which is then tested by making observation and drawing a conclusion possibly confirming the initial hypothesis. By contrast **inductive reasoning** starts with making observations, then determining a pattern which in turn leads to a hypothesis which results in theory, moving from specific observations to broader generalities.

Definition 10.3.1 Inductive Reasoning

Inductive reasoning or induction is a kind of reasoning in which a small set of observations is used to infer a larger theory.

Example 10.3.1 Number of Squares

In a grid of squares $n \times n$, how many perfect squares are there of any size, $P(n)$?

Solution

The process starts with making observations with smaller values of n.

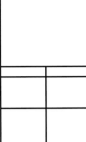

When $n=1$, we have only one square of size 1, hence $P(1)=1$.

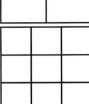

When $n=2$, we have one square of size 2 in addition to the four squares of size 1; hence $P(2)=1+4=5$.

When $n=3$, we have one square of size 3 in addition to the four squares of size 2 and nine squares of size 1; hence $P(3)=1+4+9=15$.

If you have not noticed the pattern yet, these are the sums of the square integers:

$$P(n)=1^2+2^2+\cdots+n^2=\frac{n(n+1)(2n+1)}{6}.$$

Therefore, a **4×4** square has 30 squares of various sizes.

$$P(4)=1+4+9+16=\frac{4\times5\times9}{6}=30$$

In ordering the possible number of squares as a function of the size of the largest square we have created a sequence.

Definition 10.3.2 Sequence

A **sequence** is an ordered list of objects, events or outcomes,

$$a_1, a_2, \ldots, a_n.$$

In the context of this chapter, a **sequence** will refer to an arrangement of numbers.

Sequence

A following of one outcome after another

A sequence consists of a list of outcomes, each individual outcome is a term in the sequence.

Definition 10.3.3 Term

A **term** is an individual outcome in a sequence enumerated as a function of the position in the sequence

$$a_n = P(n).$$

Example 10.3.2 Finding Terms

Identify the pattern in the sequence 1, 4, 7, 10, ... and find the next three terms and the 11th term.

$a_n = a_{n-1} + 3$
$$1$$
$$1 + 3 = 4$$
$$4 + 3 = 7$$
$$7 + 3 = 10$$

$a_n = 3n - 2$
$$3(1) - 2 = 1$$
$$3(2) - 2 = 4$$
$$3(3) - 2 = 7$$
$$3(4) - 2 = 10$$

Solution

Consider the differences between each term in the sequence: from 1 to 4, there is a difference of 3, from 4 to 7, again there is a difference of 3 and from 7 to 10 there is a difference of three. Hence, we have that any term is 3 more than the previous term:

$$a_n = a_{n-1} + 3,$$

and therefore the next three terms are 13, 16, and 19.

A sequence with common differences is called an arithmetic sequence and be written in recursive notation as above or in terms of the value n. As the difference of 3 is added $(n-1)$ times to the first term 1, we have that a_n can be expressed as

$$a_n = 1 + 3(n-1) = 3n - 2.$$

Therefore, $a_{11} = 3 \times 10 - 2 = 28$.

Definition 10.3.4 Arithmetic Sequence

An **arithmetic sequence** is a sequence of numbers such that there is a common difference between consecutive terms.
Recursive notation: $a_n = a_{n-1} + d$
Non-recursive notation: $a_n = a_1 + d(n-1)$

If instead of adding a common difference to create the next consecutive term, you multiplied, then this would be a geometric sequence.

Definition 10.3.5 Geometric Sequence

A **geometric sequence** is a sequence of numbers such that there is a common ratio between consecutive terms.
Recursive notation: $a_n = a_{n-1} \times r$
Non-recursive notation: $a_n = a_1 \times r^{n-1}$

Example 10.3.3 Finding Terms

Identify the pattern in the sequence 1, 2, 4, 8, ... and find the next three terms.

$a_n = 2 \times a_{n-1}$
$$1$$
$$2 \times 1 = 2$$
$$2 \times 2 = 4$$
$$2 \times 4 = 8$$

$a_n = 2^{n-1}$
$$2^{1-1} = 2^0 = 1$$
$$2^{2-1} = 2^1 = 2$$
$$2^{3-1} = 2^2 = 4$$
$$2^{4-1} = 2^3 = 8$$

Solution

Consider the ratio between each term in the sequence: between 2 and 1, there is a ratio of 2, between 4 and 2 there is a ratio of 2, and between 8 and 4 there is a ratio of 2. Hence, we have that any term is twice the previous term:

$$a_n = 2 \times a_{n-1},$$

and therefore the next three terms are 16, 32 and 64; or we could have written

$$a_n = 1 \times 2^{n-1} = 2^{n-1},$$

and therefore we have $a_5 = 2^4 = 16, a_6 = 2^5 = 32$ and $a_7 = 2^6 = 64$.

WARNING: There are many sequences that follow more complicated patterns; some even follow more than one.

Example 10.3.4 Double Pattern

Identify the pattern in the sequence 10, 5, 9, 4, 8, 3, ... and find the next three terms.

Solution

This pattern is neither arithmetic nor geometric as there is no common difference or ratio; however, if you view every other number, there is a pattern. That is, the differences alternate between -5 and $+4$ for the sequence, but from 10 to 9 to 8 is a difference of 1 and from 5 to 4 to 3 there is a difference of one. Therefore, the next three terms in the sequence are 7, 2, and 6.

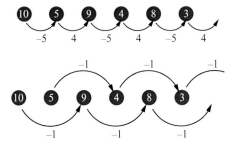

Example 10.3.5 Complex Pattern

Identify the pattern in the sequence 1, 2, 8, 22, 47, ... and find the next term.

Solution

The first differences between the terms is not common: 1, 6, 14, 25, ... and the second set of differences is not constant 5, 8, 11, ...; however, the third differences are a constant 3.

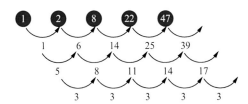

Hence, in the third row, the next number is 14, which would sum to 25 (row 2); therefore, the next number on the second row is 39 (25 + 14) and finally the next number in the first row is 86 (47 + 39).

Definition 10.3.6 Series

A **series** is the sum of the terms in a sequence,

$$s_1 = a_1, s_2 = a_1 + a_2, \ldots, s_n = a_1 + a_2 + \cdots + a_n.$$

In general, we have

$$s_n = \sum_{i=1}^{n} a_i.$$

In constructing a series by adding on term at a time, a new sequence of partial sums is generated:

$$s_1, s_2, s_3, \ldots, s_n.$$

Example 10.3.6 Series $1+2+3+\cdots$

Identify the series generated by in the sequence 1, 2, 3, 4, and 5; find the sum of the series for $n=10$.

Solution

Consider the partial sums:

$$s_1=1$$
$$s_2=1+2=3$$
$$s_3=1+2+3=5$$
$$s_4=1+2+3+4=10$$
$$s_5=1+2+3+4+5=15$$

Therefore, we see that the series for the given sequence of five numbers is 15; however, there is a pattern in that if you write the series backwards and forwards, we have

$$s_5=1+2+3+4+5$$
$$\underline{s_5=5+4+3+2+1}$$
$$2s_5=6+6+6+6+6$$
$$2s_5=5\times6=30$$
$$s_5=15.$$

Extending this idea to s_n, we have

$$s_n=1+2+3+4+5+\cdots+(n-1)+n$$
$$\underline{s_n=n+(n-1)+\cdots+5+4+3+2+1}$$
$$2s_n=(n+1)+(n+1)+\cdots+(n+1)+(n+1)$$
$$2s_5=n\times(n+1)$$
$$s_n=\frac{n(n+1)}{2}.$$

Using this formula, we have

$$s_{10}=1+2+3+4+5+6+7+8+9+10=\frac{10(10+1)}{2}=55.$$

Rule 10.3.1 Arithmetic Series

Given an **arithmetic sequence**, $a_n=a_1+d(n-1)$, the series is the average of the first and last term times the number of terms:

$$s_n=\frac{a_1+a_n}{2}\times n,$$

or equivalently, by substitution,

$$s_n=\frac{n[2a_1+(n-1)]}{2}.$$

Example 10.3.7 Series $4+2+1+\cdots$

Identify the series generated by in the sequence 4, 2, 1, 0.5, and 0.25; find the sum of the series for $n=6$.

Solution

Consider the partial sums:

$$s_1=4$$
$$s_2=4+2=6$$
$$s_3=4+2+1=7$$
$$s_4=4+2+1+0.5=7.5$$
$$s_5=4+2+1+0.5+0.25=7.75.$$

Therefore, we see that the series for the given sequence of five numbers is 7.75; however, there is a pattern in that if you write the series and subtract the series times the common ratio r, we have

$$s_5=4+2+1+0.5+0.25$$
$$-[0.5\times s_5]=-[2+1+0.5+0.25+0.125]$$
$$0.5\times s_5=4-0.125$$
$$0.5\times s_5=3.875$$
$$s_5=7.25$$

Extending this idea to s_n, we have

$$s_n=a_1+a_1r+a_1r^2+\cdots+a_1r^{(n-1)}+a_1r^n$$
$$rs_n=a_1r+a_1r^2+a_1r^3+\cdots+a_1r^n+a_1r^{n+1}$$
$$s_n-rs_n=a_1-a_1r^{n+1}$$
$$(1-r)s_n=a_1(1-r^{n+1})$$
$$s_n=\frac{a_1(1-r^{n+1})}{1-r}.$$

Using this formula, we have

$$s_6=4+2+1+0.5+0.25+0.125+0.0625=\frac{4(1-0.5^7)}{1-0.5}=7.9375.$$

Rule 10.3.2 Geometric Series

Given an geometric sequence, $a_n=a_1\times r^{n-1}$, the series is the given by

$$s_n=\frac{a_1(1-r^{n+1})}{1-r}.$$

If the absolute value of the ratio is less than 1, then the infinite sum is given by

$$s_\infty=\frac{a_1}{1-r}.$$

Example 10.3.8 Series $4+2+1+\cdots$

Identify the series generated by in the sequence 4, 2, 1, 0.5, and 0.25; find the sum of the series for $n\to\infty$.

Solution

$$s_\infty=\frac{a_1}{1-r}=\frac{4}{1-\frac{1}{2}}=8$$

EXERCISES

10.3.1. Identify the pattern and find the next three terms.

    a. 1, 6, 11, 16, …

    b. 2, 5, 4, 7, 6, 9, …

    c. 2, 1, 0.5, 0.25, …

    d. 1, 4, 9, 16, …

    e. 1, −1, 1, −1, …

10.3.2. Identify the pattern and find the next three terms.

    a. 1, 2, 4, 7, …

    b. 2, 5, 10, 17,

    c. 1, 1/2, 2/3, 3/4, 4/5, …

    d. 1, 3, 6, 10, …

    e. 1, 1, 2, 3, 5, …

10.3.3. Give the following arithmetic sequence in non-recursive form. What is the tenth term in the sequence?

    a. $a_n = a_{n-1} + 4$

    b. $a_n = a_{n-1} - 4$

    c. $a_n = a_{n-1} + 5$

    d. $a_n = a_{n-1} + \frac{1}{2}$

10.3.4. Give the following geometric sequence in non-recursive form. What are the first five terms?

    a. $a_n = 2 \times a_{n-1}$

    b. $a_n = \frac{1}{2} \times a_{n-1}$

    c. $a_n = -2 \times a_{n-1}$

    d. $a_n = 5 \times a_{n-1}$

10.3.5. Write the first five terms of the sequence and determine an alternative way of written the general term.

    a. $a_n = (1 + 2 + 3 + \cdots + n)^2$

    b. $a_n = 1 + 3 + 5 + \cdots + (2n - 1)$

10.3.6. Determine the series for the given sequence.

    a. 1, 3, 5, 7, 9

    b. 2, 8, 14, 20, 24

    c. 2, 4, 8, 16

    d. 1, −1, 1, −1, 1

10.4   ALGEBRAIC EXPRESSIONS AND RELATIONSHIPS

Notice that we abbreviate everything into a form of short hand. In the times of Plato, to express simple summation, one would write something like "five plus seven"; that is, 13 symbols to express three ideas; hence, if we use the Arabic symbol **5** for "five," the plus sign " **+** " and **7** for "seven," we have an algebraic expression **5 + 7**. If there are unknowns in a mathematical statement, we would abbreviate this to a single letter such *a* or *x*, and these single expressions can be combined to create more complicated algebraic expressions.

Definition 10.4.1 Algebraic Expression

An **algebraic expression** is a mathematical statement which contains numbers, variables and arithmetic operators.

Example 10.4.1 Algebraic Expressions

Write the algebraic expression "the sum of eleven and a number, all divided by five."

Solution

$$(11+n) \div 5$$

or

$$\frac{11+n}{5}$$

Example 10.4.2 Algebraic Expressions

Write the algebraic expression "twelve decreased by four times a number."

Solution

$$12 - 4 \times n$$

or

$$= 12 - 4n$$

Use of such notations makes it easier to determine and combine like terms.

Example 10.4.3 Apples and Oranges

Write the algebraic expression "in one box there are three apples and four pears in addition to two apples and one pear in other box" and simplify.

Solution

Let *a* represent apples and *p* represent pears, then we have

$$(3a + 4p) + (2a + p)$$

which is equivalent to

$$5a + 5p.$$

Common Algebraic Expressions

+	Plus, addition, sum, added to, increased by, more than				
−	Minus, **subtraction**, difference, less, decreased by, less than, subtracted from				
×	Product, multiplication, times, of				
÷	Quotient, division, divided by, ratio of, per				
$\square^2$	Squared, the power of two indicated by position of the two				
0	Zero, nothing, none				
1	One	2	Two	3	Three
4	Four	5	Five	6	Six
7	Seven	8	Eight	9	Nine

In general we will use 'n' for number.

Example 10.4.4 Writing Expressions

Write the algebraic expression for each of the following statements.

a. A number increased by two
b. Six less than a number
c. The product of three and a number
d. Two-thirds of a number

Solution

Write the algebraic expression for each of the following statements.

a. $n + 2$
b. $n - 6$

Note: This is not "six less a number" which would be $6 - n$.

c. $3 \times n$ or $3n$

d. $\dfrac{2}{3}n$ or $\dfrac{2n}{3}$.

These **algebraic expressions** can be broken down in to **algebraic terms** with numerical **coefficients** and unknown variables (letters).

Definition 10.4.2 Algebraic Term

An **algebraic term** is an expression where numbers and unknowns (letters) are combined in multiplication or division.

Algebraic terms come be combined in summation or differencing to create larger expressions and expressions with **like terms** can be simplified into equivalent expressions by combining the **coefficients** of like terms.

Definition 10.4.3 Coefficient

The **coefficient** in an expression is the numerical value and is written before any letters in the expression.

Example 10.4.5 Coefficients

List the numerical coefficient for each term.

a. $5x$ b. $0.2d$ c. $4xy$ d. $-3pq$

Solution

a. 5 b. 0.2 c. 4 d. -3

In the above examples, the expressions are all monomials; that is, the each contain one unique term. However, algebraic expressions can be the sum of many distinct terms. Terms that are not distinct are **like terms** which can be combined and simplified.

Definition 10.4.4 Like Terms

In an expression, **like terms** are terms that have all the same unknowns with the same exponents with different coefficients.

Example 10.4.6 Like Terms

Combine like terms in each group.

$2x$	$5y$	$-5x$
$3ab$	$5abc$	$-2ab$
$2x^2y$	$5x^2y$	$-3x^2y$
xyz	xy	z

Solution

a. $2x$ and $-5x$ are like terms:

$$2x + (-5x)$$
$$(2-5)x$$
$$-3x$$

b. $3ab$ and $-2ab$ are like terms:

$$3ab + (-2ab)$$
$$(3-2)ab$$
$$1ab$$
$$ab$$

c. $2x^2y$, $5x^2y$, and $-3x^2y$ are all like terms:

$$2x^2y + 5x^2y + (-3x^2y)$$
$$(2+5-3)x^2y$$
$$4x^2y$$

d. xyz, xy, and z are distinct terms; that is, there are no like terms.

Algebraic expressions can be evaluated for various values of the unknown or variables in each term.

Example 10.4.7 Evaluation

Evaluate the following expressions. Let $x=2$ and $y=5$.

a. $x+y$

b. $5x-y$

c. $\dfrac{y}{x}$

d. $\dfrac{x+y}{x-y}$

$$x+y$$
$$\uparrow \quad \uparrow$$
$$2 \quad 5$$

Solution

a. $x+y$ is equivalent to $2+5=7$

b. $5x-y$ is equivalent to $5\times2-5=10-5=5$

c. $\dfrac{y}{x}$ is equivalent to $\dfrac{5}{2}=2.5$

d. $\dfrac{x+y}{x-y}$ is equivalent to $\dfrac{2+5}{2-5}=\dfrac{7}{-3}=-2\dfrac{1}{3}$.

The ideas indicated by these expressions can lead to formulas for such things as perimeters and areas in geometry; as well as in computing interest and profits in finance.

Definition 10.4.5 Formula

A **formula** is an expression constructed using symbols to represent numerical values that related expressions in equality. Formulas are mathematical sentences which related underlying mathematical expressions.

Example 10.4.8 Formulations

Write an expression for the total length of lumber needed to cut **3** pieces of wood x inches in length if y inches are wasted on each cut the thickness of the saw blade.

Solution

As between each of the three pieces of wood there is a cut, there are two cuts and hence the total length, L, of the lumber needs to be

$$L = 3x + 2y.$$

Many such formulations come from **direct** or **indirect variation** between the unknowns or variables.

Definition 10.4.6 Direct Variation

Direct variation between two variables is such that their values always maintain a constant ratio:

$$\frac{y}{x} = k.$$

This relationship can also be written in terms of y **varying directly** as a multiple of x:

$$y = kx.$$

Given x and y vary directly, we might state that y is **directly proportional** to x. In this case, k is referred to as the **constant of proportionality**.

Example 10.4.9 Direct Variation

The total amount that Bob earns in a week, W, is directly proportional to the number of hours he works per week, h. Express this direct variation in a formula. Given that one week Bob earned **$147.25** for **19** hours of work, what is his hourly rate (the constant of proportionality)? How much would Bob make if he worked **40** hours a week?

Solution

The total amount of money Bob earns in a week, W, is given by

$$W = kh.$$

Using the information regarding one week's pay, we have

$$\$147.25 = k(19\,hours),$$

hence, we have that the hourly rate is

$$k = \frac{\$147.25}{19\,hours} = 7.75\$/hour.$$

Moreover, now that we have

$$W = 7.75\,hours,$$

if Bob works a 40-hour week, his total amount earned is

$$W = 7.75\frac{\$}{hour} \times 40\,hour = \$310.$$

According to the given information, we can determine that Bob gets paid **$7.75** per hour and would make **$310** in a week if he worked **40** hours that week.

Definition 10.4.7 Indirect Variation

Indirect variation between two variables is such that their values always maintain a constant product:

$$xy = k.$$

This relationship can also be written in terms of y **varying inversely** as a multiple of x:

$$y = \frac{k}{x}.$$

Give x and y vary indirectly, we might state that y is **inversely proportional** to x. In this case, k is referred to as the **constant of proportionality**.

Example 10.4.10 Indirect Variation

The total amount of time it takes a team to finish pressure washing a building, T, is inversely proportional to the number of people, n, on the team. Express this indirect variation in a formula. Given that a team of **4** can finish the job in **8** hours, what is time it would take a single worker (the constant of proportionality)? How long would it take a team of **6**?

Solution

The total amount of time, T, needed to complete the job is given by

$$T = \frac{k}{n}.$$

Using the information given, we have

$$8 = \frac{k}{4},$$

hence, we have that the constant of proportionality is

$$k = 32.$$

Moreover, now that we have

$$T = \frac{32}{n},$$

if a team of 6 works to complete the job, it would take a total of

$$T = \frac{32}{6} = 5.333 \text{ hours or 5 hours 20 minutes}.$$

According to the given information, this job would take one person 32 hours to finish whereas a team of 6 could complete it in 5 hours and 20 minutes.

EXERCISES

10.4.1. Write the algebraic expression.
   a. The sum of five and a number, quantity squared.
   b. Fifty decreased by twice a number.
   c. The difference between two and a number.
   d. Twice a number increased by nine.

10.4.2. Combine like terms.

    a. $4x + \dfrac{x}{2} - x$

    b. $5 + \dfrac{3b}{2} + 7b$

    c. $2\Delta + 4\Delta - 0.1\Delta$

    d. $9x^2 - 13x + 7x^2 + 5x - 2$

10.4.3. Write the algebraic expression.

    a. Seventeen plus twice a number.

    b. Fifteen times a number less five.

    c. The sum of ten and three times a number.

    d. Seven more than three-fourths of a number.

    e. Twice the product of two and a number squared

10.4.4. Write the statement (in words) for each of the following algebraic expressions.

    a. $n \div 12$

    b. $5x$

    c. $3(2y)$

    d. $5 - n$

    e. $5 + n$

10.4.5. Set up the following relationship.

    a. The length of a room is 5 feet longer than the width.

    b. If four tires cost d dollars, write an algebraic expression for the cost of each tire.

    c. Three shirts cost x dollars, write an algebraic expression for the cost of five shirts.

10.4.6. List the numerical coefficient for each term.

    a. $2r$

    b. $9xy$

    c. $-3abc$

    d. x

    e. $\frac{1}{2}x^2$

10.4.7. Combine like terms to simplify the expression.

    a. $5x + 7x$

    b. $6r - 3r$

    c. $-\frac{1}{2}f + 3c + 5f$

    d. $5t - 2t + 4t$

    e. $6h - 7h$

10.4.8. Combine like terms to simplify the expression.

    a. $a + a$

    b. $2e - e + 1$

    c. $5 + 2w - 1 + 3w$

    d. $4u - 0.5u + u^2$

    e. $\frac{1}{2}x + x^2 + 1 - x$

10.4.9. Evaluate the following:

    a. If $x = 12$ and $y = -2$, find $x + 2y$

    b. If $c = 2$ and $d = 3$, find $cd - 3$

    c. If $h = 1$ and $j = 5$, find $j^2 - h^2$

    d. If $n = -2$ and $m = -3$, find $nm + m^2$

10.4.10. Evaluate the following expressions when $x = -5$, $y = 2$, and $z = 3$.

    a. $x^2 + 2y - z$

    b. $2x + 3(y + z)^2$

    c. xyz

    d. $x^2 + y^2 + z^2$

    e. $\dfrac{x}{y} + z$

    f. $\dfrac{x + y}{z - x}$

10.4.11. The dimension of a room is w feet wide and l feet in length. Solve.

    a. If the **length** of a room is **5** feet longer than the **width**, write an expression for the **length** of the room in terms of the **width**, w.

    b. Evaluate the expression for **length** above for a **width** of **10** feet.

    c. Give the **area** of the room in terms of the **width** of the room.

10.4.12. **Profit** is **revenue** minus **cost**. Solve.

    a. Write an expression showing this relationship. Let R be the **revenue** and C be the **cost**.

    b. Given the revenue is **$15** per item sold and the cost is **$3** per item plus **$240** in over head, write the above expression in terms of number of items sold, x.

    c. How much **profit** is made when 50 items are sold?

    d. What is the breakeven point; that is, how many items must be sold to ensure that the **revenues** cover the **cost**, $R=C$.

10.4.13. **Distance** traveled varies directly travel **time** based on the **rate** of travel.

    a. Write an expression showing this relationship between distance, d, the constant of proportionality (rate), r, and the travel time, t.

    b. Given a car traveled **360** miles in **5** hours, what is the rate of travel, r?

    c. How far would this car travel in **7** hours?

10.5 EQUATIONS: EQUALITIES & SYSTEMS OF EQUATIONS

Formulas are distinct from expressions in that they contain an equal sign relating two or more expressions. To better understand how to use such equations, it is important that you can solve simple equalities.

Equality

✎ The state of being equal.

Definition 10.5.1 Equation

An **equation** is a mathematical sentence that asserts the equality of two expressions.

There are several properties of equality:

Property 10.51 Equality

These mathematical sentences are **reflective**, **symmetric** and **transitive**; that is,

    Reflective $a=a$

    Symmetric   If $a=b$, then $b=a$.

    Transitive   If $a=b$ and $b=c$, then $a=c$.

Example 10.5.1 Equations

Solve.

  a. $x+5=13$ **c.** $35=y+17$

  b. $y-3.7=5.9$ **d.** $22+z=12$

Solution

Step 1

$$x + 5 = 13$$

$$x + 5 = 13$$
$$x + 5 - 5 = 13 - 5$$
$$x = 8$$

Step 2

$$y - 3.7 = 5.9$$

$$y - 3.7 = 5.9$$
$$y - 3.7 + 3.7 = 5.9 + 3.7$$
$$y = 9.6$$

Step 3

$$35 = y + 17$$

$$35 = y + 17$$
$$35 - 17 = y + 17 - 17$$
$$18 = y$$
$$y = 18$$

Step 4

$$22 + z = 12$$

$$22 + z - 22 = 12 - 22$$
$$z = -10$$

Example 10.5.2 Two-Step Equations

Solve the equation in two-steps: $5x - 7 = 18$.

Solution

$$5x - 7 = 18$$
$$5x - 7 + 7 = 18 + 7$$
$$5x = 25$$

$$\frac{5x}{5} = \frac{25}{5}$$
$$x = 5$$

Example 10.5.3 Three-Step Equations

Solve the equation in three-steps: $-2x - 5 = 13x + 25$.

Solution

$$-2x - 5 = 13x + 25$$
$$+2x - 2x - 5 = +2x + 13x + 25$$
$$-5 = 15x + 25$$

$$-5 - 25 = 15x + 25 - 25$$
$$-30 = 15x$$

$$-\frac{30}{15} = \frac{15x}{15}$$
$$-2 = x$$
$$x = -2.$$

   Recall that parentheses and brackets indicate multiplication or emphasis to be distributed first.

Example 10.5.4 Solving Equations

Solve the equation: $-2(x-5)=13x+25$.

Solution

$$-2(x-5)=13x+25$$
$$-2x+10=13x+25$$
$$+2x-2x+10=2x+13x+25$$
$$10=15x+25$$

$$10-25=15x+25-25$$
$$-15=15x$$

$$-\frac{15}{15}=\frac{15x}{15}$$
$$-1=x$$
$$x=-1.$$

Definition 10.5.2 Literal Equation

A **literal equation** is an equation that contains letters such a,b,c in place of constants and letters such as x,y,z for the unknowns (variables).

Literal equations can be used to illustrate an equation in general such as a form, $ax-b=c$, or formula such as those used to define area of a rectangle in terms of the base and height, $A=bh$ or distance in terms of rate and time, $d=rt$.

Example 10.5.5 Literal Equations

Solve the equation in terms of x: $x: w=xy+z$.

Solution

$$w=xy+z$$
$$w-z=xy+z-z$$
$$w-z=xy$$

$$\frac{w-z}{y}=\frac{xy}{y}$$
$$\frac{w-z}{y}=x$$

$$x=\frac{w-z}{y}.$$

We can use the equations to solve for unknowns once a problem has been translated from English to Math.

Example 10.5.6 Solving Equations

Eight times a number is 72. Find the number.

Solution

$$\underbrace{\text{Eight times}}_{8} \; \underbrace{\text{a number}}_{n} \; \underbrace{\text{is}}_{=} \; \underbrace{72}_{72}$$

$$8 \times n = 72$$
$$8n = 72$$

$$\frac{8n}{8} = \frac{72}{8}$$
$$n = 9$$

This technique offers use a shorthand language which can be used to solve everyday problems.

Example 10.5.7 Word Problems

June has a board 15 feet long. She must cut the board so that one piece is 3 feet longer than twice the length of the other. What will be the length of each board?

Solution

Let the length of the shorter piece be represented by x; than the longer piece is 3 feet longer than twice the value of x. Then in total, there are 15 feet.

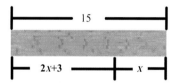

Hence, we have

$$(2x + 3) + x = 15$$
$$2x + 3 + x = 15$$
$$3x + 3 = 15$$

$$3x + 3 - 3 = 15 - 3$$
$$3x = 12$$

$$\frac{3x}{3} = \frac{12}{3}$$
$$x = 4.$$

The length of the shorter board is 4 feet and that of the longer board is 11 feet.

Definition 10.5.3 Inequality

An **inequality** is a mathematical sentence that asserts the two expressions are related, but not necessarily equal.

These mathematical sentences are less than, less than or equal to, greater than and greater than or equal to.

The strict inequalities are not **reflective** or **symmetric**; however, they are **transitive**; that is,

Not Reflective	$a \leq a$, but NOT $a < a$.
Not Symmetric	If $a < b$, then NOT $b > a$.
Transitive	If $a < b$ and $b < c$, then $a < c$.

Example 10.5.8 Inequalities

Solve the equation: $-2(x - 5) < 3x + 25$.

Solution

$$-2x+10<3x+25$$
$$-3x-2x+10<-3x+3x+25$$
$$-5x+10<25$$

$$-5x+10-10<25-10$$
$$-5x<15$$

$$\frac{-5x}{-5}>\frac{15}{-5}$$
$$x>-3.$$

Rule 10.5.1 Elementary Operations

Given an equation

$$a=b$$

the following **elementary operations** yield equivalent expressions:

$$a+c=b+c,$$
$$a-c=b-c,$$
$$a\times c=b\times c,$$
$$a\div c=b\div c, c\neq 0$$

Given an equation

$$a<b$$

the following **elementary operations** yield equivalent expressions:

$$a+c<b+c \quad \text{and} \quad a-c>b-c,$$

$$a\times c>b\times c, c>0$$
$$a\times c<b\times c, c<0$$
$$a\div c>b\div c, c>0$$
$$a\div c<b\div c, c<0$$

There are three ways to write inequalities: inequality notation, interval notation and graphic notation. When graphing inequalities on a real number line, there are several conventions. There are open intervals and there are closed intervals, and there are intervals which are closed on one end and open on the other end.

Inequality Notation	Interval Notation	Graphic Notation
$x\geq a$	$[a,\infty)$	
$x<b$	$(-\infty,b)$	
$a<x<b$	(a,b)	
$a\leq x\leq b$	$[a,b]$	
$a\leq x<b$	$[a,b)$	
$a<x\leq b$	$(a,b]$	
$x=a$		
$x\neq a$		

Example 10.5.9 Various Notations

Write the inequality in interval notation and illustrate graphically: $x > -2$.

Solution

In inequality notation,

$$x > -2$$

can be read backwards and written

$$-2 < x$$

which indicates the lower limit (not included) is –2 and has not upper bound; hence, in interval notation we have

$$(-2, \infty).$$

Graphically, -2 is open (not included) and extends to the right (to infinity).

$$-5 \ -4 \ -3 \ -2 \ -1 \ \ 0 \ \ 1 \ \ 2 \ \ 3 \ \ 4 \ \ 5$$

As this interval implies x can take on an infinitely large number (positive), this interval is unbounded.

Definition 10.5.4 Unbounded Interval

An **unbounded interval** is an interval which does not include a finite end point; there are several possible forms of an **unbounded interval**:

$(-\infty, +\infty)$,
$(-\infty, b)$,
$(-\infty, b]$,
$(a, +\infty)$,
$[a, +\infty)$.

When an **unbounded interval** is written in **inequality notation**, we have:

$(-\infty, +\infty)$ is any real number, $x \in \mathbb{R}$,
$(-\infty, b)$ is any real number less than b, $x < b$,
$(-\infty, b]$ is any real number less than or equal to b, $x \leq b$,
$(a, +\infty)$ is any real number greater than a, $x > a$,
$[a, +\infty)$ is any real number greater than a, $x \geq a$.

When **unbounded intervals** are written in inequality notation, there is only one or no boundaries on the value of x whereas **bounded intervals** are such that both ends are finite values.

Definition 10.5.5 Bounded Interval

A **bounded interval** is an interval which does have finite end points; there are several possible forms of an **bounded interval**:

a. (a, b),
b. $(a, b]$,
c. $[a, b)$,
d. $[a, b]$.

When a **bounded interval** is written in **inequality notation**, we have:

(a, b) is any real number, $a < x < b$,
(a, b) is any real number, $a < x < b$,
$(a, b]$ is any real number, $a < x \leq b$,
$[a, b)$ is any real number, $a \leq x < b$,
$[a, b]$ is any real number, $a \leq x \leq b$.

Example 10.5.10 Bounded

Write the inequality in interval notation and illustrate graphically: $x > -2$.

Solution

In inequality notation,

$$x > -2$$

can be read backwards and written

$$-2 < x$$

which indicates the lower limit (not included) is -2 and has not upper bound; hence, in interval notation we have

$$(-2, \infty).$$

Graphically, -2 is open (not included) and extends to the right (to infinity).

$$-5 \ -4 \ -3 \ -2 \ -1 \ \ 0 \ \ 1 \ \ 2 \ \ 3 \ \ 4 \ \ 5$$

Example 10.5.11 Integers

Take x to be an integer and graph the solution set for $x \geq -2$.

Solution

In inequality notation,

$$x \geq -2$$

can be read backwards and written

$$-2 \leq x$$

which indicates the lower limit (included) is -2 and has not upper bound; hence, in interval notation we have

$$(-2, \infty) \cap \mathbb{Z},$$

however, we are restricted to the integers and not the real number line in its entirety.

Graphically, -2 is close (included) and extends to the right ever integer point (to infinity).

$$-5 \ -4 \ -3 \ -2 \ -1 \ \ 0 \ \ 1 \ \ 2 \ \ 3 \ \ 4 \ \ 5$$

Recall from Chapter 3, in set notation, we can list this solution set in **roster notation**:

$$\{-2, -1, 0, 1, 2, 3, \ldots\}.$$

Other equations and inequalities include **absolute values**.

Definition 10.5.6 Absolute Value

The **absolute value** of a number x, denoted by $|x|$, is the numerical value (magnitude of the value) without the sign (direction on the real number line). It is the distance between the value on the number line and zero.

$$|x| = \begin{cases} x & x \geq 0 \\ -x & x < 0 \end{cases}$$

Example 10.5.12 Absolute Values

Take x to be a real number, solve the equation $|x|=2$.

Solution

Either $x=2$ or $-x=2$; that is, $x=2$ or $x=-2$.

Example 10.5.13 Solving Absolute Values

Take x to be a real number, solve the equation $|x+1|=5$.

Solution

Either $x+1=5$ or $-(x+1)=5$; that is, $x+1=5$ or (dividing both sides by the -1) $x+1=-5$. Solving these two equations, we have $x=4$ or $x=-7$.

Example 10.5.14 Inequalities with $|x|$

Take x to be a real number, solve the equation $|x|<3$.

Solution

First, we need to consider the boundaries: $|x|=3$. Here the solutions are 3 and -3 (not included).

$$-5\ -4\ -3\ -2\ -1\ \ 0\ \ 1\ \ 2\ \ 3\ \ 4\ \ 5$$

Now we must verify which intervals work, and in this case, it is the center interval.

$$-5\ -4\ -3\ -2\ -1\ \ 0\ \ 1\ \ 2\ \ 3\ \ 4\ \ 5$$

We can verify this as $|-4|=4<3$ is not true and therefore we do not include $(-\infty,-3]$ as with $|4|=4<3$, this is not true and therefore we do not include the interval $[3,\infty)$; however, $|2|=|-2|=2<3$, and therefore, we keep the interval $(-3,3)$.

Note: there are many points that we could have tested; however, you should test one in each interval created by the boundary points.

Rule 10.5.2 Absolute Values

In general, if a is any positive value, then we have the following equivalences with respect to **absolute values**:

$|x|=a$	$x=-a, x=a$
$|x|<a$	$-a<x<a$
$|x|\leq a$	$-a\leq x\leq a$
$|x|>a$	$x<-a$ or $x>a$
$|x|\geq a$	$x<-a$ or $x>a$

We are now in a position to discuss **linear inequalities**.

Example 10.5.15 Linear Inequalities

Find the solution set of the inequality $2x+3y>6$.

Solution

We shall first consider the inequality in terms of the boundary, when $2x + 3y = 6$, which has an infinite number of solutions. Every point on the line is a solution to the equation. However, since this boundary is not included, we use a dashed line, see Figure 10.1.

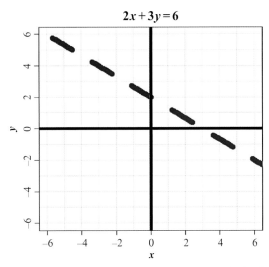

FIGURE 10.1   Boundary line for the linear inequality $2x+3y>6$.

To solve the original inequality we need to find a set consisting of values (x, y) that make the expression $2x + 3y$ greater than 6. This solution can be obtained by solving for y in the inequality to determine y relationship with the line: $y > 2 - \dfrac{2}{3}x$, so the solution set is above the line. Or you can test a point in each region to determine which set works: try $(0,0)$ in the inequality, $2 \times 0 + 3 \times 0 > 6$ is not true. Hence, the points below the line do not work; try $(3,2)$ above the line, $2 \times 3 + 3 \times 2 > 6$ is true and so the solution set is above the line, Figure 10.2.

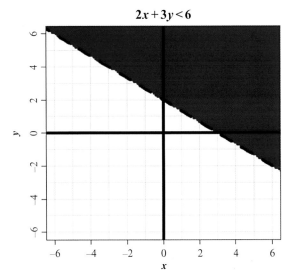

FIGURE 10.2   Scatter plot of the solution set for the linear inequality $2x+3y>6$.

Another common type of equation is the quadratic equation.

Definition 10.5.7 Quadratic Equation

A **quadratic equation** is a second-degree expression with standard form of

$$ax^2 + bx + c = 0$$

where a, b and c are real numbers and $a \neq 0$.

If the **quadratic equation** is factorable, then we can use the property that if $ab = 0$, then either $a = 0$ or $b = 0$ to solve the equation.

Example 10.5.16 Factoring Quadratics

Factor and solve the equation $2x^2 - 3x - 5 = 0$.

Solution

First, we need to factor the left hand side of the equation to be

$$2x^2 - 3x - 5 = (2x - 5)(x + 1),$$

hence, we have that either $2x - 5 = 0$ or $x + 1 = 0$. Solving these two linear equations, we have

$$x = \frac{5}{2} = 2.5 \quad \text{or} \quad x = 1.$$

The **key** to factoring this quadratic equation lies in the product ac; we need factors of the product ac that sum to be the middle term b. Here $ac = -10$ and factors of -10 that sum to -3 are -5 and 2. That is, $-5x$ is the product of the inner terms (or outer terms) and $2x$ is the product of the outer terms (or vice versa).

In fact, when the lead coefficient is one, then the key is the solution.

Example 10.5.17 Factoring Quadratics

Factor and solve the equation $x^2 - 5x - 6 = 0$.

Solution

First, the **key** to factoring this quadratic equation $ac = -6$ and factors of -6 that sum to -5 are -6 and 1. That is, $-6x$ is the product of the inner terms and $1x = x$ is the product of the outer terms:

$$x^2 - 5x - 6 = (x - 6)(x + 1),$$

hence, we have that either $x - 6 = 0$ or $x + 1 = 0$. Solving these two linear equations, we have

$$x = 6 \quad \text{or} \quad x = 1.$$

However, not all quadratics factor nicely, in which case one can use the quadratic formula. Using a technique call completing the square, we get the following formulation of the solutions.

Rule 10.5.3 Quadratic Formula

In general, the solutions to a **quadratic equation** $ax^2 + bx + c = 0$, where a, b and c are real numbers and $a \neq 0$ can be obtained from the **quadratic formula**

$$x = \frac{-b \pm \sqrt{b^2 - 4ac}}{2a}.$$

Example 10.5.18 Quadratic Inequality

Graph the solution set to the equation $x^2 - x - 6 \geq y$.

Solution

We first start with the boundary $x^2 - x - 6 = y$, which has infinitely many solutions which, in this case, are included in the solution set. To assist in graph, we can use the quadratic equation to find the zeros: $x^2 - x - 6 = 0$:

$$x = \frac{-(-1) \pm \sqrt{(-1)^2 - 4 \times 1 \times (-6)}}{2 \times 1}$$
$$\downarrow$$
$$x = \frac{1 \pm \sqrt{25}}{2}$$
$$\downarrow$$
$$x = -2, 3$$

Hence, the graph is a parabola (second-degree) in the upward direction (lead coefficient is positive) and y-intercept at $y = -6$, Figure 10.3.

$$y = x^2 - x - 6$$

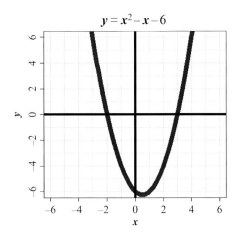

FIGURE 10.3   Scatter plot of the included boundary in the inequality $y \leq x^2 - x - 6$.

Furthermore, since $y \leq x^2 - x - 6$, it is the region below the curve that is the solution set, Figure 10.4.

$$y \leq x^2 - x - 6$$

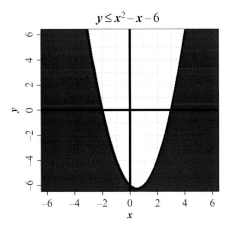

FIGURE 10.4   Scatter plot of the solution set for the inequality $y \leq x^2 - x - 6$.

Now we will study systems of equations.

Definition 10.5.8 System Equations

A **system of equations** is a set of equations to be solved simultaneously using the processes of elimination, substitution, matrix manipulation or other iterative method.

In our studies we will consider a system of two linear equations:

$$\begin{cases} ax + by = c \\ dx + ey = f \end{cases}$$

where a, b, c, d, e and f are known constants, and solve for the variables x and y.

There are three ways we will study to solve such systems are: elimination, substitution and using determinates of matrices (Cramer's Rule).

Common Solution Methods

Elimination—a process by which equations are manipulated in pairs to reduce the system by one variable; in a system of two equations and two unknowns, this is a simple equation.

Substitution—a process by which one equation is manipulated, solving for one variable that then can be substituted into the remaining equations

Determines of Matrices—a process by which the information contained in the equations are re-organized to determine the solution to the equation.

Using elimination, both equations may need to be multiplied by a constant:

Step-by-Step Elimination

Step 1. Select a variable and find the least common multiple of the coefficients.

Step 2. Determine the multiples needed for each equation and multiply the equations, respectively, except make one of the multiples the opposite sign.

Step 3. Add resulting equations reducing the number of unknowns.

Step 4. Repeat Steps 1–3 as needed and finish by plugging in the answer to the last equation to the previous equation until you have solved the system of equations.

Example 10.5.19 Elimination

Solve the system of equations below using elimination:

$$\begin{cases} 3x + 5y = 21 \\ 4x - 3y = -1 \end{cases}$$

Solution

Step 1. The least common denominator for coefficients on x is 12 and for the coefficients on y is -15; therefore, arbitrarily choosing the minimum, 12, we will eliminate x.

Step 2. The multiple needed for the first equation is 4 and the second is 3; hence, we will multiply by 4 and -3, respectively.

$$\begin{cases} (3x + 5y = 21) \times 4 \\ (4x - 3y = -1) \times -3 \end{cases}$$

Solution—cont'd

Resulting in the following:

$$\begin{cases} 12x + 20y = 84 \\ -12x + 9y = 3 \end{cases}$$

Step 3. Adding the resulting equations we have one equation with one unknown.

$$29y = 87$$

Hence, $y = 3$; and given this information and the first original equation we get

$$3x + 5 \times 3 = 21$$
$$\downarrow$$
$$3x + 15 = 21$$
$$\downarrow$$
$$3x + 15 - 15 = 21 - 15$$
$$\downarrow$$
$$3x = 6$$
$$\downarrow$$
$$\frac{3x}{3} = \frac{6}{3}$$
$$\downarrow$$
$$x = 2$$

The solution to the system of equations is $(2, 3)$.

Using substitution, only one equation initial manipulated:

Step-by-Step Elimination

Step 1. Select a variable and solve for that variable in one of the equations.
Step 2. Substitute the resulting expression in for that variable in all remaining equations.
Step 3. Repeating steps 1–2 as needed and finish by plugging in the answer to the last equation to the previous equation until you have solved the system of equations.

Example 10.5.20 Substitution

Solve the system of equations below using substitution:

$$\begin{cases} 3x + 5y = 20 \\ 4x - 3y = 17 \end{cases}$$

Solution

Step 1 The coefficients are not one and therefore, arbitrarily choosing the first equation, solve for y (as it is easier to divide by 5 that it is to divide by 3).

$$3x + 5y = 20$$
$$\downarrow$$
$$(3x + 5y) - 3x = 20 - 3x$$
$$\downarrow$$
$$5y = 20 - 3x$$
$$\downarrow$$
$$\frac{5y}{5} = \frac{20 - 3x}{5}$$
$$\downarrow$$
$$y = 4 - 0.6x$$

Continued

Solution—cont'd

Step 2   Substituting this into the second equation we get,

$$4x - 3y = 17$$
$$\downarrow$$
$$4x - 3(4 - 0.6x) = 17$$
$$\downarrow$$
$$4x - 12 + 0.6x = 17$$
$$\downarrow$$
$$4.6x - 12 = 17$$
$$\downarrow$$
$$4.6x - 12 + 12 = 17 + 12$$
$$\downarrow$$
$$4.6x = 29$$
$$\downarrow$$
$$\frac{4.6x}{4.6} = \frac{29}{4.6}$$
$$\downarrow$$
$$x = 5$$

Step 3   Plugging this into the equation found in step 1, we have

$$y = 4 - 0.6 \times 5 = 1$$

Hence, the solution to the system of equations is $(\mathbf{5}, \mathbf{1})$.

Before we cover the last procedure, we must first introduce the idea of a matrix.

Definition 10.5.9 Matrix

A **matrix** is a rectangular array of numbers that has a given number of rows and columns.

The system of equations below can be considered as two matrices:

$$\begin{cases} ax + by = c \\ dx + ey = f \end{cases},$$

first the coefficients:

$$D = \begin{bmatrix} a & b \\ d & e \end{bmatrix},$$

next are the constants that balance the equations:

$$A = \begin{bmatrix} c \\ f \end{bmatrix},$$

and the unknowns:

$$B = \begin{bmatrix} x \\ y \end{bmatrix}.$$

There are methods for solving the underlying system of equations using augmented matrices, matrix multiplications and inverse matrices; however, this is beyond the scope of this text. Here we introduce **Cramer's Rule** which uses determine to solve systems of equations.

Definition 10.5.10 Determinate of 2×2

Given a **2×2** matrix $D = \begin{bmatrix} a & b \\ c & d \end{bmatrix}$, the **determinate** is a given by

$$|D| = ad - bc.$$

Cramer's Rule uses determines as follows for a system of two equations and two unknowns:

Rule 10.5.4 Cramer's Rule

Given a system of two equations and two unknowns in the following form:

$$\begin{cases} ax+by=c \\ dx+ey=f \end{cases}$$

Let $D=\begin{bmatrix} a & b \\ d & e \end{bmatrix}$ (the coefficients on x and y), $D_x=\begin{bmatrix} c & b \\ f & e \end{bmatrix}$ (the matrix D with the balancing constants in place of the coefficients on x) and $D_y=\begin{bmatrix} a & c \\ d & f \end{bmatrix}$ (the matrix D with the balancing constants in place of the coefficients on y); then if $|D|\neq0$, by **Cramer's Rule**

$$x=\frac{|D_x|}{|D|},$$

and

$$y=\frac{|D_y|}{|D|}.$$

Example 10.5.21 Cramer's Rule

Solve the system of equations below using substitution:

$$\begin{cases} 3x-5y=-3 \\ 4x-3y=7 \end{cases}$$

Solution

Setting up the matrices we have:

$$D=\begin{bmatrix} 3 & -5 \\ 4 & -3 \end{bmatrix}, \quad D_x=\begin{bmatrix} -3 & -5 \\ 7 & -3 \end{bmatrix} \text{ and } D_y=\begin{bmatrix} 3 & -3 \\ 4 & 7 \end{bmatrix}.$$

The determinates of which are:

$$|D|=-9-(-20)=11,$$
$$|D_x|=9-(-35)=44,$$

and

$$|D_y|=21-(-12)=33.$$

Hence, the solution to the system of equations is $x=\frac{44}{11}=4$ and $y=\frac{33}{11}=3$, or $(4,3)$.

Just as there are systems of equations, there are systems of inequalities.

Example 10.5.22 System Inequalities

Graph the solution set to the following system of inequalities:

$$\begin{cases} x\geq0 \\ y\geq0 \\ x+y\leq5 \end{cases}$$

Solution

First graph the inequalities one at a time:

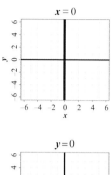

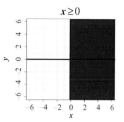

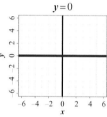

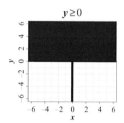

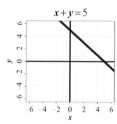

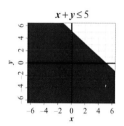

Hence the solution set is given by Figure 10.5.

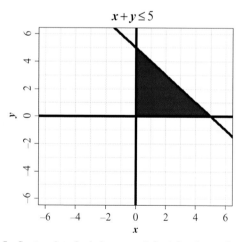

FIGURE 10.5 Scatter plot of solution set o $x \geq 0$, $y \geq 0$ and $x+y \leq 5$.

Now we introduce a function to be optimized limited to given inequalities or constrains. In optimization, our primary aim is to find a solution that will maximize or minimize the objective function.

Rule 10.5.5 Optimization

In **optimization**, the *maximum* or *minimum* of the objective function over the feasible set, assuming it exists, will be attained at least one of the vertices of the feasible set.

Example 10.5.23 Optimization

Minimize the linear function $f(x,y)=2x+3y$, subject to

$$\begin{cases} x+y\geq 1 \\ x-y\geq -1 \\ x+2y\leq 6 \\ y\geq 0 \end{cases}$$

Solution

First, we graph the solution set to the system of inequalities, Figure 10.6.

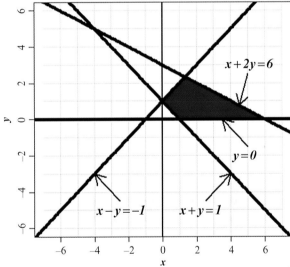

FIGURE 10.6   Scatter plot of the solution set for $x+y\geq 1$, $x-y>-1$, $x+2y\leq 6$ and $y\geq 0$.

Next, solving the pair-wise system of equations we can obtain the points of intersection: $(1,0),(6,0),\left(\dfrac{4}{3},\dfrac{7}{3}\right)$, and $(0,1)$. Now we consider each value of the objective function when it is evaluated at the extreme points of the feasible set.

$$f(1,0)=2\times 1+3\times 0=2$$
$$f(6,0)=2\times 6+3\times 0=12$$
$$f\left(\frac{4}{3},\frac{7}{3}\right)=2\times\frac{4}{3}+3\times\frac{7}{3}=\frac{29}{3}=9\frac{2}{3}$$
$$f(0,1)=2\times 0+3\times 1=3$$

Inspecting the values of the objective function at each point, we see that it will be minimized at the point (1,0) with a minimum value of 2.

EXERCISES

10.5.1. The formula for the area of a triangle is $A = \frac{1}{2}bh$, where b is the base and h is the height. Solve for h.

10.5.2. Solve.

    a. $25 = 2 + n$

    b. $x - 5 = -3$

    c. $17 = b - 17$

    d. $\frac{3}{4} + t = 1\frac{7}{8}$

    e. $m + 0.51 = 4.65$

10.5.3. Solve the following problems.

    a. Deb had $233.54 in her checking account after she deposited $54.14. What was the original balance?

    b. Jeffery spent $24.79 on a new video game. If he gave the clerk a ten and a twenty, how much change should he receive back?

    c. Jordan put 5 nickels, 3 dimes and 10 quarters in her piggy bank. She then emptied the bank and found that she had a total of $6.22. How much was in the bank before she added the change?

10.5.4. Solve these equations.

    a. $9b = 45$

    b. $-13a = 104$

    c. $-\frac{3}{5}x = 12$

    d. $\frac{z}{5} = 1.2$

    e. $4x - 3 = 7$

10.5.5. Solve these equations.

    a. $5b - 25 = -4b + 11$

    b. $-4 - 10a = 100 + 3a$

    c. $-\frac{3}{5}x + 2 = 19$

    d. $\frac{z}{5} + 0.2 = 1.2 - x$

    e. $4x - 3 = 3x + 7$

10.5.6. Solve these equations.

    a. $2(x + 5) - x = 4$

    b. $3(x - 2) = 5x - 6$

    c. $2 - 3(n - 5) = 38$

    d. $3[-2x - 6(11x)] = -6$

10.5.7. Solve these equations for the given unknown.

    a. $ax = b$ for x.

    b. $V = lwh$ for w.

    c. $d = rt$ for r.

10.5.8. Solve these equations for x.

    a. $4x + 12 = 60$

    b. $\frac{4}{7}x = 8$

    c. $\frac{4}{9}x - 9 = 11$

10.5.9. Solve.

    a. The formula used to compute the volume of a circular cylinder is $V = \pi r^2 h$ where r is the radius and h is the height. Solve for h.

    b. The formula used to compute the perimeter of a rectangle with base b and height is $P = 2b + 2h$. Solve for b.

    c. The formula used to compute the area of a circle is $A = \pi r^2$ where r is the radius. Solve for r.

10.5.10. Solve.

    a. Three times the sum of a number and 7 is 6. Find the number.

    b. The sum of three consecutive odd integers is 21. What are the numbers?

    c. The sum of three consecutive integers is 48. What are the numbers?

    d. The product of a number and 5 is equal to 9. What is the number?

    e. A number divided by 4 is equal to 7. What is the number?

10.5.11. Michael and Rachel agree to split the expenses according to their income. Michael makes twice as much as Rachel. If the expenses are $954 a month, how much does each per pay?

10.5.12. An artist mixes his own paints. For dark green he uses four times as much blue pigment as yellow. How much yellow is needed to make 20 ounces of dark green paint?

10.5.13. Solve the given inequality.
  a. $y+2 \geq 4$
  b. $2z-5 < 7$
  c. $x+2 \leq 2x+7$
  d. $x > 3-2x$

10.5.14. Take x to be a real number, solve, and graph the resulting inequality.
  a. $-2 < x+1 \leq 7$
  b. $-5 \leq x-2 < 2$
  c. $x+1=1$

10.5.15. Take x to be an integer, solve, and graph the resulting inequality.
  a. $0 < x+1 \leq 5$
  b. $-1 \leq x-7 < 6$
  c. $x+1=5$

10.5.16. Take x to be a real number, solve, and graph the resulting inequality.
  a. $x-2 > 5$
  b. $x+3 \leq -5$
  c. $x+7=9$

10.5.17. Write the given interval in inequality notation and graphic it on the real number line.
  a. $(-0.5, 5)$
  b. $(-\infty, 2)$
  c. $[4, \infty)$

10.5.18. Take x to be an integer. Write the given set in roster notation.
  a. $x \leq 5$ and $x-2 \geq -1$
  b. $x < 0$ and $x \geq -3$
  c. $x+2 \geq 5$ and $x < 8$
  d. $2x-3 < 1$ and $x \neq 0$

10.5.19. Graph the solution set, if it is non-empty.
  a. $x > 3$ and $x-5 \geq 1$
  b. $x-2 \leq 0$ and $x+2 \geq -3$
  c. $x+2 \geq 0$ and $x < 8-x$
  d. $7x-5 < 9$ and $x-2 > -5$

10.5.20. Write the given intervals in inequality and interval notation.

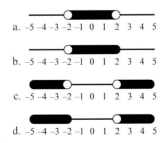

10.5.21. Use interval notation to write the following:
  a. $\{x | x \geq -2\}$
  b. $\{x\} x < 3\}$
  c. $\{x | -1 \leq x < 5\}$
  d. $\{x | x \rangle 2$ or $x < 4\}$

10.5.22. Evaluate the expression.
  a. $|-5|$
  b. $|2|$
  c. $|2-5|$
  d. $|\pi-6|$

10.5.23. Solve, given answer in both inequality and graphic notation.

   a. $|x| < 2$

   b. $|x - 2| > 3$

   c. $|2x - 1| \leq 4$

   d. $|x| > -2$

   e. $|2x| \leq 3$

10.5.24. Solve, given answer in both inequality and graphic notation.

   a. $|x| < -2$

   b. $|x + 12| \leq 3$

   c. $|2x - 11| > 5$

   d. $|x| > 2$

   e. $|2x - 3| \leq 3$

10.5.25. Factor each quadratic expression.

   a. $x^2 + 6x + 8$

   b. $x^2 + 8x + 15$

   c. $x^2 - 3x - 15$

   d. $x^2 - 13x + 12$

10.5.26. Factor each quadratic expression.

   a. $6x^2 - 5x + 1$

   b. $3x^2 + 10x + 3$

   c. $16x^2 + 4x + 1$

   d. $2x^2 - 26x + 24$

10.5.27. Solve the quadratic equation.

   a. $x^2 + 5x + 6 = 0$

   b. $x^2 - x - 6 = 0$

   c. $x^2 + x - 20 = 0$

   d. $x^2 + 4x + 8 = 5$

10.5.28. Solve each equation using the quadratic equation.

   a. $2x^2 + 7x - 5 = 0$

   b. $3x^2 + 2x = 2$

   c. $x^2 + 3x - 5 = 0$

10.5.29. Solve the system of equations using process of elimination.

   a. $\begin{cases} 3x - 2y = 23 \\ 2x + 3y = 24 \end{cases}$

   b. $\begin{cases} x + 7y = 30 \\ 4x - 3y = 5 \end{cases}$

   c. $\begin{cases} 2x - y = 23 \\ 2x + 3y = 11 \end{cases}$

10.5.30. Solve the system of equations using process of substitution.

   a. $\begin{cases} 9x - 2y = 20 \\ 5x + 3y = 7 \end{cases}$

   b. $\begin{cases} x + 7y = 22 \\ 4x - 3y = -5 \end{cases}$

   c. $\begin{cases} 2x - y = 1 \\ 2x + 3y = 21 \end{cases}$

10.5.31. Solve the system of equations using Cramer's Rule.

   a. $\begin{cases} 3x - 2y = 1 \\ 2x + 3y = 5 \end{cases}$

   b. $\begin{cases} x + y = 30 \\ x - y = 5 \end{cases}$

   c. $\begin{cases} 2x - y = 2 \\ 2x + 3y = 18 \end{cases}$

10.6   FUNCTIONS: LINEAR & QUADRATIC

Unlike the equations we solved in terms of a single unknown, a relationship is between two or more variables; when two variables, we usually denoted them x and y.

Definition 10.6.1 Relationship

A relationship or relation is a set of ordered pairs (x, y).

The largest set of real numbers the variable x takes on is referred to as the **domain** of the function and the largest set of real numbers the variable y takes on is referred to as the **range** of the function.

Example 10.6.1 Factoring Quadratics

State the **domain** and **range** in the given relationship:

$$\{(4, -1), (3, -2), (4, -3), (5, -3)\}.$$

Solution

The **domain** is $x \in \{3, 4, 5\}$ and the **range** is $y \in \{-1, -2, -3\}$.

A special type of relationship is a **function** which relates two unknowns.

Definition 10.6.2 Function

A **function** is a relation between two variables, x and y, such that for each value of the independent variable, usually selected to be x there is at most one value of the dependent variable, y. That is, "y is a function of x" when for each value of x in the domain, there is one value of y in the range.

We express this relation by $y = f(x)$.

Alternative notation for the pairing of x and y is as follows:

$$f : x \to y$$

Read as "the function is such that the value of x is mapped to the value of y"

Rule 10.6.1 Domain Restrictions

In general, there are only three **restrictions to the domain**:
1. You cannot divide by zero; that is, given an expression $\frac{\text{Top}}{\text{Bottom}}$, the bottom cannot be zero, **Bottom ≠ 0**.
2. You cannot take the square root of a negative value; that is given an expression $\sqrt{\text{Inside}}$, then the insider is greater than or equal to zero, **Inside ≥ 0**.
3. You cannot take the logarithm of a negative value or zero; that is given an expression $\log_b(\text{Inside})$, then the inside is greater than zero, **Inside > 0**.

The simplest form of a function is a **linear function**.

Definition 10.6.3 Linear Function

A **linear function** is a first-degree relationship between x and y such that

$$f(x) = mx + b,$$

where m is the **marginal change (slope)** of the line and b is the baseline or y-intercept (that is, $y = b$ when $x = 0$). The graph of this relationship is a **line**.

Continued

Definition 10.6.3 Linear Function—cont'd

Two special forms of a linear equation are the **vertical line**,

$$x=a$$

and the **horizontal line**

$$y=b.$$

Given the equation of the line, it is easiest to simply plot points. As for each value of x there is at most one y, pick values of x in your domain and evaluate the corresponding values of y.

Example 10.6.2 Linear Functions

Graph the linear function $f(x)=2x+1$.

Solution

First, as there are no restriction to the domain, we can pick various values of x, preferably near zero, and evaluate the associated values of y. Starting with the y-intercept, when $x=0$,

$$y=f(0)=2\times0=1=1.$$

Continuing in this fashion, we can recreate a t-chart with the resulting information:

x	y
0	1
1	3
2	5

Then using these three points as a guide, we can draw the linear relationship between x and y.

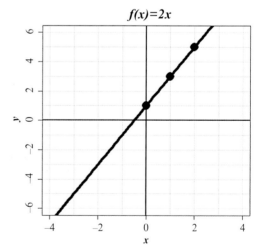

$f(x)=2x$

In the previous example, we could have plotted the line with just two points; however, it is better to graph at least three points to verify the scale of the graph. If your resulting graph does not look like a line, then your axis might be skewed. But to define a line, all that is required is a minimum of two points: (x_1,y_1) and (x_2,y_2). From these two pieces of information, we can obtain the marginal change or the slope of a line using the **slope formula**.

There are only four possibilities when discussing the slope of a line; as we read from left to right, the line either increase (moves up), decreases (moves down), does not move up or down (no movement) or movement from left to right is not possible (as the line is vertical) in which case the slope is undefined.

Solution—cont'd

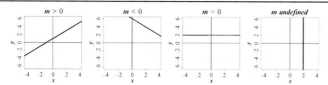

The idea of slope at first is a general assignment: upward movement will be positive, downward movement will be negative, no movement (up or down) will be zero movement, and if the line is vertical and you cannot read or move from left to right, the movement is undefined. However, the specifics need to address the magnitude of the slope; there are steep slopes and shallow slopes.

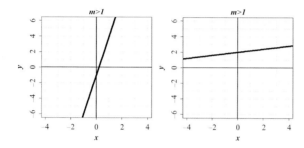

We will define steeps slopes as being greater than 1 and shallow slopes as being less than one. Therefore, one way to express the magnitude of a slope is the ratio of the rise to the run. That, the change is y over the change in x.

Definition 10.6.4 Slope Formula

The marginal change in a line is a indication of the slope of a line. Given two points on a line, (x_1, y_1) and (x_2, y_2), the **slope formula** is given by

$$m = \frac{\Delta y}{\Delta x} = \frac{y_2 - y_1}{x_2 - x_1},$$

where Δ is read "the change in."

Once you know a point and the slope of a line, we can then manipulate this equation (**slope formula**) and solve for y as a function of x.

First, given the slope m and a point (x_1, y_1), we have

$$m = \frac{y - y_1}{x - x_1}.$$

where (x, y) is an arbitrary point on the line. Note: we do not need to worry about when $x = x_1$ as we the associated value of y for this value of x and therefore, given $x \neq x_1$, we have

$$m(x - x_1) = (y - y_1)$$

or equivalently,

$$y - y_1 = m(x - x_1).$$

In this form, we need **a point** and the **slope**; hence this form is referred to as the **point-slope form** of a line.

Simplifying this further, we get

$$y = mx + mx_1 + y_1,$$

where $mx_1 + y_1$ is just one big constant, so using b or big, we have

$$y = mx + b,$$

where m is the slope and b is the y-intercept; hence, this form is referred to as the **slope-intercept form** of a line.

Once last form of a line is to manipulate the equation so that both x and y are on the same side and the coefficient on x is a positive integer. However, this is a **general form** and is not written in the form of a function.

Common Forms of a Line

Point-slope form	$y - y_1 = m(x - x_1)$
Slope-intercept form	$y = mx + b$
General form	$ax + by = c$
Horizontal line	$x = a$
Vertical line	$y = b$

Two lines are considered parallel if they have the same slope and perpendicular if the product of the slopes is negative one, or one is the negative reciprocal of the other.

Property 10.6.1 Lines

Two lines, $y = m_1 x + b_1$ and $y = m_2 x + b_2$ are **parallel** if

$$m_1 = m_2.$$

Two lines, $y = m_1 x + b_1$ and $y = m_2 x + b_2$ are **perpendicular** if

$$m_1 \times m_2 = -1$$

or

$$m_2 = -\frac{1}{m_1}.$$

Once we have an understanding of lines, we turn out attention to quadratics.

Definition 10.6.5 Quadratic Function

A **quadratic function** is a second-degree relationship between x and y such that

$$f(x) = ax^2 + bx + c,$$

where $a \neq 0$ (otherwise it is linear function) and c is the y-intercept (that is, $y = c$ when $x = 0$). The graph of this relationship is a **parabola**.

Property 10.6.2 Quadratic Function

Given $f(x) = ax^2 + bx + c$ where $a \neq 0$, when $a > 0$ then the parabola is cupping upward and when $a < 0$, the parabola is cupping downward.

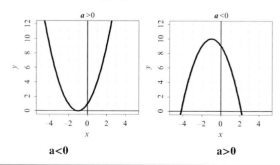

a<0   a>0

Common Forms of a Quadratic	
Standard form	$y = ax^2 + bx + c$
Graphic form	$y = a(x-h)^2 + k$
	where (h, k) is the vertex of the parabola.

EXERCISES

10.6.1. State the domain and range in the given relationship. Determine if the relationship is a function. If not, discuss why.
    a. $\{(1,2),(2,3),(3,1)\}$
    b. $\{(1,2),(1,5),(1,7)\}$
    c. $\{(2,2),(3,3),(4,4)\}$
    d. $\{(2,2),(2,3),(2,3)\}$

10.6.2. Find the slope of the line that passes through the given two points.
    a. $(1,2)$ and $(4,5)$
    b. $(-1,1)$ and $(1,4)$
    c. $(-1,2)$ and $(-1,5)$
    d. $(2,7)$ and $(3,7)$

10.6.3. Use the point-slope form of the equation of the line and simplify into slope-intercept form.
    a. Passes through $(2,1)$ with a slope of 4.
    b. Passes through $(4,2)$ with a slope of -2.
    c. Passes through $(-1,0)$ with a slope of 3.
    d. Passes through $(0,5)$ with a slope of -4.

10.6.4. Find the slope and the y-intercept of the linear function and graph the equation.
    a. $2x + y = 6$
    b. $y = -2x + 1$
    c. $x + y = 5$
    d. $2x = 8$

10.6.5. Determine if the lines are parallel, perpendicular, or neither.
    a. $y = 2x + 5$ and $2x - y = 3$
    b. $y = 2x + 4$ and $x + 2y = 3$
    c. $y = 3x + 5$ and $x - 3y = 3$
    d. $y = 0.4x - 2$ and $2x - 5y = 3$

10.6.6. Find the equation of the line parallel to $y = -2x + 5$ passing through the point $(1,5)$.

10.6.7. Find the equation of the line perpendicular to $y = x + 5$ passing through the point $(3,5)$.

CRITICAL THINKING AND BASIC EXERCISES

10.1. Determine which set each number belongs to the various sets and complete the following chart:

	Real	Rational	Irrational	Integer	Whole	Natural
2						
$\sqrt{5}$						
$\sqrt{2}$						
$\dfrac{\pi}{4}$						
e^5						
$0.\overline{112}$						
$\dfrac{\sqrt{28}}{\sqrt{7}}$						

10.2. Addition:

a.
```
  3 6
+   7
─────
□ □
```

b.
```
  1 5
+   3
─────
□ □
```

c.
```
  4 3
+   7
─────
□ □
```

d.
```
  5 8
+   7
─────
□ □
```

e.
```
  6 5 9
+   7 4
───────
□ □ □
```

10.3. Subtract:

a.
```
  3 6
−   7
─────
□ □
```

b.
```
  1 5
−   3
─────
□ □
```

c.
```
  4 3
−   7
─────
□ □
```

d.
```
  5 8
−   7
─────
□ □
```

e.
```
  6 5 9
−   7 4
───────
□ □ □
```

10.4. Multiply:

a.
```
  3 6
×   7
─────
□ □
```

b.
```
  1 5
×   3
─────
□ □
```

c.
```
  4 3
×   7
─────
□ □
```

d.
```
  5 8
×   7
─────
□ □
```

e.
```
  6 5 9
×   7 4
───────
□ □ □
```

10.5. The number 9,125,045 is divisible by which of the following: 2, 3, 4, 5, 6, or 9?

10.6. Simplify
  a. 2^0
  b. 3^2
  c. 4^{-1}
  d. $(-5)^3$
  e. $\left(\dfrac{1}{6}\right)^2$

10.7. Write the value of the underlined digit in words.
  a. $\underline{1}02.62$
  b. $3.\underline{5}$
  c. $30.0\underline{5}$
  d. $10.6\underline{2}$

10.8. Convert the given number to base 10: 1157 (base 9).

10.9. Convert the given number to base 8: 2021 (base 10).

10.10. Convert the given number to base 7: 1122 (base 3).

10.11. Give the GCF for the following sets of numbers: 15, 24, 39.

10.12. Multiply or divide.
  a. $3\dfrac{1}{2} \times 1\dfrac{3}{4}$
  b. $\dfrac{2}{3} \div \dfrac{5}{6}$
  c. $\dfrac{8}{9} \times \dfrac{3}{42}$
  d. $2\dfrac{1}{8} \div 1\dfrac{1}{3}$

10.13. Complete the chart. Reduce fraction to lowest terms.

	Fraction	Decimal	Percent
a.	$\dfrac{3}{7}$		
b.	$\dfrac{1}{4}$		
c.		0.35	
d.		2.5	
e.			19%

10.14. Solve.
  a. 5 is what percent of 45?
  b. 4 is what percent of 72?
  c. What is 12% of 300?

10.15. Identify the pattern and find the next three terms:
  a. 1, 5, 9, 13, …
  b. 1, 5, 25, 125, …
  c. 1, 8, 27, 64, …

10.16. Write the algebraic expression: Three more than five times a number.

10.17. Combine like terms: $2x^5y - 12xy + 4x^5y + xy$

10.18. Solve these equations.
  a. $5x = 45$
  b. $13b = -117$
  c. $-\dfrac{3}{5}x = 18$

10.19. Take x to be a real number, solve, and graph the resulting inequality:
  a. $2 < x + 1 \leq 7$

10.20. Solve the system of equations $\begin{cases} 3x - 2y = 23 \\ 2x + 3y = 24 \end{cases}$.

SUMMARY OF IMPORTANT CONCEPTS

Definitions:

10.1.1. **Natural numbers** are numbers used for counting and ordering objects.

$$\mathbb{N} = \{1, 2, 3, 4, \ldots\}$$

10.1.2. The word **zero** has had many derivations, but comes from the Arabic word *sifr* meaning "nothing" or "empty."

$$0$$

10.1.3. **Whole numbers** are numbers used for counting objects in a given set including the empty set, i.e., zero.

$$W = \{0, 1, 2, 3, 4, \ldots$$

10.1.4. The word **integer** comes from Latin meaning "untouched." **Integers** are positive or negative whole numbers or natural numbers including zero.

$$\mathbb{Z} = \{\ldots, -3, -2, -1, 0, 1, 2, 3, 4, \ldots\}$$

10.1.5. **Rational numbers** are the numbers which can be expressed as the quotient or fraction with integers (with the non-zero denominator) or as the quotient of an integer and a natural number.

$$\mathbb{Q} = \left\{ \frac{m}{n} \,\middle|\, m \in \mathbb{Z} \text{ and } n \in \mathbb{N} \right\}$$

10.1.6. **Irrational numbers** are the numbers which cannot be expressed as a ratio or rational number. The irrational numbers include transcendental numbers that cannot be represented as terminating or repeating decimals.

10.1.7. The set of **real numbers** are those rational and irrational numbers which constitute the continuum from $-\infty$ to ∞. That is, the union of the sets $\mathbb{R} = \mathbb{Q} \cup \overline{\mathbb{Q}}$.

10.1.8. An **operator** is a function representing a mathematical operation. This can range from **conjunction** and **disjunction** in logic, to **intersection** and **union** in sets, to **addition**, **subtraction**, **multiplication**, **division** and **exponentiation** in algebra.

10.1.9. **Addition** is the mathematical operator that represents combining collections or sets of objects together into a larger collection or set; signified by the plus sign $(+)$.

10.1.10. **Subtraction** is the inverse operation of addition; signified by the minus sign $(-)$.

10.1.11. **Multiplication** is the mathematical operation of scaling one number by another; signified by the cross symbol or multiplication sign $(\times)$.

10.1.12. **Division** is the inverse operation of addition; signified by the division sign $(\div)$ or (as the division sign is not readily available on the keyboard), the backslash (/) or more appropriately, a division line as used in fraction; for example, $1 \div 2 = \dfrac{1}{2}$.

10.1.13. **Exponentiation** is the mathematical operation involving two numbers, the base value, b, and the exponent, n. This heading covers both **powers** (n is a positive integer) and **roots** (n is one over a positive integer). Written b^n, when n is a positive integer, we have a **power** of the base, $b^n = \underbrace{b \times b \times \cdots \times b}_{n \text{ times}}$

10.2.1. **Positional notation** or **place-value notation** is a method of representing numbers. A form of coding, this is part of the number system that uses symbols or position for the different orders of magnitude. In the decimal system, we have the "ones place," "tens place," "hundreds place," etc. (all powers of base 10).

10.2.2. A number is said to be in **scientific notation** if there is one none zero number before the decimal, times a power of the base 10.

10.2.3. A **fraction** is a part of a whole or ratio of two numbers, the numerator (number of parts) divided by the denominator (how many parts in a whole). A **proper fraction** is one such that the numerator is less than the denominator. An **improper fraction** is one such that the n numerator is greater than or equal to the denominator. Proper fractions added to whole numbers are called mixed numbers or **mixed fractions**.

10.2.4. The **greatest common factor (GCF)** or **greatest common divisor (GCD)** is the largest possible positive integer that divides evenly into the numerator and the denominator.

10.2.5. A **prime number** is a natural number greater than 1 that has 1 as its only positive divisor. $\{2, 3, 5, 7, 11, 13, 17, 23, \ldots\}$. The opposite of a prime number is a **composite number**: $\{4 = 2 \times 2, 6 = 2 \times 3, 8 = 2 \times 2 \times 2, \ldots\}$. Hence, **prime factorization** is the decomposition of a composite number into smaller primes.

10.2.6. A **common denominator (CD)** is a common multiple of the denominator; the **least common denominator** is the **least common multiple (LCM)** between the given numbers.

10.2.7. A **ratio** is a comparison of two numbers (preferably whole numbers) by division. Ratios can be expressed in three distinct ways: $\dfrac{a}{b}$ or a *to* b or $a : b$, where a and b are in reduced form. A **proportion** is a statement that two ratios are equal.

10.2.8. **Percent** means "hundredth" and is denoted by the percent sign %. To change a number written in decimal form into a percentage, move the decimal point two places to the right and include the percent sign to the right of the new number. To change a number written in percentage form into a decimal, move the decimal point tow places to the left and remove the percent sign.

10.3.1. **Inductive reasoning** or induction is a kind of reasoning in which a small set of observations is used to infer a larger theory.

10.3.2. A **sequence** is an order list of objects, events or outcomes, $a_1, a_2, \ldots a_n$. In the context of this chapter, a **sequence** will refer to an arrangement of numbers.

10.3.3. A **term** is an individual outcome in a sequence enumerated as a function of the position in the sequence $a_n = P(n)$.

10.3.4. An **arithmetic sequence** is a sequence of numbers such that there is a common difference between consecutive terms. Recursive notation: $a_n = a_{n-1} + d$ Non-recursive notation: $a_n = a_1 + d(n-1)$

10.3.5. A **geometric sequence** is a sequence of numbers such that there is a common ratio between consecutive terms. Recursive notation: $a_n = a_{n-1} \times r$ Non-recursive notation: $a_n = a_1 \times r^{n-1}$

10.3.6. A **series** is the sum of the terms in a sequence, $s_1 = a_1, s_2 = a_1 + a_2, \ldots, s_n = a_1 + a_2 + \cdots + a_n$. In general, we have $s_n = \sum_{i=1}^{n} a_i$. In constructing a series by adding on term at a time, a new sequence of partial sums is generated: $s_1, s_2, s_3, \ldots, s_n$.

10.4.1. An **algebraic expression** is a mathematical statement which contains numbers, variables and arithmetic operators.

10.4.2. An **algebraic term** is an expression where numbers and unknowns (letters) are combined in multiplication or division.

10.4.3. The **coefficient** in an expression is the numerical value and is written before any letters in the expression.

10.4.4. In an expression, **like terms** are terms that have all the same unknowns with the same exponents with different coefficients.

10.4.5. A **formula** is an expression constructed using symbols to represent numerical values that related expressions in equality. Formulas are mathematical sentences which related underlying mathematical expressions.

10.4.6. **Direct variation** between two variables is such that their values always maintain a constant ratio: $\frac{y}{x} = k$. This relationship can also be written in terms of y **varying directly** as a multiple of x: $y = kx$. Give x and y vary directly, we might state that y is **directly proportional** to x. In this case, k is referred to as the **constant of proportionality**.

10.4.7. **Indirect variation** between two variables is such that their values always maintain a constant product: $xy = k$. This relationship can also be written in terms of y **varying inversely** as a multiple of x: $y = \frac{k}{x}$. Give x and y vary indirectly, we might state that y is **inversely proportional** to x. In this case, k is referred to as the **constant of proportionality**.

10.5.1. An **equation** is a mathematical sentence that asserts the equality of two expressions.

10.5.2. A **literal equation** is an equation that contains letters such a, b, c in place of constants and letters such as x, y, z for the unknowns (variables).

10.5.3. An **inequality** is a mathematical sentence that asserts the two expressions are related, but not necessarily equal. These mathematical sentences are less than, less than or equal to, greater than an greater than or equal to. The strict inequalities are not **reflective** or **symmetric**; however, they are **transitive**; that is,

Not Reflective	$a \leq a$, but NOT $a < a$.
Not Symmetric	If $a < b$, then NOT $b > a$.
Transitive	If $a < b$ and $b < c$, then $a < c$.

10.5.4. An **unbounded interval** is an interval which does not include a finite end point; there are several possible forms of an **unbounded interval**: $(-\infty, +\infty)$, $(-\infty, b)$, $-\infty, b]$, $(a, +\infty)$ and $[a, +\infty)$.

10.5.5. A **bounded interval** is an interval which does have finite end points; there are several possible forms of an **bounded interval**: (a, b), $(a, b]$, $[a, b)$, and $[a, b]$.

10.5.6. The **absolute value** of a number x, denoted by $|x|$, is the numerical value (magnitude of the value) without the sign (direction on the real number line). It is the distance between the value on the number line and zero.

$$|x| = \begin{cases} x & x \geq 0 \\ -x & x < 0 \end{cases}$$

10.5.7. A **quadratic equation** is a second-degree expression with standard form of $ax^2 + bx + c = 0$, where a, b and c are real numbers and $a \neq 0$.

10.6.1. A relationship or relation is a set of ordered pairs (x, y). The largest set of real numbers the variable x takes on is referred to as the **domain** of the function and the largest set of real numbers the variable y takes on is referred to as the **range** of function.

10.6.2. A **function** is a relation between two variables, x and y, such that for each value of the independent variable, usually selected to be x there is at most one value of the dependent variable, y. That is, "y is a function of x" when for each value of x in the domain, there is one value of y in the range. We express this relation by $y = f(x)$.

10.6.3. A **linear function** is a first-degree relationship between x and y such that

$$f(x)=mx+b,$$

where m is the **marginal change (slope)** of the line and b is the baseline or y-intercept (that is, $y=b$ when $x=0$). The graph of this relationship is a **line**.

Two special forms of a linear equation are the **vertical line**,

$$x=a$$

and the **horizontal line**

$$y=b.$$

10.6.4. The marginal change in a line is a indication of the slope of a line. Given two points on a line, (x_1, y_1) and (x_2, y_2), the **slope formula** is given by

$$m=\frac{\Delta y}{\Delta x}=\frac{y_2-y_1}{x_2-x_1},$$

where Δ is read "the change in."

10.6.5. A **quadratic function** is a second-degree relationship between x and y such that

$$f(x)=ax^2+bx+c,$$

where $a\neq0$ (otherwise it is linear function) and c is the y-intercept (that is, $y=c$ when $x=0$). The graph of this relationship is a **parabola**.

Properties

10.1.1. A set has **closure** if a and b are real numbers, then $a+b$ and $a\times b$ are also real numbers. If we let the symbol $\in$ stand for "is an element of" then this idea can be abbreviated further mathematically as follows: if $a,b\in\mathbb{R}$, then $a+b\in\mathbb{R}$ and $a\times b\in\mathbb{R}$.

10.1.2. In the additive expression: $(a+b)+c$, the number a is associating with the number b, but in the expression: $a+(b+c)$, the number b as been re-associated with the number c. And these associations are equivalent; similarly with multiplication, the expression $(a\times b)\times c$ is equivalent to $a\times(b\times c)$; the **associative** properties are mathematically: $(a+b)+c=a+(b+c)$ and $(a\times b)\times c=a\times(b\times c)$.

10.1.3. In the additive expression: $a+b$, the number a is in the first position with the number b in the second position, but in the expression: $b+a$, the number a as been commuted with the number b. And these expressions are equivalent; similarly with multiplication $a\times b$ is equivalent to $b\times a$; the **commutative** properties are written mathematically: $a+b=b+a$ and $a\times b=b\times a$.

10.1.4. In the additive expression: $a+0$, when nothing is added to the number a, the number is not changed; it is the same number a and therefore 0 is the **additive identity**; similarly with multiplication, $a\times1$ is just one number a, which is equivalent to a and therefore 1 is the **multiplicative identity**; the **identity** properties are written mathematically: $a+0=0+a=a$ and $a\times1=1\times a=a$.

10.1.5. In the additive expression: $a+(-a)$, $(-a)$ is the opposite (**additive inverse**) of a, which mean if you add both a and $(-a)$, you have in essence have added nothing. Equivalently since subtraction is the opposite of addition, in the expression $a-a$ if it is understood that you added a and then subtracted a, you have in essence have added nothing.

Hence, both of these expressions are equivalent to adding 0. Similarly with multiplication, in the expression $a\times\frac{1}{a}$, $\frac{1}{a}$ is the reciprocal (**multiplicative inverse**) of a, which means multiplying by both is in essence just identifying a given expression. Equivalently since using power notation $a=a^1$ and division is the opposite of multiplication, the reciprocal can be denoted by $\frac{1}{a}=a^{-1}$; that is in the expression $a\times a^{-1}$, if it is understood that you multiplied and divided by a, you have in essence multiplied by 1. Therefore, the **inverse** properties are written mathematically:

$$a+(-a)=a-a=0 \text{ and } a\times\frac{1}{a}=a\times a^{-1}=1.$$

10.1.6. In the expression $a \times (b + c)$ the parentheses indicate that the multiplication by the number a is affecting both the number b in addition to number c; that is, this expression is equivalent to $a \times b$ in addition to $a \times c$. The number a needs to be **distributed** to both the number b in addition to the number c. Similarly, this relation of distribution can also be seen between multiplication and powers: since multiplication leads to powers, powers can be distributed equally over multiplication. In the expression $(a \times b)^n$ the parentheses indicate that the power on n is affecting both the number a and the number b. We will learn in counting that "and" means "multiply" and therefore this expression is equivalent to a^n and b^n. The power b needs to be **distributed** to both the number a and the number b. That is, the **distributive** properties are written mathematically: $a \times (b + c) = a \times b + a \times c$ and $(a \times b)^n = a^n \times b^n$.

10.1.7. The **exponential** properties reduce multiplication between like bases to addition in the exponent: $a^x \times a^y = a^{x+y}$; reduces division between like base to subtraction: $\dfrac{a^x}{a^y} = a^{x-y}$; reduces powers of a base to a power to multiplication: $(a^x)^y = a^{x \times y}$. Moreover, $a^1 = a$ and $a^0 = 1$.

10.1.8. The **order of operation** is the order in which the operations are performed are in order of complexity from most complex operations down to addition and subtraction: **parenthesis, exponents, multiplication, division, addition**, and **subtraction**.

10.5.8. These mathematical sentences are **reflective, symmetric** and **transitive**; that is,

Reflective	$a = a$
Symmetric	If $a = b$, then $b = a$.
Transitive	If $a = b$ and $b = c$, then $a = c$.

10.6.1. Two lines, $y = m_1 x + b_1$ and $y = m_2 x + b_2$ are **parallel** if $m_1 = m_2$. Two lines, $y = m_1 x + b_1$ and $y = m_2 x + b_2$ are **perpendicular** if $m_1 \times m_2 = -1$ or $m_2 = -\dfrac{1}{m_1}$.

10.6.2. Given $f(x) = ax^2 + bx + c$ where $a \neq 0$, when $a > 0$ then the parabola is cupping upward and when $a < 0$, the parabola is cupping downward.

Rules

10.1.1. When adding real numbers the **sign effect** on positive and negative values is as follows: **Like signs**: Add and keep like sign. **Opposite signs**: Subtract and keep sign of larger absolute value

10.1.2. When multiplying positive and negatives real numbers the **sign effect** on positive and negative values is as follows: Interpreting to the right as forward and to the left as negative, multiplying or dividing by a negative is like turning round **180°**.

$$(-1) \times (+1) = -1$$

Continuing with this geometric description, multiplying by a negative twice is that same as doing a 360 turn, facing the same (positive direction) as you started.

$$(-1) \times (-1) = +1$$

Hence, an even number of negatives multiply to be positive and an odd number of negatives multiply to be negative.

10.1.3. **Subtraction is the addition of the negative**; written mathematically:

$$a - b = a + (-b)$$

10.1.4. **Division is the multiplication of the reciprocal**; written mathematically:

$$a \div b = a \times \dfrac{1}{b}.$$

10.2.1. **Operators in fractions**: When adding fractions, the fractions must have a common denominator.

$$\dfrac{a}{b} + \dfrac{c}{d} = \dfrac{ad}{bd} + \dfrac{bc}{bd} = \dfrac{ad + bc}{bd}$$

However, when multiply we multiply top to top and bottom to bottom; that is,

$$\dfrac{a}{b} \times \dfrac{c}{d} = \dfrac{ac}{bd}.$$

When divided by a fraction, we multiply by the reciprocal,

$$\frac{a}{b} \div \frac{c}{d} = \frac{a}{b} \times \frac{d}{c} = \frac{ad}{bc}.$$

10.2.2. **Operators in decimals**: When **adding** or **subtracting decimals**, first line up the decimal point underneath each other; add or subtract decimals in the same way as with whole numbers; then bring down the decimal point straight down into the answer. When **multiplication**, perform the operation without the consideration of the decimal, as if the numbers are whole and then count over the total number of decimals in each number combined in the final result. When **dividing** decimals, move the decimal place to the left the same number of places in the numerator as the dominator; or use scientific notation.

10.3.1. Given an **arithmetic sequence**, $a_n = a_1 + d(n-1)$, the series is the average of the first and last term times the number of terms:

$$s_n = \frac{a_1 + a_n}{2} \times n,$$

or equivalently, by substitution,

$$s_n = \frac{n[2a_1 + (n-1)]}{2}.$$

10.5.1. Given an equation

$$a = b$$

the following **elementary operations** yield equivalent expressions:

$$a + c = b + c,$$
$$a - c = b - c,$$
$$a \times c = b \times c,$$
$$a \div c = b \div c, c \neq 0$$

Given an equation

$$a < b$$

the following elementary operations yield equivalent expressions: $a + c < b + c$ and $a - c > b - c$,

$$a \times c > b \times c, c > 0$$
$$a \times c < b \times c, c < 0$$
$$a \div c > b \div c, c > 0$$
$$a \div c < b \div c, c < 0$$

10.5.2. In general, if a is any positive value, then we have the following equivalences with respect to **absolute values**:

$$
\begin{array}{ll}
|x| = a & x = -a, x = a \\
|x| < a & -a < x < a \\
|x| \leq a & -a \leq x \leq a \\
|x| > a & x < -a \text{ or } x > a \\
|x| \geq a & x < -a \text{ or } x > a
\end{array}
$$

10.5.3. In general, the solutions to a **quadratic equation** $ax^2 + bx + c = 0$, where a, b and c are real numbers and $a \neq 0$ can be obtained from the **quadratic formula**

$$x = \frac{-b \pm \sqrt{b^2 - 4ac}}{2a}.$$

10.5.4. Given a system of two equations and two unknowns in the following form:

$$\begin{cases} ax + by = c \\ dx + ey = f \end{cases}$$

Let $D = \begin{bmatrix} a & b \\ d & e \end{bmatrix}$ (the coefficients on x and y), $D_x = \begin{bmatrix} c & b \\ f & e \end{bmatrix}$ (the matrix D with the balancing constants in place of the coefficients on x) and $D_y = \begin{bmatrix} a & c \\ d & f \end{bmatrix}$ (the matrix D with the balancing constants in place of the coefficients on y); then if $|D| \neq 0$, by **Cramer's Rule**

$$x = \frac{|D_x|}{|D|},$$

and

$$y = \frac{|D_y|}{|D|}.$$

10.5.5. In **optimization**, the *maximum* or *minimum* of the objective function over the feasible set, assuming it exists, will be attained at least one of the vertices of the feasible set.

10.6.1. In general, there are only three **restrictions to the domain**:

1. You cannot divide by zero; that is, given an expression $\frac{\textbf{Top}}{\textbf{Bottom}}$, the bottom cannot be zero, **Bottom $\neq$ 0**.
2. You cannot take the square root of a negative value; that is given an expression $\sqrt{\textbf{Inside}}$, then the insider is greater than or equal to zero, **Inside$\geq$0**.
3. You cannot take the logarithm of a negative value or zero; that is given an expression $\log_b(\textbf{Inside})$, then the insider is greater than zero, **Inside>0**.

Common: Algebraic Expressions

+	Plus, addition, sum, added to, increased by, more than
−	Minus, subtraction, difference, less, decreased by, less than, subtracted from
×	Product, multiplication, times, of
÷	Quotient, division, divided by, ratio of, per
$\square^2$	Squared, the power of two indicated by position of the two
0	Zero, nothing, none

1	One	2	Two	3	Three
4	Four	5	Five	6	Six
7	Seven	8	Eight	9	Nine

Common: Forms of a Line

Point-slope form	$y - y_1 = m(x - x_1)$
Slope-intercept form	$y = mx + b$
General form	$ax + by = c$
Horizontal line	$x = a$
Vertical line	$y = b$

Common: Forms of a Quadratic

Standard form	$y = ax^2 + bx + c$
Graphic form	$y = a(x - h)^2 + k$

where (h, k) is the vertex of the parabola.

REVIEW TEST

1. A club sells 987 raffle tickets at \$4 each. The printing cost of the tickets cost \$27 and the prize cost \$200. What is the profit from the raffle?

2. Sort the fractions from smallest to largest: $\dfrac{2}{3}, \dfrac{5}{8}, \dfrac{3}{5}, \dfrac{3}{4}, \dfrac{5}{6}$.

3. A butcher has 27 pounds 12 ounces of meat and sells 18 pounds 15 ounces. How much meat remains?

4. Subtract: $3\dfrac{1}{2} - 1\dfrac{2}{5} =$

5. Divide: $6\dfrac{1}{3} \div 2\dfrac{1}{6} =$

6. Put the numbers in ascending order: 0.0505, 0.0408, 0.0410, 0.0590, 0.0480.

7. Give a fraction between $\dfrac{3}{7}$ and $\dfrac{6}{11}$.

8. Determine the correct inequality:

   a. $\dfrac{5}{9} \square \dfrac{26}{30}$

   b. $\dfrac{5}{9} \square \dfrac{7}{8}$

   c. $\dfrac{1}{7} \square \dfrac{5}{7}$

   d. $\dfrac{2}{3} \square \dfrac{4}{5}$

9. Simplify: $6 \times (3-8) \div 15 + 4$

10. What is 40% of 75?

11. If 45 is 25% of a number n, find n.

12. A \$125 suit is on sale for 20% off. What is the sales price?

13. If there are 45 students in a class and 5 fail, what percent passed?

14. The sales tax is given to be 7%. If a cash register has \$266.43 in the drawer including money paid for merchandize plus the tax, how much was taken in as taxes?

15. Evaluate: $5 + 9 \times 4 \div 6 - 2$

16. Evaluate: $5 \times 2^2 - 45 \div 3^2$

17. Evaluate: $-15 - (-20) + 3 - 5$

REFERENCES

Miller, D., 1942b. Popular Mathematics: The Understanding and Enjoyment of Mathematics. WM. H. Wise & CO., Inc, New York.

Hull, 1894. Elementary Arithmetic. E. H. Butler & Co.

"It takes 20 years to build a reputation and five minutes to ruin it. If you think about that, you'll do things differently."

WARREN BUFFETT

Chapter 11	Basic Finance

Warren Edward Buffett was born in Omaha, Nebraska on August 30, 1930. In 2010, his estimated worth was $45 billion. Called the Oracle of Omaha, Warren Buffett strongly adheres to the value investing philosophy and frugality despite his growing wealth. As the Chairman and CEO of Berkshire Hathaway Investor, Buffet has a yearly salary of $100,000. However, he has worked as stock broker, purchased Sinclair Texaco gas stations as a side investment, worked with Benjamin Graham in his partnership—Graham was on the board of directors for GEICO, and by 1957 had three partnerships in operations. His partnerships continued to grow and by 1962, Buffett was a millionaire, continuing to live in the house he purchased in 1957 and not spending his money frivolously, but rather investing in stocks including Coca Cola. By the 1990s, Buffett was a billionaire. In 1999, in a survey by the Carson Group, he was named the top money manager of the twentieth century. In 2007, he was listed among Time's 100 Most Influential People and in 2011, President Barack Obama award him the Presidential Medal of Freedom.

Commerce is an interchange of good or commodities such as eggs and milk.

11.1 INTRODUCTION TO BASIC FINANCING

Finance is a branch of economics concerned with resource allocation. This includes resource management, acquisition, and investment. This science of management of money and other assets started with trade.

Definition 11.1.1 Trade

Trade is the transfer of ownership of goods and services from one individual or party to another individual or party, also referred to as **commerce**, **financial transaction**, or **barter**. A network that allows trade is called a **market**, such as the **Stock Market**, **Retail Market**, **Super Market**, etc.

There are books and training videos to better teach an individual how to "trade." Based on futures and options, one invests in the desired company through the stock market and hopes that they are able to find stock worth keeping as the company is growing and share prices are increasing, or to make a quick turn over by selling the stock at a higher price than it was purchased for. However, you must know what you are looking for and be cautious with software packages which monitor the trends and try to predict when to buy and sell as these applications are often based on incorrect statistics. A wise man once advised; only purchase stock for a product you use regularly.

Commodity is an article of trade, a physical product as opposed to a service.

With the age of computers, such software packages and online banking are common place; this has made some question the need for currency. **Currency** is standardized **money** to facilitate a wider exchange of goods and services. This first stage of currency was metals that were used to represent stored value with symbols to represent commodities. Such currencies formed the basis of **trade** in the **Fertile Crescent** for over 1500 years.

A currency is a type of money and medium of exchange made from paper, cotton, and metal.

Definition 11.1.2 Money

Money is a medium of exchange, any object or note that is widely accepted as payment for goods or services rendered, for the repayment of debt. **Money** is subject to the given country or socio-economic content.

Coins and Banknotes—the two most common form of tangible currency

---

Definition 11.1.3 Debt

Debt is that which is owed; in economics, this refers to assets owed. **Debt** is created when a lender agrees to lend a specific amount or asset to the borrower or **debtor**.

If one gets overwhelmed with debt, then one might consider **bankruptcy**. Originally, **bankruptcy** was to the creditors benefit, allowing them to seize all the assets of an individual or trader who could not pay his debts. In addition to losing all of their property, the individual also when to debtor's prison. It was the burden of the family to pay the debt to obtain the debtors release.

Bankruptcy in the United States is permitted by the **United States Constitution** (Article 1, Section 8, and Clause 4) which authorizes Congress to enact "*uniform Laws on the subject of Bankruptcies throughout the United States.*" Title 11 of the United States Code is the primary source of Bankruptcy law statutes in the United States Code which is subdivided into nine remaining chapters (as many chapters have been repealed).

The most typical declaration of bankruptcy is Chapter 7 for individuals who are unable to pay their existing unsecured debt; Chapter 11 for companies or individuals over the debt limitations of Chapter 13 which for secured debts.

- Chapter 1—General Provisions
- Chapter 3—Case Administration
- Chapter 5—Creditors, The Debtor, and The Estate
- Chapter 7—Liquidation
- Chapter 9—Adjustment of Debts of a Municipality
- Chapter 11—Reorganization
- Chapter 12—Adjustment of Debts of a Family Farmer or Fisherman With Regular Annual Income
- Chapter 13—Adjustment of Debts of an Individual With Regular Income
- Chapter 15—Ancillary and Other Cross-Border Cases

> **Definition 11.1.4 Bankruptcy**
>
> **Bankruptcy** is a legally declared inability or impairment of ability of an individual or organization to pay its creditors or lenders. **Bankruptcy** is usually initiated voluntarily by the insolvent (bust) individual or organization.

There were 936,795 bankruptcy filings in 2014.

Title 11 of the United States Code—Bankruptcy

	2007 filings	2008 filings	2009 filings
Chapter 7—Liquidation	413,294	560,015	819,362
Chapter 11—Business reorganization	5199	6971	11,785
Chapter 12—Family farmers & fishermen	372	343	367
Chapter 13—Individual reorganization	276,649	334,551	370,875

Bankruptcy Facts
- For a period of time after the filing, you will normally be deprived of access to consumer loans and credit cards.
- If you own a business that files for bankruptcy, your own credit rating will be damaged.

Bankruptcy Myths
- Filing for bankruptcy will cause you to lose your job—this is not true.
- Filing for bankruptcy will cause you to lose your Social Security or other government benefits—not true; in fact, the law provides that the government cannot discriminate against you because you have filed for bankruptcy.
- You will lose all your property—not true. Although you may lose some of your property, you will not lose all of it. Many people lose little to nothing because of the bankruptcy exemptions such as your house (for example, in Florida there is no limit; however, in New Your, the limit is between $50,000 and $150,000), auto (approximately $3450 depending on if the vehicle is a tool a trade), household goods and furnishings, retirement assets, etc.
- Filing for bankruptcy will completely ruin your credit—this is not completely true, although it will affect your credit for at least 7 years and sometimes up to 10 years.

Therefore, to avoid such financial difficulties as bankruptcy, in this chapter on finance we are interested in money—how interest affects our balance and how we can earn more interest. In general, money is always abbreviated using capital letters and other information such as interest rate is represented using lower case letters.

Below is a list of several basic abbreviations.

Summary of Notation:
P	Principle
r	Interest Rate
n	Number of Periods
t	Term of Payments
i	Periodic Interest Rate
A	Future Amount
i_e	Effective Interest Rate
R	Payment

Definition 11.1.5 Principal (P)

The **principal** is the balance of the initial loan.

P That is, the borrower agrees to pay back this principal amount or proceeds and any outlined interest on this principal amount depending on the type of interested charged (simple, compound, or continuous).

The principal amount of a loan is a fixed amount. We assume no additional monies are borrowed and that a fixed **payment** will be needed to pay off the debt in periodic (number of compounds per year) payments made over a specific amount of time (**term of loan**) with interest computed at a **fixed rate**. When payments are larger than the required amount, some companies will apply the extra monies to the principal, though many companies require you specify this on the payment slip.

Definition 11.1.6 Interest Rate (r)

The **interest rate** is the percent of the initial proceeds to be charged above the principal amount borrowed.

r Nominal rate (or interest rate for one year)

The idea of simple interest has long been replaced with compound interest; that is, interest is charged every pay period which means you pay interest on previously charged interest, or in terms of savings, earn interest on interest previously earned.

Definition 11.1.7 Periods per Year (n)

The number of **periods per year** is the number of times interest is applied to the remaining principle.

n Number of times compounded per year (or number of periods per year)

Now that we know the interest rate and number of periods within each year, the questions becomes, what is the term of the loan?

Definition 11.1.8 Term of Loan (t)

The **term** of a loan is the number of years outlined in the loan agreement over which the borrower has to pay off the loan (the principal amount and accrued interest).

t The amount of time (in years)

Given a principle investment of $1, over a period of five years, see how much money you can earn:

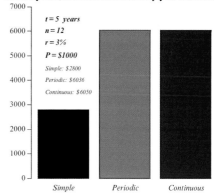

t = 5 years
n = 12
r = 3%
P = $1000
Simple: $2800
Periodic: $6036
Continuous: $6050

The first question one might be interested in is how many total payments must be made, or how many payments will it take to save a given amount.

Definition 11.1.9 Periods (m)

Number of periods is given by $m = nt$; that is, the number of periods per year times the number of years.

m The number of periods that interest is applied; that is, the total number of payments to be made.

When interest is compounded periodically over m payments, we use a periodic interest rate to determine future amount.

Definition 11.1.10 Periodic Interest (i)

The **periodic interest** rate is given by $i = \frac{r}{n}$; that is, the **interest rate per period**.

i Interest rate per period (or period interest rate).

The future amount is the amount of interest that must be paid upon interest and the interest on top of that interest, etc. over the term of the loan.

Definition 11.1.11 Future Amount (A)

The **future amount** to be paid off on a loan or a future amount desired in savings.

A Amount after time t (or future value or future amount)

Effect **rates can be evaluated** using the following formulas:

Rule 11.1.1 Future Amounts

The following formulas can be used to calculating **future amounts** under the given conditions.

$A = P(1 + rt)$ Simple Interest
$A = P(1 + i)^{nt}$ Compound Interest (compounded periodically)
$A = Pe^{rt}$ Compound Interest (compounded continuously)

As interest is accrued periodically, that is, daily, monthly, quarterly, etc., to compare these various rates, we convert them to an "effective" rate, which is the rate of yearly interest equivalent to the periodic rate; the "effective rate" is the equivalent annual rates for the given periodic rate.

Definition 11.1.12 Effective Rate (i_e)

Effective rate (annual yield) is the interest amount earned for one year per dollar amount. Effective interest rates depend on the number of compounds per year and must be formulated for each different type of interest charge. In general it will be denoted i_e.

Effect rates can be evaluated using the following formulas.

Rule 11.1.2 Effective Rates

Effective rates will depend on how the interest is compounded (if compounded).

$i_e = i$	Simple Interest
$i_e = \left(1 + \frac{r}{n}\right)^n - 1$	Compounded Periodically
$i_e = e^r - 1$	Compounded Continuously

Once all the details are specified, we can compute the necessary payment depending on the type of loan or savings plan we have.

Definition 11.1.13 Payments (R)

The periodic **payment** to be made depends on the type of interest (simple or compound), the term of the loan, and the number of periods per year and if this is a loan to be paid off or a sinking fund where monies accrue. (This cannot be abbreviated P as this notation has already been defined as principal, and R resembles P for Payment.)

R Payment per month

Understanding the minimum required payment amount is extremely important when working to save up money and when paying off a given amount of money.

Rule 11.1.3 Payments

Payments will depend on how the principal is earning interest or being charged interest: (see Section 11.2).

$R = \dfrac{Ai}{(1+i)^{nt} - 1}$	Sinking Fund when interest is earned
$R = \dfrac{Pi}{1 - (1+i)^{-nt}}$	Amortization when interest is paid
$R = P\left[1 + \dfrac{i}{1 - (1+i)^{-nt}}\right]$	Payments need to maintain a credit card

---

The effective rate depends on the number of periods per year.

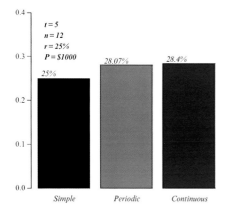

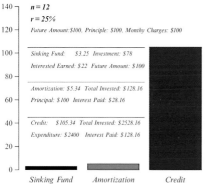

EXERCISES

Critical Thinking

11.1.1. Discuss the idea of **trade** both with and without the use of currency (**money**).

11.1.2. Is **money** universal? Discuss.

11.1.3. Is **debt** a good or bad thing? Explain. How might one avoid **debt**?

11.1.4. What is the primary source of **bankruptcy** law statutes in the United States Code? How many subdivisions?

11.1.5. Give the symbolic (shorthand) notation used to represent each of the following financial terms: **Principal, Interest Rate (Nominal Rate), Number of Periods per Year, Term of Loan, Periods, Periodic Interest, Future Amount, Effective Rates,** and **Payment**.

11.1.6. What type of funds has payments made toward a future amount where interest is earned.

11.1.7. What do you call paying off a debt with accruing interest?

11.1.8. Match the following term to its appropriate shorthanded mathematical notation:

a.	P	1)	Payment
b.	r	2)	Principal
c.	n	3)	Future amount
d.	t	4)	Total number of payments
e.	m	5)	Number of payments per year
f.	i	6)	Effective interest rate
g.	A	7)	Nominal interest rate
h.	i_e	8)	Time in years
i.	R	9)	Periodic interest rate

11.1.9. Relate in a mathematical equation the **periodic interest rate** to the **nominal interest rate** and the **number of payments per year**.

11.1.10. Relate in a mathematical equation the **total number of payments** to the **time in years** and the **number of payments per year**.

11.2 SAVINGS, SINKING FUNDS, AND AMORTIZATION

In this section, we will cover sinking funds (where interested is earned) and amortization (where interest is paid). However, we first discuss the concept of savings.

Definition 11.2.1 Savings

Savings is income not spent but rather put into a bank (deposit account) or pension fund. Savings can also be seen when the cost of commonly purchased items is reduced or moderated. Savings, as opposed to investments, are considered low-risk.

In general, there are three types of **personal savings**: emergency reserve fund, accumulation fund, and long-term investments.

Emergency Reserve Funds
- **Banks**—most convenient, but usually pay the least in interest.
- **Credit Unions**—a substitute for a bank, credit unions usually offer better returns, that is, more in interest.
- **Money Market Mutual Funds**—usually pay 1-2% interest, higher than most banks and credit unions.

Accumulation Funds
- **Certificates of Deposit (CDs)**—varying among states and nationally, CDs offer reasonable rates depending on the term. A one or two year CD would be comparable to a money market account, whereas a five year CD might offer up to 2.5%.
- **Ultra short-term Bond Funds**—these are mutual funds that offer a six-month term investment paying a reasonable rate but with higher risk than CDs.
- **Short-term Bond Fund**—offers a two to three year term with slightly better returns.
- **Mortgage-backed Bond Fund**—at least a four-year term offering attractive rates paying monthly dividends.

Long-Range Investments

- **Tier-one**—serve income investments such as Government Securities: treasury bills, treasury notes, treasury bonds, zero coupon government bonds, and U.S. Savings bonds. Such U.S. Government Securities are guaranteed by the U.S. Government and exempt from state and local income taxes.
- **Tier-two**—higher risk investments with higher rates of return.
- **Tier-three**—growth investments such as undeveloped properties, developed properties, or mutual funds.
- **Tier-four**—speculative investments such as stocks, aggressive growth mutual funds, or precious metal options.
- **Tier-five**—high risk investments, such as gold and silver, oil and gas, commodities, collectibles, precious gems, and limited partnerships.

Most Americans find it hard to build up savings; however, with discipline in spending and in maintaining a budget, having savings is the best way to prevent being fraught by debt.

Definition 11.2.2 Budgeting

Budgeting is to plan in advance of the expenditure of resources such as time and money. In personal finances, this is the total sum of monies allocated for a particular purpose or period of time.

Step 1. First gather the relevant information: principal amount, rate (annual rate), term, and periods per years, if applicable.

Step 2. Determine the appropriate formula needed to compute the future amount.

Step 3. Perform the necessary calculation.

Create a Budget:

Rent or Mortgage	$_____
Electricity	$_____
Food	$_____
Clothing	$_____
Phone	$_____
Car Payment	$_____
Insurance	$_____
Credit Card Debt	$_____
Entertainment	$_____

Once we understand the idea behind saving and budgeting, we need to discuss taxations and discounts.

Definition 11.2.3 Tax (Markup)

Tax (markup) or **taxation** is a one-time adjustment to a principal, P, amount based on a specified rate, r.

Rule 11.2.1 Tax (Markup)

The **tax** or **markup** is a percentage (rate) of the total principal:

$$T = r \cdot P = P \cdot r,$$

and the future amount (after taxations) is

$$A = P(1+r) = P + P \cdot r = P + T.$$

Example 11.2.1 State Sales Tax

The sales tax in the State of Florida is 6%. Daniel purchases a pair of rollerblades for $69.99. How much will he pay including tax?

Solution

The taxed amount is $A = \$69.99(1 + 0.06) \approx \$74.19$. Including tax, Daniel will pay $74.19 for these rollerblades.

Whereas **taxation** is a markup in the principal amount, the principal amount can also be discounted. **Discounts** are seen from sales items in department stores discounted to sell to governmental discounts when issuing treasury bills.

Definition 11.2.4 Discount (Markdown)

Discounts (depreciation) or **markdowns** is a one-time percentage decrease to a principal, P, amount based on a specified rate, r, to a principal, P, amount based on a specified annual rate, r.

Rule 11.2.2 Discount (Markdown)

The **depreciation** or **depreciated amount** is

$$D = r \cdot P = P \cdot r,$$

and the future amount (after the discount) is

$$A = P(1 - r) = P - P \cdot r = P - D.$$

Example 11.2.2 Sales Discount

Kaitlin is shopping for a new dress. She finds one marked $49.30 and the sign states 10% off all items. How much is the discount and what is the new price of the garment? What is the cost including 7% sales tax.

Solution

The discounted amount is $A = \$49.30(1 - 0.10) \approx \$44.37$, a discount of $4.93. Including tax, Kaitlin will pay $A = \$44.37(1 + 0.07) \approx \$47.48$.

Example 11.2.3 Depreciation of Value

Kendra purchased a new car worth $12,000. Given new cars on average lose 30% of their value the moment they are driven off the lot and an additional 10% each year thereafter, what is the depressed value of a car with an initial cost of $12,000 at the end of the fourth year?

Solution

After purchasing, the cars value losses 30% of its value, and is now worth $[(1 - 0.3) \times \$12,000]$ $8400. Then this amount depreciates or is discounted not just once, but four times over the next four years; that is,

$$
\begin{aligned}
A &= \$8400 \times (1 - 0.1) \times (1 - 0.1) \times (1 - 0.1) \times (1 - 0.1) \\
&= \$8400(1 - 0.1)^4 \\
&= \$5511.24.
\end{aligned}
$$

That is, after four years of use, a car that Kendra paid a total of $12,000 for is now worth $5511.24; the cost of (or money spent on) the vehicle is ($12,000 − $5511.24) $6488.76. Hence, the average yearly cost is $1622.19.

Solution—cont'd

The idea of taxations (paying an additional percent) can be extended to earning a given percent over a specified period of time. An accumulation fund is said to earn **simple interest**.

Definition 11.2.5 Simple Interest

Simple interest is interest computed once on the full amount for a given amount of time, $I = Prt$.

To evaluate the simple interest charged, the future amount or simple effective rate, use the following formulas:

Rule 11.2.3 Simple Interest

To compute the future amount in **simple interest**, first find the interest charge:

$$I = Prt$$

Therefore, the future value of a given amount earning simple interest for a given amount of time is given by

$$A = P(1 + rt) = P + I.$$

The **simple effective rate** is the actual interest rate. We write

$$i_e = r.$$

Example 11.2.4 Simple Interest

Chris invests $10,000 in principal and agreed to take 5% simple interest on this amount in two years. How much money will be earned in interest? What is the total amount that Chris will get at the end of the term?

Solution

First, we gather the information: $P = \$10,000$, $r = 0.05$, $t = 2$. The simple interest charged over these two years is given by

$$I = \$10,000 \times 0.05 \times 2 = \$1000.$$

That is, the interest on this principal amount is $1000. The total future amount is

$$A = P + I = \$10,000 + \$1000 = \$11,000,$$

that is, $11,000 is the future value for the principle of $10,000 earning simple interest at 5% annually for two years.

Definition 11.2.6 Certificate of Deposit

A **Certificate of Deposit** (CD) is a time deposit which bears a maturity date and a specified fixed interest rate. Generally they range from one to five years; formulated the same as with a tax or markup. The future amount is the principal invested plus the simple interest earned:

$$A = P(1 + r) = P + P \cdot r = P + I$$

Example 11.2.5 Investing in a CD

You purchase a $10,000 CD with an interest rate of 5% compounded annually for the term of one year. If you purchase this CD on January 1, 2012; when will it mature and how much will the CD be worth? How much is earned in interest?

Solution

This CD will mature after a single year on January 1, 2013, and will be worth

$$A = 10,000(1 + 0.05) = 10,500,$$

with $500 earned in interest.

These calculations are rather simple; however, in modern society, rarely do credit card companies or banks use the simple interest method or a rate of interest compounded once. Most charge interest on unpaid interest, that is, the interest on interest, referred to as **compound interest**:

Definition 11.2.7 Compound Interest

Compound interest is interest computed periodically (*n* times per year.)

To evaluate the future amount owed when the interest is compounded periodically, the resulting compound interest charged, or effective rate, use the following formulas:

Rule 11.2.4 Compound Interest

In **compound interest**, the amount after time *t* (or future value) is given by

$$A = P\left(1 + \frac{r}{n}\right)^{nt}.$$

The **interest** earned is given by

$$I = A - P.$$

The **effective rate** when interest is compounded periodically is

$$i_e = \left(1 + \frac{r}{n}\right)^n - 1.$$

Example 11.2.6 Compound Interest

Rebecca invests $10,000 in principal and agrees to take 5% in interest compound quarterly for two years. What is the total amount that will be due at the end of the term? How much money will be earned in interest? Compare with earning simple interest. State and interpret the effective rate.

Solution

First, we gather the information: $P = \$10,000$, $r = 0.05$, $t = 2$, $n = 4$. Note: this is similar information as in the previous example, however, here interest is earned on previously earned interest. The future value of this investment is

$$A = \$10,000\left(1 + \frac{0.05}{4}\right)^{4 \times 2} = \$11,044.86,$$

Solution—cont'd

The interest earned over these two years is given by

$$I = \$11,044.86 - \$10,000 = \$1044.86.$$

Hence, by computing compound interest quarterly over simple interest, we can earn an additional \$44.86. The effective interest rate is given by

$$i_e = \left(1 + \frac{0.04}{4}\right)^4 - 1 = 0.050945.$$

Therefore, we earn slightly more in interest when compounded quarterly; 5.0943% compared to 5% simple interest.

Example 11.2.7 Savings Account

Many banks only offer 1-2% on savings accounts. On February 17, 2011, Washington Savings Bank offered 1.5% interest compounded monthly. Assuming your average balance is \$2000, what is your expected dividend earned in a year? How much interest is earned and what is the effective rate?

Solution

First, we gather the information $P = \$2000$, $r = 0.015$, $t = 1$, $n = 12$

$$A = \$2000\left(1 + \frac{0.015}{12}\right)^{12 \times 1}$$
$$A = \$2000(1.015104)$$
$$A \approx \$2030.207$$

If you can maintain \$2000 in your savings account, you will earn \$30.20 in dividends and your final balance at the end of the year will be \$2030.20. Note: this is not conventional rounding as the fraction of the cent (0.007) is truncated; the bank will not pay you more than what they owe and our monetary system is based on the fraction of a dollar, a cent (0.01).

The interest earned over the year is given by

$$I = \$2030.20 - \$2000 = \$30.20,$$

and by computing compound interest quarterly over simple interest, we can earn an additional \$30.20. The effective interest rate is given by

$$i_e = \left(1 + \frac{0.015}{12}\right)^{12} - 1 = 0.015104.$$

Investing in a savings account paying 1.5% annually with interest compounded monthly, we earn slightly more in interest; 1.5104% compounded monthly compared to 1.5% simple interest.

CHALLENGE: Check with your bank and rework the above example with your bank's interest rate.

Example 11.2.8 Saving for a Bike

Suppose you wanted to save up enough money to put a down payment on a motor bike in a year. How much would you need to invest in a savings account paying 3% annually compounded monthly to ensure you have \$5000 in the account in a year? How much interest is earned and what is the effective rate?

Solution

$$r = 0.03,\ n = 12,\ A = \$5000,\ t = 1$$

$$5000 = P\left(1 + \frac{0.03}{12}\right)^{12 \times 1}$$

$$5000 = 1.030416P$$

$$P = \frac{5000}{1.030416} \approx 4852.41$$

You would need to invest \$4852.41 to ensure that in 12 months, you would earn \$147.59 generating a total of \$5000.

The interest earned over the year is given by

$$I = \$5000 - \$4852.41 = \$147.59,$$

Computing compound interest monthly over simple interest; the effective interest rate is given by

$$i_e = \left(1 + \frac{0.03}{12}\right)^{12} - 1 = 0.030416.$$

Investing in a savings account paying 3% annually with interest compounded monthly, we earn slightly more in interest; 3.0416% compared to 3% simple interest.

Definition 11.2.8 Compounded Continuously

Continuous interest is interest computed continuously (**n** times per year where **n** approaches ∞.)

To evaluate the future amount owed when interest is compounded continuously, the resulting interest charged, or effective rate, use the following formulas:

Rule 11.2.5 Compounded Continuously

For the future amount of a principle investment with interest **compounded continuously**, $A = Pe^{rt}$. The **interest** earned is given by $I = A - P$. The **continuous effective rate** here is $i_e = e^r - 1$.

Example 11.2.9 Compounded Continuously

Dana invests \$10,000 in principal and agrees to take 5% in interest compound continuously for two years. What is the total amount that will be due at the end of the term? How much money will be earned in interest? What is the effective rate?

Solution

First, we gather the information: $P = \$10,000,\ r = 0.05,\ t = 2,\ n = 4$. Note: this is similar information to the previous example; however, here interest is earned on previously earned interest. The future value of this investment is

$$A = \$10,000e^{0.05 \times 2} = \$11,051.71.$$

The interest earned over these two years is given by

$$I = \$11,051.71 - \$10,000 = \$1051.71.$$

Hence, by computing compound interest continuously, we can earn an additional \$44.86.

Solution—cont'd

The interest earned over the two years is given by

$$I = \$11,051.70 - \$10,000 = \$1051.70,$$

and by computing compound interest continuously, the effective interest rate is given by

$$i_e = e^{0.05} - 1 = 0.051271.$$

Investing in a savings account paying 3% annually with interest compounded continuously, we earn more in interest; 5.1271% compared to 5.0945% compounded quarterly and 5% simple interest.

The future amounts in simple interest and compound interest is determined under the assumptions that no additional amounts are invested. However, if you are seeking to save money, you might consider sinking monies into your account periodically; that is, put money in the interest-bearing account weekly, monthly, etc. Here the payments are taken to be made as often as interest is paid. Such an account is referred to as a sinking fund.

Definition 11.2.9 Sinking Fund

Sinking Fund is when periodic payments are made towards saving up to a future amount.

To evaluate the future amount owed when money is invested with interest compounded periodically, the resulting interest earned, or the time necessary to save up this amount, use the following formulas:

Rule 11.2.6 Sinking Fund

To evaluate the **payment** needed save up a given future amount and placed in an interest earning **sinking fund**, the formula is

$$R = A \left[\frac{i}{(1+i)^m - 1} \right];$$

where $i = \frac{r}{n}$ is the periodic interest rate and $m = nt$ is the total number of compounds. On the other hand, the **future amount** based on the periodic payments made is given by

$$A = R \left[\frac{(1+i)^m - 1}{i} \right].$$

The **total amount** sunk into the sinking fund is given by

$$S = mR = ntR.$$

The **interest** earned is given by

$$I = A - S.$$

If the future amount is fixed and the payment is known, then the **time** necessary to save up this future amount is given by

$$t = \frac{\ln(1+\alpha)}{n\ln(1+i)},$$

where $\alpha = \frac{Ai}{R}$ is the percent of the total amount paid in interest.

Example 11.2.10 Sinking Fund

Pete needs to have $3000 worth of work done on his truck. If his bank pays 1% interest compounded monthly, how much does Pete need to invest into his savings account to have the full amount of $3000 dollars in six months? How much is invested and how much is earned in interest?

Solution

First, we gather the information: $A = \$3000$, $r = 0.01$, $t = 0.5$, $n = 12$. The payment needed to save up this much money in six months is

$$R = \$3000 \left[\frac{\dfrac{0.01}{12}}{\left(1 + \dfrac{0.01}{12}\right)^{12 \times 0.5} - 1} \right] = \$498.96.$$

Therefore, with six payments of $498.96, a total of ($498.96 × 6) $2993.76 has been invested and ($3000 − 2993.76) $6.24 has been earned in interest.

Example 11.2.11 Christmas Fund

Deb needs to have $1000 in the bank by the end of the year to ensure she has enough money to purchase Christmas presents. If her bank pays 1.5% interest compounded monthly, how much does Deb need to invest into her Christmas Fund account to have the full amount of $1000 dollars, assuming she plans one year ahead of time? How much is invested and how much is earned in interest?

Solution

First, we gather the information: $A = \$1000$, $r = 0.015$, $t = 1$, $n = 12$. The payment needed to save up this much money in a year is

$$R = \$1000 \left[\frac{\dfrac{0.015}{12}}{\left(1 + \dfrac{0.015}{12}\right)^{12 \times 1} - 1} \right] = \$82.76.$$

Therefore, with 12 payments of $82.76, a total of ($82.72 × 12) = $993.14 has been invested and ($1000 − 993.14) $6.86 has been earned in interest.

Example 11.2.12 Sinking Fund

If Deb can afford $200 a month, how long will it take to save up the $1000 needed for Christmas assuming that her bank pays 1.5% annual interested compounded monthly.

Solution

First, we gather the information: $A = \$1000$, $r = 0.015$, $n = 12$, $R = \$200$ and therefore we have $i = \dfrac{0.015}{12} = 0.00125$ and $\alpha = \dfrac{\$1000 \times 0.00125}{\$200} = 0.00625$. The time required to save up $1000 in a year is

$$t = \frac{\ln(1 + 0.00625)}{12 \ln(1 + 0.00125)} = 0.4156,$$

Solution—cont'd

approximately 5 months. If Deb can afford $200 a month for the next five months, then she can go shopping for Christmas early or allow these monies to earn additional interest.

Now, we switch to paying interest, first on a single principal amount and then when credit is used periodically with a periodic charge, the first of which is referred to as **amortization**.

Definition 11.2.10 Amortization

Amortization is when periodic payments are made toward paying off a principal amount.

To evaluate the future amount owed when money is paid off over time with interest compounded periodically, the resulting total amount paid and interest paid, or the time necessary to pay off this amount given a specified payment, use the following formulas:

Rule 11.2.7 Amortization

To evaluate the **payment** needed pay off a principal amount which is **amortized** (**amortization**),

$$R = P\left[\frac{i}{1-(1+i)^{-m}}\right],$$

where $i = \frac{r}{n}$ is the periodic interest rate and $m = nt$ is the total number of compounds.
   The **total amount paid** to amortize the principal balance is given by

$$Q = mR = ntR.$$

The **interest paid** is given by $I = Q - A$.
   If the principle amount is fixed and the payment is known, then the **time** necessary to pay of this principal amount is given by

$$t = -\frac{\ln(1-\beta)}{n\ln(1+i)},$$

where $\beta = \frac{Pi}{R}$ is the percent of the total payment due in interest.

Example 11.2.13 Purchasing a Car

Kendra needs to purchase a new car. She needs $2000 as a down payment on a $12,000 car and the balance of which is to be paid monthly for the next 48 months (4 years) at 3.99%. What are the monthly payments? Given new cars on average lose 30% of their value and an additional 10% each year thereafter, what is the depressed value of a car with an initial cost of $12,000?

Solution

First, we gather the information: $P = \$1000$, $r = 0.0399$, $t = 4$, $n = 12$. The payment needed to pay off the unpaid balance of ($12,000 - $2000) $10,000 in four years is given by

Continued

Solution—cont'd

$$R = \$10,000 \left[\frac{\dfrac{0.0399}{12}}{1 - \left(1 + \dfrac{0.0399}{12}\right)^{-12 \times 4}} \right] = 225.75.$$

Therefore, with 48 payments of $225.75, a total of ($225.75 × 48) = $10,836.00 has been paid for the car; in addition to the $2000 down, a total of ($10,836.00 + 2000) = $12,836 has been invested into the car.

Example 11.2.14 Paying Over Time

David had a medical lien against him for $10,000, but can only afford to pay $50 a month toward this lien; assuming that he is charged 1% annual interest, how long will it take to pay off this lien? How much is eventually paid?

Solution

First, we gather the information: $P = \$10,000$, $R = \$50$, $r = 0.01$

Hence, we have $\alpha = \dfrac{10000 \times \dfrac{0.01}{12}}{50} = 0.1\overline{6}$ and

$$t = -\frac{\ln(1 - 0.1666\ldots)}{12 \ln\left(1 + \dfrac{0.01}{12}\right)} \approx 18.239.$$

At this pace, it will take 18 years and $(0.239 \times 12) = 3$ months and $(18.239 \times 50 \times 12) = \$10,943.85$ total; that is, $943.85 is paid in interest over a period of 219 months.

The payment needed to pay off an amortized loan is given by $R = \dfrac{Pi}{1 - (1+i)^{-nt}}$, where P is the principal amount borrowed, i is the periodic interest rate, n is the number of periods per year, and t is the number of years. However, this is under the assumption that no additional purchases (principal) will be added to the balance. This is not exactly how a credit card works, there is not one fixed principal amount but rather monthly purchases must be accounted for in the budget; the new balance is the previous balance including interest charged, additional purchases less payments made.

Definition 11.2.11 Credit

Credit is the ability to obtain goods, money, or services before payment based on a contract with a credit card company or simply based on the trust that payment will be made in the future.

To evaluate the future amount owed on a credit card when interest is compounded periodically and periodic charges are made, us the following formula:

Rule 11.2.8 Credit Balance

Given a starting balance, B_n; after interest is accrued compounded periodically with an periodic rate of i on the previous balance, that is, $B_i(1 + i)$, plus purchases made, P, and (minus) a payment, R, is made, the new **credit balance**, B_{n+1} is given by

$$B_{i+1} = B_i(1 + i) + P - R.$$

Example 11.2.15 New Balance

Thomas has $232.23 in debt on a credit card charging an annual rate of 21% with interest compounded monthly. He makes $141.50 worth of purchases and sent in a payment of $100 dollars. What will be the starting balance next month?

Solution

First, we gather the information:

$$B = \$232.23, \quad r = 0.21, \quad n = 12, \quad P = \$141.50, \quad R = \$100, \quad \text{and} \quad i = \frac{0.21}{12} = 0.0175.$$

Therefore, the new balance is given by

$$B = \$232.23(1 + 0.0175) + \$141.50 - \$100 = \$277.79.$$

Note here, the new balance is greater than the original balance, creating more debt. How do you maintain this; that is, constant monthly expenditures on a credit card?

Definition 11.2.12 Credit Maintenance

Credit Maintenance is when periodic payments are made toward paying off a principal balance where regular purchases (of this principal amount) are made over the same period.

To evaluate the payment necessary to maintain a credit card when interest is compounded periodically and periodic charges are made, use the following formula:

Rule 11.2.9 Credit Maintenance

Computing the **payment** needed to **maintain a credit balance** is

$$R = P\left[1 + \frac{i}{1 - (1+i)^{-nt}}\right];$$

where P is the mean periodic purchase.

The **time** necessary to pay of this principal amount is given by

$$t = -\frac{\ln\left(1 - \frac{Pi}{R-P}\right)}{n \ln(1+i)}.$$

Note that only if you pay more than your mean monthly purchases, will you ever be able to get out of debt; this is called **living within your means**.

It is extremely important to note that as the payment approaches the monthly average purchase, the time it takes to pay off the debt becomes extremely large. Moreover, as the implied domain is restricted by $1 - \frac{Pi}{R-P} > 0$, in order to pay off any debt with reoccurring expenditures requires that $R > P(1+i)$; that is, the payment is greater than the average purchases including the interest charged over a single period. If the payment made is consistently less that the average purchases (including interest paid), $R \le P(1+i)$, then the time necessary to pay off this debt is infinite (> 100 years) as the balance simply continues to grow.

In fact, if we let λ be the additional percent of the monthly purchase paid monthly, then $R = (1+\lambda)P$ and

$$t = -\frac{\ln\left(1 - \frac{i}{\lambda}\right)}{n \ln(1 + i)}.$$

Then the number of years one will be indebted is finite when $\lambda > i$; when $\lambda \leq i$ then one will be indentured (illustrated in orange in the contour plot below). The illustration below also shows that if you pay an additional 5% of the monthly purchases, you will be out of debt within five years. Even at 18% annual interest, paying an additional 5% of the monthly purchases will take two years to get out of debt.

Duration of loan by interest rate and percent payment

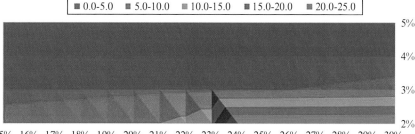

WARNING: With 18% annual interest compounded monthly, if you pay what the credit company requires, this is approximately 2% it will take 7.8 years to get out of debt.

Example 11.2.16 Minimum Payment

What is the minimum monthly payment required in order to maintain $100 worth of monthly purchases on a line of credit charging 21% annual interest compounded monthly? How long can this line of credit be maintained if this minimum amount is paid monthly? How much is paid in purchases and how much in interest?

Solution

First, we gather the information:

$$P = \$100, \quad r = 0.21, \quad n = 12 \quad \text{and} \quad i = \frac{0.21}{12} = 0.0175.$$

Therefore, the minimum required payment is

$$R > \$100(1 + 0.0175) = \$101.75,$$

and hence the minimum required payment is $101.76. Note: this is less than 2% of the monthly purchases.

$$t = -\frac{\ln\left(1 - \frac{100 \times 0.0175}{101.76 - 100}\right)}{12 \ln(1 + 0.0175)} = 24.8362.$$

As computed above, $t = 24.8362 \approx 25$; that is, if the minimum payment is made to "maintain" the line of credit, it will take nearly 25 years to pay off the debt in full. During this time, nearly 298 payments of $101.76 (the minimum payment as $R > \$101.75$) have been made for a total of $298 \times \$101.75 = \$30,327.94$ invested to maintain and payoff ($298.03395 \times \$100$) $29,903.40 of debt made in monthly purchases of $100, paying a total of ($30,327.94 − $29,903.40) $424.54 in interest.

Example 11.2.17 2%, 5%, and 10% Payment

Compare the duration in maintaining $100 worth of monthly purchases on a line of credit charging 21% annual interest compounded monthly if 2%, 5%, and 10% of the monthly purchases are made in payment, respectively, in addition to the $100 borrowed.

Solution

Given 2% of the $100 is $2, if the monthly payments are $102, then it will take

$$t = -\frac{\ln\left(1 - \frac{100 \times 0.0175}{102 - 100}\right)}{12 \ln(1 + 0.0175)} = 9.9885.$$

Therefore, if 102% of the monthly purchases are paid each month, $100 in purchases can be maintained for 10 years before the entire debt is paid off, approximately one-fifth of the time as when the minimum payments are made. During this time, 120 payments of $102 have been made for a total of $12,225.92 to maintain and pay off $12,086.19 of debt made in purchases and $239.72 in interest.

Given 5% of the $100 is $5, if the monthly payments are $105, then it will take

$$t = -\frac{\ln\left(1 - \frac{100 \times 0.0175}{105 - 100}\right)}{12 \ln(1 + 0.0175)} = 2.0692$$

years to pay off this accruing debt. If 105% of the monthly purchases are paid each month, $100 in monthly purchases can be maintained for 2 years before the entire debt is paid off, approximately one-fifth of the time as when the 2% above the monthly purchases are made in payment. During this time, 25 payments of $105 have been made for a total of $2607.25 to maintain and payoff $2583.09 of debt made in purchases and $124.15 in interest.

Given 10% of the $100 is $10, if the monthly payments are $110, then it will take

$$t = -\frac{\ln\left(1 - \frac{100 \times 0.0175}{110 - 100}\right)}{12 \ln(1 + 0.0175)} = 0.9240$$

years to pay off this accruing debt. Therefore, if 110% of the monthly purchases are paid each month, $100 in purchases can be maintained 11 months before the entire debt is paid off, approximately half of the time as when the 5% above the monthly purchases are made in payment. During this time, 11 payments of $110 have been made for a total of $1219.74 to maintain and payoff $1208.86 of debt made in purchases and $10.89 in interest.

Example 11.2.18 Getting Out of Debt

Rebecca strives to maintain a zero balance on her credit card; however, she presently has a $500 balance and will need to spend $500 a month on her credit over the next three months. The interest rate on her credit is 24%, what should be the monthly payment be to ensure she is out of debt in the fourth month?

Solution

First, we gather the information: $P = \$500$, $r = 0.24$, $t = 0.25$, $n = 12$ and $i = 0.02$.

Continued

Solution—cont'd

$$R = 500\left[1 + \frac{0.02}{1 - (1 + 0.02)^{-0.25 \times 12}}\right] \approx 673.38.$$

That is, if you spend ($3 \times \$500$) $1500 on this credit card in these three months, you need to make monthly payments of $673.38 to ensure a zero (or negative) balance at the end of the third month.

Payment number	Previous balance	Purchases	Payment amount	New balance
1	$500.00	$500.00	$673.38	$336.62
2	$336.62	$500.00	$673.38	$169.98
3	$169.98	$500.00	$673.38	$(0.00)

Note: in the above chart, the new balance is the previous balance including interest plus the purchases made, minus the payment.

$$\$500.00 \times (1 + 0.02) + \$500.00 - \$673.38 = \$336.62$$
$$\$336.62 \times (1 + 0.02) + \$500.00 - \$673.38 = \$169.98$$
$$\$169.98 \times (1 + 0.02) + \$500.00 - \$673.38 = \$0$$

Example 11.2.19 Getting Out of Debt

In the previous example, how long would it take to pay off $500 worth of monthly charges if a minimum of $600 is paid monthly? How much interest is accrued during this time?

Solution

First, we gather the information: $P = \$500$, $r = 0.24$, $R = \$100$, $n = 12$ and $i = 0.02$.

$$t = -\frac{\ln\left(1 - \dfrac{500 \times 0.02}{600 - 500}\right)}{12\ln(1 + 0.02)} = 0.443377$$

That is, it will take approximately 6 ($12 \times 0.443377 = 5.32$) months to reach a zero balance.

Payment number	Previous balance	Purchases	Payment amount	New balance
1	$500.00	$500.00	$589.26	$420.74
2	$420.74	$500.00	$589.26	$339.89
3	$339.89	$500.00	$589.26	$257.42
4	$257.42	$500.00	$589.26	$173.31
5	$173.31	$500.00	$589.26	$87.51
6	$87.51	$500.00	$589.26	$0.00

That is, to determine the accounting for the payments made we have the following calculations:

Solution—cont'd

$$\$500.00 \times (1 + 0.02) + \$500.00 - \$589.26 = \$420.74$$
$$\$420.74 \times (1 + 0.02) + \$500.00 - \$589.26 = \$339.89$$
$$\$339.89 \times (1 + 0.02) + \$500.00 - \$589.26 = \$257.42$$
$$\$257.42 \times (1 + 0.02) + \$500.00 - \$589.26 = \$173.31$$
$$\$173.31 \times (1 + 0.02) + \$500.00 - \$589.26 = \$87.51$$
$$\$87.51 \times (1 + 0.02) + \$500.00 - \$673.38 = \$0$$

WARNING: Once I paid off the balance of my debt on the first, the statement stating a payment was due the 20th was not sent out; however, my husband made a purchase on the 16th and then, because now there was a balance and I did not send in a payment (as I was not sent a statement), the credit card company tried to charge me a late fee. That is, it is sometimes necessary to call in and verify balances regularly (especially when no statement is received) or avoid this by not paying off a credit card in full (maybe all but the minimum payment) as otherwise they might not send you a billing that month.

The primary question, therefore, is *"**How do you budget your money?**"*

EXERCISES

Critical Thinking

11.2.1. Distinguish between **debt** and **savings**.

11.2.2. What are the three types of investments usually used for **emergency reserve funds**?

11.2.3. What are the three steps in budgeting for a desired item?

11.2.4. Distinguish between a **tax** and a **discount**.

Basic Exercises

11.2.5. Compute the sales tax when a $203 item is taxed at 7%.

11.2.6. Compute the sales tax and the final cost of an item on sale for $349 with a 6% sales tax.

11.2.7. What is the discounted price of a $200 bag which is 15% off? What is the discount?

11.2.8. Compute the sales price of an item worth $459, discounted 20%.

11.2.9. Compute the sales price of a pair of jeans marked 25% off the selling price of $120.

11.2.10. Compute the total purchasing cost for a dress marked down 10% off the selling price of $239 with a 7% sales tax.

11.2.11. Bob invests $1200 in principal into an account paying 4% simple interest for six years. How much interest is earned? What is the future amount in this account?

11.2.12. Fred invests $10,100 in principal into an account paying 1.4% simple interest for two years. How much interest is earned? What is the future amount in this account?

11.2.13. Greg invests $4200 in principal into an account paying 3.3% simple interest for three years. How much interest is earned? What is the future amount in this account?

11.2.14. Jerry invests $5000 in principal into an account paying 5% simple interest for one year. How much interest is earned? What is the future amount in this account?

11.2.15. Frances invests $8000 in principal into an account paying 6.1% simple interest for ten years. How much interest is earned? What is the future amount in this account?

11.2.16. Jeremiah invests $200 into an account paying 1.5% interest compounded monthly. What is the future amount in the account after two years? How much interest is earned? What is the effective rate?

11.2.17. Zeke invests $20,000 into an account paying 1.1% interest compounded quarterly. What is the future amount in the account after three years? How much interest is earned? What is the effective rate?

11.2.18. Joel invests $500 into an account paying 2.1% interest compounded daily. What is the future amount in the account after four years? How much interest is earned? What is the effective rate?

11.2.19. Wanda invests $1000 into an account paying 1.3% interest compounded weekly. What is the future amount in the account after a year and a half? How much interest is earned? What is the effective rate?

11.2.20. Betty invests $200 into an account paying 1.5% interest compounded continuously. What is the future amount in the account after two years? How much interest is earned? What is the effective rate?

11.2.21. Dyanne invests $200 into an account paying 2.1% interest compounded continuously. What is the future amount in the account after two years? How much interest is earned? What is the effective rate?

11.2.22. Pete invests $2000 into an account paying 1.3% interest compounded continuously. What is the future amount in the account after two years? How much interest is earned? What is the effective rate?

11.2.23. Billy invests $100 into an account paying 2.5% interest compounded continuously. What is the future amount in the account after two years? How much interest is earned? What is the effective rate?

11.2.24. Craig is saving up for a new computer that cost $2480. He is budgeting to ensure that he has the full amount in his savings in four months. If his savings account pays 1% in interest, how much does Craig need to pay into this account each month to have enough money saved? How much is invested into the account by Craig? How much is earned in interest?

11.2.25. In the previous exercise, if Craig could only afford to put $400 in this account each month, how long will it take to save up for the new computer?

11.2.26. Geoff is saving up for new business equipment that costs $15,400. He is budgeting to ensure that he has the full amount in his savings in two year. If his savings account pays 1.7% in interest, how much does Geoff need to pay into this account each month to have enough money saved? How much is invested into the account by Geoff? How much is earned in interest?

11.2.27. In the previous exercise, what would be the future amount after two years if Geoff could only afford $600 payments each month? How short is he of his goal?

11.2.28. In exercise 10.26, how long would it take to save up this money if Geoff can only afford $500 monthly?

11.2.29. Kaitlin wants a long engagement and plans to get married in three years. She estimates her wedding will cost $25,000 and does not plan on putting anything on credit. How much will Kaitlin have to put into an account paying 1.9% each month to accumulate this amount for her wedding?

11.2.30. In the previous exercise, if Kaitlin can only afford $400 a month, how much will she have in this account for her wedding? How short is Kaitlin of her goal?

11.2.31. In exercise 10.29, if Kaitlin insists on the $25,000 wedding, how long will it take to save this amount of money if she can only afford $400 a month to put in her sinking fund?

11.2.32. In exercise 10.29, if Kaitlin makes $400 monthly payments into a sinking fund, charges the remaining amount on a credit card charging 16% in annual interest (compounded monthly) and pays this card off (assuming no additional purchases) over the next year, compare this to postponing the wedding for a year and sinking monies into the fund for that additional year.

11.2.33. Wanda has three children and one grandchild, and wants to have saved $100 for each by Christmas, only nine months away. How much money needs to be sunk into an account paying 1.2% interest compounded monthly to ensure that the account has $400 by this time?

11.2.34. Donny needs to save $1000 in five months to be able to afford to move when his lease is up. How much money needs to be sunk into an account paying 0.9% interest compounded monthly to ensure that the account has $1000 at the end of the fifth month?

11.2.35. If you want to save $10,000 for a down payment and can only afford to pay $250 into a savings account paying 0.8% annual interest compounded monthly, how long will it take to save up this down payment?

11.2.36. In the previous exercise, how much time can you save if you can increase your monthly payments to $300; that is, if you budget an additional $50 toward getting a house.

11.2.37. Given you have been able to save up the $10,000 needed for the down payment on a $125,000 house, if there is no closing cost and you are able to get a 30-year fixed mortgage at 5.3%, how much are your monthly payments? How much is paid in interest?

11.2.38. In the previous exercise, if you can afford to put twice as much down up front, how much money can be saved in interest?

11.2.39. In exercise 11.2.37, how would your monthly payments change if the term of the loan is decreased to a 15-year term loan? How much would you save in interest?

11.2.40. In exercise 11.2.37, how would your monthly payments change if the term of the loan is increased to a 40-year term loan? How much additional interest is paid?

11.2.41. Otto needs to fix his roof and takes out a loan for $4000 at 2.1% fixed interest. He makes arrangements to pay off this loan $200 a month. How long will it take Otto to pay off this loan?

11.2.42. Jackson takes out a loan for $500 to go on vacation. If the loan charges 3% annual interest compounded monthly and Jackson can afford only $50 a month to pay off this debt, how long will it take Jackson to pay off this loan?

11.2.43. Pete has a balance of $432 on his credit card which charges 27% annual interest compounded monthly. He makes $200 in purchases and makes a $300 payment. What is the new balance of this account?

11.2.44. Natalie has a balance of $32.15 on her credit card which charges 22% annual interest compounded monthly. She makes $50 in purchases and makes a $30 payment. What is the new balance of this account?

11.2.45. Mary has a balance of $751 on her credit card which charges 18% annual interest compounded monthly. She makes $100 in purchases and makes a $250 payment. What is the new balance of this account?

11.2.46. Dillon has a balance of $77 on his credit card which charges 17% annual interest compounded monthly. He makes $52 in purchases and makes a $30 payment. What is the new balance of this account?

11.2.47. Kristin's average monthly purchases are $225 on a credit card charging 25% annual interest compounded monthly. Assuming that her previous balance was $225, what should her monthly payments be to maintain this credit card and pay off all the charges in six months? How much money is paid out? How much for purchases and how much in interest?

11.2.48. Dana's average monthly purchases are $150 on a credit card charging 19% annual interest compounded monthly. Assuming that her previous balance was $150, what should her monthly payments be to maintain this credit card and pay off all the charges in four months? How much money is paid out? How much for purchases and how much in interest?

11.2.49. When maintaining credit, it is assumed the average monthly purchases are constant and the starting balance is this average amount. Discuss the effects of having a starting balance greater than that of the average monthly purchases. Hint: you need to consider this as credit maintenance in addition to amortizing the difference in the balances, that is, the starting balances minus the average monthly purchases.

11.2.50. When paying off a credit card charging 15% annual interest compounded monthly where the average monthly purchases are $100, how long will it take to pay off this amount if the monthly payment is $105? How long will it take to pay off this amount if the monthly payments are $110?

11.2.51. Discuss what happens at the monthly payment approaches the average monthly purchases.

11.3 COMPARISON SHOPPING

In modern day society, comparison shopping is a necessity.

Budgeting Scheme I: Cash versus Credit

First consider purchasing an item with cash (either making a principal deposit or saving up for it using a sinking fund) or credit (buy now, pay later). In general, if you do not have enough cash to pay the incoming bill in its entirety, then using a credit card will cost you more in the long run.

Definition 11.3.1 Cash

Cash refers to money in the physical form of currency, such as banknotes and coins. In bookkeeping and finance, **cash** refers to current liquid assets.

Whereas using cash is good, carrying (excess) cash might not be a good way to save money as money can "burn a hole in your pocket," that is, you are more likely to spend what cash you have on you. However, carrying no cash might cause you to use your credit more often than you should. Therefore, if you make such small periodic purchases, you need to compare the cost of cash versus credit.

Definition 11.3.2 Credit and Debt

Credit is based on trust which allows one party (**creditor** or **lender**) to provide resources to another party (**debtor** or **borrower**) where that second party reimburses the first party in regular payments; all unpaid proportion of the amount provided is referred to as debt.

Example 11.3.1 Cash vs. Credit

Pete has determined that he will need $498.96 paid into a sinking fund paying 1% for six months to save up the $3000 for work to be done on his truck. That is an investment of $2993.76; compare this with charging this amount, the $3000 to a credit card charging 21% interest with no additional charges for six months.

Solution

First, we gather the information: $A = \$3000$, $r = 0.21$, $t = 0.5$, $n = 12$. As no additional purchases will be made over the next year, this is amortization.

$$R = 3000 \left[\frac{\frac{0.21}{12}}{1 - \left(1 + \frac{0.21}{12}\right)^{-12 \times 0.5}} \right] = 531.07$$

Month	Previous balance	Purchases	Payment	New balance
1	$3000.00	$–	$531.07	$2521.43
2	$2521.43	$–	$531.07	$2034.49
3	$2034.49	$–	$531.07	$1539.03
4	$1539.03	$–	$531.07	$1034.89
5	$1034.89	$–	$531.07	$521.93
6	$521.93	$–	$531.07	$0.00

Hence, in the first case where we saved up cash, we paid out $2993.76 and earned $6.24 in interest. However, we had to wait six months to work on the truck. In the second case (assuming we can refrain from using this credit card for a full six months), we will pay out $3186.41 for the $3000 borrowed and $186.41 in interest.

Example 11.3.2 Christmas Shopping

Deb plans on spending $1000 for Christmas; either by savig up the money over the next year or using credit and paying the balance off over the next year. Her first option is to put money into a Christmas Fund. Assuming her bank pays 1.5% interest compounded monthly, we have determined Deb would need to deposit $82.76 dollars a month for a year to have the full $1000 savings in the bank. Compare this with charging this amount to a credit card charging 18% interest with no additional charges for a full year.

Solution

First, we gather the information: $A = \$1000$, $r = 0.18$, $t = 1$, $n = 12$. The payment needed to pay off $1000 in a year at 18% interest is

$$R = 1000 \left[\frac{\dfrac{0.18}{12}}{1 - \left(1 + \dfrac{0.18}{12}\right)^{-12 \times 1}} \right] = 91.68.$$

Therefore, with 12 payments of $91.68, a total of ($91.68 × 12) $1100.16 has been paid and ($1100.16 − 1000) = $100.16 has been paid in interest.

Month	Previous balance	Purchases	Payment	New balance
1	$1000.00	$–	$91.68	$923.32
2	$923.32	$–	$91.68	$845.49
3	$845.49	$–	$91.68	$766.49
4	$766.49	$–	$91.68	$686.31
5	$686.31	$–	$91.68	$604.92
6	$604.92	$–	$91.68	$522.32
7	$522.32	$–	$91.68	$438.47
8	$438.47	$–	$91.68	$353.37
9	$353.37	$–	$91.68	$266.99
10	$266.99	$–	$91.68	$179.32
11	$179.32	$–	$91.68	$90.33
12	$90.33	$–	$91.68	$–

Hence, in the first case where we saved up cash, we paid out $993.12 and earned $6.88 in interest. However, we had to wait months until Christmas. In the second case (assuming we can refrain from using this credit card for a full year), we will pay out $1100.16 for the $1000 borrowed and $100.16 in interest.

Example 11.3.3 Planning a Wedding

Bethany is getting married in two years and has budgeted $6000 for the wedding. Compare sinking monies into a fund that pays 1.15% for the next two years versus saving $200 a month and charging the balance onto a credit card charging 23% interest for two additional years (assuming that no previous balance exists and no additional purchases will be made for these additional two years).

Solution

Case I: Sinking Fund

First, we gather the information for the sinking fund: $A = \$6000$, $r = 0.0115$, $t = 2$, $n = 12$.

$$R = \$6000 \left[\frac{\dfrac{0.23}{12}}{\left(1 + \dfrac{0.23}{12}\right)^{12 \times 2} - 1} \right] = \$247.26$$

Therefore, if you deposit $247.26 in an account each month, then you will have deposited a total of $5934.14 into savings and with $65.86 earned in interest for a grand total of $6000.

Continued

Solution—cont'd

Case II: Partial Sinking Fund and Partial Charge

If we can only afford $200 a month, then can solve the equation below to determine the future value of the account:

$$A \left[\frac{\frac{0.23}{12}}{\left(1 + \frac{0.23}{12}\right)^{12 \times 2} - 1} \right] = \$200$$

$$A = \$200 \left[\frac{\left(1 + \frac{0.23}{12}\right)^{12 \times 2} - 1}{\frac{0.23}{12}} \right] = \$4853.27$$

That is, depositing only $200 per month will allow this bride to save $4853.27 toward the wedding. Therefore, the payment need to pay off the remaining $1146.73 over a two year period at 23% interest is

$$R = 1146.73 \left[\frac{\frac{0.23}{12}}{1 - \left(1 + \frac{0.23}{12}\right)^{-12 \times 2}} \right] = \$60.06.$$

Therefore, with twenty-four payments of $60.06, at total of ($60.06 × 24) = $ 1441.39 has been paid and ($1441.39 − 1146.73) = $294.66 has been paid in interest. Including the ($200 × 24) $4800 deposited plus the $1441.39 to pay off the credit card; a total of $6241.39 will be spent during a period of four years, the ($4853.27 + $1146.73) $6000 spent on the wedding and a loss of $241.39 as while we earned $53.27 the first two years of savings, we paid $294.66 taking two years to pay off the credit card debt.

	Case I	Case II
Sinking Fund	$5934.24	$4800.00
Interest Earned	$65.86	$–
Credit Card	$–	$1146.73
Interest Paid	$–	$294.66
TOTAL PAID	$5934.24	$6241.39

Example 11.3.4 Plasma TV

Dana wants to purchase a new Panasonic "VIERA TC-P42C2 42" Plasma TV for $1332. Assume this price includes sales tax. She can afford to make $300 payments a month. Compare how long it would take to save up for this item in a savings account paying 1% versus how long it would take to pay this amount off on a credit card charging 27% interest.

Solution

First, to determine how long it would take to save up this amount of money, we use the gathered information $A = \$1332$, $r = 0.01$, $R = \$300$, $n = 12$ and

$$\alpha = \frac{\$1332 \times \frac{0.01}{12}}{\$300} = 0.004255. \text{ Therefore, we have}$$

Solution—cont'd

$$t = \frac{\ln(1 + 0.004255)}{12 \ln\left(1 + \dfrac{0.01}{12}\right)} = 0.36939.$$

At this rate, it will take 0.36939 years or approximately 5 (4.43) months before enough money is saved to simply purchase this item. However, only $1329.81 is deposited and the remaining $2.19 is earned in interest.

On the other hand, if we had purchased this item today planning to pay it off $300 a month, then (assuming no additional purchases are made on this credit card) we have the gathered information $P = \$1332$, $r = 0.27$, $R = \$300$, $n = 12$ and

$\alpha = \dfrac{\$1332 \times \dfrac{0.23}{12}}{\$300} = 0.0851$. Therefore, we have

$$t = \frac{\ln(1 - 0.0851)}{12 \ln\left(1 + \dfrac{0.23}{12}\right)} = 0.390392.$$

At this rate, it will take 0.390392 years or approximately 5 (4.68) months before enough money is saved to simply purchase this item. However, a total of $(4.68 \times \$300) = \$1405.41$ is paid out, the $1332 for the TV and $73.41 is paid in interest.

This is a difference of $(\$1405.41 - \$1329.81) = \$75.60$. Thus, the question becomes, "Is it worth the additional $75.60 to have the TV today and are you able to (1) maintain your payments of $300 a month and (2) not make any additional charges to this account?"

Here it is more efficient to wait and save up for a fixed period of time than it is to make payments toward a principal amount borrowed. This is also true when looking at leases. Unless this expense is a tax write off or deductable as a business expense, it is usually cheaper to purchase rather than lease.

Budgeting Scheme II: Leasing versus Purchasing

Historically, leases have served many purposes with just as many varied legal policies. One example is in agriculture, during the late 18th and early 19th centuries, where leasing, an important form of landholding, for industrialized countries to affect less developed regions and urban areas. The modern laws retain a laissez-faire philosophy that still dominates the law of contract and property law. The term *laissez-faire* (English pronunciation: /ˌlɛseɪˈfɛər/, French: [lɛsefɛʁ] "describes an environment in which transactions between private parties are free from state intervention, including restrictive regulations, taxes, tariffs, and enforced monopolies."

Definition 11.3.3 Contract

A **contract** is a legally binding agreement between two or more parties with mutual obligations.

The term **contract** is best defined as "agreement to be kept" and a breach of contract can lead to "damages" or monetary compensation detailed in the contract. The contract of interest here is that of a lease. The idea of a lease is that you only pay the "depreciated value" of the car; however, one you add in finance charges and possible fees which affect the end cost. On the other hand, when you buy the car, you pay the full price included the finance charges and possible fees; unlike with a lease, you also retain possession of the car which means you still possess value.

Definition 11.3.4 Lease

A **lease** is a contract calling for the user (**lessee**) to pay the owner (**lessor**) for use of assets; a rental agreement is a lease in which the asset is a tangible property. Other such leases on intangible property include licensing such as with computer programs or use of a radio frequency.

Example 11.3.5 Leasing a Honda

In January of 2011, Honda's current offer is for a 2011 Accord Coupe 5 Speed Automatic LX-S Featured Special Lease with monthly payments of $199 per month with $3499 due at lease signing for 36 months. How much will you have paid at the end of this lease before you return this car? What are the retained value and the mean monthly cost?

Solution

The total payments, T, is $199 per month for 36 months with $3499 at signing is

$$T = \$3499 + 36 \times \$199 = \$10,663.$$

With a $0 retained value as car has not been purchased, the average monthly expense is $296.19.

Example 11.3.6 Alternative Lease

In January of 2011, an alternative offer for the 2011 Accord Coupe 5 Speed Automatic LX-S Featured Special Lease with $0 due at lease signing and $0 first month's payment and then $300 per month for the remaining 35 months. How much will you have paid at the end of this lease before you return this car? What is the monthly cost and what is the retained value?

Solution

The total payments, T, is $300 per month for 35 months, for a total of

$$T = \$0 + 35 \times \$300 = \$10,500.$$

With a $0 retained value as car has not been purchased, the average monthly expense is $291.67.

WARNING: Some leases charge additional for excess mileage or wear on the car.

Definition 11.3.5 Purchase

To **purchase** is to obtain ownership of a security or other asset in exchange for money or value; to buy.

The question now is, "Is it better to lease a car or purchase the car?"

Example 11.3.7 Finance a Honda

The same Honda outlined in Example 11.3.5 and 11.3.6, a 2011 Accord Coupe 5 Speed Automatic LX-S Special APR (annual percentage rate) at 0.9% annual interest, for

Example 11.3.7 Finance a Honda—cont'd

36 months costs $28.16 a month for every $1000 financed of the dealer's price. With options, a 2.4 liter engine, 4 cylinder, front wheel drive, 5-speed automatic is $21,383. How much will you have paid at the end of this lease before you return this car assuming cars depreciate 30% the first year (70% retained) and 10% thereafter (90% retained)?

Solution

Monthly payments are $R = \$21,383 \times \dfrac{\$28.16}{\$1000} = \$602.15$. Therefore, the total paid is

$$T = 36 \times \$602.15 = \$21,677.$$

However, as this is a purchase, there is retained value, V, in the car: 70% the first year and 90% thereafter for the remaining two years.

$$V = \$21,383 \times 0.70 \times 0.90^2 = \$12,124.16.$$

Hence, the total expenditure, E, is

$$E = \$21,383 - 12,124.16 = \$9258.84,$$

and, therefore, the mean monthly expenditure is $257.19.

Budgeting Scheme III: Renting versus Owning

You can rent many things, from movies on VHS or DVD to games for PC or Game Boy; you may rent a hotel room or an umbrella at the beach. **Renting** is typically an implied, explicit, or written contract involving specific terms which regulate and manage the property or asset.

In modern day society, a rite of passage before you get into your own home may be the rental of an apartment.

Definition 11.3.6 Renting

Renting is an agreement where a payment is made for a temporary use of a good, service, or property owned by another.

Example 11.3.8 Renting in Tampa

In the college community in Tampa, an apartment with two bedrooms on average will cost **$874**. See Figure 11.1 for other rates. How much would it cost to rent this apartment yearly while you earn a four year degree?

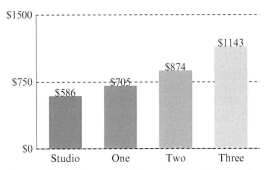

FIGURE 11.1 Average Rental Rates near University of South Florida, January 16, 2011, as reported by www.myapartmentmap.com.

Solution

Four years of monthly payments gives a total of 48 months. Therefore the total amount paid in rent is

$$T = 48 \times \$874 = \$20,976.$$

However, instead of renting one may purchase a house in an effort to obtain ownership, whether it is a car or a house, to create wealth or establish an estate.

Ownership is the starting point for many other concepts that form the foundations of ancient and modern societies such as **trade, money, debt, bankruptcy**, the **criminality of theft** and **private vs. public property**. Ownership is the key building block in the development of the capitalist socio-economic system.

Definition 11.3.7 Ownership

Ownership is the state or fact of exclusive rights and control over property; either land/real estate, tangible (objects), or intangible property (intellectual property).

What is the cost of ownership? There is the yard maintenance, plumbing, electrical, pool maintenance, pest control, home insurance, flood insurance, water and sewer, etc. and this does not include the purchasing of the home. Moreover, there are fees and expenses, over and above the price of the property itself such as inspection fees and closing cost. Closing cost, including the lawyer's fees, survey charges, and deed filing fees or settlement costs which are usually 2-6% of the mortgage amount.

Example 11.3.9 Buying a Home in Tampa

House prices in Tampa, Florida have dropped significantly over the past ten years and it appears to be a buyer's market. Rebecca finds a house for $94,500 and has $5000 to put down. What are the monthly payments (not including home owners insurance), with 3% in closing costs on a 30 year fixed mortgage at 5.125%.

Solution

First, we need to determine the amount of the mortgage: the purchasing price is $94,500, minus the $5000 down payment, plus the 3% ($2835) in closing cost—the amount of the mortgage is

$$\$94,500 - \$5000 + 0.03 \times \$94,500 = \$92,335.$$

As this is a single amount to be paid off in monthly installments, this is amortization of a mortgage loan:

$$R = 92,335 \left[\frac{\dfrac{0.05125}{12}}{1 - \left(1 + \dfrac{0.05125}{12}\right)^{-12 \times 30}} \right] = \$502.75.$$

That is, for a 30 year fixed mortgage on a $94,500 down, the monthly payments will be $502.75 (without property tax and insurance). That is a total of $180,990.74 in addition to the $7835 paid in closing costs and down payment. That is a grand total of $188,825.74, nearly twice what the house is worth.

Example 11.3.10 Extra Payments

In the previous example, how much time and interest would you save by paying an extra $100 a month toward your balance?

Solution

First, we gather the information:

$R = \$602.75$, $P = \$92,335$, $r = 0.05125$, $t = 30$, $n = 12$ and therefore,

$$\alpha = \frac{92,335 \times 0.05125}{602.75} \approx 0.65425 \text{ and } i = \frac{0.05125}{12} \approx 0.00427.$$

$$t = -\frac{ln(1 - 0.65425)}{12 ln(1 + 0.00427)} \approx 20.77.$$

By paying an additional $100 per month, Rebecca can save approximately 10 years of payments, paying only $125,688.01 total. Even with the down payment, this is nearly $50,000 in savings.

Many people wonder if they can afford a house and don't realize that sometimes purchasing a home costs less than renting an apartment; it is usually fear of commitment of making payments for 30 years than the actual cost.

Example 11.3.11 House Payments vs. Rent

Dana pays $885 in rent a month for a two bedroom, two bath apartment. Compare living in this apartment compared to purchasing a comparable house for $200,000 with 5% down and 2% closing cost for a 30 year mortgage. Assuming the interest rate for the mortgage is 6% and closing costs can be included and not paid in cash above and beyond the initial amount; compare the yearly cost for renting versus owning.

Solution

If you pay $885 a month for 30 years, then you will pay $885 \times 12 \times 30 = \$318,600$; that is, $10,620 a year. Whereas if you purchase a house for $200,000 plus the 5% down ($10,000) and 2% closing cost ($4000), you will need to take out a mortgage for $194,000 (including the closing cost). The monthly payments work out to be

$$R = 194,000 \left[\frac{\frac{0.06}{12}}{1 - \left(1 + \frac{0.06}{12}\right)^{-12 \times 30}} \right] = \$1163.13.$$

Therefore, we would spend $432,726.09 total including the down payment over 360 payments, that is, $1202.02 on average monthly. This is more than the monthly cost of renting; however, you do not have any rights to the apartment and you would own the house. If you take into account the home's worth, even if it did not appreciate, this average cost is significantly reduced. Assuming the house maintains its worth, after the 30 year mortgage is paid off, you are out ($432,726.09 - $200,000) = $232,726.09; however, it is significantly less than the $318,600 paid in rent and reduces the average monthly cost of owning down to $646.46.

EXERCISES

11.3.1. Distinguish between **cash** and **credit**.

11.3.2. Joel needs to purchase a $400 lawn mower. Compare sinking money into an account paying 1.3% annual interest compounded quarterly for 6 months and purchasing the mower on credit at 19% annual interest compounded monthly for 6 months.

11.3.3. Tara plans on spending $1400 on her kids for Christmas next year. Compare sinking money into an account paying 1.5% annual interest compounded daily for a year and purchasing the presents on credit at 17% annual interest compounded monthly for the year.

11.3.4. Kendra needs $250 for a new cell phone. Compare sinking money into an account paying 0.7% annual interest compounded monthly for four months and purchasing the phone on credit at 27% annual interest compounded monthly for the year.

11.3.5. Bridget wants to purchase a used car and estimates it will cost $3000. She can only afford to make $300 payments a month. Compare how long it would take to save up for the car putting monies into a sinking fund earning 2% interest and how long it would take to pay this amount off on a line of credit charging 16% annual interest compounded monthly.

11.3.6. Distinguish between leasing and purchasing.

11.3.7. A new car offers a three year lease for $99 dollars a month with $1000 due at signing. If the car would cost $7995 at 7% annual interest compounded monthly, compare leasing versus purchasing. Assume the car depreciates 30% upon leaving the lot and an additional 10% each year.

11.3.8. Distinguish between renting and owning.

11.3.9. Compare paying rent of $895 a month in an apartment to purchasing a $225,000 house paying 10% down, $0 closing cost at a fixed rate of 9% for a term of 30 years.

11.4   COMPARISON USING EFFECTIVE RATES

This section concentrates on the **effective rates** as a way of comparing two or more accounts or account types. For example, when comparing CD and Savings; **effective rate** (annual yield) is the interest amount earned for one year per dollar amount depending on the number of compounds per year. In general, there are three types of simple interest used for CDs.

Recall the **effective rates** are:

$i_e = i$	Simple Interest
$i_e = 1 - \left(1 + \frac{r}{n}\right)^n$	Compounded Periodically
$i_e = 1 - e^r$	Compounded Continuously

Example 11.4.1 CD vs. Savings

You purchase a $5000 CD with an interest rate of 3% compounded annually for the term of five years. If you purchase this CD on January 1, 2012, using effective rates, compare this to a savings account paying 1.2% compounded monthly.

Solution

This CD will mature after five years on January 1, 2017, and will be worth

$$A = \$5000(1 + 0.03 \times 5) = \$5750,$$

Solution—cont'd

with $750 earned in interest. The effective interest rate is the stated 3% annual rate given. Whereas, for the saving account where interest is earned on interest, for the same five years, the future amount is given by

$$A = 5000\left(1 + \frac{0.012}{12}\right)^{12 \times 5} = \$5309.02,$$

with only $309.02 earned in interest.

Hence, a CD can earn an additional $440.98 over the savings account; however, there are penalties if money is withdrawn from the CD whereas the monies available in the savings account are liquid (meaning that they can be withdrawn at any time without penalty).

Example 11.4.2 Comparing Banks

You have researched three banks: the first bank pays 1.5% compounded monthly, the second pays 1.3% quarterly, and the third pays 1.2% continuously. Use effective rates to compare these three banks.

Solution

For the first bank, $r = 0.015$ and $n = 12$. Therefore the effective annual rate is

$$i_e = \left(1 + \frac{0.015}{12}\right)^{12} - 1 = 0.015104.$$

For the second bank, $r = 0.013$ and $n = 4$. Therefore, the effective annual rate is

$$i_e = \left(1 + \frac{0.013}{4}\right)^{4} - 1 = 0.013064.$$

For the third bank, $r = 0.012$ and $n \to \infty$. Therefore, the effective annual rate is

$$i_e = 1 - e^{0.012} = 0.012072.$$

Therefore, the effective interest rates depend more on the interest rate and less on the number of compounds per year. In general, it is usually best to select a higher interest rate over a larger number of compounds per year. The effective rates are equivalent to rounding.

Example 11.4.3 Comparing Compounds

Use effective rates to compare the interest rate of 4% compounded quarterly, monthly, and continuously to four decimals.

Solution

Note: $r = 0.04$ for all three effective rates.

For the first number of compounds: $n = 4$, the effective annual rate is

$$i_e = \left(1 + \frac{0.04}{4}\right)^{4} - 1 = 0.0406.$$

For the second number of compounds: $n = 12$, the effective annual rate is

Continued

Solution—cont'd

$$i_e = \left(1 + \frac{0.04}{12}\right)^{12} - 1 = 0.0407.$$

For the third number of compounds: compounded continuously, $n \to \infty$, the effective annual rate is

$$i_e = 1 - e^{0.0124} = 0.0408.$$

These rates differ only in the fourth decimal range of 0.0002 between the effective rates when compounding four times a year and when interest is compounded continuously.

EXERCISES

Critical Thinking

11.4.1 Distinguish between **annual interest rate**, **periodic interest rate**, and **effective interest rate**.

11.4.2 Distinguish between **sinking fund** and **amortization**.

11.4.3 Distinguish between **renting**, **leasing,** and **owning**.

Basic Problems

11.4.4. Find the effective interest rates when interest is compounded periodically given:
   (a) $r = 0.24$, $n = 12$
   (b) $r = 0.12$, $n = 12$
   (c) $r = 0.24$, $n = 24$
   (d) $r = 0.12$, $n = 24$

11.4.5. Find the effective interest rates when interest is compounded continuously given:
   (a) $r = 0.24$
   (b) $r = 0.12$
   (c) $r = 0.06$
   (d) $r = 0.01$

11.4.6. Compare the effective interest rates compounded monthly versus compounded continuously given:
   (a) $r = 0.27$
   (b) $r = 0.18$
   (c) $r = 0.16$
   (d) $r = 0.01$

11.5 PERSONAL FINANCE

Important questions you should know the answers to:

- Do you know if you earn interest on your savings account, and if so how much?
- Do you have any credit cards? How many? What are your average monthly purchases?
- Do you know what your interest rate is/are on each? Do you know your current balance?
- Did you know that to pay off a credit card by making the minimum payment you could be making those small payments for up to eight years? And what do you think about all these new cars offering you a lease. When would it be feasible to lease or is it just more of a convenient or tax deductable.

- Have you considered buying a house? How does this compare to renting? Or should you just live with Mom and Dad? And then there are the necessities: water, food, and the luxuries such as phone, cable, etc.
- How do we budget what we want and need into what we have?
- Are you insured?

These are questions all adults face. And we have not even mentioned retirement.

Definition 11.5.1 Insurance

Insurance is a form of risk management used to hedge against the risk of uncertain loss. The **insurance rate** is the factored weight used to determine the amount to be charged for the insurance coverage. This amount is referred to as the **premium**.

Types of insurance:

Auto insurance protects the policy holder against financial loss in the event of an incident involving their automotive vehicle. While this type of insurance is required by law, it is only the third most recommended insurance behind health and life.

Property Coverage for theft of or damage to one automobile.

Liability Coverage for bodily injury or property damage to others.

Medical Coverage for medical bills accumulated for treatment or rehabilitation; this may also include coverage of lost wages or funeral expenses.

Home insurance provides coverage for destruction or damage of the policy holder's home.

Health insurance covers the cost of medical treatments. In 2011, nearly 43 million Americans didn't have health insurance, yet it is the number one recommended insurance as medical costs are expected to increase faster than inflation.

Dental insurance, like medical insurance, covers cost of dental work.

Funeral insurance covers funeral expenses of the insured, a type of "health" insurance.

Casualty insurance providing coverage for accidents such as career hazards, sometimes a supplemental medical insurance.

Life insurance provides monetary support to the surviving members of the family. While this does provide assistance to the descendants, special provisions may apply. Many life insurance companies have a suicide clause. The policy may not be worth its face value if the policy matures at a specified age (which could be as high as 100). Life insurance is the second most recommended insurance after health insurance.

Property insurance provides protection to property, covering fire, theft, or weather damage. Specialized forms include: flood insurance, earthquake insurance, etc.

Liability insurance such as public or professional liability insurance.

Credit insurance repays some or all of a loan when the borrower is unable to pay due to unemployment, injury, disability, or even death.

Definition 11.5.2 Budget

A **budget** is an itemized list of expected income and expenses for a given period in the future. This plan of operations can be based on estimates such as for food or clothing, or fixed such as car payments, rent, insurances, etc.

Archival Note

Funeral Insurance was introduced by the Greeks and Romans circa 600 AD when *benevolent societies* paid the funeral expenses of the members upon death and cared for the surviving family.

Example 11.5.1 First Budget

Kendra has just moved into her new appartment paying $640 a month, has a $225.75 car payment, $199 for car insurance every four out of six months, and the following monthly expenses: $400 for food, $50 for water, $142 for electricity, $52 for a mobile phone, $210 for entertainment, and $150 in clothes. What is Kendra's average monthly expense? Working for $10 an hour with a 20% federal income tax, how many hours does Kendra need to work to make this budget work?

Solution

We can estimate this assuming that the insurance is due monthly and not just eight months out of the year. This would be $640 + $225.75 + $199 + $400 + $50 + $142 + $210 + $150, which is a total of $2068.75 a month.

More accurately, we have to pay $1869.75 a month not including insurance and $199 for 8 out of 12 months:

$$12 \times \$1869.75 + 8 \times \$199 = \$24,029$$

a year total, for an average of

$$\frac{\$24,029}{12} = \$2002.42$$

a month on average.

This amount, $2002.42, needs to be the take home pay; that is, after the federal income tax is taken out. Hence, the amount needed to be earned is the solution to the equation

$$P(1 - 0.2) = \$2002.42,$$

and therefore

$$P = \frac{\$2002.42}{0.8} = \$2503.02.$$

Kendra needs to earn $2503.02 a month to clear $2002.42 a month take home pay, and at $10 per hour, she will need to work

$$h = \frac{\$2503.02}{\$10} = 250.302.$$

Therefore, at this pay rate, Kendra needs to work approximately 250 hours a month or between 62 and 63 hours a week.

This is rather high as a usual work week is 40 hours. Kendra needs to better budget her monies.

Example 11.5.2 Living on a Budget

In the previous example, how many hours a week would Kendra need to work if she cut her entertainment and clothing expenses in half?

Solution

Without the additional expenses for entertainment and clothes, we have to pay $1689.75 a month not including insurance and $199 for 8 out 12 months:

$$12 \times \$1689.75 + 8 \times \$199 = \$21,869$$

a year total, for an average of

$$\frac{\$21,869}{12} = \$1822.42$$

a month on average.

Solution—cont'd

This amount, $1822.42, needs to be the take home pay, that is, after the federal income tax is taken out. Hence, the amount needed to be earned is the solution to the equation

$$P(1-0.2) = \$1822.42,$$

and therefore,

$$P = \frac{\$1822.42}{0.8} = \$2278.02.$$

Kendra needs to earn $2278.02 a month to clear $1822.42 a month take home pay, and at $10 per hour, she will need to work

$$h = \frac{\$2278.02}{\$10} = 227.8.$$

Therefore, at this pay rate, Kendra needs to work approximately 228 hours a month or approximately 57 hours a week.

In fact, if Kendra completely removes entertainment and clothes from her budget, she will still have to work 51.4 hours a week.

Example 11.5.3 Living on a Budget

In the previous example, assuming Kendra removes entertainment and clothes from her budget and has a roommate move in to cut her part of the rent in half, how many hours will Kendra need to work to make budget?

Solution

Without any expenses for entertainment and clothes, we have to pay $1509.75 a month not including insurance; however, if she finds a roommate willing to pay half the rent, this amount drops to $1189.75 and $199 for 8 out 12 months:

$$12 \times \$1189.75 + 8 \times \$199 = \$15,869$$

a year total, for an average of

$$\frac{\$15,869}{12} = \$1322.42$$

a month on average.

This amount, $1822.42, needs to be the take home pay, that is, after the federal income tax is taken out. Hence, the amount needed to be earned is the solution to the equation

$$P(1-0.2) = \$1322.42,$$

and therefore,

$$P = \frac{\$1822.42}{0.8} = \$1653.02.$$

Kendra needs to earn $1653.02 a month to clear $1322.42 a month take home pay, and at $10 per hour, she will need to work

$$h = \frac{\$1653.02}{\$10} = 165.3.$$

Therefore, at this pay rate, Kendra needs to work approximately 166 hours a month or approximately 42 hours a week.

Creating a Personal Budget

Should include:

- Housing: Rent or House Payment
- Car Payment
- Auto Insurance
- Gas
- Renter or Home Owner's Insurance
- Water
- Electricity
- Food
- Clothes
- Cable
- Home Phone
- Cellular Phone
- Internet
- Creditors

Can also include:

- Entertainment
- Tuition
- Books
- School Fees
- Computer Expense

It is important to be on top of your budget, especially if you use credit cards. Everything you do in the financial world is recorded and used to evaluate your credit in the form of a credit report.

Definition 11.5.3 Credit Report

A **credit report** is a report containing detailed information on a person's credit history (who and how much they owe as well as their payment history: the individual's past borrowing and repaying behavior; this report lists personal information, credit lines currently in that person's name and other risk factors such as late payments or recent bankruptcy.

There are three bureaus that provide these scores: Experian, Equifax, and Trans Union. The U.S. average credit score on February 19, 2011, was 693. The highest possible FICO score is 850; however, scores usually do not exceed 825. These scores are based on the following six pieces of information:

1. Personal data including current and previous addresses, your social security number, and employment history.
2. Summary of credit history including the number and type of accounts that are in good standing or are past-due as well as your payment history.
3. Detailed account information: how long you have had the account, with which you have the account and balance.
4. The number and type of inquiries into applicant's credit history report. That is, having someone check your score can, in and of itself, lower your score.
5. Details of any accounts turned over to credit agency such as information about liens, wages garnishments and other federal, state and county records.

For this reason, it is important to live within your means. If you make purchases you cannot afford whether or not you take out a loan or purchase it on credit, you are probably not living within your means. First used in the 1930s, the **Means Test** is an examination into the financial state of an individual to determine the eligibility for public assistance. However, with a good budget, such assistance can be avoided.

Definition 11.5.4 Means

Means in finance, refers to the resources available for disposal, such as the purchase of material goods.

The main ways to live within your means is to:

1. Spend less money than you earn or bring into your household.
2. Stop using credit cards, and know how to maintain them when used.
3. Use a sinking fund to save up for large purchases instead of using credit.
4. Don't compete with the Jones—don't give into pressure from people around you.
5. Have a savings account or emergency fund.
6. Find additional work to increase your income.

EXERCISES

11.5.1. Define insurance.

11.5.2. Create a budget for yourself that includes housing, food, and any other bill you pay periodically (say monthly.)

11.5.3. Create a budget for Todd and determine how much he can put in savings a month given Todd makes $3400 monthly and has the following expenses:

House Payment	$885
Car Payment	$230
Auto Insurance	$200
Gas	$40
Home Owners Insurance	$150
Water	$50
Electric	$125
Food	$400
Clothes	$150
Cable	$65
Cellular Phone	$56
Internet	$35
Creditors	$250
Entertainment	$500

11.5.4. Create a budget for Jennifer and determine how much she can put in savings a month given Jennifer makes $5200 monthly and has the following expenses:

House Payment	$1275
Car Payment	$350
Auto Insurance	$220
Gas	$ 80
Home Owners Insurance	$250
Water	$ 78
Electric	$154
Food	$800
Clothes	$250
Cable	$85
Home Phone	$25
Cellular Phone	$45
Internet	$45
Creditors	$500
Entertainment	$1000

11.5.5. Create a budget for Shirley and determine how much more she needs to earn each month to break even given Shirley makes $1900 monthly and has the following expenses:

Rent	$565
Car Payment	$275
Auto Insurance	$175
Gas	$ 40
Water	$ 50
Electric	$75
Food	$300
Clothes	$150
Cable	$65
Cellular Phone	$56
Internet	$35
Creditors	$275
Entertainment	$100

CRITICAL THINKING AND BASIC EXERCISES

Critical Thinking

11.1 Define the following terms: **trade, money, debt, bankruptcy, principal, interest rate, term, period, future amount, effective rate,** and **payment**.

11.2 Distinguish between **annual interest rate** and **periodic interest rate**.

11.3 Distinguish between **effective interest rate** and **periodic interest rate**.

11.4 Distinguish between **effective interest rate** and **annual interest rate**.

11.5 Define the following terms: **savings, budgeting, tax, discount, simple interest, compound interest, CD, compound interest, compounded continuously, sinking fund, amortization,** and **credit**.

11.6 Distinguish between **simple interest** and **compound interest**.

11.7 Distinguish between **interest compound periodically** and **interest compounded continuously**.

11.8 Define the following terms: **cash**, **credit**, **contract**, **lease**, **purchase**, **rent**, **own,** and **insurance**.

Basic Exercises

11.9 Find the **future amount** when interest is compounded periodically given
  (a) $P = \$100$, $r = 0.23$, $n = 12$, $t = 2$
  (b) $P = \$200$, $r = 0.12$, $n = 12$, $t = 3$
  (c) $P = \$1000$, $r = 0.17$, $n = 24$, $t = 5$
  (d) $P = \$12{,}000$, $r = 0.08$, $n = 12$, $t = 5$

11.10 In the previous problem, determine the following for each part: the **period interest rate**, the **effective interest rate**, the total **number of payments**, the total **amount paid,** and the amount paid in **interest**.

11.11 Find the future amount when interest is compounded periodically given
  (a) $P = \$500$, $r = 0.23$, $n = 12$, $t = 1$
  (b) $P = \$500$, $r = 0.23$, $n = 12$, $t = 2$
  (c) $P = \$500$, $r = 0.23$, $n = 12$, $t = 5$
  (d) $P = \$500$, $r = 0.23$, $n = 12$, $t = 10$
  (e) Discuss the various amounts earned over time as computed in parts a-d.

11.12 Find the future amount when interest is compounded periodically given
  (a) $P = \$100$, $r = 0.08$, $n = 12$, $t = 2$
  (b) $P = \$100$, $r = 0.12$, $n = 12$, $t = 2$
  (c) $P = \$100$, $r = 0.18$, $n = 12$, $t = 2$
  (d) $P = \$100$, $r = 0.24$, $n = 12$, $t = 2$
  (e) Discuss the various amounts earned as computed in parts a-d as a function of interest rate.

SUMMARY OF IMPORTANT CONCEPTS

11.1.1 **Trade** is the transfer of ownership of goods and services from one individual or party to another individual or party, also referred to as **commerce, financial transaction,** or **barter**. A network that allows trade is called a **market,** such as the **Stock Market, Retail Market, Super Market,** etc.

11.1.2 **Money** is a medium of exchange, any object, or note that is widely accepted as payment for goods or services rendered, for the repayment of debt. **Money** is subject to the given country or socio-economic content.

11.1.3 **Debt** is that which is owed; in economics, this refers to assets owed. **Debt** is created when a lender agrees to lend a specific amount or asset to the borrower or **debtor**.

11.1.4 **Bankruptcy** is a legally declared inability or impairment of ability of an individual or organization to pay its creditors or lenders. **Bankruptcy** is usually initiated by the debtor voluntarily by the insolvent (bust) individual or organization.

11.1.5 The **principal** is the balance of the initial loan. Denoted P; that is, the borrower agrees to pay back this principal amount or proceeds and any outlined interest on this principal amount depending on the type of interested charged (simple, compound, or continuous).

11.1.6 The **interest rate** is the percent of the initial proceeds to be charged above the principal amount borrowed. Denoted r; this is the nominal rate (or interest rate for one year)

11.1.7 The number of **periods per year** is the number of times interest is applied to the remaining principal; denoted n.

11.1.8 The **term** of a loan is the number of years outlined in the loan agreement over which the borrower has to pay off the loan (the principal amount and accrued interest). Denoted t for the amount of time (in years).

11.1.9 **Number of periods** is given by $m = nt$; that is, the number of periods per year times the number of years; denoted m.

11.1.10 The **periodic interest** rate is given by $i = \frac{r}{n}$; that is, the **interest rate per period**.

11.1.11 The **future amount** to be paid off on a loan or a future amount desired in savings.

11.1.12 **Effective rate** (annual yield) is the interest amount earned for one year per dollar amount. Effective interest rates depend on the number of compounds per year and must be formulated for each different type of interest charge. In general it will be denoted i_e.

11.1.13 The periodic **payment** to be made depends on the type of interest (simple or compound), the term of the loan, and the number of periods per year and if this is a loan to be paid off or a sinking fund where monies accrue. (This cannot be abbreviated P as this notation has already been defined as principal, and R resembles P for Payment.)

11.2.1 **Savings** is income not spent but rather put into a bank (deposit account) or pension fund. Savings can also be seen when the cost of commonly purchased items is reduced or moderated. Savings, as opposed to investments, are considered low-risk.

11.2.2 **Budgeting** is to plan in advance of the expenditure of resources such as time and money. In personal finances, this is the total sum of monies allocated for a particular purpose or period of time.

11.2.3 **Tax (markup)** or **taxation** is a one-time adjustment to a principal, P, amount based on a specified rate, r.

11.2.4 **Discounts (depreciation)** it is a one-time percentage decrease to a principal, P, amount based on a specified rate, r, to a principal, P, amount based on a specified annual rate, r.

11.2.5 **Simple interest** is interest computed once on the full amount for a given amount of time, or written mathematically: $I = Prt$.

11.2.6 A **Certificate of Deposit** (CD) is a time deposit which bears a maturity date, a specified fixed interest rate. Generally range from one to five years; formulated the same as with a tax or markup. The future amount is the principal invested plus the simple interest earned: $A = P(1+r) = P + P \cdot r = P + I$

11.2.7 **Compound interest** is interest computed periodically (n times per year.)

11.2.8 **Continuous interest** is interest computed continuously (n times per year where n approaches ∞.)

11.2.9 **Sinking Fund** is when periodic payments are made towards saving up to a future amount.

11.2.10 **Amortization** is when periodic payments are made toward paying off a principal amount.

11.2.11 Given a starting balance, B_n; after interest is accrued compounded periodically with an periodic rate of i on the previous balance, that is, $B_i(1+i)$, plus purchases made, P, and (minus) a payment, R, is made, the new **credit balance**, B_{n+1} is given by $B_{i+1} = B_i(1+i) + P - R$.

11.2.12 **Credit Maintenance** is when periodic payments are made toward paying off a principal balance where regular purchases (of this principal amount) are made over the same period.

11.3.1 **Cash** refers to money in the physical form of currency, such as banknotes and coins. In bookkeeping and finance, **cash** refers to current liquid assets.

11.3.2 **Credit** is based on trust which allows one party (**creditor** or **lender**) to provide resources to another party (**debtor** or **borrower**) where that second party reimburses the first party in regular payments; all unpaid proportion of the amount provided is referred to as debt.

11.3.3 A **contract** is a legally binding agreement between two or more parties with mutual obligations.

11.3.4 A **lease** is a contract calling for the user (**lessee**) to pay the owner (**lessor**) for use of assets; a rental agreement is a lease in which the asset is a tangible property. Other such leases on intangible property include licensing such as with computer programs or use of a radio frequency.

11.3.5 To **purchase** is to obtain ownership of a security or other asset in exchange for money or value; to buy.

11.3.6 **Renting** is an agreement where a payment is made for a temporary use of a good, service, or property owned by another.

11.3.7 **Ownership** is the state or fact of exclusive rights and control over property; either land/real estate, tangible (objects), or intangible property (intellectual property).

11.5.1 **Insurance** is a form of risk management used to hedge against the risk of uncertain loss. The **insurance rate** is the factored weight used to determine the amount to be charged for the insurance coverage. This amount is referred to as the **premium**.

11.5.2 A **budget** is an itemized list of expected income and expenses for a given period in the future. This plan of operations can be based on estimates such as for food or clothing, or fixed such as car payments, rent, insurances, etc.

11.5.3 A **credit report** is a report containing detailed information on a person's credit history (who and how much they owe as well as their payment history: the individual's past borrowing and repaying behavior; this report lists personal information, credit lines currently in that person's name and other risk factors such as late payments or recent bankruptcy.

11.5.4 **Means** in finance, refers to the resources available for disposal, such as the purchase of material goods.

Rules

11.1.1 The following formulas can be used to calculating **future amounts** under the given conditions.

$A = P(1 + rt)$	Simple Interest
$A = P(1 + i)^{nt}$	Compound Interest (compounded periodically)
$A = Pe^{rt}$	Compound Interest (compounded continuously)

11.1.2 **Effective rates** will depend on how the interest is compounded (if compounded).

$i_e = i$	Simple Interest
$i_e = \left(1 + \frac{r}{n}\right)^n - 1$	Compounded Periodically
$i_e = e^r - 1$	Compounded Continuously

11.1.3 **Payments** will depend on how the principle is earning interest or being charged interest:

$R = \frac{Ai}{(1+i)^{nt} - 1}$	Sinking Fund when interest is earned
$R = \frac{Pi}{1 - (1+i)^{-nt}}$	Amortization when interest is paid
$R = P\left[1 + \frac{i}{1 - (1+i)^{-nt}}\right]$	Maintaining

11.2.1. The **tax** or **markup** is a percentage (rate) of the total principal: $T = r \cdot P = P \cdot r$, and the future amount (after taxations) is $A = P(1 + r) = P + P \cdot r = P + T$.

11.2.2. The **depreciation** or **depreciated amount** is $D = r \cdot P = P \cdot r$, and the future amount (after the discount) is $A = P(1 - r) = P - P \cdot r = P - D$.

11.2.3. To compute the future amount in **simple interest**, first find the interest charge: $I = Prt$. Therefore, the future value of a given amount earning simple interest for a given amount of time is given by $A = P(1 + rt) = P + I$. The **simple effective rate** is the actual interest rate. We write $i_e = r$.

11.2.4. Therefore, the amount after time t (or future value) is given by $A = P\left(1 + \frac{r}{n}\right)^{nt}$. The **interest** earned is given by $I = A - P$. The **effective rate** when interest is compounded periodically is $i_e = \left(1 + \frac{r}{n}\right)^n - 1$.

11.2.5. For the future amount of a principle investment with interest **compounded continuously**, $A = Pe^{rt}$. The **interest** earned is given by $I = A - P$. The **continuous effective rate** here is $i_e = e^r - 1$.

11.2.6. To evaluate the **payment** needed save up a given future amount and placed in an interest earning sinking fund, the formula is $R = A\left[\frac{i}{(1+i)^m - 1}\right]$; where $i = \frac{r}{n}$ is the periodic interest rate and $m = nt$ is the total number of compounds. On the other hand, the **future amount** based on the periodic payments made is given by $A = R\left[\frac{(1+i)^m - 1}{i}\right]$. The **total amount** sunk into the **sinking fund** is given by $S = mR = ntR$. The **interest** earned is given by $I = A - S$. If the future amount is fixed and the payment is known, then the **time** necessary to save up this future amount is given by $t = \frac{\ln(1+\alpha)}{n\ln(1+i)}$, where $\alpha = \frac{Ai}{R}$ is the percent of the total amount paid in interest.

Continued

Rules—cont'd

11.2.7. To evaluate the **payment** needed pay off a principal amount which is **amortized (amortization)**, $R = P\left[\frac{i}{1-(1+i)^{-m}}\right]$, where $i = \frac{r}{n}$ is the periodic interest rate and $m = nt$ is the total number of compounds. The total amount paid to amortize the principal balance is given by $Q = mR = ntR$. The interest paid is given by $I = Q - A$. If the principal amount is fixed and the payment is known, then the time necessary to pay of this principal amount is given by $t = -\frac{\ln(1-\beta)}{n\ln(1+i)}$, where $\beta = \frac{Pi}{R}$ is the percent of the total payment due in interest.

11.2.8. Given a starting balance, B_n; after interest is accrued compounded periodically with an periodic rate of i on the previous balance, that is, $B_i(1+i)$, plus purchases made, P, and (minus) a payment, R, is made, the new **credit balance**, B_{n+1} is given by $B_{i+1} = B_i(1+i) + P - R$.

11.2.9. Computing the **payment** needed to **maintain a credit balance** is $R = P\left[1 + \frac{i}{1-(1+i)^{-nt}}\right]$; where P is the mean periodic purchase. The **time** necessary to pay of this principal amount is given by $t = -\frac{\ln\left(1 - \frac{Pi}{R-P}\right)}{n\ln(1+i)}$. Note that only if you pay more than your mean monthly purchases, will you ever be able to get out of debt; this is called, **living within your means**.

REVIEW TEST

1. What is the transfer of ownership of goods and services from one individual or party to another individual or party?
2. Name a medium of exchange, any object or note that is widely accepted as payment for goods or services rendered, for the repayment of debt.
3. Another word for that which is owed.
4. How much is a 7% tax on an item selling for $15.45?
5. What is the discount rate for a television set selling for $975 with an original price of $1250?
6. Compute the effective interest rate when interest is compounded monthly with an annual rate of 7%.
7. Compute the effective interest rate when interest is compounded continuously with an annual rate of 12%.
8. Which is the better interest rate on a loan: 5% compounded monthly or 4.95% compounded daily?
9. A principal of $100 is deposited into an account paying 3% interest compounded monthly. What is the future amount after 2 years?
10. A principal of $100 is deposited into an account paying 3% interest compounded continuously. What is the future amount after 2 years?
11. A principal of $100 is deposited into an account paying 3% interest compounded monthly. How much time would it take for this amount to double?
12. How much money needs to be invested into a sinking fund weekly to save $1000 in six months if the fund pays 2.7% interest?
13. What are the monthly payments for a 30 year mortgage with a principle amount of $165,000 and 9% annual interest rate?
14. A credit card company charges 17% annual interest compounded monthly. With a balance of $180 at the end of this cycle (month) and new purchases totaling $45, what is next month's balance?
15. A credit card company charges 17% annual interest compounded monthly. What is the minimum payment necessary to maintain this credit card when $115 of purchases is made monthly?

REFERENCE

Buffett, M., Clark, D., 2008. Warren Buffett and the Interpretation of Financial Statements: The Search for the Company with a Durable Competitive Advantage. Scribner, New York.

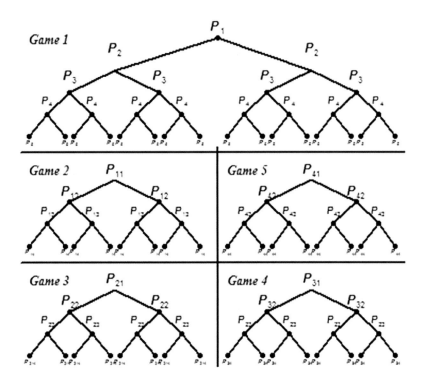

The Beautiful Mind—John Forbes Nash, Jr.—His works in Game Theory, geometry, etc. have provided insight into the factors that govern chance and events in our daily lives.

Chapter 12   Game Theory

JOHN VON NEUMANN (1903-1957)

John von Neumann, one of the foremost twentieth-century mathematicians, was born in Budapest, Hungary on December 28, 1903. He was a general scientific prodigy, in the mode of some of his great predecessors. He received his formal education at the University of Berlin, the Technical Institute in Zurich, and the University of Budapest, where he earned his Ph.D. in mathematics in 1926 at the age of twenty.

Von Neumann taught at the University of Hamburg from 1926 until 1929. He left Germany in 1930, just before the Second World War, taking refuge in this country, where he accepted a professorship in mathematical physics at Princeton University. He became professor of mathematics at the Institute for Advanced Study at Princeton, New Jersey, when it was founded in 1933, retaining this position until his death in 1957.

Von Neumann was one of the founders of game theory. In 1944 he collaborated with Oskar Morgenstern on a book, Theory of Games and Economic Behavior. This important and initial comprehensive treatment of game theory offered a new approach to the study of economic behavior through the use of game-theoretic methods.

The development of the first electronic computers at the institute was directed by von Neumann. He initiated the concept of stored programs in the computer—that programs and data should be treated similarly. This was one of the major break-throughs in computer development. Among the computers for which von Neumann is credited are the MANIAC, NORC, and ORDVAC.

Von Neumann served on the U.S. Atomic Energy Commission and consulted on the Atomic Bomb Project at Los Alamos. In 1956, he received the $50,000 Enrico Fermi award for his outstanding work on computer theory, design, and construction.

12.1 INTRODUCTION TO GAME THEORY

Game theory is a relatively new branch of mathematics; it is the study of the rational behavior of people in conflict situations. The conflict may be between individuals involved in a game of chance, between teams engaged in an athletic contest, between nations engaged in war, or between firms engaged in competition for a share of the market for a certain product. In our study of game theory, we shall refer to each contestant as a player and restrict our investigations to games employing two players. In general, however, we need not restrict ourselves to conflicts generated by two players only; such games can involve any number of players.

The two players involved in a game have precisely opposite interests. Each player has a specified number of actions from which to choose. However, the action taken by one of the players at a particular stage of the game is not known to the other player. Each of the players has a certain objective in mind, and he attempts to choose his actions so as to achieve this aim.

Definition 12.1.1 Game Theory

Game theory is concerned with the analysis of human behavior in conflict situations. In our brief introduction to the subject, we have considered only games between two persons (players); however, they may involve any number of players, teams, companies and so forth.

The initial study of game theory was conducted independently by John von Neumann and Emil Borel during the 1920s. After World War II, von Neumann and Oskar Morgenstern formulated the subject area as an independent branch of mathematics. The close relationship between systems of equations and game theory will become apparent as we proceed with our study.

12.2   THE MATRIX GAME

Consider a modified version of the game "matching pennies." Player P_1 puts a coin either heads up, H, or tails up, T. A second Player P_2 without knowing P_1's choice calls either "heads" or "tails." Player P_1 will pay Player P_2 \$5 if P_1 shows H and P_2 chooses H; Player P_1 will pay Player P_2 \$3 if both have chosen T. However, if Player P_2 guesses incorrectly, he must pay P_1 \$4 (that is, if P_1 shows H and P_2 guesses T, or if P_1 shows T and P_2 guesses H Player P_2 must pay P_1 \$4).

This game is a two-person game since there are two players involved. It can be displayed in the form of a matrix as follows:

$$P_1 \begin{array}{c} \\ H \\ T \end{array} \begin{array}{cc} & \overset{\displaystyle P_2}{\overset{\displaystyle H \quad T}{}} \\ \begin{bmatrix} -5 & 4 \\ 4 & -3 \end{bmatrix} \end{array}$$

The positive entries of the matrix denote the gains of Player P_1, and his losses to Player P_2 are the negative entries of the array. We designate P_1, the role player and P_2 the column player. Player P_1's two choices are each associated with a row of the matrix. Player P_2's choices are each associated with a column of the matrix. Thus, if P_1 chooses row one and P_2 chooses column one, then P_1 pays P_2 \$5; instead, if P_2 selects column two, then P_2 pays P_1 \$4.

Definition 12.2.1 Game Theory/Payoff Matrix

The states of the game are given in the form of a matrix called the **game matrix** or **payoff matrix**.

This game is a *zero-sum game* because whatever is lost or gained by P_1 is gained or lost by P_2. The matrix representation of the game is called the *matrix game*. Each entry of the matrix game represents the payoff to either player P_1 or P_2. Thus a matrix game is also referred to as the payoff matrix.

Definition 12.2.2 Two-person Zero Game

A **two-person zero-sum game** is a game played by two opponents with opposing interest and such that the payoff to one player is equal to the loss of the other.

The problem facing each player is what choice to make so that it will be in his best interest. That is, should P_1 select row one or row two? Should P_2, not knowing P_1's choice, select column one or column two? Before we proceed to discuss the method for choosing optimally, we shall summarize the basic concept and terminology constituting a two-person zero-sum game:

(a) Two players, P_1 and P_2, are engaged in a conflict of interest. For example, in the game of "matching pennies," P_1 wants to maximize his winnings while P_2 wants to minimize his losses.

(b) Each player has at his disposal a set of instructions regarding the action to take in each conceivable position of the game. We shall refer to these instructions as *strategies*. (In the game given above each player has two strategies H and T.)

(c) Associated with each strategy employed by P_1, there is a certain payoff. The set of all strategies will result in an array of payoffs known as the matrix game or payoff matrix.

For instance, in "matching pennies," the payoff matrix was

$$\textbf{Strategies of } P_2$$

$$\overbrace{\begin{array}{cc} H & T \end{array}}$$

$$\textbf{Strategies of } \{P_1 \quad \begin{array}{c} H \\ T \end{array} \begin{bmatrix} -5 & 4 \\ 4 & -3 \end{bmatrix}$$

(d) The objective of P_1 is to utilize his strategies so as to maximize his winnings; P_2's objective is to select his strategies so as to minimize his losses.

(e) Such games are called zero-sum games or strictly competitive games, since the sum of the amounts won by the two players is always zero.

Note: The winnings of one player are equal to the losses of the other.

Like all mathematics, game theory is a tautology whose conclusions are true because they are contained in the premises.

R.A. Epstein

Example 12.2.1 Payoff Matrix

Consider a two-person zero-sum game, the payoff matrix of which is given by

$$P_2$$

$$\begin{array}{ccc} \gamma_1 & \gamma_2 & \gamma_3 \end{array}$$

$$P_1 \begin{array}{c} \beta_1 \\ \beta_2 \end{array} \begin{bmatrix} -2 & 1 & 3 \\ 6 & -4 & 5 \end{bmatrix}$$

Here, there are two strategies available to Player P_1; namely β_1 and β_2, and three strategies, $\gamma_1, \gamma_2, \gamma_3$, available to Player P_2. The entries in the matrix denote winnings of Player P_1. Recall that negative entries mean that Player P_1 will pay P_2. In the game P_1 chooses one of his two strategies, β_1 or β_2, and simultaneously, Player P_2, without knowing P_1's choice selects $\gamma_1, \gamma_2, \gamma_3$. The intersection of the row corresponding to P_1's choice and the column corresponding to P_2's selection gives the payoff of this play in the game. For example, if P_1 chooses strategy β_2 and P_2 selects γ_3, then P_1 wins \$5. If P_1 chooses β_1 and P_2 selects γ_1, then P_1 receives \$-2; that is, he must pay Player P_2 \$2.

Here, we have the question of importance: What strategy should P_1 choose to maximize his winnings? At the same time, what should P_2's choice be to minimize his losses? The selection of strategies will be the focus of the remaining discussions in this chapter.

12.3 STRICTLY DETERMINED GAME: THE SADDLE POINT

Let us consider a two-person game, the payoff matrix of which is given by

$$
\begin{array}{c}
 & \quad\quad P_2 \\
 & \begin{array}{ccc} \gamma_1 & \gamma_2 & \gamma_3 \end{array} \\
\begin{array}{cc} & \beta_1 \\ P_1 & \beta_2 \\ & \beta_3 \end{array}
& \left[\begin{array}{ccc} 3 & 7 & 2 \\ 4 & 5 & 8 \\ 2 & 3 & 4 \end{array}\right]
\end{array}
$$

Here, P_1 wants to maximize his payoff, while P_2 wishes to minimize it. Before we proceed to determine the optimal strategy for the game, we shall employ the concept of row and column domination to simplify the game. That is, each of the elements of row 2 is greater than the corresponding element of row three ($4 > 2$, $5 > 3$, $8 > 4$). In view of P_1's objectives of maximizing his payoff, he can always do better by choosing β_2 instead of β_3. Thus, we can eliminate row 3 of the game matrix. P_2's aim is to minimize his payoff to P_1. Assuming that P_1 will be playing rationally, it is clear, looking at the three strategies available to P_2, that he will not choose strategy γ_2 (second column). This is because no matter what strategy P_1 uses, P_2 can do better than γ_2 by selecting γ_1, since each entry in the first column is smaller than the corresponding entry in the second column. Thus, the second column can be eliminated from the game matrix since it is dominated by column one. That is, if Player P_2 plays intelligently, he will never consider strategy γ_2.

At this point, the payoff matrix is reduced to

$$
\begin{array}{c}
 & \quad\quad P_2 \\
 & \begin{array}{cc} \gamma_1 & \gamma_2 \end{array} \\
\begin{array}{cc} & \beta_1 \\ P_1 & \beta_2 \end{array}
& \left[\begin{array}{cc} 3 & 2 \\ 4 & 8 \end{array}\right]
\end{array}
$$

We shall not attempt to analyze the game. If P_1 uses strategy β_1, the worst that can happen is that P_2 will select strategy γ_3 and P_1 will only gain \$2. If P_1 uses strategy β_2, the worst that can happen is that P_2 will select strategy γ_1 and P_1 will only gain \$4. The best approach to the game by Player P_1 is to aim at the larger of these minimum payoffs. Thus, P_1 must choose strategy β_2, which will result in the *maximum* of the *minimum* amounts he stands to gain.

The method of selecting the best strategy for P_1 is called the *maximin decision*, which is simply the maximum of the row minimum.

Rule 12.3.1 Maximin/Minimax

The search for the saddle point of a matrix game, if one indeed exists, can be made in three steps:

Step 1 Write the minimum entry of each row at the end of that row and write the maximum entry or each column at the end of the column.

Step 2 Identify the maximin strategy for Player P_1 by circling the largest of the row minimum. Similarly, circle the smallest of the column maximum to identify the minimax strategy for Player P_2.

Step 3 The game has a saddle point if and only if the maximin and minimax are equal to the same number, and that number also appears at the intersection in the matrix of maximin (row) and minimax (column).

Now let us consider P_2. If he uses strategy γ_1, the worst that can happen is that he will lose \$4 if P_1 employs strategy β_2. If he selects strategy γ_3, the worst that can happen is that he will lose \$8. Thus, it is to his advantage to choose γ_1

in that this choice will minimize his maximum loss to P_1. This method of selecting the strategy for Player P_2 is called the minimax decision, which is simply the minimum of the column maximum.

The selection of the preceding strategies can be displayed in the form of a table as follows:

		P_2		
		γ_1	γ_3	*Row minimum*
P_1	β_1	3	2	2
	β_2	(4)	8	(4)
	Column maximum	(4)	8	

Thus, the maximum decision for P_1 is the strategy resulting in the maximum of the row minimum, which is β_2, and the minimax decision for P_2 is the strategy resulting in the minimum of the column maximum, which is γ_1. In the present example, P_1's maximum decision results in a payoff of \$4, which coincides with P_2's minimax decision, also resulting in a payoff of \$4. When this fact occurs, we say that the matrix game possesses a *saddle point or point of equilibrium*. We call this entry in the game matrix the *value of the game*. A matrix game possessing a point of equilibrium is said to be *strictly determined*. We shall summarize these important terms in the following definition:

Definition 12.3.1 Strictly Determined

A matrix game is said to be **strictly determined** if and only if there is an entry in the payoff matrix that is the smallest entry in its row and also the largest entry in its column. This entry is called the **saddle point** or **equilibrium point** and is the value of the game.

The name *saddle point* is derived from the property of the point as the minimum in its row and the maximum in its column, as visualized by the point (β_i, γ_j) in Figure 12.1.

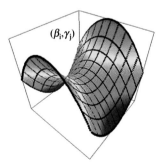

FIGURE 12.1   The saddle point.

We shall illustrate our discussion by considering some examples.

Example 12.3.1 Saddle Point

Obtain the value of the following matrix game:

$$
\begin{array}{c}
 \\
P_1 \begin{array}{c} \beta_1 \\ \beta_2 \\ \beta_3 \end{array}
\end{array}
\begin{array}{c}
P_2 \\
\begin{array}{ccc} \gamma_1 & \gamma_2 & \gamma_3 \end{array} \\
\left[\begin{array}{ccc}
\$-3 & \$1 & \$4 \\
\$4 & \$-3 & \$5 \\
\$6 & \$2 & \$7
\end{array} \right]
\end{array}
$$

Solution

Here again, the aim of P_1 is to maximize his winnings and that of P_2, to minimize his losses. Inspecting the payoff matrix, we see that the second row is dominated by the third row. This is, each entry in the third row is larger than the corresponding entry in the second row. Thus, if P_1 plays intelligently, he will never employ strategy β_2, so we can eliminate row 2. Similarly, column 3 is dominated by column 2. Thus, if P_2 plays intelligently, he will not use strategy γ_3. Hence, the payoff matrix reduces to

$$
P_1 \begin{array}{c} \beta_1 \\ \beta_2 \end{array}
\begin{bmatrix} \$-3 & \$1 \\ \$6 & \$2 \end{bmatrix}
$$

$$
\begin{array}{cc} & P_2 \\ & \gamma_1 \quad \gamma_2 \end{array}
$$

Applying the *maximin* and *minimax* method we have

		γ_1	γ_2	Row minimum
	β_1	$-3	$-1	$-3
P_1	β_3	$6	($2)	($2)
	Column maximum	$6	($2)	

(Header above table: P_2)

Thus, the best strategy for P_1 to employ is β_3. This will assure him a gain of $2, provided player P_2 plays intelligently and uses strategy γ_2 which will cost him only $2. Since the *maximin and minimax* decisions result in the same amount of payoff; namely, $2, this is a *saddle point*. The game is *strictly determined* since it possesses a *point of equilibrium*. The *value of the game* is $2.

In attempting to find the best strategies of the players, we need not employ the row and column dominance technique to reduce the payoff matrix. We can directly employ the maximin-minimax procedure.

Game theory is a study of strategic decision making.

Example 12.3.2 Best Strategies

Find the best strategies for Players P_1 and P_2 which will lead to the value of the game if the payoff matrix is given by

$$
P_1 \begin{array}{c} \beta_1 \\ \beta_2 \\ \beta_3 \\ \beta_4 \end{array}
\begin{bmatrix}
\$10 & \$0 & \$24 & \$-14 \\
\$12 & \$5 & \$6 & \$18 \\
\$-4 & \$-3 & \$ & \$-2 \\
\$12 & \$-7 & \$6 & \$5
\end{bmatrix}
$$

$$
\begin{array}{ccccc} & & P_2 & & \\ \gamma_1 & \gamma_2 & \gamma_3 & \gamma_4 \end{array}
$$

Solution

Again, the aim of P_1 is to find the best strategy to maximize his winnings. P_2 wishes to employ the strategy to minimize his losses. Utilizing the maximin-mimimax procedure, we construct the table below. Thus, Player P_1 should employ strategy β_2 to maximize the minimum amount he stands to win. Player P_2 should employ strategy γ_2 to minimize the maximum amount he stands to lose. It is clear that the value of the game is $5.

Continued

Solution—cont'd

		P_2				
		γ_1	γ_2	γ_3	γ_4	*Row minimum*
P_1	β_1	$10	$0	$24	$–14	$–14
	β_2	$12	($5)	$6	$18	($5)
	β_3	$–4	$–3	$16	$–2	$–4
	β_4	$12	$–7	$6	$5	$–7
	Column maximum	$12	($5)	$24	$18	

The solutions of strictly determined games possess certain properties that are optimal to both players provided that they play intelligently. If one of the players deviates from the procedure leading to the saddle point, the other player may increase his profit. For instance, suppose that in Example 12.3.2, Player P_1 uses his best strategy, β_2, but that P_2 plays irrationally by using strategy γ_4. This will result in P_1 increasing his payoff from $5 to $18. If both players deviate from their optimal strategies, then, needless to say, one of them will suffer.

In this section, we have discussed two-person games that are strictly determined; however, there are games that do not possess a point of equilibrium. Such games will be studied in Section 12.4.

PROBLEMS

12.3.1 Find the best strategies for players P_1 and P_2 in the matrix games given by

(a)
$$
\begin{array}{c}
& P_2 \\
& \gamma_1 \ \ \gamma_2 \\
P_1 \ \begin{array}{c} \beta_1 \\ \beta_2 \end{array}
\begin{bmatrix} \$4 & \$6 \\ \$5 & \$8 \end{bmatrix}
\end{array}
$$

(b)
$$
\begin{array}{c}
& P_2 \\
& \gamma_1 \ \ \ \gamma_2 \ \ \ \gamma_3 \\
P_1 \ \begin{array}{c} \beta_1 \\ \beta_2 \end{array}
\begin{bmatrix} \$4 & \$10 & \$6 \\ \$-3 & \$5 & \$-2 \end{bmatrix}
\end{array}
$$

The aim of P_1 is to maximize his payoff, P_2's objective is to minimize his losses.

12.3.2 Find the saddle point of the following matrix games and interpret their meanings:

(a)
$$
\begin{bmatrix}
\$6 & \$0 & \$-5 \\
\$3 & \$2 & \$6 \\
\$-7 & \$1 & \$-11
\end{bmatrix}
$$
(b)
$$
\begin{bmatrix}
\$-3 & \$4 & \$-2 \\
\$0 & \$1 & \$5 \\
\$-2 & \$16 & \$-13
\end{bmatrix}
$$

12.3.3 Determine whether or not the following matrix games are strictly determined:

(a)
$$
\begin{bmatrix}
\$-2 & \$3 & \$-1 \\
\$-5 & \$1 & \$-16
\end{bmatrix},
$$
(b)
$$
\begin{bmatrix}
\$9 & \$2 & \$4 \\
\$7 & \$1 & \$7 \\
\$-1 & \$3 & \$14
\end{bmatrix}
$$

12.3.4 Construct a 3×3 matrix game, the value of which is $10.

12.3.5 Construct a two-person zero-sum matrix game in which the payoff to the maximizing player is negative.

12.4 GAMES WITH MIXED STRATEGIES

As we mentioned in the previous section, not all two-person zero-sum games possess a saddle point. We shall now turn our attention to those games that are not strictly determined. Consider the following two-player game:

Evolution and the Theory of Games
J. Maynard Smith

$$P_1 \begin{array}{c} \\ \beta_1 \\ \beta_2 \end{array} \begin{array}{cc} \gamma_1 & \gamma_2 \\ \left[\begin{array}{cc} \$8 & \$3 \\ \$4 & \$16 \end{array}\right] \end{array}$$

Applying the procedure we discussed in our search for a saddle point, we can write the game as

		P_2		
		γ_1	γ_2	*Row minimum*
P_1	β_1	$8	$3	$3
	β_2	$4	$16	($4)
	Column maximum	($8)	$16	

If Player P_1 chooses his maximin strategy, then his maximin payoff will be $4. If Player P_2 employs his minimax strategy, then his minimax payoff will be $8. Thus, the maximin payoff of P_1 and the minimax payoff of P_2 are not equal—the matrix game does not possess a point of equilibrium. For such a game, it is difficult to see what either player should do.

Let us consider the case in which Player P_2 employs his minimax strategy, γ_1. If Player P_1 suspects P_2's choice, he can then select strategy β_1 instead of his maximin strategy, β_2, to increase his payoff from $4 to $8. However, if Player P_1 continues to employ strategy β_1, P_2 will detect this and can shift from his minimax strategy, γ_1, to strategy γ_2 decreasing his losses from $8 to $3. After a while, Player P_1 will likely become aware of P_2's strategy and can shift his choice from β_1 to β_2, thus increasing his payoff from $3 to $16. It is clear, then, in games without a saddle point, that a player should not stick with a particular strategy, but rather mix his strategies in such a way that the opposite player will not easily be able to predict his choice.

Mixing strategies can be done by some random mechanism to assure that the opposing player in the game will not discover the pattern of moves. For example, the random mechanism can be a fair coin. Player P_1 flips a coin; if a head appears, he chooses strategy β_1 with probability p; if a tail occurs, he chooses strategy β_2 with probability $1-p$. Note that $p+(1-p)=1$; that is, we are certain that at each move one of the two strategies will be chosen. Thus, the random mechanism mixes the strategies of the players. Games in which each player's strategies are mixed for lack of a saddle point are called *mixed-strategy games*.

Now, suppose that in mixed-strategy games, we assign to each player a certain probability for choosing each of his strategies. What we want to know here is how we can characterize the payoff of a game? In strictly determined games the payoff of the game was defined by the maximin and minimax strategies. However, in the present case we do not know which strategies are being used by the players and we cannot define the payoff if the game is played only once. However, if the game is played a number of times, by utilizing the frequency with which each strategy is used, we can obtain the expected payoff or central tendency of the matrix game. For example, suppose that in the matrix game on page 504 Player P_1 chooses his strategies β_1 and β_2 with probabilities 0.6 and 0.4, respectively. Similarly, player P_2 selects his strategies γ_1 and γ_2 with probabilities 0.7 and 0.3, respectively. (Shortly we shall be discussing the best way of arriving at these probabilities.) The expected payoff of the matrix game is given by evaluating the following expression:

$$[0.6 \quad 0.4] \cdot \begin{bmatrix} 8 & 3 \\ 4 & 10 \end{bmatrix} \begin{bmatrix} 0.7 \\ 0.3 \end{bmatrix}$$

First, we multiply the vector $[0.6\ 0.4]$ and the 2×2 matrix game $\begin{bmatrix} 8 & 3 \\ 4 & 16 \end{bmatrix}$ that is,

$$[0.6 \quad 0.4] \cdot \begin{bmatrix} 8 & 3 \\ 4 & 16 \end{bmatrix} = [4.8 + 1.6 \quad 1.8 + 6.4] = [6.4 \quad 8.2].$$

Next, we multiply the row vector $[6.4\ 8.2]$ by the column vector $\begin{bmatrix} 0.7 \\ 0.3 \end{bmatrix}$ to obtain the expected payoff,

$$[0.6 \quad 0.4 \quad 8.2] \cdot \begin{bmatrix} 0.7 \\ 0.3 \end{bmatrix} = (6.4)(0.7) + (8.2)(0.3)$$
$$= 4.48 + 2.46$$
$$= \$6.94$$

Thus, the expected payoff which favors Player P_1, is \$6.94. That is, in the long run and under the specified probabilities of selecting the strategies, P_1's expected gain is \$6.94.

In general, the expected payoff of a 2×2 matrix game is obtained as follows: Let P_1 and P_2 be the probabilities that Player P_1 selects his strategies, β_1 and β_2 respectively. We shall refer to these probabilities as *strategy probabilities* of Player P_1 and write them as a row vector $[p_1 p_2]$. Similarly, we shall denote the strategy probabilities for Player P_2 by the column vector $\begin{bmatrix} q_1 \\ q_2 \end{bmatrix}$. Note that $p_1 + p_2 = 1$ and $q_1 + q_2 = 1$. The expected payoff of the matrix game $\begin{bmatrix} a_{11} & a_{12} \\ a_{21} & a_{22} \end{bmatrix}$ denoted by E is defined by

$$E = [p_1 \quad p_2] \begin{bmatrix} a_{11} & a_{12} \\ a_{21} & a_{22} \end{bmatrix} \begin{bmatrix} q_1 \\ q_2 \end{bmatrix}$$

Example 12.4.1   Expected Payoff

Obtain the expected payoff of the matrix game,

$$\begin{array}{c} \textbf{\textit{P}}_2 \\ \begin{array}{cc} \gamma_1 & \gamma_2 \end{array} \\ P_1 \begin{array}{c} \beta_1 \\ \beta_2 \end{array} \begin{bmatrix} 4 & -3 \\ -5 & 7 \end{bmatrix} \end{array}$$

If Player P_1 and P_2 decide on selecting their strategies with probabilities [1/2 ½] and [1/3 2/3], respectively.

Solution

The expected payoff of this game is

$$E = \begin{bmatrix} \dfrac{1}{2} & \dfrac{1}{2} \end{bmatrix} \begin{bmatrix} 4 & -3 \\ -5 & 7 \end{bmatrix} \begin{bmatrix} \dfrac{1}{3} \\ \dfrac{2}{3} \end{bmatrix}$$

$$= \begin{bmatrix} -\dfrac{1}{2} & \dfrac{4}{2} \end{bmatrix} \begin{bmatrix} \dfrac{1}{3} \\ \dfrac{2}{3} \end{bmatrix}$$

$$= \dfrac{7}{6}$$

Thus, the expected payoff is in favor of P_1. Note that if the actual value of E had been negative, then the game would have been in favor of Player P_2.

For higher order matrix games, say $k \times k$, given by

$$\begin{array}{c} & P_2 \\ & \begin{array}{cccc} \gamma_1 & \gamma_2 & \cdots & \gamma_k \end{array} \\ \begin{array}{c} \beta_1 \\ \beta_2 \\ \vdots \\ \beta_k \end{array} & \begin{bmatrix} a_{11} & a_{12} & \cdots & a_{1k} \\ a_{21} & a_{22} & \ldots & a_{2k} \\ \vdots & \vdots & & \vdots \\ a_{k1} & a_{k2} & \cdots & a_{kk} \end{bmatrix} \end{array}$$

with strategy probabilities for P_1 and P_2 given by

$$\begin{bmatrix} p_1 & p_2 & \cdots & p_k \end{bmatrix} \quad \text{and} \quad \begin{bmatrix} q_1 \\ q_2 \\ \vdots \\ q_k \end{bmatrix}$$

respectively, the expected payoff of the game is

$$E = \begin{bmatrix} p_1 & p_2 & \cdots & p_k \end{bmatrix} \begin{bmatrix} a_{11} & a_{12} & \ldots & a_{1k} \\ a_{21} & a_{22} & \ldots & a_{2k} \\ \cdot & \cdot & & \cdot \\ a_{21} & a_{22} & \ldots & a_{2k} \end{bmatrix} \begin{bmatrix} q_1 \\ q_2 \\ \vdots \\ q_k \end{bmatrix}$$

with $p_1 + p_2 + \cdots + p_k = 1$ and $q_1 + q_2 + q_k = 1$.

Example 12.4.2 Expected Payoff

Obtain the expected payoff of the matrix game,

$$\begin{array}{c} & & P_2 \\ & & \begin{array}{ccc} \gamma_1 & \gamma_2 & \gamma_3 \end{array} \\ P_1 & \begin{array}{c} \beta_1 \\ \beta_2 \\ \beta_3 \end{array} & \begin{bmatrix} 4 & -3 & 0 \\ 2 & 1 & -4 \\ -10 & -3 & 1 \end{bmatrix} \end{array}$$

if the strategy probabilities of P_1 and P_2 are given by

$$\begin{bmatrix} \dfrac{1}{3} & \dfrac{1}{3} & \dfrac{1}{3} \end{bmatrix} \quad \text{and} \quad \begin{bmatrix} \dfrac{1}{2} \\ \dfrac{1}{4} \\ \dfrac{1}{4} \end{bmatrix}$$

respectively.

Solution

The expected payoff of the 3×3 matrix game is

$$E = \begin{bmatrix} \dfrac{1}{3} & \dfrac{1}{3} & \dfrac{1}{3} \end{bmatrix} \cdot \begin{bmatrix} 4 & -3 & 0 \\ 2 & 1 & -4 \\ -10 & -3 & 1 \end{bmatrix} \begin{bmatrix} \dfrac{1}{2} \\ \dfrac{1}{4} \\ \dfrac{1}{4} \end{bmatrix}$$

$$= \begin{bmatrix} -\dfrac{4}{3} & -\dfrac{5}{3} & -1 \end{bmatrix} \begin{bmatrix} \dfrac{1}{2} \\ \dfrac{1}{4} \\ \dfrac{1}{4} \end{bmatrix}$$

$$= -\dfrac{4}{3}$$

Thus, the expected payoff of the game is $-4/3$ and it favors Player P_2.

There is still a basic question remaining: What is the best way for Player P_1 to choose his strategy probabilities so that his expected payoff will be maximum? To answer this question we proceed as follows. Consider a two-person zero-sum game given by

$$\begin{array}{cc} & P_2 \\ & \begin{array}{cc} \gamma_1 & \gamma_2 \end{array} \\ P_1 \begin{array}{c} \beta_1 \\ \beta_2 \end{array} & \begin{bmatrix} a_{11} & a_{12} \\ a_{21} & a_{22} \end{bmatrix} \end{array}$$

and P_1's strategy probabilities by $[p_1 \, p_2]$. Player P_1 wants to choose the probabilities p_1 and p_2 to select his strategies so as to maximize the expected payoff E.

That is,

$$E \leq \begin{bmatrix} p_1 & p_2 \end{bmatrix} \begin{bmatrix} a_{11} & a_{12} \\ a_{21} & a_{22} \end{bmatrix}$$

Which can be written as

$$a_{11}p_1 + a_{21}p_2 \geq E$$
$$a_{12}p_1 + a_{22}p_2 \geq E,$$

such that

$$p_1 + p_2 = 1.$$

Treating these inequalities as equalities and solving the three equations we obtain

$$p_1 = \frac{a_{22} - a_{21}}{a_{11} - a_{12} - a_{21} + a_{22}}$$

and

$$p_2 = \frac{a_{11} - a_{12}}{a_{11} - a_{12} - a_{21} + a_{22}}$$

Similarly, the aim of P_2 is to select his strategy probabilities, q_1 *and* q_2 in such a manner as to minimize the expected payoff. That is,

$$E = \begin{bmatrix} a_{11} & a_{12} \\ a_{21} & a_{22} \end{bmatrix} \begin{bmatrix} q_1 \\ q_2 \end{bmatrix}$$

which can be stated as

$$a_{11}q_1 + a_{12}q_2 \geq E$$
$$a_{21}q_1 + a_{22}q_2 \leq E,$$

such that

$$q_1 + q = 1.$$

Solution—cont'd

Solving the inequalities as if they were equalities, we have

$$q_1 = \frac{a_{22}-a_{12}}{a_{11}-a_{12}-a_{21}+a_{22}}$$

$$q_2 = \frac{a_{11}-a_{21}}{a_{11}-a_{12}-a_{21}+a_{22}}$$

and

$$E = [p_1 \quad p_2] \begin{bmatrix} a_{11} & a_{12} \\ a_{21} & a_{22} \end{bmatrix} [q_1 \quad q_2]$$

$$= \frac{a_{11}-a_{21}}{a_{11}-a_{12}-a_{21}+a_{22}}$$

Baseball, it is said, is only a game. True, … and the Grand Canyon is only a hole in Arizona. Not all holes, or games are created equal.

George Will

Definition 12.4.1 Optimal Strategies

The strategy probabilities $\{p_1, p_2\}$ and $\{q_1, q_2\}$, as defined above are called the **optimal strategies** for Players P_1 and P_2, respectively. The expected payoff E is called the value of the game.

Example 12.4.3 Expected Payoff

Consider "matching pennies" under the following rules. If Player P_1 has turned up H and P_2 calls H, then P_1 must pay P_2 \$6. However, if P_2 calls T, then P_2 must pay P_1 \$4. If P_1 has turned up T and P_2 calls H, then P_2 must pay P_1 \$8, and if P_2 calls T, then P_1 pays P_2 \$2. Find the optimal strategies for P_1 and P_2 and the value of the game.

Solution

The game matrix is given by

$$P_1 \begin{matrix} & P_2 \\ & \begin{bmatrix} \$-6 & \$4 \\ \$8 & \$-2 \end{bmatrix} \end{matrix}$$

It can be seen that the game does not have a saddle point. Thus, we must employ the mixed-strategies procedure. The optimal strategy for P_1 is

$$p_1 = \frac{a_{22}-a_{21}}{a_{11}-a_{12}-a_{21}+a_{22}} = \frac{-2-8}{-6-4-8-2} = \frac{1}{2}$$

and

$$p_2 = \frac{a_{11}-a_{12}}{a_{11}-a_{12}-a_{21}+a_{22}} = \frac{-6-4}{-6-4-8-2} = \frac{1}{2}$$

The optimal strategy for P_2 is

$$q_1 = \frac{a_{22}-a_{12}}{a_{11}-a_{12}-a_{21}+a_{22}} = \frac{-2-4}{-6-4-8-2} = \frac{3}{10}$$

and

$$q_2 = \frac{a_{11}-a_{21}}{a_{11}-a_{12}-a_{21}+a_{22}} = \frac{-6-8}{-6-4-8-2} = \frac{7}{10}$$

The expected payoff of the game corresponding to these optimal strategies is

$$E = \frac{a_{11}a_{22}-a_{12}a_{21}}{a_{11}-a_{12}-a_{21}+a_{22}} = \frac{(-6)(-2)-(4)(8)}{-6-4-8-2}$$

$$= \$1.00.$$

Continued

Solution—cont'd

Thus, Player P_1's optimal strategy is to choose row one with probability 1/2 and row two with the same probability. Player P_2's optimal strategy is to select column one with probability 3/10 and column two with probability 7/10. The value of the game is $1.00. That is, since E is positive, the game is favorable to Player P_1 in the long run.

Example 12.4.4 Expected Payoff

Find the optimal strategies for Players P_1 and P_2 for the matrix game given in Example 12.3.2. That is,

$$P_1 \begin{array}{c} P_2 \\ \begin{bmatrix} 4 & -3 \\ -5 & 7 \end{bmatrix} \end{array}$$

Also obtain the value of the game.

Solution

We should mention here that in Example 12.4.1, we stated the strategy probabilities of [1/2 1/2] and [1/3 2/3] for P_1 and P_2, respectively, for purpose of illustrating the concept of expected payoff. Here we shall obtain the optimal strategies for each of the players.

The optimal strategy for P_1 is

$$p_1 = \frac{a_{22} - a_{21}}{a_{11} - a_{12} - a_{21} + a_{22}} = \frac{7+5}{4+3+5+7} = \frac{12}{19}$$

and

$$p_2 = \frac{a_{11} - a_{12}}{a_{11} - a_{12} - a_{21} + a_{22}} = \frac{4+3}{4+3+5+7} = \frac{7}{19}$$

The optimal strategy for P_2 is

$$q_1 = \frac{a_{22} - a_{12}}{a_{11} - a_{12} - a_{21} + a_{22}} = \frac{7+3}{4+3+5+7} = \frac{10}{19}$$

and

$$q_2 = \frac{a_{11} - a_{21}}{a_{11} - a_{12} - a_{21} + a_{22}} = \frac{4+5}{4+3+5+7} = \frac{9}{19}$$

The expected payoff of the game corresponding to these optimal strategies is

$$E = \frac{a_{11}a_{22} - a_{12}a_{21}}{a_{11} - a_{12} - a_{21} + a_{22}} = \frac{(4)(7) - (-3)(-5)}{4+3-5+7} = \frac{13}{19}$$

Thus, the optimal strategies for P_1 and P_2 are [12/19 7/19] and [10/19 9/19], respectively. The value of the game is 13/19 and since it is positive it favors Player P_1. That is, in the long run P_1 is expected to gain 13/19.

We should mention here that we can sometimes employ the row and column dominance relation to reduce a higher order matrix game so that the preceding formulas would be applicable. For example, consider the 3×3 matrix game given by

$$P_1 \begin{array}{c} P_2 \\ \begin{bmatrix} 2 & 1 & 3 \\ 12 & 3 & 11 \\ 5 & 6 & 10 \end{bmatrix} \end{array}$$

Here, row one of the matrix is dominated by the second row, since each element of row two is greater than the corresponding element of row one. Similarly, since each element of column two is smaller than the corresponding element of column three, we conclude that column three is dominated by column two. Thus, the matrix game reduces to

Solution—cont'd

$$P_1 \begin{array}{c} P_2 \\ \begin{bmatrix} 12 & 3 \\ 5 & 6 \end{bmatrix} \end{array}$$

and we can apply the mixed-strategies procedure to obtain optimal strategies for P_1 and P_2.

The optimal strategy for P_1 is

$$p_1 = \frac{a_{22} - a_{21}}{a_{11} - a_{12} - a_{21} + a_{22}} = \frac{6 - 5}{12 - 3 - 5 + 6} = \frac{1}{10}$$

and

$$p_2 = \frac{a_{11} - a_{12}}{a_{11} - a_{12} - a_{21} + a_{22}} = \frac{12 - 3}{12 - 3 - 5 + 6} = \frac{9}{10}$$

The optimal strategy for P_2 is

$$q_1 = \frac{a_{22} - a_{12}}{a_{11} - a_{12} - a_{21} + a_{22}} = \frac{6 - 3}{12 - 3 - 5 + 6} = \frac{3}{10}$$

and

$$q_2 = \frac{a_{11} - a_{21}}{a_{11} - a_{12} - a_{21} + a_{22}} = \frac{12 - 5}{12 - 3 - 5 + 6} = \frac{7}{10}$$

The expected payoff of the game corresponding to these optimal strategies is

$$E = \frac{a_{11} a_{22} - a_{12} a_{21}}{a_{11} - a_{12} - a_{21} + a_{22}} = \frac{(12)(6) - (3)(5)}{12 - 3 - 5 + 6} = \frac{57}{10}$$

The value of the game is 57/10 and it favors Player P_1 because it is positive. This means that in the long run Player P_1 is expected to gain 57/10.

PROBLEMS

12.4.1 Find the expected payoff of the matrix game given by

$$P_1 \begin{array}{c} \begin{array}{c} P_2 \\ \gamma_1 \quad \gamma_2 \end{array} \\ \begin{array}{c} \beta_1 \\ \beta_2 \end{array} \begin{bmatrix} -6 & 12 \\ 4 & -10 \end{bmatrix} \end{array}$$

If Player P_1 chooses strategy β_1 60% of the time and Player P_2 selects strategy γ_2 70% of the time. Which player does the game favor?

12.4.2 Determine the expected payoff for the game

$$P_1 \begin{array}{c} P_2 \\ \begin{bmatrix} 10 & -6 & 2 \\ -12 & 8 & 4 \\ 16 & -14 & -8 \end{bmatrix} \end{array}$$

for the strategy probabilities $[\mathbf{1/3 \ 1/3 \ 1/3}]$ and $\begin{bmatrix} 1/4 \\ 1/4 \\ 1/2 \end{bmatrix}$ for Players P_1 and P_2, respectively.

12.4.3 Find the optimal strategies for Players P_1 and P_2, and the value of the following games:

$$P_1 \begin{array}{c} \begin{bmatrix} -4 & 6 \\ 8 & -2 \end{bmatrix} \end{array} \quad \text{and} \quad P_1 \begin{array}{c} P_2 \\ \begin{bmatrix} 10 & -12 \\ -16 & 14 \end{bmatrix} \end{array}$$

12.4.4 In the game of matching pennies, consider the following situation: If Player P_1 turns up T and P_2 calls T, P_1 pays P_2 $5. However, if Player P_2 calls H, then P_2 must pay P_1 $8. Now if P_1 turns up H and P_2 calls T, then P_2 must pay P_1 $10, but if P_2 calls H, then P_1 must pay P_2 $12.

(a) Obtain the matrix of the game.
(b) Determine the optimal strategies for Players P_1 and P_2.
(c) Find the value of the game.
(d) Does the game favor Player P_1?

12.4.5 Find the optimal strategies for Players P_1 and P_2 and the value of the game

$$P_1 \begin{array}{c} P_2 \\ \begin{bmatrix} 3 & 7 & 1 \\ 2 & 6 & 8 \\ 1 & 3 & 4 \end{bmatrix} \end{array}$$

12.4.6 Determine the optimal strategies for Players P_1 and P_2 and the value of the following game, if possible:

$$P_1 \begin{array}{c} P_2 \\ \begin{bmatrix} 3 & 4 & -4 & 6 & 0 \\ -3 & 5 & -2 & 7 & 10 \end{bmatrix} \end{array}$$

12.5   REDUCING MATRIX GAMES TO SYSTEMS OF EQUATIONS

The aim of the present section is to show that a matrix game can be reduced to a problem in linear programming. We can then solve matrix games by applying the techniques introduced in Chapter 10. We shall illustrate the procedure by considering a specific example. Suppose that we have a two-person zero-sum matrix game given by

$$P_1 \begin{array}{c} P_2 \\ \begin{bmatrix} 4 & -2 \\ -6 & 8 \end{bmatrix} \end{array}$$

Player P_1 wants to utilize the best strategy to maximize his expected payoff E. Player P_2 wants to choose his strategies so that he will minimize the expected payoff. Thus P_1 and P_2 are the maximizer and minimizer of E, respectively. Let $[p_1 \ \ p_2]$ and $\begin{bmatrix} q_1 \\ q_2 \end{bmatrix}$ be the strategy probabilities of Players P_1 and P_2, respectively. Proceeding as in the previous section, we can write

$$[p_1 \ \ p_2] \begin{bmatrix} 4 & -2 \\ -6 & 8 \end{bmatrix} \geq E$$

which can be stated as

$$4p_1 - 6P_2 \geq E$$
$$-2p_1 + 8P_2 \geq E$$

restricted by the fact that we must have

$$p_1 \geq 0. \quad p_2 \geq 0, \quad Ep_1 + p_2 = 1.$$

There is new restriction here that E must be positive for this procedure to work. However, if E is not positive, we can make it so by adding an appropriate constant to every term in the inequalities without changing the character of the problem.

Thus, P_1 must select P_1 and P_2 to maximize his expected payoff **E** subject to the following constraints:

$$4p_1 - 6P_2 \geq E$$
$$-2p_1 + 8P_2 \geq E$$
$$p_1 + p_2 = 1.$$
$$p_1 \geq 0, \quad p_2 \geq 0, \quad E > 0.$$

To simplify the problem further, we divide the preceding inequalities and equality by E; that is,

$$4\frac{p_1}{E} - 6\frac{p_2}{E} \geq 1$$
$$-2\frac{p_1}{E} + 8\frac{p_2}{E} \geq 1$$
$$\frac{p_1}{E} - 6\frac{p_2}{E} = \frac{1}{E}$$
$$\frac{p_1}{E} \geq 0, \quad \frac{p_2}{E} \geq 0.$$

Let $z_1 = p_1/E$ and $z_2 = p_2/E$. Now, the object of the preceding inequalities is to minimize

$$z_1 + z_2 = \frac{1'}{E}$$

subject to the constraints

$$4z_1 - 6z_2 \geq 1$$
$$-2z_1 + 8z_2 \geq 1$$
$$z_1 \geq 0, \quad z_2 \geq 0,$$

which is simply in the form of a linear program. Note that when we maximize E, it means that we minimize $1/E$; that is, increasing E implies that we are decreasing the value of $1/E$.

We begin by treating the constraints as if they were equalities. Solving them for z_1 and z_2 we obtain

$$z_1 = \frac{7}{10} \quad \text{and} \quad z_2 = \frac{3}{10}.$$

You recall that

$$z_1 = \frac{p_1}{E} \quad \text{and} \quad z_2 = \frac{p_2}{E}.$$

subject to

$$z_1 + z_2 = \frac{1}{E}.$$

Substituting the values of z_1 and z_2 in the above equation we have

$$\frac{7}{10} + \frac{3}{10} = \frac{1}{E}.$$

Solving this expression for the expected payoff, we obtain **E = 1.0**. Thus, substituting this value of **E** in

$$p_1 = z_1 E \quad \text{and} \quad p_2 = z_2 E,$$

we have

$$p_1 = \frac{7}{10}(1) = \frac{7}{10} \quad \text{and} \quad p_2 = \frac{3}{10}(1) = \frac{3}{10}$$

Thus, Player P_1 must select the first row of the matrix game with probability 7/10 and the second row with probability 3/10 so as to maximize his gain.

Similarly, Player P_2's objective to select the strategy probabilities q_1 and q_2 so as to minimize the expected payoff. That is,

$$\begin{bmatrix} 4 & -2 \\ -6 & 8 \end{bmatrix} \begin{bmatrix} q_1 \\ q_2 \end{bmatrix} \leq E,$$

which can be written as

$$4q_1 - 2q_2 \leq E$$
$$-6q_1 + 8q_2 \leq E$$

along with the restrictions

$$q_1 \geq 0, \quad q_2 \geq 0, \quad q_1 + q_2 = 1, \quad E > 0.$$

P_2's problem, then, is to choose q_1 and q_2 that will minimize the expected payoff E subject to the constraints

$$4q_1 - 2q_2 \leq E$$
$$-6q_1 + 8q_2 \leq E$$
$$q_1 - + q_2 = 1$$
$$q_1 \geq 0 \quad q_2 \geq 0.$$

Using arguments similar to those previously mentioned, and letting $\gamma_1 = q_1/E$ and $\gamma_2 = q_2/E$, we have reduced the game into the following linear **equations**:

$$4\gamma_1 - 2\gamma_2 \leq 1$$
$$-6\gamma_1 + 8\gamma_2 \leq 1$$
$$\gamma_1 \geq 0, \quad \gamma_2 \geq 0.$$

subject to minimizing $\gamma_1 + \gamma_2 = 1/E$. Treating the first two inequalities as if they were equalities and solving them, we obtain $y_1 = 1/2$ and $y_2 = 1/2$.

Since we know that $E = 1$, we can obtain the strategy probabilities:

$$q_1 = y_1 E \quad \text{or} \quad q_1 = \frac{1}{2}(1) = \frac{1}{2}$$

and

$$q_2 = y_2 E \quad \text{or} \quad q_2 = \frac{1}{2}(1) = \frac{1}{2}.$$

Note that the restriction that $y_1 + y_2 = \frac{1}{E}$ or $\frac{1}{2} + \frac{1}{2} = 1$ is satisfied. Thus, Player P_2 must select columns one and two of the matrix game with equal probability.

Furthermore, we check our results by obtaining the value of the game as follows:

$$E = \begin{bmatrix} \dfrac{7}{10} & \dfrac{3}{10} \end{bmatrix} \begin{bmatrix} 4 & -2 \\ -6 & 8 \end{bmatrix} \begin{bmatrix} \dfrac{1}{2} \\ \dfrac{1}{2} \end{bmatrix}$$

$$= \begin{bmatrix} 1 & 1 \end{bmatrix} \begin{bmatrix} \dfrac{1}{2} \\ \dfrac{1}{2} \end{bmatrix}$$

$$= 1.$$

Therefore, the optimal strategies for Players P_1 and P_2 are [7/10 3/10] and [1/2 1/2], respectively. The corresponding value of the game is 1 and, since it is positive, it favors P_1.

PROBLEMS

12.5.1 Find the optimal strategies for Players P_1 and P_2, in the following games:

(a) $P_1 \begin{bmatrix} -2 & 8 \\ 10 & 6 \end{bmatrix}$ (b) $P_1 \begin{bmatrix} 4 & -5 \\ -3 & 2 \end{bmatrix}$

with P_2 labeling the columns.

12.5.2. Reduce the matching pennies game given below into a system of linear equations and find the optimal stratiegies for Players P_1 and P_2. In the game of matching pennies, consider the following situation: If Player P_1 turns up T and P_2 calls T, P_1 pays P_2 \$5. However, if Player P_2 calls H, then P_2 must pay P_1 \$8. Now if P_1 turns up H and P_2 calls T, then P_2 must pay P_1 \$10, but if P_2 calls H, then P_1 must pay P_2 \$12.

12.5.3. Solve the following matrix game:

$$P_1 \begin{bmatrix} 6 & 8 & -8 & 12 & 0 \\ -6 & 10 & -4 & 14 & 20 \end{bmatrix}$$

with P_2 labeling the columns.

12.5.4. Find the optimal strategies for Players P_1 and P_2, of the matrix game

$$P_1 \begin{bmatrix} 4 & -3 & 2 \\ 5 & 4 & 3 \\ 8 & -2 & 8 \end{bmatrix}$$

with P_2 labeling the columns.

by reducing it into a system of linear equations.

Hint: Begin by employing the row and column dominance relation.

CRITICAL THINKING AND BASIC EXERCISE

12.1 Find the best strategies for Players P_1 and P_2, in the matrix games given by:

(a)
$$P_1 \begin{matrix} \beta_1 \\ \beta_2 \end{matrix} \begin{bmatrix} -1 & 4 \\ -3 & 2 \end{bmatrix}$$
with P_2 columns γ_1 γ_2

(b)
$$P_1 \begin{matrix} \beta_1 \\ \beta_2 \\ \beta_3 \end{matrix} \begin{bmatrix} -10 & 5 & -8 \\ 4 & 2 & 1 \\ 0 & -4 & -1 \end{bmatrix}$$
with P_2 columns γ_1 γ_2 γ_3

12.2 Find the point of equilibrium of the following matrix games:

(a) $\begin{bmatrix} 4 & -4 & 4 \\ -3 & -3 & -3 \\ 5 & -5 & 5 \end{bmatrix}$ (b) $\begin{bmatrix} 2 & -4 & 1 & -3 \\ 0 & 1 & 6 & -5 \\ 7 & 1 & 4 & -1 \end{bmatrix}$

12.3 Are the following matrix games strictly determined?

(a) $\begin{bmatrix} 4 & 12 & 10 \\ 14 & 8 & 2 \\ -6 & 4 & -8 \end{bmatrix}$ (b) $\begin{bmatrix} 2 & 3 & 4 & 5 & 6 \\ -1 & -2 & 1 & 2 & 3 \end{bmatrix}$

12.4 Construct a 3×3 matrix game, the point of equilibrium of which is 5.

12.5 Construct a zero-sum two-person matrix game in which the payoff to P_1 is negative; that is, a gain for Player P_2.

12.6 Find the expected payoff of the matrix game given by

$$P_1 \begin{matrix} \beta_1 \\ \beta_2 \end{matrix} \begin{bmatrix} -10 & 8 \\ 5 & -8 \end{bmatrix}$$
with P_2 columns γ_1 γ_2

Player P_1 chooses his strategies with equal probability and Player P_2 selects strategy y_1 three-fourths of the time and strategy y_2 one-fourth of the time.

12.7 Determine the expected payoff for the matrix game given by

$$P_1 \begin{array}{c} P_2 \\ \begin{bmatrix} 7 & 3 & -2 \\ -4 & 5 & 4 \\ 6 & 8 & 10 \end{bmatrix} \end{array}$$

for the strategy probabilities $[1/2\ 1/4\ 1/4]$ and $\begin{bmatrix} 1/6 \\ 1/3 \\ 1/2 \end{bmatrix}$ for Players P_1 and P_2, respectively.

12.8 Determine the optimal strategies for Players P_1 and P_2, and the value of the following games:

$$\text{(a)}\ P_1 \begin{array}{c} P_2 \\ \begin{bmatrix} 4 & -3 \\ -2 & 10 \end{bmatrix} \end{array} \quad \text{(b)}\ P_1 \begin{array}{c} P_2 \\ \begin{bmatrix} 1 & -2 \\ -1 & 2 \end{bmatrix} \end{array}$$

12.9 What are the optimal strategies for Players P_1 and P_2, and the value of the game:

$$P_1 \begin{array}{c} P_2 \\ \begin{bmatrix} -8 & 4 & 5 \\ 10 & -6 & -5 \\ -9 & -7 & -6 \end{bmatrix} \end{array} ?$$

12.10 Find the optimal strategies for Players P_1 and P_2, and the value of the following matrix game:

$$P_1 \begin{array}{c} P_2 \\ \begin{bmatrix} 7 & -1 & 6 & 10 & 12 \\ 14 & 5 & -2 & 6 & 8 \end{bmatrix} \end{array}$$

12.11 Find the optimal strategies for Players P_1 and P_2, in the following matrix games by reducing the games into linear programming problems:

$$\text{(a)}\ P_1 \begin{array}{c} P_2 \\ \begin{bmatrix} -8 & 4 \\ 10 & -6 \end{bmatrix} \end{array} \quad \text{(b)}\ P_1 \begin{array}{c} P_2 \\ \begin{bmatrix} -8 & 12 \\ 14 & -6 \end{bmatrix} \end{array}$$

12.12 Reduce the following matrix game into a linear programming problem and then obtain the optimal strategies and value of the game:

$$P_1 \begin{array}{c} P_2 \\ \begin{bmatrix} a_{11} & a_{12} \\ a_{21} & a_{22} \end{bmatrix} \end{array}$$

SUMMARY OF IMPORTANT CONCEPTS

Definitions:

12.1.1. **Game theory** is concerned with the analysis of human behavior in conflict situations. In our brief introduction to the subject, we have considered only games between two persons (players); however, they may involve any number of players, teams, companies and so forth.

12.1.2. The states of the game are given in the form of a matrix called the **game matrix** or **payoff matrix**.

12.2.1 A **two-person zero-sum game** is a game played by two opponents with opposing interest and such that the payoff to one player is equal to the loss of the other.

12.2.2. A **zero-sum** games such that the loss or gain by Player P_1 is equal to the gain or loss by Player P_2.

12.3.1 A matrix game is said to be **strictly determined** if and only if there is an entry in the payoff matrix that is the smallest entry in its row and also the largest entry in its column. This entry is called the **saddle point** or **equilibrium point** and is the value of the game.

12.4.1. The strategy probabilities $\{p_1, p_2\}$ and $\{q_1, q_2\}$, as defined above are called the **optimal strategies** for Players P_1 and P_2, respectively. The expected payoff E is called the value of the game.

Rules:

12.3.1 The search for the saddle point of a matrix game, if one indeed exists, can be made in three steps:

   Step 1 Write the minimum entry of each row at the end of that row and write the maximum entry of each column at the end of the column.

   Step 2 Identify the *maximin strategy* for Player P_1 by circling the largest of the row minimum. Similarly, circle the smallest of the column maximum to identify the *minimax strategy* for Player P_2.

   Step 3 The game has a saddle point if and only if the maximin and minimax are equal to the same number, and that number also appears at the intersection in the matrix of maximin (row) and minimax (column).

REVIEW TEST

1. Define **game theory**.
2. What is another name for **game matrix**?
3. A game played by two opponents with opposing interest and such that the payoff to one player is equal to the loss of the other is referred to by what name?
4. What is meant by **strictly determined**?
5. What is the value of the two-person zero-sum game, the payoff matrix of which is given by $\begin{bmatrix} -2 & 3 & 4 \\ 6 & -1 & 5 \end{bmatrix}$
6. Find the best strategies for Players P_1 and P_2 which will lead to the value of the game if the payoff matrix is given by $\begin{bmatrix} -4 & 1 & 5 \\ 5 & -4 & 6 \\ 7 & 3 & 8 \end{bmatrix}$.
7. Construct a **3×3** matrix game with a value of \$5.
8. Construct a **3×3** matrix game with a negative value.
9. Obtain the expected payoff of the matrix game: $\begin{bmatrix} -5 & 7 \\ 4 & -3 \end{bmatrix}$

REFERENCES

Luce, R. Duncan, Raiffa, Howard, 1957. Games and Decisions. Wiley, New York.

May, F.B., 1970. Introduction to Games of Strategy. Allyn & Bacon, Boston.

McKinsey, J.C.C., 1957. Introduction to the Theory of Games. Wiley, New York.

von Neumann, J., Morgenstern, O., 1954. Theory of Games and Economic Behavior, 2nd ed. Princeton University Press, Princeton, NJ.

Owens, G., 1970. Finite Mathematics. W. B. Saunders, Philadelphia, PA.

Williams, J.D., 1964. The Compleat Strategyst. McGraw-Hill, New York.

Tsokos, C.P., 1978h. Mainstreams of Finite Mathematics with Application. Charles E. Merrill Publishing Company, A Bell & Howell Company.

Appendix: Standard Normal

TABLE A1: Standard Normal Probability Distribution: To the Center $P(0 \leq Z \leq z)$

	0.00	0.01	0.02	0.03	0.04	0.05	0.06	0.07	0.08	0.09
0.0	0.0000	0.0040	0.0080	0.0120	0.0160	0.0199	0.0239	0.0279	0.0319	0.0359
0.1	0.0398	0.0438	0.0478	0.0517	0.0557	0.0596	0.0636	0.0675	0.0714	0.0753
0.2	0.0793	0.0832	0.0871	0.0910	0.0948	0.0987	0.1026	0.1064	0.1103	0.1141
0.3	0.1179	0.1217	0.1255	0.1293	0.1331	0.1368	0.1406	0.1443	0.1480	0.1517
0.4	0.1554	0.1591	0.1628	0.1664	0.1700	0.1736	0.1772	0.1808	0.1844	0.1879
0.5	0.1915	0.1950	0.1985	0.2019	0.2054	0.2088	0.2123	0.2157	0.2190	0.2224
0.6	0.2257	0.2291	0.2324	0.2357	0.2389	0.2422	0.2454	0.2486	0.2517	0.2549
0.7	0.2580	0.2611	0.2642	0.2673	0.2704	0.2734	0.2764	0.2794	0.2823	0.2852
0.8	0.2881	0.2910	0.2939	0.2967	0.2995	0.3023	0.3051	0.3078	0.3106	0.3133
0.9	0.3159	0.3186	0.3212	0.3238	0.3264	0.3289	0.3315	0.3340	0.3365	0.3389
1.0	0.3413	0.3438	0.3461	0.3485	0.3508	0.3531	0.3554	0.3577	0.3599	0.3621
1.1	0.3643	0.3665	0.3686	0.3708	0.3729	0.3749	0.3770	0.3790	0.3810	0.3830
1.2	0.3849	0.3869	0.3888	0.3907	0.3925	0.3944	0.3962	0.3980	0.3997	0.4015
1.3	0.4032	0.4049	0.4066	0.4082	0.4099	0.4115	0.4131	0.4147	0.4162	0.4177
1.4	0.4192	0.4207	0.4222	0.4236	0.4251	0.4265	0.4279	0.4292	0.4306	0.4319
1.5	0.4332	0.4345	0.4357	0.4370	0.4382	0.4394	0.4406	0.4418	0.4429	0.4441
1.6	0.4452	0.4463	0.4474	0.4484	0.4495	0.4505	0.4515	0.4525	0.4535	0.4545
1.7	0.4554	0.4564	0.4573	0.4582	0.4591	0.4599	0.4608	0.4616	0.4625	0.4633
1.8	0.4641	0.4649	0.4656	0.4664	0.4671	0.4678	0.4686	0.4693	0.4699	0.4706
1.9	0.4713	0.4719	0.4726	0.4732	0.4738	0.4744	0.4750	0.4756	0.4761	0.4767
2.0	0.4772	0.4778	0.4783	0.4788	0.4793	0.4798	0.4803	0.4808	0.4812	0.4817
2.1	0.4821	0.4826	0.4830	0.4834	0.4838	0.4842	0.4846	0.4850	0.4854	0.4857
2.2	0.4861	0.4864	0.4868	0.4871	0.4875	0.4878	0.4881	0.4884	0.4887	0.4890
2.3	0.4893	0.4896	0.4898	0.4901	0.4904	0.4906	0.4909	0.4911	0.4913	0.4916
2.4	0.4918	0.4920	0.4922	0.4925	0.4927	0.4929	0.4931	0.4932	0.4934	0.4936
2.5	0.4938	0.4940	0.4941	0.4943	0.4945	0.4946	0.4948	0.4949	0.4951	0.4952
2.6	0.4953	0.4955	0.4956	0.4957	0.4959	0.4960	0.4961	0.4962	0.4963	0.4964
2.7	0.4965	0.4966	0.4967	0.4968	0.4969	0.4970	0.4971	0.4972	0.4973	0.4974
2.8	0.4974	0.4975	0.4976	0.4977	0.4977	0.4978	0.4979	0.4979	0.4980	0.4981
2.9	0.4981	0.4982	0.4982	0.4983	0.4984	0.4984	0.4985	0.4985	0.4986	0.4986
3.0	0.4987	0.4987	0.4987	0.4988	0.4988	0.4989	0.4989	0.4989	0.4990	0.4990
3.1	0.4990	0.4991	0.4991	0.4991	0.4992	0.4992	0.4992	0.4992	0.4993	0.4993
3.2	0.4993	0.4993	0.4994	0.4994	0.4994	0.4994	0.4994	0.4995	0.4995	0.4995
3.3	0.4995	0.4995	0.4995	0.4996	0.4996	0.4996	0.4996	0.4996	0.4996	0.4997
3.4	0.4997	0.4997	0.4997	0.4997	0.4997	0.4997	0.4997	0.4997	0.4997	0.4998

TABLE A2: Standard Normal Probability Distribution: To the Left $P(Z \leq z)$

	0.00	0.01	0.02	0.03	0.04	0.05	0.06	0.07	0.08	0.09
0.0	0.5000	0.5040	0.5080	0.5120	0.5160	0.5199	0.5239	0.5279	0.5319	0.5359
0.1	0.5398	0.5438	0.5478	0.5517	0.5557	0.5596	0.5636	0.5675	0.5714	0.5753
0.2	0.5793	0.5832	0.5871	0.5910	0.5948	0.5987	0.6026	0.6064	0.6103	0.6141
0.3	0.6179	0.6217	0.6255	0.6293	0.6331	0.6368	0.6406	0.6443	0.6480	0.6517
0.4	0.6554	0.6591	0.6628	0.6664	0.6700	0.6736	0.6772	0.6808	0.6844	0.6879
0.5	0.6915	0.6950	0.6985	0.7019	0.7054	0.7088	0.7123	0.7157	0.7190	0.7224
0.6	0.7257	0.7291	0.7324	0.7357	0.7389	0.7422	0.7454	0.7486	0.7517	0.7549
0.7	0.7580	0.7611	0.7642	0.7673	0.7704	0.7734	0.7764	0.7794	0.7823	0.7852
0.8	0.7881	0.7910	0.7939	0.7967	0.7995	0.8023	0.8051	0.8078	0.8106	0.8133
0.9	0.8159	0.8186	0.8212	0.8238	0.8264	0.8289	0.8315	0.8340	0.8365	0.8389
1.0	0.8413	0.8438	0.8461	0.8485	0.8508	0.8531	0.8554	0.8577	0.8599	0.8621
1.1	0.8643	0.8665	0.8686	0.8708	0.8729	0.8749	0.8770	0.8790	0.8810	0.8830
1.2	0.8849	0.8869	0.8888	0.8907	0.8925	0.8944	0.8962	0.8980	0.8997	0.9015
1.3	0.9032	0.9049	0.9066	0.9082	0.9099	0.9115	0.9131	0.9147	0.9162	0.9177
1.4	0.9192	0.9207	0.9222	0.9236	0.9251	0.9265	0.9279	0.9292	0.9306	0.9319
1.5	0.9332	0.9345	0.9357	0.9370	0.9382	0.9394	0.9406	0.9418	0.9429	0.9441
1.6	0.9452	0.9463	0.9474	0.9484	0.9495	0.9505	0.9515	0.9525	0.9535	0.9545
1.7	0.9554	0.9564	0.9573	0.9582	0.9591	0.9599	0.9608	0.9616	0.9625	0.9633
1.8	0.9641	0.9649	0.9656	0.9664	0.9671	0.9678	0.9686	0.9693	0.9699	0.9706
1.9	0.9713	0.9719	0.9726	0.9732	0.9738	0.9744	0.9750	0.9756	0.9761	0.9767
2.0	0.9772	0.9778	0.9783	0.9788	0.9793	0.9798	0.9803	0.9808	0.9812	0.9817
2.1	0.9821	0.9826	0.9830	0.9834	0.9838	0.9842	0.9846	0.9850	0.9854	0.9857
2.2	0.9861	0.9864	0.9868	0.9871	0.9875	0.9878	0.9881	0.9884	0.9887	0.9890
2.3	0.9893	0.9896	0.9898	0.9901	0.9904	0.9906	0.9909	0.9911	0.9913	0.9916
2.4	0.9918	0.9920	0.9922	0.9925	0.9927	0.9929	0.9931	0.9932	0.9934	0.9936
2.5	0.9938	0.9940	0.9941	0.9943	0.9945	0.9946	0.9948	0.9949	0.9951	0.9952
2.6	0.9953	0.9955	0.9956	0.9957	0.9959	0.9960	0.9961	0.9962	0.9963	0.9964
2.7	0.9965	0.9966	0.9967	0.9968	0.9969	0.9970	0.9971	0.9972	0.9973	0.9974
2.8	0.9974	0.9975	0.9976	0.9977	0.9977	0.9978	0.9979	0.9979	0.9980	0.9981
2.9	0.9981	0.9982	0.9982	0.9983	0.9984	0.9984	0.9985	0.9985	0.9986	0.9986
3.0	0.9987	0.9987	0.9987	0.9988	0.9988	0.9989	0.9989	0.9989	0.9990	0.9990
3.1	0.9990	0.9991	0.9991	0.9991	0.9992	0.9992	0.9992	0.9992	0.9993	0.9993
3.2	0.9993	0.9993	0.9994	0.9994	0.9994	0.9994	0.9994	0.9995	0.9995	0.9995
3.3	0.9995	0.9995	0.9995	0.9996	0.9996	0.9996	0.9996	0.9996	0.9996	0.9997
3.4	0.9997	0.9997	0.9997	0.9997	0.9997	0.9997	0.9997	0.9997	0.9997	0.9998

TABLE A3: Standard Normal Probability Distribution: To the Left $P(Z \leq z)$

	0.00	0.01	0.02	0.03	0.04	0.05	0.06	0.07	0.08	0.09
−0.0	0.5000	0.4960	0.4920	0.4880	0.4840	0.4801	0.4761	0.4721	0.4681	0.4641
−0.1	0.4602	0.4562	0.4522	0.4483	0.4443	0.4404	0.4364	0.4325	0.4286	0.4247
−0.2	0.4207	0.4168	0.4129	0.4090	0.4052	0.4013	0.3974	0.3936	0.3897	0.3859
−0.3	0.3821	0.3783	0.3745	0.3707	0.3669	0.3632	0.3594	0.3557	0.3520	0.3483
−0.4	0.3446	0.3409	0.3372	0.3336	0.3300	0.3264	0.3228	0.3192	0.3156	0.3121
−0.5	0.3085	0.3050	0.3015	0.2981	0.2946	0.2912	0.2877	0.2843	0.2810	0.2776
−0.6	0.2743	0.2709	0.2676	0.2643	0.2611	0.2578	0.2546	0.2514	0.2483	0.2451
−0.7	0.2420	0.2389	0.2358	0.2327	0.2296	0.2266	0.2236	0.2206	0.2177	0.2148
−0.8	0.2119	0.2090	0.2061	0.2033	0.2005	0.1977	0.1949	0.1922	0.1894	0.1867
−0.9	0.1841	0.1814	0.1788	0.1762	0.1736	0.1711	0.1685	0.1660	0.1635	0.1611
−1.0	0.1587	0.1562	0.1539	0.1515	0.1492	0.1469	0.1446	0.1423	0.1401	0.1379
−1.1	0.1357	0.1335	0.1314	0.1292	0.1271	0.1251	0.1230	0.1210	0.1190	0.1170
−1.2	0.1151	0.1131	0.1112	0.1093	0.1075	0.1056	0.1038	0.1020	0.1003	0.0985
−1.3	0.0968	0.0951	0.0934	0.0918	0.0901	0.0885	0.0869	0.0853	0.0838	0.0823
−1.4	0.0808	0.0793	0.0778	0.0764	0.0749	0.0735	0.0721	0.0708	0.0694	0.0681
−1.5	0.0668	0.0655	0.0643	0.0630	0.0618	0.0606	0.0594	0.0582	0.0571	0.0559
−1.6	0.0548	0.0537	0.0526	0.0516	0.0505	0.0495	0.0485	0.0475	0.0465	0.0455
−1.7	0.0446	0.0436	0.0427	0.0418	0.0409	0.0401	0.0392	0.0384	0.0375	0.0367
−1.8	0.0359	0.0351	0.0344	0.0336	0.0329	0.0322	0.0314	0.0307	0.0301	0.0294
−1.9	0.0287	0.0281	0.0274	0.0268	0.0262	0.0256	0.0250	0.0244	0.0239	0.0233
−2.0	0.0228	0.0222	0.0217	0.0212	0.0207	0.0202	0.0197	0.0192	0.0188	0.0183
−2.1	0.0179	0.0174	0.0170	0.0166	0.0162	0.0158	0.0154	0.0150	0.0146	0.0143
−2.2	0.0139	0.0136	0.0132	0.0129	0.0125	0.0122	0.0119	0.0116	0.0113	0.0110
−2.3	0.0107	0.0104	0.0102	0.0099	0.0096	0.0094	0.0091	0.0089	0.0087	0.0084
−2.4	0.0082	0.0080	0.0078	0.0075	0.0073	0.0071	0.0069	0.0068	0.0066	0.0064
−2.5	0.0062	0.0060	0.0059	0.0057	0.0055	0.0054	0.0052	0.0051	0.0049	0.0048
−2.6	0.0047	0.0045	0.0044	0.0043	0.0041	0.0040	0.0039	0.0038	0.0037	0.0036
−2.7	0.0035	0.0034	0.0033	0.0032	0.0031	0.0030	0.0029	0.0028	0.0027	0.0026
−2.8	0.0026	0.0025	0.0024	0.0023	0.0023	0.0022	0.0021	0.0021	0.0020	0.0019
−2.9	0.0019	0.0018	0.0018	0.0017	0.0016	0.0016	0.0015	0.0015	0.0014	0.0014
−3.0	0.0013	0.0013	0.0013	0.0012	0.0012	0.0011	0.0011	0.0011	0.0010	0.0010
−3.1	0.0010	0.0009	0.0009	0.0009	0.0008	0.0008	0.0008	0.0008	0.0007	0.0007
−3.2	0.0007	0.0007	0.0006	0.0006	0.0006	0.0006	0.0006	0.0005	0.0005	0.0005
−3.3	0.0005	0.0005	0.0005	0.0004	0.0004	0.0004	0.0004	0.0004	0.0004	0.0003
−3.4	0.0003	0.0003	0.0003	0.0003	0.0003	0.0003	0.0003	0.0003	0.0003	0.0002

Index

NOTE: Page numbers followed by *b* indicates boxes, *f* indicates figures and *t* indicates tables.

Edwards Brothers Malloy
Ann Arbor MI. USA
August 11, 2016